Advances in Penicillium and Aspergillus Systematics

NATO ASI Series

Advanced Science Institutes Series

A series presenting the results of activities sponsored by the NATO Science Committee, which aims at the dissemination of advanced scientific and technological knowledge, with a view to strengthening links between scientific communities.

The series is published by an international board of publishers in conjunction with the NATO Scientific Affairs Division

A	**Life Sciences**	Plenum Publishing Corporation
B	**Physics**	New York and London
C	**Mathematical and Physical Sciences**	D. Reidel Publishing Company Dordrecht, Boston, and Lancaster
D	**Behavioral and Social Sciences**	Martinus Nijhoff Publishers
E	**Engineering and Materials Sciences**	The Hague, Boston, and Lancaster
F	**Computer and Systems Sciences**	Springer-Verlag
G	**Ecological Sciences**	Berlin, Heidelberg, New York, and Tokyo

Recent Volumes in this Series

Volume 95—Drugs Affecting Leukotrienes and Other Eicosanoid Pathways
edited by B. Samuelsson, F. Berti, G. C. Folco, and G. P. Velo

Volume 96—Epidemiology and Quantitation of Environmental Risk in Humans from Radiation and Other Agents
edited by A. Castellani

Volume 97—Interactions between Electromagnetic Fields and Cells
edited by A. Chiabrera, C. Nicolini, and H. P. Schwan

Volume 98—Structure and Function of the Genetic Apparatus
edited by Claudio Nicolini and Paul O. P. Ts'o

Volume 99—Cellular and Molecular Control of Direct Cell Interactions
edited by H.-J. Marthy

Volume 100—Recent Advances in Nervous System Toxicology
edited by Corrado L. Galli, Liugi Manzo, and Peter S. Spencer

Volume 101—Chromosomal Proteins and Gene Expression
edited by Gerald R. Reeck, Graham H. Goodwin, and Pedro Puigdomènech

Volume 102—Advances in Penicillium and Aspergillus Systematics
edited by Robert A. Samson and John I. Pitt

Series A: Life Sciences

Advances in Penicillium and Aspergillus Systematics

Edited by

Robert A. Samson

Centraalbureau voor Schimmelcultures
Baarn, The Netherlands

and

John I. Pitt

CSIRO Division of Food Research
North Ryde, NSW, Australia

Plenum Press
New York and London
Published in cooperation with NATO Scientific Affairs Divison

Proceedings of the First International Penicillium and Aspergillus Workshop, held May 6–10, 1985,
at the Trippenhuis of the Royal Dutch Academy of Sciences and Letters in Amsterdam, The Netherlands

Library of Congress Cataloging in Publication Data

International Penicillium and Aspergillus Workshop (1st: 1985: Trippenhuis Royal Dutch Academy of Sciences and Letters)
Advances in penicillium and aspergillus systematics.

(NATO ASI series. Series A, Life sciences; v. 102)
"Proceedings of the First International Penicillium and Aspergillus Workshop, held May 6–10, 1985, at the Trippenhuis of the Royal Dutch Academy of Sciences and Letters in Amsterdam, The Netherlands"—T.p. verso.
"Published in cooperation with NATO Scientific Affairs Division."
Includes bibliographies and index.
1. Penicillium—Classification—Congresses. 2. Aspergillus—Classification—Congresses. 3. Fungi—Classification—Congresses. I. Samson, Robert A. II. Pitt, John I. III. North Atlantic Treaty Organization. Scientific Affairs Division. IV. Title. V. Series. [DNLM: 1. Aspergillus—classification—congresses. 2. Penicillium—classiciation—congresses. W3 IN833 1st 1985a / QW 180.5.D38 I61 1985a]
QK625.M7I58 1985 589.2′3 85-28329
ISBN 0-306-42222-0

A Division of Plenum Publishing Corporation
233 Spring Street, New York, N.Y. 10013

Printed in the United States of America

PREFACE

It is difficult to overestimate the importance of the mould genera *Penicillium* and *Aspergillus* in the natural world and in the affairs of man. These are large and ubiquitous genera; several species are beneficial to mankind because of their role in food fermentation or the production of useful antibiotics and enzymes. Others are important as agents of food spoilage. In recent years, these moulds have gained attention because a number of species produce toxic metabolites. The great impact of mycotoxicology on food problems, particularly in developing countries has generated a vast body of research. The ecological roles of *Aspergillus* and *Penicillium* species in nature is also of importance.

For the past several decades, the taxonomic treatments of Raper and Thom (1949) and Raper and Fennell (1965) were the only standard descriptive monographs of *Penicillium* and *Aspergillus*. However, in practice, the identification of these economically important moulds was problematic and caused considerable confusion in the literature and subsequently in applied research. In the past ten years, several laboratories have carried out taxonomic research resulting in a number of valuable publications. However, the differences in methodology and species concepts made comparison between the different classifications difficult. This confusion in the taxonomy and nomenclature of common species is a great handicap to applied and industrial research. While we do not suggest that a consensus on the classification of these genera will solve all industrial and agricultural problems, such a step forward would, indeed, be progress in this direction.

The idea of a workshop on *Penicillium* and *Aspergillus* systematics was born during the Third International Mycological Congress in Tokyo in August 1983. At this congress, a symposium on "Taxonomy of food-borne fungi" organised by Drs S. Udagawa and R. A. Samson, included papers by several mycologists who had recently proposed a variety of taxonomic schemes in *Penicillium* and *Aspergillus*. At the conclusion of the symposium, it was agreed that a specialists' workshop was desirable. Specifically, standardization of methodology, terminology and typification was necessary to allow future research to be conducted with uniform principals. Additionally, the introduction of new techniques into the taxonomy of these fungi, such as secondary metabolite profiles, serology and DNA and RNA homology should be considered as possible means of supplimenting morphological data. Following the initiative presented at Tokyo, R. A. Samson undertook the responsibility for organizing, together with Mr. K. A. Seifert and Mrs. Ellen van Reenen-Hoekstra, the First International Penicillium and Aspergillus NATO Workshop.

The workshop was held in the Netherlands on May 6-10, 1985 at the "Trippenhuis" of the Royal Academy of Science and Letters, Amsterdam with some sessions at the Hotel Royal in Baarn. It was a historic occasion. Never before had such a group of specialists met as a small, defined group. The enthusiastic response from scientists around the world was a clear indication both of the importance they attached to the study of these genera, and the great desire for dialoque and, where possible, consensus. A very friendly atmosphere prevailed, coupled with a sense of urgency that there was much to be done if the aims set out by Dr. Samson were to be achieved.

The workshop was centred around the formal presentation of papers, which form the bulk of this volume. Following each paper, the results, observations and ideas were discussed. Salient points from these discussions are presented following the printed text of each. Several additional sessions for more general discussion were also held, which resulted in the acceptance of a series of recommendations for future taxonomic practise. These recommendations comprise the final section of this book. The editors were perhaps surprised, and certainly delighted, at the large amount of agreement evident among the participants and we hope that this will be apparent to the reader.

The success of the workshop and the subsequent publication of the proceedings could not be achieved without the help of various individuals. Mrs. Ans Spaapen, Tineke van den Berg, Ellen Mul, Francis Claus, Gisela Samson and Mr. Bert Gerrits van den Ende helped with typing and preparing the manuscript. We are grateful to Prof. D. L. Hawksworth who prepared the initial text for the recommendations and to Mr. Keith Seifert, who spent many hours listening to tapes with us and typing the salient points of the discussions. Finally, all authors are acknowledged for their readiness to prepare their contributions for publication rapidly, in order to ensure that the proceedings would appear soon after the meeting.

The First International Penicillium and Aspergillus Workshop was sponsored by the NATO ARW programme. Co-sponsors were: the "Programma-commissie Biotechnologie" (The Haque), the Royal Academy of Sciences and Letters (Amsterdam), Gist-Brocades (Delft) and Bayer-Nederland (Mijdrecht).

The Editors, Baarn - July, 1985

CONTENTS

Chapter 3: ADVANCES IN PENICILLIUM TAXONOMY

Chapter 4. ADVANCES IN ASPERGILLUS TAXONOMY

Chapter 5: NEW TECHNIQUES AND APPROACHES TO THE SYSTEMATICS OF PENICILLIUM AND ASPERGILLUS

Chapter 6: TELEOMORPHS AND THEIR RELATIONSHIPS

Chapter 7: RECOMMENDATIONS

Chapter 1:

TYPIFICATION AND NOMENCLATURE

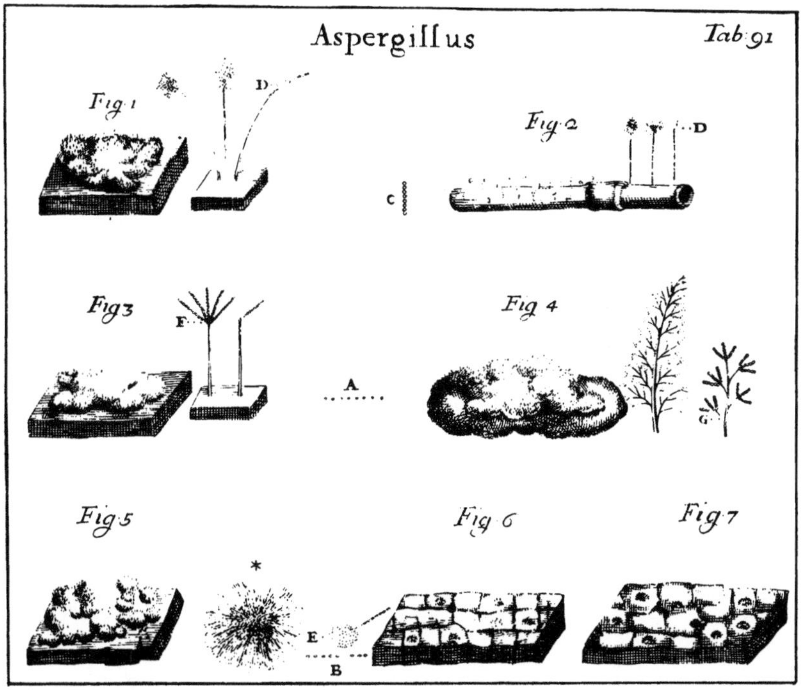

THE TYPIFICATION AND CITATION OF THE GENERIC NAME PENICILLIUM

David L. HAWKSWORTH

Commonwealth Mycological Institute
Kew, Surrey TW9 3AF, United Kingdom

SUMMARY

The typification and citation of the generic name Penicillium is reviewed in the light of changes in the International Code of Botanical Nomenclature made in 1981. The correct citation is now Penicillium Link, not "Link ex Gray" or "Fr."; "Link:Fr." can be used to indicate its sanctioned status but this form is unnecessary for general use. The species name Penicillium expansum can be attributed to "Link" alone, and is acceptable as lectotype for the generic name. The name P. crustaceum must now be attributed to "(L.)Fr." and does not affect the use of P. expansum as it is not conspecific (or congeneric) with that species. The need for conservation and the debate as to the name of the type species have both been avoided as a result of the changes enacted in the Code.

The question of the typification and citation of the generic name Penicillium was addressed by Hawksworth et al. (1976) who considered that in order to maintain current usage it was necessary to conserve Penicillium Link ex Gray (1821; lectotype P. expansum Link ex Gray) against Penicillium Fr. (1832; holotype Mucor crustaceus L.) This problem arose because Fries (1832:407) stated "Typus: Mucor crustaceus L." The only recognizable element in the protologue of Linnaeus' (1753:1186) name was based on Botrytis non ramosa alba... Micheli (1729:213, tab. 91 (below) fig. 3), a coprophilous Botrytis-like fungus and not a Penicillium. However, Jørgensen and Gunnerbeck (1977) did not consider conservation necessary as (1) there was no evidence that Fries had a type concept in the sense of the then operative Code, and (2) the interpretation of the then Art. 10 on generic typifications which they regarded as resting on the species concept of the author rather than the name he used for it. The view of the Special Committee for Fungi and Lichens was that conservation was not necessary (Petersen, 1980).

At that time two proposals to change the International Code of Botanical Nomenclature were under consideration which could affect the situation with *Penicillium* if adopted. These were to change the starting point date for the nomenclature of Fungi from 1 January 1821 to 1 May 1753 (van Warmelo, 1979; Demoulin et al., 1981) and to clarify the ambiquity in Art. 10 with respect to the typification of generic names (McNeill, 1981). In view of these uncertainties, it was not appropriate to re-open the discussion on *Penicillium* until decisions on these key nomenclatural matters had been taken at the International Botanical Congress in Sydney in 1981. At that Congress, many changes in the rules pertinent to the nomenclature of fungi were agreed to and incorporated into the new edition of the Code (Voss et al., 1983); a summary of these changes is provided by Hawksworth (1984). It is consequently now timely to re-address the question of the citation and typification of the generic name *Penicillium* in the light of the actions taken at Sydney.

The generic name *Penicillium* was first introduced by Link (1809:16) for the three species *P. glaucum* Link, *P. candidum* Link, and *P. expansum* Link. As the change in starting point date to 1753 was agreed, the generic name and these three species names are now all regarded as validly published by Link in 1809 (Art. 13 (d)) and can be attributed to Link alone. Further, in the revised Art. 10.2, it was made clear that the type of the genus must be selected from amongst the types of the names of species mentioned when the genus name was introduced. *P. expansum* has been consistently accepted as the lectotype for the genus since Thom (1910) and can be retained as such under the revised rules. In the absence of any extant material of Link, a dried culture (IMI 39761, isoneotypes BPI, K, UPS; - living cultures ATCC 7861, CBS 325.48, IMI 39761) has been designated as neotype for the species name *P. expansum* (Hawksworth et al., 1976; Samson et al., 1976; Pitt, 1979*).

Under the revised Art. 13(d), however, in order to minimize changes occasioned by the alteration in the starting point date for the names of fungi from 1821 to 1753, names of Hyphomycetes adopted in Fries (1821-32) are sanctioned; i.e. are treated as if conserved against earlier homonyms and synonyms. Fries (1832:406) clearly accepted *Penicillium* and, as he attributed the name to Link, he can be regarded as having sanctioned the generic name. Under Rec. 50E.2, a ":" can be used to indicate this status (i.e. *Penicillium* Link:Fr.), but, as with citations such as "nom. cons." (treated in Rec. 50E.1) this is only pertinent as a form of short-hand in nomenclatural discussions. Such citations are confusing to those who are not specialists in nomenclature and cannot be recommended for general use (Hawksworth, 1982; Santesson, 1984), especially as proposals to modify their use are already being made (Gams and Kuyper, 1984). In the interests of the majority of their users, taxonomists have an obligation to keep changes in names *and* citations to a minimum.

---

*actual publicaton date in 1980

Link's original three species were all included within Fries' concept of Penicillium. As Fries clearly attributed the genus to Link there seems to be no obstacle to the acceptance of one of these as lectotype, i.e. P. expansum as generally accepted (see above). Fries' use of "Types" (see above) was interpreted as "typical of" by Jørgensen and Gunnerbeck (1977). If this had been a new genus, most nomenclatural specialists would disagree with that argument. However, as it was (1) not a new genus and all Link's taxa were still included; (2) the name Fries cited did not accord with the generic diagnosis; and further (3) as the sanctioning provisions do not specifically refer to typification by the sanctioning author, but rather everything associated with the name (Art.7.17) Fries' citation of "Typus: Mucor crustaceus L." looses its nomenclatural relevance.

The name Penicillium expansum Link was treated by Fries (1832:407) as a synonym of "P. crustaceum". Fries clearly based this name on Mucor crustaceus L., first making the transfer to Penicillium in his contribution to the flora edited by Liljeblad (1816: 678), from which date that combination is now regarded as being validly published. Jørgensen and Gunnerbeck (1977) considered that Fries' name had to be treated as newly introduced by Fries (1832) under the then operative Code and cited it as "P. crustaceum Fr.". Their view was contrary to the practice of most mycologists at that time who would have cited it as "P. crustaceum L. ex Fr." and based its typification on the Linnaean protologue. However, with the revised starting point date Linnaeus' name is treated as validly published from 1753, and Fries' transfer to Penicillum is regarded as a new combination, i.e. P. crustaceum (L.) Fr. There is now no alternative but to typify that combination by the Botrytis-like figure published by Micheli (see above). The name P. crustaceum cannot therefore be taken up for P. expansum as it was not based on a nomenclatural type conspecific (or even congeneric) with the neotype of P. expansum.

In summary, as a result of the changes in the Code enacted in 1981, the need to conserve Penicillium to maintain usage in its present sense has been avoided, its citation has been simplified, and the debate as to whether the name of its type species should be P. expansum or P. crustaceum has been unequivocably resolved.

Acknowledgements

This study was undertaken as a part of Science and Engineering Research Council (SERC) Contract No. SO/17/84. I am grateful to my colleagues Drs A.H.S. Onions and B.C. Sutton for reading and commenting on a draft of this paper.

REFERENCES

DEMOULIN, V., HAWKSWORTH, D.L., KORF, R.P. and POUZAR, Z. 1981. A solution to the starting point problem in the nomenclature of fungi. Taxon 30: 52-63.

FRIES, E.M. 1821-32. "Systema mycologicum". 3 vols. Gryphiswald: E. Mauritius.

GAMS, W. and KUYPER, T. W. 1984. Problems involved in the sanction of fungal names. Mycotaxon 20: 619-631.
GRAY, S.F. 1821. A natural arrangement of British Plants. 2 vols London: Baldwin, Cradock and Joy.
HAWKSWORTH, D.L. 1982. Changes to the British checklist arising from the abolition of later fungal starting points. Lichenologist 14: 131-137.
HAWKSWORTH, D.L. 1984. Recent changes in the international rules affecting the nomenclature of fungi. Microbiol. Sci. 1: 18-21.
HAWKSWORTH, D.L., PITT, J.I. and SUTTON, B.C. 1976. Typification of the genus Penicillium. Taxon 25: 665-670.
JØRGENSEN, P.M. and GUNNERBECK, E. 1977. The nomenclature of Penicillium. Taxon 26: 581-582.
LILJEBLAD, S. 1816. Utkast till en Svensk Flora, 3rd edition. Uppsala: Zeiil and Palmblad.
LINK, H.F. 1809. Observations in ordines plantarum naturales. Dissertatio 1ma. Berl. NatKde 3: 1-42.
LINNAEUS, C. 1753. Species plantarum. 2 vols. Stockholm: L. Salvii.
MCNIELL, J. 1981. Report of the Committee on Generic Typification.Taxon 30: 200-207.
MICHELI, P.A. 1729. Nova plantarum genera ivxta Tovrnefortii methodvm disposita. Florence.
PETERSEN, R,H, 1980. Report of the Special Committee for Fungi and Lichens. Taxon 29: 148-149.
PITT, J.I. 1979. The genus Penicillium and its teleomorphic states Eupenicillium and Talaromyces. London: Academic Press.
SAMSON, R. A., STOLK, A. C. and HADLOK, R. 1976. Revision of the subsection Fasciculata of Penicillium and some allied species. Stud. Mycol. 11: 1-47.
SANTESSON,R. 1984. Lichens of Norway and Sweden. Stockholm: Swedish Museum of Natural History.
THOM, C. 1910. Cultural studies of species of Penicillium. Bur. Anim. Ind. Dep. Agric. 118: 1-109.
VOSS, E.G. et al., eds. 1983. International Code of Botanical Nomenclature adopted by the Thirteenth International Botanical Congress, Sydney, August 1981. (Regnum veg. 111.) Utrecht: Bohn, Scheltema and Holkema.
VAN WARMELO, K.T. 1979. Proposals for modifications of the Code of Botanical Nomenclature: IMC2 proposals. Taxon 28: 424-431.

DIALOGUE FOLLOWING PROF. HAWKSWORTH'S PAPER

PITT: What do you and other nomenclaturalists think about the use ": Fr.", particularly with Penicillium for routine use?

HAWKSWORTH: I would prefer that you say "Penicillium" Link unless you are dealing with a formal nomenclatural discussion. I think the fewer barriers we can have between nomenclaturalists and the consumers of taxonomy the better.

GAMS: I would use ":Fr." in papers dealing with nomenclatural matters but not necessarily in citing strains from a culture collection. I would also use it in formal generic revisions.

HAWKSWORTH: The problem here, which was not appreciated by the Sydney Botanical Congress in 1981, is that this is the only case in the whole botanical nomenclature where citation of the author has anything to do with the nomenclatural status of a name. All other author citations are concerned with the valid publication of the name. Illegitimacy, conservation and other indications of status are not normally mentioned as part of citations. The use of ":Fr." is a new concept. It is very interesting to note where it was placed by the editors of the new Code: it was positioned in Recommendation 50E immediately after the paragraph stating that "nom.cons." could be used after the citation of conserved names. This use of "nom. cons.", of course, is also a rather unusual practice, except in very formal taxonomic treatments or nomenclatural discussions where it is a useful form of short-hand.

DIERCKX' CONTRIBUTION TO THE GENUS PENICILLIUM

Grégoire L. HENNEBERT

Laboratoire de Mycologie systématic et appliquée
Université Catholique de Louvain, Place de Croix du Sud 3
B-1348 Louvain-la-Neuve, Belgique

SUMMARY

Original line drawings of Dierckx' species of *Penicillium* validly described in 1901 have been recovered and constitute iconotype material. Dierckx' contribution to *Penicillium* taxonomy is reappraised in light of its history and methodology.

INTRODUCTION

Twenty-four of the twenty-five line drawings illustrating the 25 species of *Penicillium*, described by Dierckx in 1901, were recovered recently. Apart from a single culture recovered by Biourge from Dierckx, *P. griseo-roseum* Dierckx, Dierckx' drawings are the only extant authentic material of Dierckx. They typify twenty-two of Dierckx's twenty-three new species. In this paper, the history and the characteristics of Dierckx' contribution to *Penicillium* taxonomy is analysed and two of Dierckx's drawings are presented.

History

After the publication of his controversial paper about polymorphism in fungi (Carnoy, 1870), which most probably was based on mixed cultures, when Pasteur was demonstrating the necessity of pure culture in the study of beer yeast fermentation, Professor Jean-Baptiste Carnoy realised that study of fungi or bacteria had to be made with extreme care, in pure culture and in the laboratory. When he had started teaching and research in microbiology at the Ecole Superieure de Brasserie de l'Université de Louvain in 1888, he decided to develop experimental microbiology by sending Philibert Biourge on a two-year government grant "to learn physiological chemistry and the method of pure culture" with Pasteur, in 1891.

P. Biourge was borne on April 8, 1864, became a priest in 1888, a doctor of natural sciences in botany in 1891, and was currently one of Carnoy's Ph.D. students. Biourge stayed at Pasteur's laboratory in Paris, working with Dr. Roux and Dr. Duclaux from 1891 to 1892, then worked with Prof. Hoppeseyler in Strasbourg and with Dr. Emil Hansen at the Carlsberg Laboratory in Copenhagen. There he became acquainted with single spore culture methods. On his return to the Louvain University in February 1893, Biourge was appointed assistant to Carnoy. He was promoted to Professor of Microbiology in 1894 (Hennebert, 1979).

In 1893, Biourge initiated research on the bacterial "diseases" of beer. He purified the yeasts used in Belgian breweries and started the maintenance of a pure culture collection of beer yeasts together with contaminant strains of bacteria, wild yeasts and moulds. In investigating germinating barley, malt infusions, water, cheese, fruit and fresh products Biourge also collected a large number of moulds, mainly Penicillium (Biourge, 1920).

When in 1897 Jean Ray published his paper on "Variations des Champignons inferieurs sous l'influence du milieu", which dealt with the variations of Penicillium glaucum in culture, Carnoy suggested to Biourge that he should investigate the same subject. Biourge objected that nobody could tell what Link's Penicillium glaucum was exactly, and that he had in his collection a number of Penicillium isolates, which although all glaucus in colour were surely distinct species. Carnoy then suggested imperatively "then start making a monograph of the genus". Biourge started.

However, in December 1897, when Carnoy was obliged to retire through illness, Biourge became too busy with lectures and demonstrations to have much time left for the monograph. The work slowed until the next year, when in July 1898, Francois Dierckx, a student in botany, chose to prepare a PhD thesis in microbiology in Biourge's Laboratory.

Born in Turnhout on 14 January 1863, François Dierckx chose to become a Jesuit in 1880. In 1887 he studied natural sciences at the University of Lille and, from 1888 to 1892, zoology at the University of Louvain, working in Professor Gilson's Laboratory. In 1895, after three years of theology, he was ordained as a priest. He was awarded his bachelor's degree in natural sciences (botany) at the University of Louvain in November 1897. In July 1898 Biourge asked Dierckx to contribute to the Penicillium monograph. Biourge taught him monospore isolation, pure culture techniques under standarized conditions, and macroscopic and microscopic observations and drawings. All the Penicillium isolates in Biourge's collection were handed to him.

We can imagine Biourge and Dierckx, the professor and the student, of approximately the same age (34 and 35 respectively), both priests, both educated and well trained scientists, working together with the same devotion on these colourful Penicillium moulds. Surely Biourge, with his six years experience in microbiology, and aware of the problems of Penicillium, inspired Dierckx's research work, his descriptive approach and classification. After two years, on July 17, 1900, Dierckx presented his thesis, an illustrated contribution to a monograph of the genus Penicillium, and graduated with the highest grade.

In October 1900, Dierckx read a summary of his monograph at the Societe Scientifique de Bruxelles, which was published in 1901 as "Essai de revision du genre Penicillium Link". It was an all too brief summary, which, however, was considered important enough by Lindau to publish a translation of the description of Dierckx' 23 new species into German (Lindau 1907). Saccardo (1902) gave only a mention, in the Sylloge Fungorum, looking for the publication of Dierckx' full account. But this was to be the only published material, as Dierckx was not able to afford the cost of full publication, with illustrations, of his thesis. In the meantime Dierckx published a memoir (La Cellule, 16, 116 p., 1899) and prepared another one (ibid., 18, 56 p., 1901) on comparitive anatomy of Coleoptera. For those contributions, he obtained a two years government travel grant on 29 December, 1900.

Dierckx went to Indonesia in May 1901, where he spent most of his time in the Botanical Garden of Bogor (Buitenzorg) with Treub. On his way to Indonesia, Dierckx visited Saccardo in Padua and showed him his thesis with drawings, aquarelles and cultures of Penicillum. Saccardo recalled that visit in the "Sylloge Fungorum" (Saccardo 1902, p. 1020). On his way back to Belgium, Dierckx spent a while in Gosio's Laboratory of Hygiene at Rome, where Penicillia were being studied. Unfortunately Dierckx lost all his Penicillium cultures during his travels, while of duplicate cultures in Biourge's laboratory, only one survived. When Dierckx rejoined Biourge's laboratory at the end of 1902, he had to start again. He isolated new strains, recovered some but not all of the species he had previously described, and discovered many new ones.

Dierckx was appointed Professor in the Faculty of Sciences at the University Faculties in Namur in December 1903. He intended to continue with Penicillium descriptions in his free time, but without any success. Finally he abandoned the project and gave to Biourge his cultures, his thesis, colour plates (aquarelles) and drawings, and his laboratory notes from Louvain and Namur. Included were descriptions of unnamed species dated 1904, eventually published by Biourge in 1923.

From 1904 to 1914, Biourge slowly reorganized, renumbered and developed his culture collection but his yearly charge of 465 hours of lectures was too much to let him carry on with the Penicillium monograph. In 1914, when the University closed because of the war and the destruction of the city of Louvain, Biourge had the unexpected opportunity to retire with his culture collection in his home at Feluy (Brabant) where he stayed up to 1919. In addition to P. griseo-roseum Dierckx which survived from before 1900, Biourge was able to reidentify 21 of Dierckx' species from the isolates found by Dierckx and himself after the end of 1902, using as his basis Dierckx' drawings, paintings and unedited descriptions. When P. duclauxii Delacr. was replaced from Bainier's collection, only P. congolense Dierckx and P. biourgei Dierckx were not reidentified.

On 3 April 1916, at a conference at the University of Louvain, Biourge gave the outlines of his monograph, "Les moisissures du groupe Penicillium Link: étude monographique". Published after the war, in 1920, that paper is a preliminary conspectus of Biourge's monograph published with the same title in 1923. There Biourge described 133 species of Penicillium, including taxa described by Dierckx. He accepted 18 of Dierckx' 23 new species, synonymized P. elongatum with P. leucopus(Pers.) Biourge, excluded P. atroviride as heterogenious, mentioned P. congolense and P. biourgei as possible species and ignored P. rubropunctatum. Each species description was based on a single fungus culture, indicated by a strain number selected from Biourge's culture collection. These strains have been considered as type or neotype isolates of Biourge's or Dierckx' species, respectively.

Biourge subsequently sent his type and neotype cultures to Thom in 1924, 1926, 1927 and 1928, and these were integrated into Thom's monograph (1930). Biourge also deposited many of his cultures in the CBS in 1929 and a few more in 1930. In 1934, Thom visited Biourge at the Carnoy Institute in Louvain and received further subcultures. In 1936, Professor P. Simonart, in charge of the collection since 1934, went to Thom's Laboratory in Washington carrying other duplicates of Biourge's cultures with him. In 1939, when Biourge retired from work at the age of 75 years, Simonart named the culture collection "Mycotheque Philibert Biourge". The Collection was placed in the care of the author in 1969 and in 1972 was integrated with Meyer's collection and the author's collection to form the "Mycotheque de l'Université Catholique de Louvain" (MUCL).

Origin of Dierckx' collection of Penicilllium isolates

In 1897, Dierckx received from Biourge all the Penicillium isolates from Biourge's collection, which Biourge himself had isolated from beer yeasts, barley, malt, beer wort, hops, brewery water and other products such as cheese, fruits, vegetables, and the air in the laboratory and classrooms. Dierckx added more strains isolated by himself or received or purchased from abroad. He received the type isolate of P. duclauxii Delacr. both from Delacroix and from Duclaux in the Pasteur Institute. From Wehmer came P. olivaceum and P. italicum. He also purchased several isolates from Kral's Culture Bureau in Prague, such as a strain named "Penicillium roseum Link" that he described as P. citreoroseum Dierckx. He also requested herbarium material from several sources on loan and attempted to revive the material in culture. He made drawings and descriptions from such herbarium material in any case. His thesis was based on 61 isolates. When working again at Biourge's Laboratory in Louvain from 1902 to 1903 and his laboratory in Namur he developed again a collection, numbered at least up to no 404 according to records in Biourge (1923).

Pure culture methods

Trained by Biourge in pure culture methods which consisted of plating 3 to 4 successive dilutions of spore suspensions and in single spore transfer, Dierckx applied these techniques both to his own and acquired isolates. The single spore cultures were then processed through three or more succesive transfers on a series of culture media in order to check the stability of characteristics as a demonstration of purity. After receiving the type of P. italicum Wehmer from Kral's collection, Dierckx tested it for purity and isolated from it, a second fungus, similar to his P. hirsutum.

Dierckx' culture media for comparative studies

Before 1898, Biourge and his student Semal (1897) had already tested several culture media to help differentiate his Penicillium isolates. From 1893 Biourge used beer wort with or without gelatine,the "milieu classique des Zymotechniciens, comme Hansen, Will, Lindau" (Biourge, 1923) . Gelatine (10%) was used to solidify the medium, or a mixture of gelatine (5%) and agar (0.75%) during summer. He also modified Raulin's and Hayduk's formulae which contained ammonium nitrate and asparagine, respectively, as nitrogen sources. In both cases he substituted sucrose for the original glucose.

Dierckx used seven culture media, but four of them were most useful in his comparative study of Penicillia. He used beer wort or malt extract gelatine agar of 12% sugar (12-14° Balling) containing 5% gelatine and 0.75% agar without pH adjustment. He modified Raulin-sucrose medium from Biourge's recipe to make it neutral. This neutral Raulin-sucrose medium, called Raulin-Dierckx medium by Biourge (1923), contains 4.6% sucrose. The formula is given by Biourge (1923), Thom (1930) and Raper and Thom (1949).

Dierckx used neutral Raulin sucrose as a liquid medium and solidified by 5% gelatine and 0.75% agar. He also used Hayduk's sucrose asparagine medium, as modified by Biourge, in liquid culture. Biourge (1923) used it both as a liquid and solidified with 10% gelatine or 5% gelatine and 0.75% agar. Hayduk's asparagine sucrose medium contains 8% sucrose, 0.08% sparagine, 0.1% dipotassium phosphate and 0.032% hydrated magnesium sulfate. To those three culture media, Dierckx added a fourth one, a bacteriological peptone with both glucose and glycerol, slightly acidified and solidified by 5% gelatine and 0.75% agar. On that medium cultures aged rapidly and their conidiophore tips and conidia showed abnormal swellings, like those recorded as "Corda's phenomenon" by Biourge (1923) and Thom (1930).

Less defined media were also used by Dierckx in his parallel series of cultures on 7 substrates. Autoclaved dried bread, like that previously used by Brefeld and Biourge, favoured sclerotial and cleistothecial development as well as the formation of coremia in species such as P. griseofulvum, P. aeruginosum and P. elongatum. Dierckx initiated the use of cheese (fresh Holland cheese) as melted autoclaved slant in tubes, which caused development of particular colours in some strains. Dierckx also used liquid sterilized beer for the production of coremia.

Macroscopic and microscopic description

According to Dierckx (1901), and his original 1904 descriptions published by Biourge (1923), Dierckx recorded the following macroscopic characteristics of Penicillium isolates: growth rates, zonation, radial or annular striation, surface texture, colour of sporulating areas and margins, colour of reverse, medium pigmentation, exudate colours, odour, gelatine liquefaction, and formation of coremia and sclerotia. Colours were taken from Klincksieck and Valette's chromotaxy. The obverse and reverse of each strain in culture on seven or more media were then painted in water colours, in an artistic way. Dierckx indicated having prepared about 15 aquarelles after "nature", of each species.

Dierckx carefully applied strictly standardized conditions of nutrition, temperature and light in the cultivation of his isolates. He emphasized that only in such conditions could the purity and identification of cultures be assured. Microscopic morphological characteristics were also used by Dierxck in species diagnoses, except in a few cases where the noted this. Slides were prepared from conidiophores taken at the zone between the white margin and the coloured sporulating area, where conidiophores are fully developed but only starting to sporulate. Slides were stained in methylene blue. Line drawings were made with Abbe's camera lucida, at 500 x magnification.

Dierckx' unpublished Ph.D. thesis

In 1920, Biourge stated about Dierckx' Ph.D. thesis: "Après trois mille cultures de comparaison entre 61 souches, il (Dierckx) reconnut 25 espèces nettement distinctes. Sa thèse (32 pages, in folio, 3 planches en noir et 13 cartons d'aquarelles) a été recue en juillet 1900. Elle est resteé inédite, parce que les procédés de reproduction des aquarelles étaient d'un prix inabordable ou encore trop imparfaits" and, in 1923 (p.34), "La thèse doctorale, inédite, de Dierckx etablissait les caractères culturaux des 25 espèces. J'en ai retrouvé 22 ou 24" and (p. 27), "J'ai déjà pu identifier une grosse partie des espèces de Dierckx. De ses 57 cartons, 45 representaient des espèces qui sont actuellement en ma collection".

Dierckx' thesis was thus a folio manuscript of 32 pages describing 25 Penicillium species, 23 of them being named as new, on the basis of 61 isolates. The text was accompanied by 16 cardboard plates, numbered I to XVI. The 13 first plates were bearing 32 aquarelles made from cultures. The last three plates, no XIV to XVI, were covered by 25 sets of line drawings, one of each species. Neither the text of the thesis nor the 13 plates of aquarelles have been recovered up to now. However, the three plates n° XIV, XV and XVI, bearing 24 of the 25 sets of line drawings by Dierckx, have been recovered recently, together with some of Biourge's correspondence, notes and aquarelles, by Miss Marie-Louise Biourge at Biourge's house in Feluy (Brabant). These have been identified by the author.

The recovered plates, made of cardboard, 24 x 33 cm, and numbered XIV, XV and XVI, consist of 8 (with the first one missing), 9 and 8 drawings, respectively. The drawings, made on light bristol cardboard, are of different sizes and bear black-inked line drawings of 1 to 2 conidiophores and a few conidia. Dierckx' pencil written strain number appears in the upper left corner of each, the corresponding Biourge strain number in the upper right corner, the species name at the bottom and, near each group of conidia, the size, length by width, in micrometers. Some other notations, in Dierckx' handwriting are still apparent. At the top of each drawing is a notation of 2 numbers and one or two capital letters that possibly might indicate a colour standard. Near each conidiophore there are two numbers with decimals disposed as a fraction, but their significance is not known so far. On two drawings, Dierckx added the year 1900. Some other annotations are in Biourge's hand; e.g. "(B!)" near the Biourge collection number as confirmation of the existence of a similar strain in the last Biourge collection.

The drawings are disposed on the 3 plates exactly in the same order as that of the species presented in Dierckx' 1901 paper, i.e. the same as they appear in the 32 pages of descriptions in the thesis. Just one drawing, the first one, that of P. rubro-punctatum, is missing. It is this species that is ignored in Biourge's monograph (1923). It is noticeable that one drawing of P. congolense, has not been inked, but water-painted in shades of light blue, with dark blue stained inclusions in the conidiophore stipes and branches as seen by Dierckx after methylene blue staining of the slides.

It is worth mentioning here that Biourge, anxious to know whether Dierckx' work would provide an answer to the question, what is Penicillium glaucum Link, sent slides and cultures of each of Dierckx' species to Saccardo before Dierckx' thesis was presented. Saccardo answered indicating that 6 strains might possibly be candidates for identification as P. glaucum and that all the other species appeared to be to him new to science. This is why the single strain identified by Dierckx as P. glaucum was indicated with a question mark (Biourge, 1920).

Validity of Dierckx' names of Penicillium species

Saccardo (1902, p. 1090) who saw the full thesis in the hands of Dierckx while he was visiting him, realised the importance of Dierckx' paper ("opusculo gravis ponderis"), cited the species names without description in the Sylloge Fungorum and mentioned the shortness of the published descriptions and the lack of the "locus natalis" for each of the species. Lindau (1907, p. 170) also appreciated the novelty of Dierckx' taxonomy of the genus Penicillium based on cultural investigations but regretted the shortness of the descriptions, which might increase confusion. However, he provided a full standardized translation of Dierckx' descriptions.

Westling (1911) recognized and accepted Dierckx' subdivisions of the genus, but rejected all of his new species as hardly recognizable. Thom (1930), although he was greatly enlighted by Biourge's (1923) redescriptions of Dierckx' species and the study of Biourge's strains, refused to recognize the validity of Dierckx' species as published in 1901. He attributed them to Dierckx in Biourge (1923), and gave priority to other names published before 1923 for the same species.

Recently, the 1983 CBS catalogue listed P. candidofulvum and P. aurantiocandidum as "Dierckx ex Biourge" as if those names were validated by Biourge, which is incorrect. In accordance with the Code of International Botanical Nomenclature, it must be asserted that Dierckx' Penicillium names are validly published with Latin names and descriptions, even if descriptions are brief. Latin diagnosis and type designation are not required for the validity of taxa published in 1901.

Typification of Dierckx' Penicillium species

There is no doubt that the recovered line drawings are the originals made by Dierckx. They therefore constitute the iconotype material of 22 of his 23 species, that of Penicillium rubropunctatum being lost. The only cultural type material preserved of Dierckx' species is that of Penicillium griseoroseum. It came from the Pasteur Institute, before 1898, as no. 85, was registered as old series Biourge 401, renumbered by Dierckx as no. 3, and by Biourge in 1914 as new no 29. That strain was renumbered by Thom as 4733.70 in 1927 and eventually as NRRL 820. It was accepted as type of P. griseoroseum by Pitt (1979) with IMI 92220 i as lectotype herbarium material.

All other isolates identified by Biourge as Dierckx' species are new, dating from after 1902. Biourge's isolate no. 146, identified by Biourge as P. citreonigrum Dierckx, is not the Westling's type strain of P. subcinereum Westling, separately received by Biourge from Dr. Westerdijk before 1923, but his own isolate, according to Biourge's handwritten list of cultures (see Thom, 1930; Pitt 1979). It is now designated as neotype of P. citreonigrum Dierckx (Pitt, 1979). Delacroix' type of P. duclauxii was received by Dierckx in 1898 from Delacroix as no. 84, given no. 351 in Biourge's collection, and sent to Thom who numbered it 4733-53. It was lost in Louvain, and replaced later by a subculture of Thom's 4733-53, under the same number 351.

Table 1 lists the original numbers of the strains selected by Dierckx for description and drawing of his new species and others, and the numbers of Biourge's strains in Biourge's and other culture collections*.

As illustrations in this paper, two sets of drawings by Dierckx are reproduced at original size, P. duclauxii Delacr. and P. brevicompactum Dierckx with the original published descriptions (1901). Drawings of the same species from Biourge (1923) are reproduced for comparison (Fig. 1-4).

*Most strains are kept in the Laboratoire de Mycologie at Leuven.

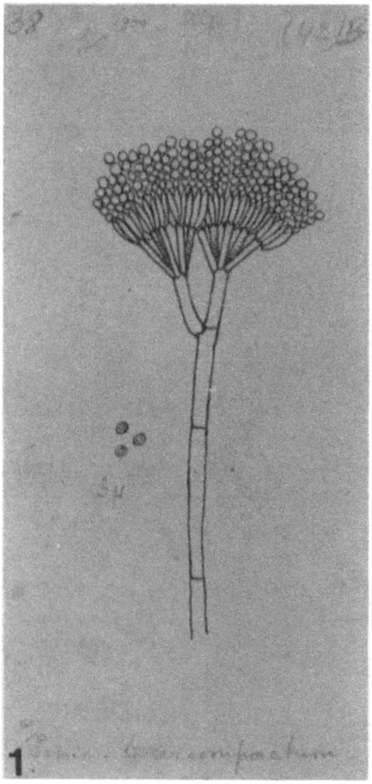

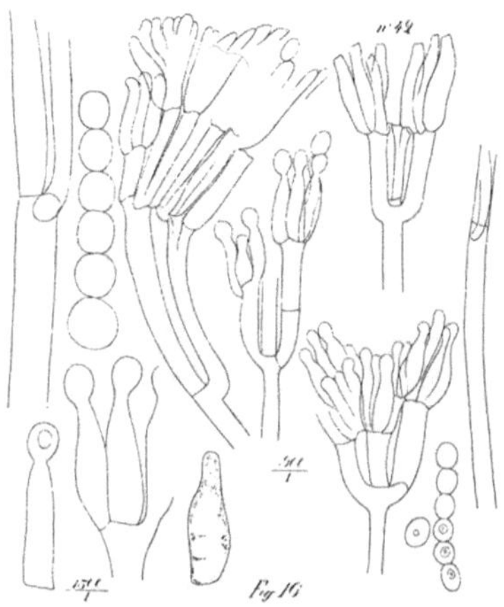

Fig. 1-2. Penicillium brevicompactum, 1. Dierckx' drawing (iconotype) and description of type isolate Dierckx-38, 2. Biourge drawing of the same from neotype strain, Biourge-42 = CBS 257.29.

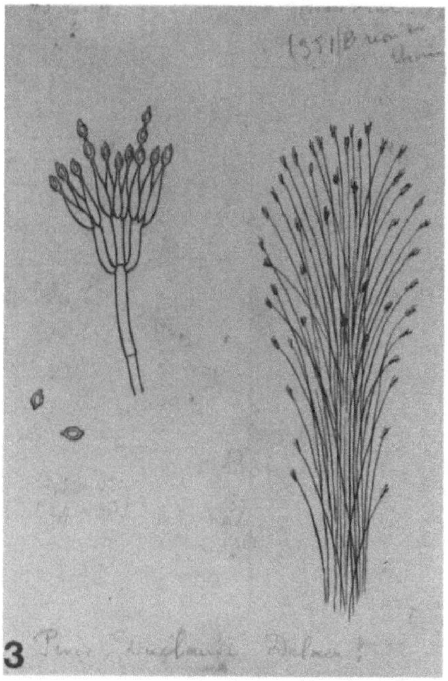

Fig. 3-4. Penicillium duclauxii, 3. Dierckx' drawing and description of type isolate, Delacroix 84 = Dierckx-1= Biourge 351. 4. Biourge's drawing of the same from Biourge 351.

Table 1. History of Dierckx' *Penicillium* species and type strains and Biourge's strains

Dierckx (1901)	before 1901	after 1904
Subgenus *Aspergilloides*		
P. rubropunctatum Dierckx	Bi-343b Dx-23	----
P. candidofulvum	Bi-330 → Dx-29	Bi-46 → Thom 4733.25 → NRRL 769 → CBS 254.37 → IMI 92201 ↘ CBS st. Biourge
P. aurantiobrunneum	Dx-53	Bi-145 → Thom 4733.5 → NRRL 766 = IMI 92195 = ATCC 10103 ↘ CBS st. Biourge
P. citreoroseum	Krâl → DX-54	Bi-18 → Thom 4733.36 → NRRL 834 = IMI 92210 ↘ CBS st. Biourge
P. carminoviolaceum	DX-18	Bi-176 → Thom 4733.28 → NRRL ↘ CBS st. Biourge NRRL 2066
P. roseopurpureum	Bi-344 → Dx-8	Bi-8 → Thom 4733.108 ↘ CBS st. Biourge 266.29 → NRRL 2064 = IMI 40573= ATCC 10492
P. citreonigrum	Bi-342 → Dx-10	→ Bi-146 → Thom 4733.35 → → NRRL 76
P. corylophilum	Dx-45	Bi-78 → Thom 4733.42 → NRRL 802 → CBS 312.48 = IMI 39754 = ATCC 9784

Dierckx (1901)	before 1901	after 1904
Subgenus <u>Eupenicillium</u>		
P. duclauxii	Del.-84→Dx-1	Bi-351→Thom 4733.53→NRRL 1030→CBS 322.48 = IMI 40044 = ATCC 10439 ↘ CBS st. Biourge
P. aeruginosum	Wehmer→Dx-59b	Bi-123
P. minioluteum	Krâl →Dx-47	Bi-60 ↗ Thom 4733.89→NRRL → CBS st. Biourge ↘ LSTM Smith P-44→IMI 89377→CBS 642.68
P. congolense	Dx-57	----
P. elongatum	Dx-2	Bi-40 as P. juglandis = P. leucopus→Thom 4733.55→NRRL Bi-49 as P. leucopus→Thom 4733.81→NRRL 978
P. glaucum Link sensu Dierckx	Dx-9	st. Westling. P. solitum, T, Bi-3→Thom 4733.114→NRRL 950 Thom 2546→NRRL 937→CBS 288.36 = IMI 39810 = ATCC 9923
P. atroviride	Dx-5	----
P. verrucosum	Dx-12	Bi-44→Thom 4733-125→NRRL 965→CBS 603.74 = IMI 200310
P. griseobrunneum	Dx-31	Bi-148→Thom 4733.68→NRRL 867 ↘ CBS st. Biourge→NRRL→LSTM = IMI 92219
P. brevicompactum	Dx-38	Bi-42 ↗ Thom 4733.21 NRRL 864 → Simonart NRRL 863 ↘ CBS st. Biourge 257-29 NRRL 2011 = IMI 40225 = ATCC 10448
P. griseofulvum	Dx-25	Bi-34 ↗ Thom 4733.69 → LSTM Smith→NRRL 2152 ↘ CBS st. Biourge 185.27 = IMI 75832 ↘ LSTM Raistrick→Thom 5112.2→NRRL 734 ↘ P.W. Brian I.C.I.→NRRL 2300
P. biourgei	Dx-5	----
P. brunneorubrum	Bi-331→Dx-28	Bi-48→Thom 4733.22→NRRL 841 = IMI 92198 ↘ CBS st. Biourge→Thom
P. aurantiocandidum	Bi- →Dx-26	Bi-11→Thom 4733.6 NRRL 884 ↘ CBS st. Biourge→NRRL→CBS 292.48 = IMI 39814 = ATCC 10411
P. aurantiogriseum	Krâl →Dx-55	Bi-63→Thom 4733.7→NRRL 971 ↘ CBS st. Biourge
P. hirsutum	IP-84 →Dx-4	Bi-186→Thom 4733.71
P. griseoroseum	IP-85→Bi-401 Dx-3	Bi-29→Thom 4733.70→NRRL 820→NSHB p/39→IMI 92220

Dierckx' classification of Penicillium species

In the same paper, Dierckx (1901) proposed a new classification. He divided the genus in two subgenera: *Aspergilloides* Dierckx and *Eupenicillium* Dierckx (non teleomorph genus *Eupenicillium* Ludwig), on the basis of the conidiophore branching.

Subgenus *Aspergilloides* is characterized by unbranched or loosely branched conidiophores ending in single verticils of phialides. It orginally included 8 species and has been accepted by Biourge (1920), but later renamed *Monoverticillium* Biourge (1923). It is reinstated by Pitt (1979), with *P. aurantiobrunneum* Dierckx as lectotype species (actually *P. glabrum* (Wehmer) Westling), instead of the section *Monoverticillata* Raper & Thom.

Subgenus *Eupenicillium* is characterized by branched conidiophores, either with a single lateral (asymmetric) branch or two to five symmetric, verticillate primary branches, with secondary branches ending with a verticil of phialides. Subgenus *Eupenicillium* has been accepted by Biourge (1920, 1923), who divided it in two sections: *Bulliardium* Biourge and *Biverticillium* Dierckx in Biourge, in accordance to handwritten notes left by Dierckx (Biourge, 1920). Making formal sections Biourge, however, proposed to raise them at subgeneric rank. Pitt(1979) renamed subgenus *Eupenicillium* sect. *Bulliardum* Biourge as subgenus *Penicillium*, in accordance to the International Code of Botanical Nomenclature, instead of the section *Asymmetrica* Raper & Thom, pp. Pitt (1979) also recognized the subgeneric rank to the section *Monoverticillatum* Dierckx in Biourge, instead of the section *Biverticillata-symmetrica* Raper & Thom.

CONCLUSIONS

The use of Dierckx' line drawings as type material of his species will allow checking strains identified by Biourge and distributed by him to culture collections, and designated as neotypes by Pitt (1979). Such identification checks must be made on pure strains and in standardized conditions as defined by Dierckx and Biourge respectively, and compared with the original descriptions. Such an investigation is intended at the MUCL culture collection.

REFERENCES

BIOURGE, Ph. 1920. Les moisissures du groupe *Penicillium* Link. Etude monographique. Conference au Cercle professoral de l'Université de Louvain, le 3 avril 1916. Association des Anc. Elèves de l'Ecole Supérieure de Brasserie de l'Université de Louvain, 3: 1-33.

BIOURGE, Ph. 1923. Les moisissures du groupe *Penicillium* Link. Étude monographique. La Cellule, 33: 322 p. ill. Louvain.

CARNOY, J.B. 1870. Recherches morphologiques et physiologiques sur les champignons. Bull. Soc. Roy. bot. Belgique 9: 157-321.

DIERCKX, Fr. 1901. Essai de Revision du genre *Penicillium*

Link. Note preliminaire. Annales de la Société Scientifique de Bruxelles, 25: 83-88.
HENNEBERT, G. L. 1979. Philibert Biourge, microbiologiste et mycologue (1864-1924). Colloque d'Histoire des Sciences III, 1977. Les Sciences exactes et naturelles à l'Université de Louvain de 1835 à 1940. Recueil de travaux d'histoire et de philologie, Louvain 15: 61-98.
HENRARD, P. 1937. Le Pere François Dierckx, s.j. Professeur à la Faculte des Sciences à Namur, 1863-1937. Annuaire des Facultés Notre-Dame de la Paix, 1936-1937, 8p, Namur
LINDAU, G. 1907. Fungi imperfecti. Hyphomycetes in Dr. L. Rabenhorst's Kryptogamen flora von Deutschland, Oesterreich und der Schweir. Leipzig.
PITT, J. I. 1979. The genus Penicillium and its teleomorphic states Eupenicilluum and Talaromyces. London: Academic Press.
RAPER, K. B. and THOM, C. 1949. A manual of the Penicillia. Baltimore:Williams & Wilkins.
SACCARDO, P.A. 1902. Sylloge Fungorum, vol XVI Supplementum universale. Pars. 5, Padova, 1291 p.
SEMAL, O. 1897. Recherches sur la fermentation ammoniacale due aux Mucédinées simples. La Cellule 13: 284-312.
THOM, C. 1930. The Penicillia. Baltimore:Williams and Wilkins.
WESTLING, R. 1911. Über die grüne Species der Gattung Penicillium. Arch. Bot. 11: 1-156.

TYPIFICATION OF ASPERGILLUS AND RELATED TELEOMORPH GENERA

W. GAMS and R.A. SAMSON

Centraalbureau voor Schimmelcultures
3740 AG Baarn, Netherlands

SUMMARY

After a historical survey, the genus Aspergillus Mich. ex Link : Fr. is lectotypified by Aspergillus glaucus (Mich. ex L. : Fr.) Link. The monotype of the associated teleomorph-genus Eurotium Link : Fr. is Eurotium herbariorum (Wiggers : Fr.) Link. Two specimens preserved in Persoon's herbarium in Leiden as Mucor herbariorum agree with the present concept of this species but none of them can be considered as a holotype of the species. Hence Malloch & Cain's designation of a neotype (DAOM 137960) with an associated living culture (CBS 516.65 = ATCC 16469) can be upheld. For the anamorph two specimens are preserved in Persoon's herbarium too, which are not sufficiently representative, so that the above-mentioned specimen listed as type for the anamorph A. glaucus by Blaser can be recognized. Typification of the remaining teleomorph genera does not present problems.

INTRODUCTION

There is considerable consensus concerning species concepts and epithets used in Aspergillus and related teleomorphs. Nevertheless, typification of the genus and the recognized species still shows many deficiencies, which must be resolved before nomenclature of the genus can be regarded as stable.

Aspergillus, according to all protologues, is an anamorphic genus. Unlike other groups of the genus, the connection of Eurotium with anamorphs of the A. glaucus group has never been in doubt. Clarification of relationships in other groups came from Subramanian (1972) and Malloch and Cain (1972, 1973), as summarized by Samson (1979).

1. VALIDATION AND TYPIFICATION OF THE PRINCIPAL GENERA

Anamorphic genera

The genus Aspergillus was described by Micheli (1729) with nine original species designated with polynomial names. According to Art. 13.1d of the International Code of Botanical Nomenclature (Voss, 1983), a validation of the genus by Linnaeus or a post-Linnean author is required. Linnaeus (1753, p. 1185) treated three of Micheli's fungi as Mucor leprosus (= Aspergillus cespitosus densissimus, initio niveus, deinde aureus Mich.), M. glaucus (= Aspergillus capitatus, capitulo glauco, seminibus rotundis Mich.) and M. cespitosus (= Aspergillus albus tenuissimus Mich.), and hence he did not validate the genus but only these specific names. In 1797 he again cited these species in Mucor and referred to the illustration of M. glaucus ... by O.F. Müller in Flora Danica (1777: tab. 777, fig. 2).

Gleditsch (1753, considered as post-Linnean) used the generic name Byssus with polynomial combinations, including species VI. based on Micheli's fungus.

As far as we can ascertain, the first post-Linnean author to use the name Aspergillus was Albrecht von Haller (1768), but he did not use it at generic level and maintained a polynomial nomenclature and his work can not therefore be accepted as a validation of the genus.

Persoon (1801, p. 691) treated the species known at that time in Monilia, including Monilia glauca, which was again based on Micheli's "Aspergillus capitatus capitulo glauco". Sowerby (1803) illustrated Mucor glaucus in vol.III, plate 378, fig. 9 of his English Fungi.

Link (1809: 16) is usually cited as the first author who "revalidated" Aspergillus. He recognized six species, including A. glaucus Link, with the possible synonymy of Monilia glauca Pers.

Fries (1832, p. 385) recognized (= sanctioned) all of Link's species of Aspergillus and explicitly stated (p. 383): "Typus Mucor glaucus Linné". More references of that period were given by Fries (1832, p. 383) and Corda (1838), but they are not relevant to the present discussion.

The genus Sterigmatocystis was introduced by Cramer (1859) with the new species S. antacustica, which is now considered a possible synonym of A. niger (Raper and Fennell, 1965). Fresenius (1863, p. 83) redescribed the genus as Sterigmatocystis Fresen., with the single species S. sulphurea Fresen., which he considered congeneric with S. antacustica and distinct from Aspergillus because of what is presently called the biseriate conidiophore structure.

Saccardo (1880, p. 16) listed A. glaucus "(Sow.) Link" as "exemplum" for Aspergillus and was followed in this selection by Clements and Shear (1931) who changed "exemplum" to "type". In addition, Saccardo retained Sterigmatocystis Fresen. as a distinct genus, but most authors of the twentieth century have regarded these fungi as congeneric.

b) Teleomorphic genera:

Link (1809, p. 31) described the genus Eurotium with a single species, Eurotium herbariorum (Pers.) Link (based on Mucor herbariorum Pers.), independently from Aspergillus. Only later it was realized that it applied to cleistothecial teleomorphs correlated with Aspergillus anamorphs. M. herbariorum was, however, attributed by Persoon (1801, p. 202) to Wiggers (1780), von Haller (1768, vol.3, No. 1257) and Gleditsch (1753). Among these, only Wiggers used a binary nomenclature consistently and can thus be considered as author of the basionym Mucor herbariorum. His text consists only of the four words "Mucor herbariorum, sessilis luteus" without citation of a possible basionym or synonym. Haller's corresponding fungus is probably the number 2151 (as correctly cited by Fries, 1832, and not 1257) "Mucor sessilis globosus luteus", corresponding with Gleditsch's "VIc. Mucor; luteus, globosus, in basi tomentosa sessilis"; Gleditsch in turn referred to "Mucor globosus flavus" v.Haller, Enum. meth. Stirp. Helv. 8. 1742. Eurotium Persoon was then sanctioned by Fries (1832, p. 331) among the Mucorini.

An important post-Friesian author dealing with Aspergillus and Eurotium was Corda. In Sturm Deutschlands Flora (1831, p. 13, Tafel 7) he described and illustrated only the anamorph as Aspergillus glaucus "Link". In Icones 2 (1838, p. 18) and Icones 4 (1840, p. 31) he described four species, including A. glaucus Link, but none of the other species recognized by Link. In Icones 4: 36 he also described Eurotium herbariorum among the Physarales, noticing its growth together with, amongst others, Aspergillus glaucus. In Plate 7, fig. 99 he clearly illustrated this association.

The ascal nature of E. herbariorum was recognized by Fresenius (1853). De Bary (1854) firmly established the connection between Aspergillus anamorphs and Eurotium teleomorphs and in 1870 he decided to distinguish Eurotium repens de Bary (=E. herbariorum sensu Fuckel) from E. aspergillus-glaucus de Bary (=E. herbariorum Link) because of considerably smaller conidia and ascospores. Collecting the fungus frequently on jam, he considered the epithet 'herbariorum' inappropriate. He also transferred three anamorphic taxa to Eurotium.

All the other recognized teleomorphic genera with Aspergillus anamorphs are post-Friesian and type species are unequivocally designated. Their type specimens are listed by Samson & Gams (this publication).

Chaetosartorya Subram. 1972 = Harpezomyces Malloch & Cain 1972.
Holotype C. chrysella (Kwon-Chung & Fennell) Subram.

Dichlaena Mont. & Dur. 1848-49.
Holotype D. lentisci Mont. & Dur.

Emericella Berk. & Br. 1857
Holotype E. variecolor Berk. & Br.

Fennellia Wiley & Simmons 1973
Holotype F. flavipes Wiley & Simmons

Hemicarpenteles Sarbhoy & Elphick 1968 = Sclerocleista Subram. 1972
Holotype H. paradoxa Sarbhoy & Elphick

Neosartorya Malloch & Cain 1972
Holotype N. fischeri (Wehmer) Malloch & Cain

Petromyces Malloch & Cain 1972 = Syncleistostroma Subram. 1972
Holotype P. alliaceus Malloch & Cain

Warcupiella Subram. 1972 = Sporophormis Malloch & Cain 1972
Holotype W. spinulosa (Warcup) Subram.

2. NOMENCLATOR OF Aspergillus glaucus AND Eurotium herbariorum

A. glaucus is obviously the lectotype or, if we follow the sanctioning authority of Fries, even the holotype of the genus and it may be cited with the following nomenclator.

ASPERGILLUS GLAUCUS (Mich. ex L. : Fr.) Link

Aspergillus capitatus capitulo glauco Mich. -- Nova Plantarum Genera: 212, Tab. 91:1. 1729.

= Mucor glaucus Mich. ex Linné -- Sp. Pl. 2: 1186. 1753.
= Monilia glauca (Mich. ex L.) Pers. -- Syn. meth. Fung.: 691. 1801.
= Aspergillus glaucus (Mich. ex L. : Fr.) Link -- Mag. Ges. naturf. Fr. Berlin 3: 16. 1809 (synonymy with a question mark) : Fries -- Syst. mycol. 3: 385. 1832.

The holotype (monotype) of Eurotium is E. herbariorum.

EUROTIUM HERBARIORUM (Wiggers : Fr.) Link -- Mag. Ges. naturf. Fr. Berlin 3: 31. 1809. : Fries -- Syst. mycol. 3: 332. 1832.

?= Mucor; luteus, globosus, in basi tomentosa sessilis Gleditsch -- Meth. Fung. p. 162. 1753 = Mucor sessilis globosus luteus v.Haller -- Hist. Stirp. indig. Helv. 3: 413, No. 2151. 1768.
= Mucor herbariorum Wiggers -- Primitiae Florae Holsatiae, p. 111. 1780. Pers., Syn. meth. Fung.: 202. 1801.
= Eurotium aspergillus-glaucus de Bary -- Abh. Senckenberg. naturf. Ges. 7: 379. 1870.
= Eurotium herbariorum ser. minor Mangin 1909 = Aspergillus minor (Mangin) Thom & Raper 1941 = A. mangini Thom & Raper 1944.

3. TYPE SPECIMENS FOR THE TYPE SPECIES OF THE OLD GENERA

The specific identity of A. glaucus has been doubtful since its original description, but at least since Corda (1840) it was usually associated with little-differentiated anamorphs of Eurotium species and no mycologist ever contested this connection. The original description, however, indicates "In mali limonii atque cidonii & pyri, nec non in cucurbitae frustulis, & aliis similibus semiputridis rebus oriri plerumque solet". This habitat indication does not render identification with the osmophilic anamorphs of Eurotium very likely. Linnaeus restricted the substrate to "in Citris, Melonibus, Pomis aliisque corruptis". Only the connection with herbarized plants from Corda onwards renders the identification according to the present concept highly likely.

It has repeatedly been ascertained that no type material is preserved from Micheli and Link (Malloch & Cain, 1972). From De Bary's work some slides were preserved at BM, but they are not at the moment accessible. Study of this material would be interesting, as De Bary (1870) mentioned having seen Link's material, which is no longer available.

Therefore Malloch & Cain (1972) designated the living and dried culture DAOM 137960 (originating from ATCC 16469 = CBS 516.65) as neotype of Eurotium herbariorum. Blaser (1975, p.33) listed A. glaucus only as the anamorph of E. herbariorum and indicated Malloch & Cain's neotype specimen apparently for both the teleomorph and the anamorph. This expansion of the typification was subsequently explicitly taken up by Domsch et al. (1980).

Our own inquiries also resulted in negative information from Uppsala (Herb. Fries) and Prague (herb. Corda). No herbarium is known to be preserved from Wiggers. In Leiden (herb. Persoon), however, two specimens are preserved, L 910.255-494 as "Monilia glauca Pers.?" on withered stems and L 910.255-342 as "Mycoderma glaucum" (both in Persoon's handwriting). The former contains a few uncharacteristic heads of an Aspergillus, the latter only a few penicilli probably belonging to P. roqueforti.

In addition, two specimens, L 910.262.99 "Erotium herbarum" (sic!) on a pressed rose flower, and 910.262-96 "Erotium herbariorum Syn. Fung. p. 202" on an old agaric (both in Persoon's handwriting) are preserved. Both contain ascomata of Eurotium herbariorum in the sense of present usage and no anamorph. While the second specimen obviously dates after 1801, the former undated specimen may by older. We had suggested to consider this as the most authentic neotype specimen to be traced. But subsequent discussion made it clear that there is no need to assume that automatically all of Persoon's specimens have a type value and that it is by far preferable to stick to Malloch & Cain's neotypification, so that both teleomorph and anamorph are present on the same neotype specimen.

The name A. glaucus cannot be used as indication of a group only. It is needed to designate the anamorph of E. herbariorum specifically. The neotypification with L 910.255-494 is not satisfactory and Blaser's choice of DAOM 137960 can stand to preserve present usage.

No herbarium material is known to exist of Sterigmatocystis antacustica Cramer. Neither is there any material preserved of the second species, as Fresenius' collections were destroyed at Frankfurt a.M. during World War II. Raper & Fennell (1965) based their description of Aspergillus sulphureus (Fresen.) Thom & Church = A. fresenii Subram. on a living culture WB 4077 = CBS 550.65, received from E. Yuill, which fits the protologue. This strain was dried down as herb. IMI 211397 and can serve as neotype. In any case there is no doubt about the fact that Sterigmatocystis can be regarded as congeneric with Aspergillus.

Acknowledgements

We are greatly indebted to the curators of the herbaria IMI, PC, PRM, UPS and L for information on and loan of herbarium specimens in their keeping.

REFERENCES

DE BARY, A. (1854). Ueber die Entwicklung und den Zusammenhang von Aspergillus glaucus und Eurotium. Bot. Ztg. 12: 425-434, 441-451, 465-471.

DE BARY, A. & Woronin, M. (1870). Beiträge zur Morphologie und Physiologie der Pilze 3.(1870). Eurotium, Erysiphe, Cicinnobolus. Abh. Senckenb. naturforsch. Ges. 7: 361-382.

BENJAMIN, C.R. (1955). Ascocarps of Aspergillus and Penicillium. Mycologia 47: 669-687.

BLASER, P. (1975). Taxonomische und physiologische Untersuchungen über die Gattung Eurotium Link ex Fr. Sydowia 28: 1-49.

CLEMENTS, F.E. & SHEAR, C.L. (1931). The Genera of Fungi. (Reprint Hafner, New York, 1957).

CORDA, A.C.J. (1831). In J. Sturm Deutschlands Flora. III. Abth., 11. Heft, Nürnberg.

CORDA, A.C.J. (1838). Icones Fungorum hucusque cognitorum. Vol. 2. Pragae.

CORDA, A.C.J. (1840). Icones Fungorum hucusque cognitorum. Vol. 4. Pragae.

CRAMER, C. (1859). Über eine neue Fadenpilzgattung im Ohr, Sterigmatocystis antacustica. Vierteljahresschr. naturf. Ges. Zürich 4: 323; also Bot. Ztg 18: 133. 1860.

DOMSCH, K.H., GAMS, W. & ANDERSON, T.-H. (1980). Compendium of soil fungi. Academic Press, London.

FISCHER, E. (1897). Plectascineae. In Engler & Prantl, Natuerl. Pflanzenfam. 1(1): 290-320. Leipzig.

FRESENIUS, G. (1853). Mykologische Notizen. Bot. Ztg. 11: 474.

FRESENIUS, G. (1850-1863). Beiträge zur Mykologie. Frankfurt a.M.

FRIES, E.M. (1832). Systema mycologicum. Vol. 3(2). Gryphiswaldae.

GLEDISCH, J. G. (1753). Methodus Fungorum. Berolini.

HALLER, A. von (1768). Historia stirpum indigenarum Helvetiae inchoata. Vol. 3: 113-114. and p. 413 (nr. 1257).

LINNÉ, C. von. (1753). Species Plantarum, Tomus II. Holmiae.

LINNÉ, C. von. (1797). Systema Vegetabilium 15: 1020.

MALLOCH, D. & CAIN, R.F. (1972). The Trichocomataceae: Ascomycetes with Aspergillus, Paecilomyces and Penicillium imperfect states. Can. J. Bot. 50: 2613-2628.

MALLOCH, D. & CAIN, R.F. (1973). The Trichocomacae (Ascomycetes): synonyms in recent publications. Can. J. Bot. 51: 1647-1648.

MICHELI, P.A. (1729). Nova Plantarum Genera. Florentiae.

MÜLLER, O.F. (1777). Flora Danica, vol. V, fasc. 13. Kopenha-

gen. (date fide Stafleu F.A. & Cowan, R.S. 1981: Taxonomic literature III: 805, The Hague).
PERSOON, C.H. (1801). Synopsis methodica Fungorum.
PITT, J.I. (1979). The genus Penicillium and its teleomorphic states Eupenicillium and Talaromyces. Academic Press, London (published 1980).
RAPER, K.B. & Fennell, D.I. (1965). The genus Aspergillus. Williams & Wilkins, Baltimore.
SACCARDO, P.A. (1880). Conspectus generum fungorum Italiae inferiorum. Michelia 2: 1-38.
SAMSON, R.A. (1979). A compilation of the Aspergilli described since 1965. Stud. Mycol. 18: 38 pp.
SOWERBY, J. (1797-1809). Coloured figures of English Fungi or Mushrooms. London. Vol. III. 1803.
SUBRAMANIAN, C.V. (1971). Hyphomycetes, an account of Indian species except Cercosporae. ICAR, New Delhi.
SUBRAMANIAN, C.V. (1972). The perfect states of Aspergillus. Current Science 41: 755-761.
WIGGERS, F.H. (1780). Primitiae Florae Holsatiae. Kiliae.

DIALOGUE FOLLOWING DR. GAMS' PAPER

WICKLOW: Is it my understanding that there is no attempt here to name anamorphs because there are no examples of these occurring alone in nature?

GAMS: In cultures, we have no problems because they are both found together. In nature this is not always so. Link, for example, was not aware that Aspergillus and Eurotium were linked.

WICKLOW: S. T. Hill in England has shown that mites which colonize cereals, will eat the conidial heads of the Aspergillus, but will not eat the cleistothecia of the Eurotium.

PITT: In nature you do find the Aspergillus stage of Eurotium by itself, on substrates of low water activities. In the dry habitats where these organisms really grow, the Eurotium stage is simply not found. There is a practical reason for giving names to these Aspergillus anamorphs and that is that these fungi occur in nature without going through the teleomorph stage at all.

WICKLOW: These strains have not lost the ability to produce ascocarps though, because when we isolate them they always produce cleistothecia.

PITT: In nature these fungi can go through thousands of generations without ever producing an ascocarp. In the reduced water activity environments where these fungi thrive, they live there day to day lives as anamorphs.

TYPIFICATION OF THE SPECIES OF ASPERGILLUS AND ASSOCIATED TELEOMORPHS

Robert A. SAMSON and Walter GAMS

Centraalbureau voor Schimmelcultures
Baarn, Netherlands

SUMMARY

The typification of 190 taxa of Aspergillus and associated teleomorphs is investigated and adjusted to meet the rules of the International Code of Botanical Nomenclature. When no dried type specimen was indicated by the original author(s), a lectotype is designated. In 24 cases where the original names apply to the teleomorph (in combination with Aspergillus or other generic names) and no other names are available for the anamorph, new names are proposed for the anamorph.

INTRODUCTION

In the following pages we investigate the typification of the remaining taxa described in Aspergillus, Sterigmatocystis and their associated teleomorph genera and designate types in cases where typification was not appropriate or lacking. Although it would be useful to sort out the typification for all taxa ever described in Aspergillus, we confine this study to those taxa accepted by Raper and Fennell (1965) and Samson (1979) and some species described more recently.

Most taxonomic studies were based on observations of cultures and recent new taxa were always based on living type cultures. The Sydney edition of the International Code of Botanical Nomenclature again confirmed the requirement for dried type specimens to satisfy Art. 37, while living type cultures are considered insufficient. In the case of Aspergillus the available living type cultures are generally stable enough to document the identity of a species several decades after its description. Hence we accept species described after 1958 (Art. 37) for which only living type cultures were designated, and cite them with their original date of publication, but ensure that dried type material is preserved too (see Recommmendation 3, chapter 7).

Aspergillus, according to all protologues, is an anamorphic genus. In this century, following the precedent of Fischer (1897), Aspergillus was repeatedly treated with the explicit inclusion of teleomorphs. The necessity for recognizing distinct genera for groups of teleomorphs within the Trichocomaceae (=Eurotiaceae) was clearly established by Benjamin (1955). When teleomorphic fungi were originally described as Aspergillus, their epithets must be transferred into a teleomorphic genus and they then apply to the holomorphic fungus (Malloch & Cain, 1972). These epithets are no longer available for use in combination with Aspergillus to designate the anamorph only. Although we feel that there is a limited need for different anamorph names for many such species, we follow the recommendation (Recommendation 5, see chapter 7) made in subsequent discussions during this workshop and introduce new epithets for such fungi.

For the species of Eurotium Blaser (1975) designated living type cultures; dried type material has still to be designated based on his suggestions. Further decisions about species delimitation in this genus made by Pitt (this publication) are followed. For the typification of taxa of the A. niger group some typifications were proposed by Al-Musallam (1980). Kozakiewicz (1982) clarified the typification of A. parasiticus. The cultures designated as representative for the species of the A. ochraceus group by Christensen (1982) agree with those cited here.

An investigation of other Aspergillus types showed that dried reference cultures of many improperly typified taxa are preserved in the herbarium of the Commonwealth Mycological Institute in Kew (IMI). Because these specimens were dried down from subcultures of original type cultures at a date later than the original publications, these specimens are considered lectotypes. If no dried IMI specimen is present, dried specimens from CBS subcultures were prepared and designated lectotypes (in herb. CBS). All these accession numbers are identical with those of available living CBS or IMI cultures.

Referring to Pitt's paper on Eurotium (see chapter 6) we take the view that apparently identical teleomorphs with distinguishable anamorphs should have different names (as they apply to the holomorph), whereas taxa differing in minor details of the teleomorph only and to date distinguished as varieties, need not have separate anamorph names.

For all names the protologues have been checked and the precise nomenclature and author's citation are corrected where applicable.

Hemicarpenteles acanthosporus Udagawa & Takada--Bull. Nat. Sci. Mus. Tokyo 14: 503. 1971.
Anam: A. acanthosporus Udagawa & Takada -- ibid.
Holotype NHL 22462, derived from CBS 558.71, ex soil, Bougainville Island.

A. aeneus Sappa -- Allionia 2: 84, 1954 (no type indicated, no dried type specimen).
Original culture WB 4769 = CBS 128.54 = IMI 69855, ex forest soil, Somalia. Herb. CBS 128.58, derived from the type culture, is here designated as lectotype.

A. allahabadii Mehrotra & Agnihotri -- Mycologia 54: 400. 1962 (no dried type specimen).
Original culture WB 4539 = CBS 164.63 = IMI 139273, ex garden soil, Allahabad, India. Herb. CBS 164.63, derived from the type culture, is here designated as lectotype.

Petromyces alliaceus Malloch & Cain -- Can. J. Bot. 50: 2623. 1972 [= Syncleistostroma alliaceum (Thom & Church emend. Fennell & Warcup) Subram. -- Curr. Sci. 41: 756. 1972 (nomen anamorphosis)].
Anamorph: A. alliaceus Thom & Church -- The Aspergilli p. 163. 1926.
Holotype of teleomorph designated by Malloch & Cain (1972) and neotype of anamorph: TRTC 46232, derived from culture ATCC 16891 = CBS 542.65, ex soil Mt. Riddock, Australia.

A. alutaceus Berk. & Curt. -- Grevillea 3(25): 108. 1875. Syn. A. ochraceus Wilhelm -- "Beiträge zur Kenntniss der Pilzgattung Aspergillus". Diss. Strassburg, p. 66. 1877.
Holotype in herb. K, on decaying maize.
No type culture of A. ochraceus available but WB 398 = CBS 108.08 received from Blochwitz is considered representative.

A. ambiguus Sappa -- Allionia 2: 254. 1955 (no type indication, no dried type specimen).
Original culture WB 4737 = CBS 117.58 = IMI 139274, ex savannah soil, Somalia. Herb. CBS 117.58, derived from the type culture, is here designated as lectotype.

Eurotium amstelodami Mangin -- Annls Sci. nat., Bot., Sér. 9, 10: 360. 1909 = A. amstelodami (Mangin) Thom & Church -- The Aspergilli, p. 113. 1926.
As no epithet is available for the anamorph, we propose
ASPERGILLUS HOLLANDICUS st. nov.
Anamorphosis Eurotii amstelodami. Stipites conidiophororum hyalini vel dilute luteovirides, 275-350 µm alti, vesiculae subglobosae, 18-25 µm diam., phialides 5.0-6.5 x 2.5-3.5 µm. Conidia (sub)globosa, spinulosa, plerumque 4.5-5.0 x 3.5-4.0 µm, catenis radiantibus vel columnaribus disposita, olivaceo-viridia. Holotypus CBS 518.65.
Neotype culture CBS 518.65 (designated by Blaser), received from C. Thom. Herb. CBS 518.65, derived from this culture, is designated as neotype for the teleomorph.

A. amylovorus Panasenko ex Samson -- Stud. Mycol., Baarn 18: 28, 1979.
Holotype herb. CBS 600.67, derived from the original culture, ex wheat starch, Ukraine.

A. anthodesmis Bartoli & Maggi -- Trans. Br. mycol. Soc. 71: 386. 1978.
Holotype 103 S in herb. RO, derived from culture CBS 522.77, ex soil in Ivory Coast.

Eurotium appendiculatum Blaser -- Sydowia 28: 38. 1975.
Anam. A. appendiculatus Blaser -- ibid.
Holotypus ZT 8286, isolated from smoked sausage, Switzerland. Derived from living culture CBS 374.75 = IMI 278374.

A. arenarius Raper & Fennell -- Genus Aspergillus p. 475. 1965 (no dried type specimen).
Type culture WB 5012 = CBS 463.65 = IMI 55632, ex soil in Mysore, India. Herb. IMI 55632, derived from the type culture, is here designated as lectotype.

A. asperescens Stolk -- Antonie van Leeuwenhoek 20: 303. 1954.
Type culture CBS 110.51 = WB 4770 = IMI 46813, ex soil from caves, Poland. Dried type specimens were deposited in herb. K and L. The specimen in herb K. is here designated as holotype.

Edyuillia athecia (Raper & Fennell) Subram. -- Curr. Sci. 41: 756. 1982 = Gymneurotium athecium (Raper & Fennell) Malloch & Cain -- Can. J. Bot. 50: 2619. 1972 = Eurotium athecium (Raper & Fennell) v. Arx -- = A. athecius Raper & Fennell -- Gen. Aspergillus, p. 183, 1965 (no dried type specimen).
As no epithet is available for the anamorph, we propose
ASPERGILLUS ATHECIELLUS st. nov.
Anamorphosis Edyuilliae atheciae. Diagnosis Latina in "Genus Aspergillus" (Raper & Fennell) p. 184. 1965 continetur. Typus IMI 32048.
Original culture WB 5000 = CBS 464.65 = IMI 32048, ex coffee beans. Herb. IMI 32048, derived from the type culture, is here also designated as lectotype of the teleomorph.

A. aureolatus Muntanjola-Cvetković & Bata -- Bull. Inst. Jardin bot. Univ. Beograd 1: 196. 1965 (no dried type specimen).
Original culture BKZ = WB 5126 = CBS 190.65 = IMI 136527, ex air, Yugoslavia. Herb. CBS 190.65, derived from the type culture, is here designated as lectotype.

Neosartorya aureola (Fennell & Raper) Malloch & Cain -- Can. J. Bot. 50: 2620. 1972 = A. aureolus Fennell & Raper -- Mycologia 47: 71. 1955 (no dried type specimen).
As no epithet is available for the anamorph, we propose
ASPERGILLUS AUREOLUTEUS st. nov.
Anamorphosis Neosartoryae aureolae. Diagnosis Latina in Mycologia 47: 71. 1955. continetur. Typus CBS 105.55.
The original cultures WB 2244 and 2391, ex soil Gold Coast, Africa, were listed as representative by Raper & Fennell (1965). WB 2244 = CBS 105.55 = IMI 61451 is selected and Herb. CBS 105.55 designated as lectotype.

A. auricomus (Guégen) Saito -- J. Ferment. Technol. Osaka 17: 3. 1939 = Sterigmatocystis auricoma Guegen -- Bull. Soc. mycol. Fr. 15: 171. 1899.
No type material is available. The culture WB 391 = CBS 467.65, received from P. Biourge as A. auricomus, was considered as representative by Raper and Fennell (1965). Herb. CBS 467.65 is hence designated as neotype.

A. avenaceus G. Smith -- Trans. Br. mycol. Soc. 25: 24. 1943 (no dried type specimen).
Type culture WB 517 = CBS 109.46 = IMI 16140, ex pea seed, UK. Herb. CBS 109.46, derived from the type culture, is here designated as lectotype.

A. awamori Nakazawa -- Rep. Inst. Govt. Res. Formosa 1, 1907.
Al-Musallam (1980) designated herb. CBS 557.65 as the neotype. The culture (WB 4948) was originally received from Instituto Ozwaldo Cruz.

Emericella bicolor M. Christensen & States -- Mycologia 70: 337. 1978.
Anam. A. bicolor M. Christensen & States -- ibid.
Holotype RMF 2058 in herb. NY, derived from living culture CBS 425.77 = IMI 216612, ex Artemisia grassland, Wyoming, USA.

A. biplanus Raper & Fennell -- Gen. Aspergillus p. 434. 1965 (no dried type specimen).
Raper and Fennell (1965) mentioned two isolates, WB 5071 en 5072, ex soil in Costa Rica. The first isolate is preserved as herb. IMI 235602 and is here designated as lectotype.

A. bisporus Kwon-Chung & Fennell -- Mycologia 63: 481. 1971.
Dried type specimen in BPI, derived from culture CBS 707.71 = NRRL 3693, ex mouse infected by soil-passage technique, USA.

A. brevipes G. Smith -- Trans. Br. mycol. Soc. 35: 241 1952.
Dried holotype specimen in herb. K, derived from culture WB 4078 = CBS 118.53 = IMI 16034 = IMI 51494, ex soil in Australia.

A. bridgeri M. Christensen -- Mycologia 74: 210. 1982.
Holotype JB 26-1-2 in herb. NY, derived from culture CBS 350.81 = IMI 259098, ex soil under Atriplex gardneri, Sweetwater County, Wyoming, USA.

A. brunneo-uniseriatus Singh & Bakshi -- Trans. Br. mycol. Soc. 44: 160. 1961 (no dried type specimen).
Type culture WB 4273 = CBS 127.61 = IMI 227677, ex soil under Shisham trees, India. Herb. IMI 227677, derived from the type culture, is here designated as lectotype.

A. brunneus Delacr., see Eurotium echinulatum

A. caesiellus Saito -- J. Fac. Sci. Coll. Imp. Univ. Tokyo 18: 49-50. 1904 (no dried type specimen).
Type culture WB 5061 = CBS 470.65 = IMI 172278, received from IFO Saito's type strain. Herb. IMI 172278, derived from the type culture, is here designated as lectotype.

A. caespitosus Raper & Thom -- Mycologia 36: 563. 1944 (no dried type specimen).
Type culture WB 1929 = CBS 103.45 = IMI 16034, ex soil, Fayetteville, Arkansas, USA. Herb. IMI 16034ii, derived from the type culture, is here designated as lectotype.

A. campestris M. Christensen -- Mycologia 74: 212. 1982.
Holotype ST 2-3-1 in herb. NY, derived from CBS 348.81 = IMI 259099, ex top soil from native prairie, near Zap, North Dakota, USA.

A. candidus Link -- Mag. Ges. naturf. Fr. Berlin 3: 16. 1809.
No type specimen could be traced in herb. B. The culture WB 303 = CBS 567.65 = IMI 91889, supplied by J. Westerdijk, was listed as representative by Raper & Fennell (1965). Consequently herb. CBS 567.65, derived from this isolate, is here designated as neotype.

A. carbonarius (Bain.) Thom -- J. agric. Res. 7: 12. 1916 = Sterigmatocystis carbonaria Bain. -- Bull. Soc. bot. Fr. 27: 27. 1880.

As no original material is available, Al-Musallam (1980) designated CBS 556.65 = Thom's 4030-1, originating from A. F. Blakeslee, as neotype culture and specimen.

A. carneus Blochw. -- Annls mycol. 31: 81. 1933.
No type material is known to exist. The culture WB 527 = CBS 494.65 = IMI 135818, ex air, USA, was listed as representative by Raper & Fennell (1965). Herb. IMI 135818, derived from this culture, is here designated as neotype.
Note: Blochwitz described this as a new species and considered it different from Sterigmatocystis carnea van Tieghem. Hence it must not be cited as A. carneus (van Tiegh.) Blochw. (as done by Raper & Fennell, 1965).

Eurotium carnoyi Malloch & Cain -- Can. J. Bot. 50: 63. 1972.
Anam. A. carnoyi Biourge ex Thom & Raper -- Misc. Publ. U.S. Dep. Agric. 426: 34. 1941.
Type culture WB 126 = CBS 471.65 = IMI 17227, obtained by G. Smith from P. Biourge. Herb. IMI 17227, derived from the type culture, is here designated as lectotype.

A. cervinus Massee -- Kew Bull., Misc. Inf. 4: 158. 1914.
Christensen & Fennell (1964) designated a neotype specimen WT 540 (herb. WISC) because Massee's holotype has apparently been lost. This specimen is derived from WB 5025 = CBS 537.65, ex soil from rain forest near Kuala Lumpur, Malaysia.

Eurotium chevalieri Mangin -- Annls Sci. nat., Bot., Sér. 9, 10: 361. 1909 = A. chevalieri (Mangin) Thom & Church -- The Aspergilli, p. 111. 1926.
As no epithet is available for the anamorph, we propose
ASPERGILLUS EQUITIS st. nov.
Anamorphosis Eurotii chevalieri. Stipites conidiophororum 700-850 µm alti, vesiculae fere globosae, 25-35 µm diam., phialides 5-7 x 3.0-3.5 µm. Conidia ovalia vel ellipsoidea, spinulosa, plerumque 4.5-5.5 µm longa. Typus IMI 211382.
Neotype culture designated by Blaser (1975) WB 78 = CBS 522.65 = IMI 211382, isolated by C. Thom from coffee beans in 1916. Herb. IMI 211382, derived from the neotype culture, is here designated as neotype for both the teleomorph and the anamorph.

Chaetosartorya chrysella (Kwon & Fennell) Subram. -- Curr. Sci. 41: 761. 1972 = Harpezomyces chrysellus (Kwon & Fennell) Malloch & Cain -- Can. J. Bot. 50: 2619. 1972 = A. chrysellus Kwon & Fennell -- Gen. Aspergillus, p. 424, 1965 (no dried type specimen).
As no epithet is available for the anamorph, we propose
ASPERGILLUS CHRYSEIDES st. nov.
Anamorphosis Chaetosartoryae chrysellae. Diagnosis in "Gen. Aspergillus" (Raper & Fennell) p. 424 continetur. Typus IMI 238612.
Type culture NRRL 5084 = CBS 472.65 = IMI 238612, ex forest soil, Costa Rica. Herb. IMI 238612, derived from the type culture, is here designated as lectotype.

A. citrisporus Höhnel, see Sclerocleista thaxteri

A. clavato-flavus Raper & Fennell -- Gen. Aspergillus p. 378. 1965 (no dried type specimen).
Type culture WB 5113 = CBS 473.65 = IMI 124937, ex forest soil, Australia. Herb. IMI 124937, derived from the type culture, is here designated as lectotype.

A. clavato-nanicus Batista, Maia & Alecrim -- Anais Fac. med. Univ. Recife 15: 197. 1955 (no dried type specimen).
Type culture WB 4741 = CBS 474.65 = IMI 235352, ex man. Herb. IMI 235352, derived from the type culture, is here designated as lectotype.

A. clavatus Dezm. -- Annls Sci. nat., Bot., Sér. 2, 2: 71. 1834.
No type specimen could be traced in PC.
The culture WB 1 = CBS 513.65 = IMI 15949, (received by C. Thom from J. Westerdijk in 1909), was listed as representative by Raper & Fennell (1965). Herb. IMI 15949, derived from this culture, is here designated as lectotype.

A. conicus Blochwitz apud Dale -- Annls mycol. 12: 38. 1914 and Annls mycol. 27: 206. 1929 (no dried type specimen).
No type material available. WB 149 = CBS 475.65 = IMI 172281, received from P. Biourge in 1924, was considered representative by Raper and Fennell (1965). Herb. IMI 172281, derived from this culture, is here designated as neotype.

A. conjunctus Kwon & Fennell apud Raper & Fennell -- Gen. Aspergillus p. 552. 1965 (no dried type specimen).
Type culture WB 5080 = CBS 476.65 = IMI 135421, ex forest soil, Costa Rica. Herb. IMI 135421, derived from the type culture, is here designated as lectotype.

A. coremiiformis Bartoli & Maggi -- Trans. Br. mycol. Soc. 71: 386. 1978.
Holotype 102 S in herb. RO, derived from CBS 553.77 = IMI 22302, ex tropical rain forest soil, Ivory Coast.

Emericella corrugata Udagawa & Horie -- Mycotaxon 4: 535. 1976.
Anam. A. corrugatus Udagawa & Horie -- ibid.
Holotype NHL 2763, ex sugar cane field, Thailand. Corresponding culture CBS 191.77.

Chaetosartorya cremea (Kwon & Fennell) Subram. -- Curr. Sci. 41: 761. 1972 = Harpezomyces cremeus (Kwon & Fennell) Malloch & Cain, Can. J. Bot. 50: 2620. 1972 = A. cremeus Kwon & Fennell apud Raper & Fennell -- Gen. Aspergillus p. 418. 1965 (no dried type specimen).
As no epithet is available for the anamorph, we propose
ASPERGILLUS CREMEOFLAVUS st. nov.
Anamorphosis Chaetosartoryae cremeae. Diagnosis in "Gen. Aspergillus" (Raper & Fennell) p. 418 continetur. Typus IMI 123749ii.
Type culture WB 5081 = CBS 477.65 = 123749, ex forest soil, Costa Rica. Herb. IMI 123749ii, derived from the type culture, is here designated as lectotype.

Eurotium cristatum (Raper & Fennell) Malloch & Cain -- Can. J. Bot. 50: 64. 1972 = A. cristatus Raper & Fennell -- "Gen. Aspergillus" p. 169. 1965.
Anam. A. spiculosus Blaser -- Sydowia 28: 42. 1975. Connection made by Pitt (this publication).
Type culture for E. cristatum WB 4222 = CBS 123.53 = IMI 172280, received from H. J. Swart, S.Africa. IMI 172278, derived from the type culture, is here designated as lectotype. Type culture for A. spiculosus and dried lectotype specimen CBS 377.75, ex soil, Spain.

A. crustosus Raper & Fennell -- Gen. Aspergillus, p. 532. 1965 (no dried type specimen).
Type culture WB 4988 = CBS 478.65 = IMI 135819, ex human skin,

Illinois, U.S.A. Herb. IMI 135819, derived from the type culture, is here designated as lectotype.

A. curviformis Chowdhery & Rai -- Nova Hedwigia 32: 229. 1980.
Holotype, ex mangrove mud, Kagh Isl. (West Bengal), India (in herb. Lucknow). Living culture not seen.

A. crystallinus Kwon & Fennell apud Raper & Fennell -- Gen. Aspergillus, p. 471. 1965 (no dried type specimen).
Type culture WB 5082 = CBS 479.65 = IMI 139270, ex forest soil, Costa Rica. Herb. IMI 139270, derived from the type culture, is here designated as lectotype.

A. deflectus Fennell & Raper -- Mycologia 47: 83. 1955 (no dried type specimen).
Type culture NRRL 2206 = CBS 109.55 = IMI 61448, ex soil in Brazil. Herb. IMI 61448, derived from the type culture, is here designated as lectotype.

Emericella desertorum Samson & Mouchacca -- Antonie van Leeuwenhoek 40: 121. 1974. Has no anamorph.
Holotype specimen and culture CBS 653.73, ex soil, Kharga Oasis, Egypt.

A. dimorphicus Mehrotra & Prasad -- Trans. Br. mycol. Soc. 52: 331. 1969 (no dried type specimen).
Type culture CBS 649.74 = IMI 131533, ex garden soil, India. Herb. IMI 131533, derived from the type culture, is here designated as lectotype.

A. diversus Raper & Fennell -- Gen. Aspergillus, p. 437. 1965 (no dried type specimen).
The cultures WB 5074 and 5075, ex soil in Costa Rica, were listed as representative by Raper & Fennell (1965). WB 5074 is preserved as CBS 480.65 = IMI 232882. Herb. IMI 232882, derived from this authentic culture, is here designated as lectotype.

A. duricaulis Raper & Fennell -- Gen. Aspergillus, p. 249. 1965 (no dried type specimen).
Type culture WB 4021 = CBS 481.65 = IMI 172282, ex soil in Argentina. Herb. IMI 172282, derived from the type culture, is here designated as lectotype.

A. eburneo-cremeus Sappa -- Allionia 2: 87. 1954 (no indication of type, no dried type specimen).
Type culture WB 4773 = CBS 130.54 = IMI 69856, ex forest soil, Somalia. Herb. IMI 69856, derived from the type culture, is here designated as lectotype.

Eurotium echinulatum Delacr. -- Bull. Soc. mycol. Fr. 9: 266. 1893 = A. echinulatus (Delacr.) Thom & Church -- The Aspergilli, p. 107. 1926.
Anam. A. brunneus Delacr. -- Bull. Soc. mycol. Fr. 9: 185. 1893.
Neotype culture designated by Blaser (1975) for both states WB 131 = CBS 524.65 = IMI 211378, ex Californian figs. Herb. IMI 211378, derived from the neotype culture, is here designated as neotype.

A. egyptiacus Moubasher & Moustafa -- Egypt J. Bot. 15: 153. 1972 (no dried type specimen).
Type culture CBS 656.73 = IMI 141415, ex sandy soil, Egypt. Herb. IMI 141415, derived from the type culture, is here designated as lectotype.

A. *elegans* Gasperini -- Atti Soc. toscana Sci. nat. Pisa, Mem. 8: 328. 1887.
No type material available. The culture WB 4850 = CBS 102.14 = IMI 133962, received from A. Blochwitz, was listed as representative by Raper & Fennell (1965). Herb. CBS 102.14 derived from this culture, is here designated as neotype.

A. *ellipticus* Raper & Fennell -- Gen. Aspergillus p. 319. 1965 (no dried type specimen).
Type culture WB 5120 = CBS 482.65 = IMI 172283, ex soil in Costa Rica. Al-Musallam (1980) found that this culture was mixed with *A. helicothrix* and purified it as CBS 707.79. Herb. CBS 707.79, derived from this culture, is here designated as lectotype.

A. *elongatus* Rai & Agarwal -- Can. J. Bot. 48: 791. 1970 (no dried type specimen).
Type culture WB 5495 = CBS 387.75, ex alkaline soil, India. Herb. CBS 387.75, derived from the type culture, is here designated as lectotype.

A. *flaschentraegeri* Stolk -- Trans. Br. mycol. Soc. 47: 123. 1964.
Stolk (1965) deposited dried specimens of CBS 108.63 = WB 5042 = IMI 101651 in herb. K and L. The specimen in herb K is here designated as the lectotype.

Neosartorya fischeri (Wehmer) Malloch & Cain var. *fischeri* -- Can J. Bot. 50: 2621. 1972.= *A. fischeri* Wehmer -- Centbl. Bakt. ParasitKde, Abt. 2, 18: 390. 1907.

As there is no epithet available for the anamorph, we propose
ASPERGILLUS FISCHERIANUS stat. nov.
Anamorphosis *Neosartoryae fischeri*. Stipites conidiophororum 300-500(-magis quam 1000) µm alti, vesiculae lageniformes, 12-18 µm diam., griseo-virides. Phialides viridescentes, 5.5-7.0 x 2.0-2.5 µm. Conidia subglobosa, asperulata, dilute pigmentata, 2.0-2.5 µm diam., sed ad 3.0 µm longa. Typus IMI 21139ii.
Type culture WB 181 = CBS 101.12 = IMI 21139, received from Wehmer in 1923. Herb. IMI 21139ii, derived from the type culture, is here designated as lectotype.

Neosartorya fischeri var. *glabra* (Fennell & Raper) Malloch & Cain -- Can. J. Bot. 50: 2621. 1972 = *A. fischeri* var. *glaber* Fennell & Raper -- Mycologia 47: 74. 1955.
Type culture WB 2163 = NRRL 2163 = CBS 111.55 = IMI 61447, isolated from rubber scrap, by G.W. Martin, Iowa. Herb. IMI 61447, derived from the type culture, is designated as lectotype here.

Neosartorya fischeri var. *spinosa* (Raper & Fennell) Malloch & Cain -- Can. J. Bot. 50: 2621. 1972 = *A. fischeri* var. *spinosus* Raper & Fennell -- Gen. Aspergillus, p. 256. 1965.
Type culture WB 5034 = CBS 483.65 = IMI 211390, ex soil in Nicaragua. Herb IMI 211390, derived from the type culture, is designated as lectotype here.

Fennellia flavipes Wiley & Simmons -- Mycologia 65: 937. 1973.
Anam. *A. flavipes* (Bain. & Sart.) Thom & Church -- The Aspergilli, p. 155. 1926 = *Sterigmatocystis flavipes* Bain. & Sart. -- Bull. trimest. Soc. mycol. Fr. 27: 90. 1911.
Type of the teleomorph herb. QM 9131. Type culture for the anamorph WB 302 = CBS 103.22 = IMI 171885, received by Da Fonseca from the

Bainier collection. Herb. IMI 171885, derived from the type culture, is here designated as lectotype.

A. flavo-furcatus Batista & Maia ('flavo-furcatis') -- Anais Soc. biol. Pernambuco 8: 94-96. 1955 (no dried type specimen).
Type culture WB 4911 = CBS 484.65 = IMI 124938, ex culture contaminant. Herb. IMI 124938, derived from the type culture, is here designated as lectotype.

A. flavus Link var. flavus -- Mag. ges. naturf. Fr. Berlin 3: 16. 1809 : Fries -- Syst. mycol. 3: 386. 1832
Neotype designated by Kozakiewicz (1982): IMI 124930, derived from NRRL 1957 = CBS 569.65, isolated from cellophane.

A. flavus var. columnaris Raper & Fennell -- Gen. Aspergillus, p. 366. 1965.
Type culture WB 4818 = CBS 485.65 = IMI 124932, ex butter in Japan. Herb. IMI 124932, derived from the type culture, is here designated as lectotype.

A. floriformis Samson & Mouchacca -- Antonie van Leeuwenhoek 40: 343. 1975.
Holotype in herb. CBS, derived from CBS 937.73, ex desert soil, Western Desert, Egypt.

Emericella foeniculicola Udagawa -- Trans. mycol. Soc. Japan 20: 13. 1979.
Anamorph A. foeniculicola Udagawa -- ibid.
Holotype NHL 2777, ex Foeniculum vulgare seed, China; corresponding culture CBS 156.80.

A. foetidus Thom & Raper -- Man. Aspergilli, p. 219. 1945 = A. aureus Nakazawa 1907 non Berk. 1836.
Al-Musallam (1980) pointed out that CBS 121.28 = WB 341, received in 1930 from Yokoyama as A. aureus Nakazawa, is the type culture of this species. Herb. CBS 121.28 is here designated as lectotype.

Emericella foveolata Horie -- Trans. mycol. Soc. Japan 19: 313. 1978.
Anamorph A. foveolatus Horie -- ibid.
Holotype IFM 4547, corresponding culture NHL 2839 = CBS 279.81, ex Indian herbal drug, Japan.

A. fresenii Subram. = A. sulphureus (Fres.) Wehmer -- Pilzgatt. Aspergillus, p. 113. 1901; Thom & Church -- The Aspergilli, p. 185. 1926 = Sterigmatocystis sulphurea Fres. -- Beiträge zur Mykologie 3:
No type material is preserved from Fresenius. The culture WB 4077 = CBS 550.65 = IMI 211397, ex soil from Mysore, India, was listed as representative by Raper & Fennell (1965). Herb. IMI 211397, derived from this culture, is here designated as neotype.

Emericella fruticulosa (Raper & Fennell) Malloch & Cain -- Can. J. Bot. 50: 61. 1972 = A. fruticulosus Raper & Fennell - Gen. Aspergillus, p. 506. 1965 (no dried type specimen).
As no epithet is available for the anamorph, we propose
ASPERGILLUS FRUTICANS st. nov.
Anamorphosis Emericellae fruticulosae. Diagnosis Latina in "Gen. Aspergillus" (Raper & Fennell) p. 506. 1965 continetur. Typus IMI 139279.
Type culture WB 4903 = CBS 468.65 = IMI 139279, ex soil from California. Herb. IMI 139279, derived from the type culture, is here designated as lectotype.

A. fumigatus Fres. var. fumigatus -- Beiträge zur Mykologie: 81. 1863
Fresenius's drawing is perfectly recognizable and can serve as iconotype. In addition, a representative culture is WB 163 = CBS 136.61 = IMI 16152, ex chicken lung. The dried culture IMI 16152, derived from this culture, is here designated as neotype.

A. fumigatus var. ellipticus Raper & Fennell -- Gen. Aspergillus, p. 246. 1965.
Type culture WB 5109 = CBS 487.65 = IMI 172286, ex man. Herb. CBS IMI 172286, derived from the type culture, is here designated as lectotype.

A. funiculosus G. Smith -- Trans. Br. mycol. Soc. 39: 111-114. 1956 (no dried type specimen designated).
Type culture WB 4744 = CBS 116.56 = IMI 44397, ex soil, Nigeria. Herb. IMI 44397, derived from the type culture, is here designated as lectotype.

A. giganteus Wehmer -- Pilzgatt. Aspergillus, p. 85-87. 1901.
No type material is known to exist. The culture WB 10 = CBS 526.65 = IMI 227678, isolated ex bat dung in Yucatan caves by F.A. Wolf, was listed as representative by Raper & Fennell (1965). Herb. IMI 227678, derived from this culture, is here designated as neotype.

Eurotium glabrum Blaser -- Sydowia 28: 35. 1975.
Anam. A. glaber Blaser -- ibid.
Holotype ZT 8218T, ex leaves of Vaccinium myrtillus, Switzerland, derived from culture CBS 379.75.

A. glaucus (Mich. ex L. : Fr.) Link, see Gams & Samson, see Chapter 2.

A. globosus Chowdhery & Rai -- Nova Hedwigia 32: 229. 1980.
Holotype in herb. Lucknow, ex mangrove mud, Kagh Island, India. Culture not seen.

A. gorakhpurensis Kamal & Bhargava -- Trans. Br. mycol. Soc. 52: 338. 1969 (no dried type specimen).
Type culture CBS 684.74 = IMI 130728, ex forest soil under Tectona grandis, India. Herb. IMI 130728, derived from the type culture, is here designated as lectotype.

A. gracilis Bain. -- Bull. trimest. Soc. mycol. Fr. 23: 90. 1907.
No type material available. A representative culture is WB 4962 = CBS 539.65 = IMI 211393, received from R. Emerson, isolated ex gun-firing mechanism returning from the South Pacific. Herb. IMI 211393, derived from this culture, is here designated as neotype.

A. granulosus Raper & Thom -- Mycologia 36: 565. 1944 (no dried type specimen).
Type culture WB 1932 = CBS 588.65 = IMI 172278, ex soil, Fayetteville, Arkansas, USA. Herb. IMI 172278, derived from the type culture, is here designated as lectotype.

Eurotium halophilicum C.M. Christensen, Papavizas & C.R. Benjamin -- Mycologia 51: 636. 1959.
Anam. A. halophilicus C.M. Christensen et al. -- ibid.
Type culture NRRL 2739 = WB 4679 = CBS 122.62 = IMI 211802, ex Triticum sativum. The dried specimen from the type culture deposited

at BPI, which was listed first among the syntypes for both states, is here designated as lectotype.

A. helicothrix Al-Musallam -- Revision of the black Aspergilli, Diss. Utrecht, p. 22. 1980.
Holotype in herb. CBS, derived from culture CBS 677.79, originally a culture contaminant in CBS 482.65 (A. ellipticus).

Eurotium herbariorum (Wiggers : Fr.) Link, see Gams & Samson, this publication.

Eurotium heterocaryoticum C. M. Christensen et al. -- Mycologia 57: 535. 1965
Anam. A. heterocaryoticus C. M. Christensen et al. - ibid.
Holotype NCF C-100 (herb. BPI) derived from CBS 410.65, isolated from rough stored rice in Mexico.

A. heteromorphus Batista & Maia -- Anais Soc. biol. Pernambuco 15: 200. 1955 (no dried type specimen).
Type culture WB 4747 = CBS 117.55 = IMI 172288, ex culture contaminant. Herb. IMI 172288, derived from the type culture, is here designated as lectotype.

Emericella heterothallica (Kwon et al.) Malloch & Cain -- Can. J. Bot. 50: 62. 1972 = A. heterothallicus Kwon, Fennell & Raper apud Raper & Fennell - Gen. Aspergillus, p. 502. 1965.
As no epithet is available for the anamorph, we propose
ASPERGILLUS COMPATIBILIS, st. nov.
Anamorphosis Emericellae heterothallicae. Diagnosis Latina in "Gen. Aspergillus" (Raper & Fennell), p.502. 1965 continetur. Typus CBS 488.65.
Type cultures indicated WB 5096 = CBS 488.65 = IMI 139277 and WB 4981 = CBS 489.65 = IMI 139278, both ex soil, Costa Rica. A cross of the two cultures in herb CBS, representing the teleomorph, is here designated as lectotype.

A. insulicola Montemayor & Santiago -- Mycopathologia 55: 130. 1975 (no dried type specimen).
Type culture CBS 382.75, ex soil, Aves Isl. north of Venezuela. Herb. CBS 382.75, derived from the type culture, is here designated as lectotype.

A. itaconicus Kinoshita -- Bot. Mag. Tokyo 45: 60. 1931 (no dried type specimen).
Type culture WB 161 = CBS 115.32 = IMI 16119, ex juice of salted plums. Herb.IMI 16119, derived from the type culture, is here designated as lectotype.

A. ivoriensis Bartoli & Maggi -- Trans. Br. mycol. Soc. 71: 383. 1978.
Holotype 101 S in herb. RO, derived from CBS 551.77 = IMI 223068, ex tropical rain forest soil, Ivory Coast.

A. janus Raper & Thom -- Mycologia 36: 556. 1944 (no dried type specimen).
The two cultures, WB 1787 and 1936, were listed as representative by Raper & Fennell (1965). Herb. IMI 16065, derived from WB 1787 = CBS 118.45 = IMI 16065, ex soil, Panama, is here designated as lectotype.

A. janus var. brevis Raper & Thom -- Mycologia 36: 561. 1944 (no dried type specimen).
Type culture WB 1935 = CBS 111.46 = IMI 16066, ex soil, Alameda,

Mexico. Herb. IMI 16066, derived from the type culture, is here designated as lectotype.

A. japonicus Saito -- Bot. Mag. Tokyo 20: 61. 1906.
Type culture CBS 114.51, received as Saito 5087 from K. Kominami. Herb. CBS 114.51, derived from the type culture, is here designated as lectotype.

A. kanagawaensis Nehira -- J. Jap. Bot. 26: 109. 1951 (no dried type specimen).
Type culture HUT 2045 = WB 4774 = CBS 538.65 = IMI 126690, ex soil, Kanagawa Pref., Japan. Herb. IMI 126690, derived from the type culture, is here designated as lectotype.

A. lanosus Kamal & Bhargava -- Trans Br. Mycol. Soc. 52: 336. 1969 (no dried type specimen).
Type culture CBS 650.74 = IMI 130727, ex forest soil under Tectona grandis, India. Herb. IMI 130727, derived from the type culture, is here designated as lectotype.

Dichlaena lentisci Mont. & Dur. -- Fl. Alg., 1: 405. 1846-49.
Anam. Aspergillus sp., described by Malloch & Cain (1972: 2616).
Holotype, on Pistacia lentiscus, in v.Höhnel's herb. (FH). Culture of anamorph is lost and the anamorph cannot yet be properly described and named therefore.

A. leporis States & M. Christensen -- Mycologia 58: 738. 1966. Holotype RMF 99 (herb. NY) derived from CBS 151.66 = IMI 25910, ex hare dung (Lepus townsendii), USA.

Eurotium leucocarpum Hadlok & Stolk -- Antonie van Leeuwenhoek 35: 9. 1969.
Anam. A. leucocarpus Hadlok & Stolk -- ibid.
Holotype herb. CBS 353.68, derived from culture CBS 353.68, ex dried sausage, W. Germany.

A. longivesica Huang & Raper -- Mycologia 63: 53. 1971.
Holotype N1129 in Herb WIS, derived from culture CBS 530.71 = IMI 156966, ex rain forest soil, Nigeria.

A. malodoratus Kwon & Fennell apud Raper & Fennell -- Gen. Aspergillus, p. 468. 1965 (no dried type specimen).
Type culture WB 5083 = CBS 490.65 = IMI 172289, ex forest soil, Costa Rica. Herb. IMI 172289, derived from the type culture, is here designated as lectotype.

Hemisartorya maritima Rai & Chowdheri -- Kavaka 3: 73. 1975.
Anam. ASPERGILLUS MARITIMUS st. nov., named here.
Anamorphosis Hemisartoryae maritimae. Diagnosis Latina in Kavaka 3: 73. 1975 continetur. Typus in herb. Lucknow.
Holotypus ex mangrove mud, Kagh Islands, India, in herb. Lucknow. Corresponding culture CBS 186.77.

Eurotium medium Meissner -- Bot. Ztg 55: 536. 1897.
Anam. A. medius Meissner -- ibid.
Probable type culture WB 124 = CBS 113.27, received from Bierberg via Raistrick. Herb. CBS 113.27, derived from this culture, is here designated as lectotype.

A. melleus Yukawa. J. Coll. Agric. Imp. Univ. Tokyo. 1: 366. 1911.
No type culture known, but representative culture WB 5103 = CBS

546.65, received from B. S. Mehrotra, India. Herb. CBS 546.65 derived from this culture, is here designated as neotype.

A. microcysticus Sappa -- Allionia 2: 251.1955 (no type indication, no dried type specimen).
Type culture WB 4749 = CBS 120.58 = IMI 139275, ex savannah soil, Somalia. Herb. IMI 139275, derived from the type culture, is here designated as lectotype.

A. multicolor Sappa -- Allionia 2: 87. 1954 (no indication of type, no dried type specimen).
Type culture WB 4775 = CBS 133.54 = IMI 69875 ex forest soil, Somalia. Herb. IMI 69875, derived from the type culture, is here designated as lectotype.

Emericella navahoensis M. Christensen & States -- Mycologia 74: 226. 1982.
Anam. A. navahoensis M. Christensen & States -- ibid.
Holotype SD-5 in herb. NY, derived from culture CBS 351.81, ex soil near Page, Arizona, USA.

Emericella nidulans (Eidam) Vuill. var. nidulans -- C. r. hebd. Séanc. Acad. sci. Paris 184: 137. 1927 = Sterigmatocystis nidulans Eidam in Cohn -- Beitr. Biol. Pfl. 3: 392. 1883 = A. nidulans (Eidam) Winter -- Rabenh. Krypt. Fl. 1(2): 62. 1883.
As no epithet is available for the anamorph, we propose
ASPERGILLUS NIDULELLUS st. nov.
Anamorphosis Emericellae nidulantis. Stipites conidiophororum saepe sinuosi, leves, cinnamomeo-brunnei, saepe 75-100 x 2.5-3 µm, sursum ad 3.5-5 µm dilatati. Vesiculae 8-10 µm diam. Metulae 5-6 x 2-3 µm, phialides 5-6 x 2.0-2.5 µm. Conidia globosa rugulosa, 3-3.5 µm diam., in massa viridia. Typus IMI 86806.
No type material known to exist. Representative culture WB 187= CBS 589.65 = IMI 86806, obtained from Bainier's collection. Herb. IMI 86806, derived from this culture, is here designated as neotype.

Emericella nidulans var. acristata (Fennell & Raper) Subram. -- Curr. Sci. 41: 758. 1972 = A. nidulans var. acristatus Fennell & Raper, Mycologia 47: 79. 1955 (no dried type specimen).
Type culture WB 2394 = CBS 119.55 = IMI 61453, isolated from fabric exposed in New Mexico. Herb. IMI 86806, derived from the type culture, is here designated as lectotype.

Emericella nidulans var. dentata (Sandhu & Sandhu) Subram. -- Curr. Sci. 41: 758. 1972 = A. nidulans var. dentatus Sandhu & Sandhu -- Mycologia 55: 297. 1963.
Type culture WB 4908 = CBS 114.63 = IMI 126693, isolated from diseased nails in India. Herb. IMI 126693, derived from the type culture, is here designated as lectotype.

Emericella nidulans var. echinulata (Fennell & Raper) Subram. -- Curr. Sci. 41: 758. 1972 = A. nidulans var. echinulatus Fennell & Raper -- Mycologia 47: 79. 1955.
Type culture WB 2395 = CBS 120.55 = IMI 61454, ex soil, Argentina. Herb.IMI 61454, derived from the type culture, is here designated as lectotype.

Emericella nidulans var. lata (Thom & Raper) Subram. -- Curr. Sci. 41: 758. 1972 = A. nidulans var. latus Thom & Raper -- Mycologia 31: 657. 1939.
Type culture WB 200 = CBS 492.65 = IMI 45625. Herb. IMI 45625,

derived from the type culture, is here designated as lectotype.

A. niger van Tieghem -- Annls Sci. nat., Bot., Sér. 5, 8: 240. 1867 = Sterigmatocystis nigra (van Tieghem) van Tieghem -- Bull. Soc. bot. Fr. 24: 102. 1877.
Al-Musallam (1980) designated WB 326 = CBS 554.65, ex tannin-gallic fermentation, USA, as neotype, but several older epithets predate the species and the conservation of the epithet will be necessary.

Eurotium niveoglaucum (Thom & Raper) Malloch & Cain, Can. J. Bot. 50: 64. 1972 = A. niveoglaucus Thom & Raper -- Misc. Publ. U.S. Dep. Agric. 426: 35. 1941.
As no epithet is available for the anamorph, we propose
ASPERGILLUS GLAUCONIVEUS st. nov.
Anamorphosis Eurotii niveoglauci. Stipites conidiophororum leves, hyalini vel brunnei, 1000-1500 µm alti, vesiculam versus dilatati, vesiculae subglobosae, 40-50 µm diam. Phialides 8-10 x 3-4 µm. Conidia ellipsoidea ad ovalia, hyalina, spinulosa, plerumque 7-8 x 6.0-6.5 µm. Typus IMI 32050ii.
Type culture WB 127 = CBS 114.27 = IMI 32050, received from CBS as A. glaucus mut. alba Blochwitz. Herb. IMI 32050ii, derived from the type culture, is here designated as lectotype.

Fennellia nivea (Wiley & Simmons) Samson -- Stud. Mycol. Baarn 18: 5. 1979 = Emericella nivea Wiley & Simmons - Mycologia 65: 934. 1973.
Anam. A. niveus Blochwitz -- Annls mycol. 27: 205. 1929.
Holotype of the teleomorph herb QM 8942, ex soil, Thailand.
Type culture for the anamorph NRRL 5505 = CBS 115.27 = IMI 171878, received from A. Blochwitz. Herb. IMI 171878, derived from this culture, is here designated as neotype.

A. nutans McLennan & Ducker -- Aust. J. Bot. 2: 355. 1954 (no dried type specimen).
Type culture WB 4364 = CBS 121.56 = IMI 62874, ex soil, Australia. Herb.IMI 62874ii, derived from the type culture, is here designated as lectotype.

A. ochraceoroseus Bartoli & Maggi -- Trans. Br. mycol. Soc. 71: 393. 1978.
Holotype 104 S in herb. RO, derived from CBS 550.77 = IMI 223071, ex tropical rain forest soil, Ivory Coast.

A. ochraceus Wilhelm, see A. alutaceus

Sclerocleista ornata (Raper, Fennell & Tresner) Subram. -- Curr. Sci. 41: 757. 1972 = A. ornatus Raper, Fennell & Tresner -- Mycologia 45: 678. 1953.
As no epithet is available for the anamorph, we propose
ASPERGILLUS ORNATULUS st. nov.
Anamorphosis Sclerocleistae ornatae. Diagnosis Latina in Mycologia 45: 678. 1953 continetur. Typus IMI 55295.
Type culture NRRL 2256 = CBS 124.53 = IMI 55295, ex soil in oakwood, Madison, Wisc., U.S.A. Herb. IMI 55295, derived from the type culture, is here designated as lectotype.

A. oryzae (Ahlburg 1878) Cohn var. oryzae -- Jber. schles. ges. vaterl. Kult. 61: 226. 1884 = Eurotium oryzae Ahlburg apud Korschelt -- Dingler polytechn. J. 230: 330. 1878.
Representative culture WB 447 = CBS 102.07 = IMI 16266, received from CBS in 1909. Herb. IMI 167226 derived from this culture, is here designated as neotype.

A. oryzae var. brunneus Murakami -- J. gen. appl. Microbiol. Tokyo 17: 304. 1972 (no dried type specimen)
Type culture CBS 817.72, ex koji. Herb. CBS 817.72, derived from the type culture, is here designated as lectotype.

A. oryzae var. effusus (Tirab.) Ohara -- Res. Bull. Fac. Agric. Gifu Univ. No. 1: 81. 1951 = A. effusus Tirab. -- Annali Bot. Roma 7: 16. 1908.
Representative culture WB 506 = CBS 574.65 = IMI 124935 ex Zea mays. Herb. IMI 124935, derived from this culture, is here designated as neotype.

A. oryzae var. viridis Murakami -- J. Gen. appl. Microbiol. 17. 303. 1972 (no dried type specimen).
Type culture CBS 819.72, ex tane koji. Herb. CBS 819.72, derived from the type culture, is here designated as lectotype.

A. ostianus Wehmer -- Bot. Centbl. 80: 449-461. 1899 (no dried type specimen) =? Sterigmatocystis butyracea Bain. -- Bull. Soc. bot. Fr. 27: 29. 1880 = A. butyraceus (Bain.) Thom & Raper -- Man. Aspergilli, p. 282. 1945.
Representative culture WB 420 = CBS 103.07 = IMI 15960, received from R. Westling as A. ostianus. Herb. IMI 15960, derived from this culture, is here designated as neotype for A. ostianus. A presumably authentic isolate for A. butyraceus, WB 422 = CBS 627.78, deviates from the original diagnosis of that species, but fits the concept of A. ostianus.

A. pallidus Kamyschko -- Nov. Sist. niz. Rast. 1963: 93. 1963.
Type culture VKM F-1136 = CBS 344.67 = IMI 129967, ex soil in Moldavian SSR. Herb. IMI 129967, derived from the type culture, is here designated as lectotype.

A. panamensis Raper & Thom -- Mycologia 36: 568. 1944 (no dried type specimen).
Type culture WB 1785 = CBS 120.45 = IMI 19393, ex soil, Panama. Herb. IMI 19393, derived from the type culture, is here designated as lectotype.

Hemicarpenteles paradoxus Sarbhoy & Elphick -- Trans. Br. mycol. Soc. 51: 156. 1968.
Anam. A. paradoxus Fennell & Raper -- Mycologia 47: 69. 1955 (no dried type specimen).
Type culture for anamorph WB 2162 = CBS 527.65 = IMI 117502, isolated by J. Warcup from possum dung in New Zealand. Herb. IMI 117502, derived from this culture, is here designated as lectotype. Type culture for the teleomorph IMI 61446 = CBS 793.68. Herb IMI 61446 (microscopical slide), derived from the type culture, is here designated as lectotype.

A. parasiticus Speare var. parasiticus -- Bull. Hawaii. Sug. Planters' Assoc. Exp. Stn, Pathol. & Physiol. Ser. 12: 38. 1912.
Type culture WB 502 = CBS 103.13 = IMI 15957(ix), ex Pseudococcus calceolariae. Herb. IMI 15958ix, derived from the type culture was designated as lectotype by Kozakiewicz (1982).

Emericella parvathecia (Raper & Fennell) Malloch & Cain -- Can. J. Bot. 50: 62. 1972 = A. parvathecius Raper & Fennell - Gen. Aspergillus, p. 509. 1965.
As no epithet is available for the anamorph, we propose
ASPERGILLUS MICROTHECIUS st. nov.

Anamorphosis Emericellae parvatheciae. Diagnosis Latina in "Genus Aspergillus" (Raper & Fennell), p. 509. 1965 continetur. Typus IMI 139280.
Type culture WB 4904 = CBS 493.65 = IMI 139280, ex human skin in California. Herb. IMI 139280, derived from the type culture, is here designated as lectotype.

A. parvulus G. Smith -- Trans. Br. mycol. Soc. 44: 45. 1961.
Holotype IMI 86558, derived from culture WB 4753 = CBS 136.61 = IMI 86558, ex forest soil, USA.

A. peniciIloides Speg. -- Revta La Plata Univ. Fac. Agron. Vet. 2: 246. 1896.
No type material obtained from herb. LPS, but representative culture WB 4548 = CBS 540.65 = IMI 211342, ex human skin, Brazil. Herb. IMI 211342, derived from this culture, is here designated as neotype.

A. petrakii Vörös -- Sydowia Beih. 1: 62. 1957 (no type indication, no dried type specimen).
Type culture WB 4369 = CBS 105.57 = IMI 172291, ex Leptinotarsa decemlineata. Herb. IMI 172291, derived from the type culture, is here designated as lectotype.

A. peyronelii Sappa -- Allionia 2: 248. 1955 (no type indication, no dried type specimen).
Type culture WB 4754 = CBS 122.58 = IMI 139271, ex savannah soil, Somalia. Herb. IMI 139271, derived from the type culture, is here designated as lectotype.

A. phoenicis (Corda) Thom -- J. agric. Res. 7: 14. 1916 = Ustilago phoenicis Corda -- Icon. Fung. 4: 9. 1840.
Holotype U. phoenicis on Phoenix dactylifera fruits in PRM, see Al-Musallam (1980). Representative living culture WB 4757 = CBS 552.65 ex kuro-koji.

A. proliferans G. Smith -- Trans. Br. mycol. Soc. 26: 25. 1943.
Type culture WB 1908 = CBS 121.45 = IMI 16105, ex cotton fabric. Herb. IMI 16105iii, derived from the type culture, is here designated as lectotype.

A. protuberus Muntanjola-Cvetkovič -- Mikrobiologija 5: 119. 1968 (no dried type specimen).
Type culture CBS 602.74, ex rubber-coated electric cables. Herb. CBS 602.74, derived from the type culture, is here designated as lectotype.

A. pseudodeflectus Samson & Mouchacca -- Antonie van Leeuwenhoek 40: 345. 1975.
Holotype and living culture CBS 756.74, ex desert soil, Egypt.

Eurotium pseudoglaucum (Blochwitz) Malloch & Cain -- Can. J. Bot. 50: 64. 1972 = A. pseudoglaucus Blochwitz -- Annls mycol. 27: 207. 1929.
As no epithet is available for the anamorph alone, we propose ASPERGILLUS GLAUCOAFFINIS st.nov.
Anamorphosis Eurotii pseudoglauci. Stipites conidiophororum 150-300 µm alti, vesiculae 12-20 µm diam, phialides 6-8 x 3-4 µm. Conidia subglobosa, spinulosa, 5 -7.5 µm diam. Typus IMI 16122ii.
Type culture WB 40 = CBS 123.28 = IMI 16122, received from A. Blochwitz. Herb. IMI 16122ii, derived from the type culture, is here designated as lectotype.

A. pulverulentus (McAlpine) Wehmer -- Centbl. Bakt. ParasitKde, Abt. 2, 18: 385. 1907 = Sterigmatocystis pulverulenta McAlpine -- Agric. Gaz. N.Z.Wales 7: 302. 1897.
Holotype on French beans in herb. VPRI (Al-Musallam, 1980). Representative culture WB 4851 = CBS 558.65, received as A. elatior Mosseray.

A. pulvinus Kwon & Fennell apud Raper & Fennell -- Gen. Aspergillus, p. 455. 1965 (no dried type specimen).
Type culture WB 5078 = CBS 578.65 = IMI 139628, ex forest soil, Costa Rica. Herb. IMI 139628, derived from the type culture, is here designated as lectotype.

A. puniceus Kwon & Fennell apud Raper & Fennell -- Gen. Aspergillus, p. 547. 1965 (no dried type specimen).
Type culture WB 5077 = CBS 495.65 = IMI 126692, ex soil, Costa Rica. Herb. IMI 126692, derived from the type culture, is here designated as lectotype.

Emericella quadrilineata (Thom & Raper) C.R. Benjamin -- Mycologia 47: 680. 1955 = A. quadrilineatus Thom & Raper -- Mycologia 31: 660. 1939.
As no epithet is available for the anamorph, we propose
ASPERGILLUS TETRAZONUS st. nov.
Anamorphosis Emericella quadrilineatae. Diagnosis Latina in Mycologia 47: 680. 1955 continetur. Typus IMI 89351.
Type culture WB 201 = CBS 591.65 = IMI 89351, ex soil in New Jersey. Herb. IMI 89351, derived from the type culture, is here designated as lectotype.

A. raperi Stolk -- Trans. Br. mycol. Soc. 40: 190. 1957.
Holotype specimen in herb. K, derived from culture WB 4221 = CBS 123.56 = IMI 70949, ex grassland soil, Zaïre.

A. recurvatus Raper & Fennell -- Gen. Aspergillus, p. 529. 1965 (no dried type specimen).
Type culture WB 4902 = CBS 496.65 = IMI 36528, ex lizard dung, California. Herb. CBS IMI 36528, derived from the type culture, is here designated as lectotype.

Eurotium repens de Bary -- Abh. Senckenberg. naturf. Ges. 7: 379. 1870 = A. repens (de Bary) Fischer -- Engler & Prantl, Natürl. PflFam. 1(1): 302. 1897.
Anam. A. glaucus var. repens Corda -- Icon. Fung. 5: 53. 1839.
As the binomial A. repens for the anamorph is preoccupied by Fischer's incorrect combination for the teleomorph, the new name
ASPERGILLUS REPTANS nom. nov.
is substituted for Corda's fungus.
Holotype material possibly preserved among De Bary's slides at BM, but at present not accessible. A neotype culture was designated by Blaser (1975), CBS 529.65 = WB 13, received from Da Fonseca. Herb. CBS 529.65, derived from this culture, is designated as neotype here.

A. restrictus G. Smith -- J. Textile Inst. 22. 115. 1931 (no dried type specimen).
Type culture WB 154 = CBS 117.33 = CBS 541.65 = IMI 16267, ex cotton goods, UK. Herb. IMI 16267, derived from the type culture, is here designated as lectotype.

A. robustus M. Christensen & Raper -- Mycologia 70: 200. 1978.
Holotype WB 5286 (NY) derived from CBS 428.77 = IMI 216610, ex soil in thorn forest, Kenya.

Eurotium rubrum König, Spieckermann & Bremer -- Z. Unters. Nahr- u. Genussm. 4: 726. 1901 = A. ruber (König et al.) Thom & Church -- The Aspergilli, p. 112. 1926. Syn. A. sejunctus Bain. & Sart. -- Bull. trimest. Soc. mycol. Fr. 27: 361. 1911.

As no epithet is available for the anamorph only, we propose

ASPERGILLUS RUBROBRUNNEUS st. nov.

Anamorphosis Eurotii rubri. Stipites conidiophororum leves, hyalini vel aurantiobrunnei, 500-750 µm alti, vesiculae subglobosae, 25-35 µm diam. Phialides 7-9 x 4-5 µm. Conidia ellipsoidea, ovalia vel subglobosa, dense spinulosa, plerumque 5-7.5 µm diam., capitula griseo-brunnea formantia. Typus CBS 530.65.

A. ferrugineus (Sow.) Link (L. Sp. Pl. Ed. 4, 6(1): 68. 1824) was suggested by Raper & Fennell (1965) to be an available name for the anamorph. Sowerby's (Col. Fig. Engl. Fungi 3, pl. 378, fig. 10. 1803) figure of Mucor ferrugineus renders this interpretation unlikely; its classification as Nematogonium ferrugineum (Sow.)Hughes was based on Persoon's specimen

As no type material for E. rubrum is known to be in existence, the representative culture WB 52 = CBS 530.65 is selected and a dried culture of it is designated as neotype.

Emericella rugulosa (Thom & Raper) C.R. Benjamin -- Mycologia 47: 680. 1955 = A. rugulosus Thom & Raper -- Mycologia 31: 660. 1939.

As no epithet is available for the anamorph, we propose

ASPERGILLUS RUGULOVALVUS st. nov.

Anamorphosis Emericellae rugulosae. Stipites conidiophororum sinuosi, leves, dilute brunnei, 50-80 µm alti, vesiculae hemiphaericae, 8-10 µm diam. Metulae 7-8 x 3-3.5 µm, phialides 6-7 x 2.5-3.0 µm. Conidia globosa, viridia, rugulosa, 3-4 µm diam. Typus IMI 136775.

Type culture Thom 4138.T11 = WB 206 = CBS 133.60 = IMI 136775, ex soil in New Jersey, U.S.A. Herb. IMI 136775, derived from the type culture, is here designated as lectotype.

A. sclerotiorum Huber -- Phytopathology 23: 306. 1933 (no dried type).

Type culture WB 415 = CBS 549.65 = IMI 56673, ex Pyrus malus. Herb. IMI 55673, derived from the type culture, is here designated as lectotype.

A. silvaticus Fennell & Raper -- Mycologia 47: 83. 1955 (no dried type specimen).

Type culture WB 2398 = CBS 128.55 = IMI 61456, ex soil in Tafo, Gold Coast, Africa. Herb.IMI 61456, derived from the type culture, is here designated as lectotype.

A. sparsus Raper & Thom -- Mycologia 36: 572. 1944 (no dried type specimen).

Type culture WB 1933 = CBS 139.61 = IMI 19394, ex soil, Honduras. Herb. IMI 19394, derived from the type culture, is here designated as lectotype.

Emericella spectabilis M. Christensen & Raper -- Mycologia 70: 333.1978.

Anam.A. spectabilis M. Christensen & Raper -- ibid.

Holotype RMFH 429 in herb. NY, derived from culture CBS 429.77 = IMI 21661, ex coal mine soils, USA.

A. speluneus Raper & Fennell -- Gen. Aspergillus, p. 457. 1965 (no dried type specimen).

The cultures WB 4989 and 4990 were listed as representative by Raper & Fennell (1965). WB 4989 = CBS 497.65 = IMI 211389, ex dead cane crickets, is selected and Herb. IMI 211389, derived from this culture, is here designated as lectotype.

A. spiculosus Blaser, see Eurotium cristatum

Warcupiella spinulosa (Warcup) Subramanian -- Curr. Sci. 41: 757. 1972 = Sporophormis spinulosa (Warcup) Malloch & Cain -- Can. J. Bot. 50: 2624. 1972 = A. spinulosus Warcup in Raper & Fennell -- Gen. Aspergillus, p. 204. 1965.

As no epithet is available for the anamorph only, we propose ASPERGILLUS WARCUPII st.nov.

Anamorphosis Warcupiellae spinulosae. Diagnosis Latina in "Gen. Aspergillus" (Raper & Fennell) p. 204. 1965 continetur. Typus IMI 75885.

Type culture WB 4376 = CBS 512.65 = IMI 75885. Herb. IMI 75885 is designated here as lectotype.

A. stellatus Curzi, see Emericella variecolor

Neosartorya stramenia (Novak & Raper) Malloch & Cain -- Can. J. Bot. 50: 2622. 1972 = A. stramenius Novak & Raper in Raper & Fennell -- Gen. Aspergillus, p. 260. 1965.

As no epithet is available for the anamorph alone, we propose ASPERGILLUS PALEACEUS st. nov.

Anamorphosis Neosartoryae strameniae. Diagnosis Latina in "Gen. Aspergillus" (Raper & Fennell) p. 260 continetur. Typus IMI 172293.

Type culture WB 4652 = CBS 498.65 = IMI 172293, ex forest soil, Wisconsin, USA. Herb. IMI 172293, derived from the type culture, is here designated as lectotype.

Emericella striata (Rai, Tewari & Mukerji) Malloch & Cain -- Can. J. Bot. 50: 62. 1972 = A. striatus Rai et al., Can. J. Bot. 42: 1521. 1964.

As no epithet is available for the anamorph alone, we propose ASPERGILLUS STRIATULUS st. nov.

Anamorphosis Emericellae striatae. Diagnosis Latina in Can. J. Bot. 42: 1521. 1972 continetur. Typus IMI 96679.

Type culture WB 4699 = CBS 592.65 = IMI 96679, ex mangrove mud, Kagh Islands, India. Herb.IMI 96679, derived from the type culture, is here designated as lectotype.

Chaetosartorya stromatoides Wiley & Simmons -- Mycologia 65. 935. 1973.

Anam. A. stromatoides Raper & Fennell -- Gen. Aspergillus, p. 421. 1965 (no dried type specimen).

Holotype for the teleomorph herb. QM 8944, ex soil, Thailand, derived from QM 8944 = CBS 265.73. Type culture of the anamorph is WB 4519 = CBS 500.65 = IMI 123750, ex soil, Panama. Herb. IMI 123750, derived from the type culture, is here designated as lectotype.

Emericella sublata Horie -- Trans. mycol. Soc. Japan 20: 481. 1979.

Anam. A. sublatus Horie -- ibid.

Holotype IFM 4553, ex Japanese herbal drug, Japan. Correlated cultures in NHL and IFO (not seen).

A. subolivaceus Raper & Fennell -- Gen. Aspergillus, p. 385. 1965 (no dried type specimen).

Type culture WB 4998 = CBS 501.65 = IMI 44882, ex Lintafelt cotton, England. Herb. IMI 44882, derived from the type culture, is here designated as lectotype.

A. subsessilis Raper & Fennell -- Gen. Aspergillus, p. 530. 1965 (no dried type specimen).

Type culture WB 4905 = CBS 502.65 = IMI 135820, ex desert soil, USA. Herb. IMI 135820, derived from the type culture, is here designated as lectotype.

A. sulphureus (Fres.) Thom & Church, see A. fresenii

A. sydowii (Bain. & Sart.) Thom & Church -- The Aspergilli, p. 147. 1926 = Sterigmatocystis sydowii Bain. & Sart. -- Annls mycol. 11: 25. 1913. No type material is known to be extant. Representative culture WB 250 = CBS 593.65 = IMI 211384. Herb. IMI 211384, derived from this culture, is here designated as neotype.

A. tamarii Kita -- Centbl. Bakt. ParasitKde, Abt. 2, 37: 433. 1913. Authentic culture CBS 104.13, received from G. Kita, ex koji, Japan. Herb. CBS 104.13, derived from this culture, is here designated as lectotype.

A. terreus Thom apud Thom & Church var. terreus -- Am. J. Bot. 5: 85. 1918.
Type culture WB 255 = CBS 601.65 = IMI 17294, ex soil, USA.
Herb. IMI 17294, derived from the type culture, is here designated as lectotype.

A. terreus var. africanus Fennell & Raper -- Mycologia 47: 86. 1965 (no dried type specimen).
Type culture WB 2399 = CBS 130.55 = IMI 61457, ex soil, Tafo, Gold Coast, Africa. Herb. IMI 61457, derived from the type culture, is here designated as lectotype.

A. terreus var. aureus Thom & Raper -- Man. Aspergilli, p. 198. 1945 (no dried type specimen).
Type culture WB 1923 = CBS 503.65 = IMI 82431, ex soil, Texas. Herb. IMI 82431, derived from the type culture, is here designated as lectotype.

A. terricola Marchal var. terricola -- Rev. mycol. 15: 101. 1893. No type known to be preserved, but representative culture WB 426 = CBS 579.65 = IMI 172294, received from C. Thom. Herb. IMI 172294, derived from this culture, is here designated as neotype.

A. terricola var. americanus Thom & Church -- Am. J. Bot. 8: 125. 1921 (No dried type specimen).
Type culture WB 424 = CBS 580.65 = IMI 16127, ex soil, Georgia. Herb. IMI 16127, derived from the type culture, is here designated as lectotype.

A. terricola var. indicus (Mehrotra & Agnihotri) Raper & Fennell -- Gen. Aspergillus, p. 412. 1965 = A. indicus Mehrotra & Agnihotri -- Mycologia 54. 403. 1962.
Type culture WB 4680 = CBS 167.73 = IMI 172295, ex bread. Herb. IMI 172295, derived from the type culture, is here designated as lectotype.

Sclerocleista thaxteri Subram. -- Curr. Sci. 41: 757. 1972 = Neosartorya citrispora Malloch & Cain -- Can. J. Bot. 50: 2620. 1972.
Anam. A. citrisporus Höhnel -- Sber. K. Akad. Wiss. Wien, Math.-naturw. Cl. 111: 1036 (Fragm. Mykol. 45). 1902.
Subramanian (1972) designated a dried specimen (herb. FH) of WB 2292 = CBS 105.25 = IMI 55296, ex caterpillar dung, Kittery Point, USA, as the holotype. Living representative culture WB 2292 =

CBS 105.25, isolated by R. Thaxter and maintained at CBS, probably originating from the same source.

A. thomii G. Smith. -- Trans. Br. mycol. Soc. 34: 17. 1951 (no dried type specimen).
Type culture WB 2097 = CBS 120.51 = IMI 45664, culture contaminant. Herb. IMI 45664, derived from the type culture, is here designated as lectotype.

Eurotium tonophilum Ohtsuki -- Bot. Mag. Tokyo 75: 438. 1962.
Anam. A. tonophilus Ohtsuki -- ibid.
Type culture WB 5124 = CBS 405.65 = IMI 108299, ex optical lens, Japan. Herb. IMI 108299, derived from the type culture, is here designated as lectotype.

A. toxicarius Murakami -- J. gen. appl. Microbiol. Tokyo 17: 307. 1971. (no dried type specimen).
Type culture CBS 822.72 = IMI 89717, ex Uganda groundnuts. Herb. CBS 822.72, derived from the type culture, is here designated as lectotype.

Emericella unguis Malloch & Cain, Can. J. Bot. 50: 62. 1972.
Anam. A. unguis (Emile-Weil & Gaudin) Thom & Raper -- Mycologia 31: 667. 1939 = Sterigmatocystis unguis Emile-Weil & Gaudin -- Arch. Méd. exp. Anal. Pathol. Paris 28: 463. 1918.
Type culture of teleomorph WB 2393 = QM 25b = CBS 132.55 = IMI 61452, ex shoe leather. Herb. IMI 61452, derived from the type culture, is designated as lectotype.
Type culture for the anamorph WB 216 = CBS 595.65 = IMI 136525 ex man. Herb. IMI 136525, derived from the type culture, is here designated as lectotype.

A. unilateralis Thrower -- Aust. J. Bot. 2: 355. 1954.
Type culture WB 4366 = CBS 126.56 = IMI 62876. Herb. IMI 62876, derived from the type culture, is here designated as lectotype.

A. ustus (Bain.) Thom. & Church -- The Aspergilli, p. 152. 1926 = Sterigmatocystis usta Bain. -- Bull. Soc. bot. Fr. 28: 78. 1881.
No type material is known to be preserved. Representative culture WB 275 = CBS 261.67 = IMI 211805. Herb. IMI 211805, derived from this culture, is designated here as neotype.

A. varians Wehmer -- Bot. Centbl. 80: 460. 1899. emend. Raper & Fennell, Gen. Aspergillus p. 467. 1965.
Type culture Thom 115 = WB 4793 = CBS 505.65 = IMI 172297. Herb. IMI 172297, derived from the type culture, is designated here as lectotype.

Emericella variecolor Berk. & Br. apud Berk. var. variecolor -- Introd. Crypt. Bot., p. 340. 1857 = A. variecolor (Berk. & Br.) Thom & Raper -- Mycologia 31: 663. 139. Syn. A. stellatus Curzi -- Rendic. Accad. naz. Lincei 19: 424. 1934.
As no epithet is available for the anamorph alone, we propose
ASPERGILLUS STELLIFER st. nov.
Anamorphosis Emericellae variecoloris. Stipites conidiophororum leves, cinnamomeo-brunnei, 140-200 µm alti, vesiculae hemisphaericae, 8-10 µm diam. Metulae 7-8 x 3-4 µm, phialides 8-9 x 2.5-3.0 µm. Conidia globosa, rugulosa, 2.5-3.5 µm diam. Typus CBS 598.65.
Holotype material in herb. K, "ad Bowenpilly, prope Secundarabad". Representative culture CBS 598.65 = IMI 136778, ex soil, Panama.

Emericella variecolor var. astellata (Fennell & Raper) C.R. Benjamin -- Mycologia 47: 679. 1955 = A. variecolor var. astellatus Fennell & Raper -- Mycologia 47: 81. 1955.
Type culture WB 2396 = CBS 134.55 = IMI 61455, isolated from leaves of Ilex sp., Baltra Island, Galápagos, by G.W. Martin. Herb. IMI 61455, derived from the type culture, is designated as lectotype here.

A. versicolor (Vuill.) Tiraboschi -- Annali Bot. Roma 7: 9. 1908-9; apud Thom & Church, The Aspergilli, p. 142. 1926 = Sterigmatocystis versicolor Vuill. apud B. Mirsky, Thse Nancy, p. 15. 1903.
Type material not known to be preserved. Representative culture WB 238 = CBS 583.65. Herb. CBS 538.65, derived from this culture, is designated here as neotype.

Emericella violacea (Fennell & Raper) Malloch & Cain -- Can. J. Bot. 50: 62. 1972 = A. violaceus Fennell & Raper -- Mycologia 47: 75. 1955.
As no epithet is available for the anamorph alone, we propose
ASPERGILLUS VIOLACEO-BRUNNEUS st.nov.
Anamorphosis Emericellae violaceae. Diagnosis Latina in Mycologia 47: 75. 1972 continetur. Typus IMI 61449.
Type culture WB 2240 = CBS 138.55 = IMI 61449, ex soil, Tafo, Gold Coast, Africa. Herb. IMI 61449, derived from the type culture, is here designated as lectotype.

A. violaceofuscus Gasperini -- Atti Soc. Toscana Sci. nat. Pisa 8: 326. 1887.
No type material known to be preserved. Representative culture is WB 360 = CBS 123.27, received from M. Church. Herb. CBS 123.27, derived from this culture, is designated as lectotype.

A. viridinutans Ducker & Thrower -- Aust. J. Bot. 2: 355. 1954 (no dried type specimen).
Type culture WB 4365 = CBS 127.56 = IMI 62875, ex rabbit dung, Frameston, Australia. Herb. IMI 62875, derived from the type culture, is here designated as lectotype.

A. wentii Wehmer -- Centbl. Bakt. ParasitKde, Abt. 2, 2: 150. 1896.
Representative, possibly original culture WB 375 = CBS 104.07 = IMI 17295, ex soybeans, Java. Herb. IMI 17295, derived from this culture, is here designated as neotype.

A. zonatus Kwon & Fennell apud Raper & Fennell -- Gen. Aspergillus, p. 377. 1965 (no dried type specimen).
Type culture WB 5079 = CBS 506.65 = IMI 124936, ex forest soil, Costa Rica. Herb. CBS 506.65, derived from the type culture, is here designated as lectotype.

REFERENCES

AL-MUSALLAM, A. (1980). Revison of the black Aspergillus species. Ph.D. Thesis, University of Utrecht.

BENJAMIN, C. R. (1955). Ascocarps of Aspergillus and Penicillium. Mycologia 47: 669-687.

BLASER, P. (1975). Taxonomische und physiologische Untersuchungen über die Gattung Eurotium Link ex Fries. Sydowia 28: 1-49.

CHRISTENSEN, M. (1982). The Aspergillus ochraceus group: Two

new species from Western soils and a synoptic key. Mycologia 74: 210-225.
CHRISTENSEN, M. & FENNELL, D.I. (1964). The rediscovery of Aspergillus cervinus. Mycologia 56: 350-361.
FISCHER, E. (1897). Plectascineae. In: Engler & Prantl, Natürl. PflFam. 1(1): 290-320.
KOZAKIEWICZ, Z. (1982). The identity and typification of Aspergillus parasiticus. Mycotaxon 16: 293-305.
MALLOCH, D. & CAIN, R.F. (1972). The Trichocomaceae: Ascomycetes with Aspergillus, Paecilomyces, and Penicillium imperfect states. Can. J. Bot. 50: 2613-2628.
MALLOCH D. & CAIN, R.F. (1973). The Trichocomaceae (Ascomycetes): synonyms in recent publications. Can. J. Bot. 51: 1647-1648.
RAPER, K.B. & FENNELL, D.I. (1965). The genus Aspergillus. Williams & Wilkins, Baltimore.
SAMSON, R.A. (1979). A compilation of the Aspergilli described since 1965. Stud. Mycol. 18: 38 pp.
SUBRAMANIAN, C.V. (1972). The perfect states of Aspergillus. Curr. Sci. 41: 755-761.

INFRAGENERIC TAXA OF ASPERGILLUS

Walter GAMS, Martha CHRISTENSEN
Agnes H. ONIONS, John I. PITT
and Robert A. SAMSON

The species of Aspergillus fall into distinct clusters, which have been widely accepted. Since Thom and Church (1926), Thom and Raper (1945) and Raper and Fennell (1965), these clusters have been called "groups", a category without nomenclatural standing. In a few cases "series" were named, which do have a nomenclatural status (when described before 1935 even without a Latin diagnosis), but they do not compete for priority unless at the same rank. For the sake of simplicity in citation, we prefer to introduce new sections rather than new combinations from previous series. In contrast, in Penicillium subgenera and sections were named at the beginning of the century and a complete scheme was formally devised by Pitt (1979).

In discussions during this workshop the need was felt to formalize infrageneric taxa in Aspergillus (see Recommendation 6). The only problem to be solved was to establish a hierarchy between few subgenera [possibly correlated with teleomorph genera, which all belong to Malloch's (1985) subfamiliy Dichlaenoideae of the Trichocomaceae] and a larger number of sections equalling the present "groups". Our discussions resulted in the following scheme. The new taxa are formalized below *.

1. Subgenus ASPERGILLUS (automatically generated)

Xerophilic species. Conidiophore stipes smooth-walled, hyaline or brownish or greenish. Vesicles slightly inflated to subglobose, fertile in the upper half. Metulae absent. Conidial masses mostly in shades of green.
Type species A. glaucus (Mich. ex L. : Fr.) Link.
Teleomorphs: Eurotium Link : Fr., ?Dichlaena Mont. & Durieu, Edyuillia Subram..

---

* See also Samson and Seifert, p. 412 of this publication.

a. Section ASPERGILLUS (automatically generated)

A. glaucus group Thom & Church 1926.

Xerophilic species. Conidiophore stipes smooth-walled, hyaline or brownish. Vesicles dome-like. Metulae absent. Conidial heads radiate to somewhat columnar, typically in shades of green but light brown in one species.
Type species A. glaucus (Mich. ex L. : Fr.) Link.
Teleomorphs: Eurotium Link : Fr., Edyuillia Subram.

b. Section RESTRICTI, sect. nov.

A. restrictus series Thom & Raper 1945; A. restrictus group Raper & Fennell 1965.

Species xerophilicae. Stipites leves, hyalini, nonnumquam in summo viriduli. Vesiculae modice inflatae ad subglobosae. Metulae absentes. Capitula conidiorum plus minusve manifeste columnaria. Conidia primum cylindrica, deinde ellipsoidea ad globosa, in massa dilute viridia. Species typica A. restrictus G. Smith.

Xerophilic species. Conidiophore stipes smooth-walled, hyaline, sometimes greenish in apical part. Vesicles slightly inflated to subglobose. Metulae absent. Conidial heads definitely or loosely columnar. Conidia cylindrical when young, later ellipsoidal to globose. Conidial masses in pale green shades.
Type species A. restrictus G. Smith.
Teleomorph unknown.

2. Subgenus FUMIGATI subgen. nov

Stipites leves, hyalini vel viriduli vel flavo-brunnei. Vesiculae lageniformes vel clavatae. Metulae absentes. Conidiorum massa seu dilute griseobrunnea ad obscure coeruleo-viridis, seu roseo-bubalina. Species typica A. fumigatus Fresen.

Conidiophore stipes smooth-walled, hyaline, greenish or yellow-brown. Vesicles flask-shaped or clavate. Metulae absent. Conidial masses either in pale grey-green to dark blue-green or in pinkish fawn shades.
Type species A. fumigatus Fresen.
Teleomorph: Neosartorya Malloch & Cain.

a. Section FUMIGATI (automatically generated)

A. fumigatus group Thom & Church 1926.

Conidiophore stipes smooth-walled, often greenish. Vesicles flask-shaped to clavate. Metulae absent. Phialides confined to the apical part, parallel. Conidial masses compactly columnar, commonly grey-green to dark blue-green.
Type species A. fumigatus Fresen.
Teleomorph: Neosartorya Malloch & Cain.

b. Section CERVINI sect. nov.

A. cervinus group Raper & Fennell 1965.

Stipites leves, flavo-brunnei. Vesiculae globosae ad lageniformes. Metulae absentes. Conidiorum capitula radiantia vel breviter columnaria, conidiorum massa roseo-bubalina. Species typica A. cervinus Massee.

Conidiophore stipes smooth-walled, yellow-brown. Vesicles globose to flask-shaped. Metulae absent. Conidial heads radiate or short columnar, conidial masses in pinkish fawn shades.
Type species A. cervinus Massee.
Teleomorph unknown.

3. Subgenus ORNATI subgen. nov
A. ornatus group Raper & Fennell 1965 (pro parte, to exclude species with hülle cells).

Stipites leves ad inconspicue asperulati. Vesiculae clavatae ad lageniformes. Metulae absentes. Conidiorum capitula radiantia ad laxe columnaria, conidiorum massa griseola vel flavido-viridis ad olivaceo-brunnea, deinde fere nigra. Species typica A. ornatus Raper, Fennell & Tresner.

Conidiophore stipes smooth-walled to slightly roughened. Vesicles clavate to flask-shaped. Metulae absent. Conidial heads radiate to loosely columnar, conidial masses in greyish or yellowish green to olive-brown shades, mostly becoming almost black in age.
Type species: A. ornatus Raper, Fennell & Tresner.
Teleomorphs: Warcupiella Subram., Sclerocleista Subram., Hemicarpenteles Sarbhoy & Elphick.
Note: Because of the teleomorphs, this is still a heterogeneous assemblage.

4. Subgenus CLAVATI subgen nov

Stipites alti, crassitunicati, leves, hyalini. Vesiculae longae clavatae. Metulae absentes, phialides totam phialidem obtegentes. Conidiorum capitula magna, in complures columnas divisa, conidiorum massa coeruleo-viridis. Species typica A. clavatus Desm.

Conidiophore stipes tall, thick-walled, smooth-walled, hyaline. Vesicles long clavate. Metulae absent. Phialides covering the entire vesicle. Conidial heads large, splitting into compact columns, conidial masses in blue-green shades.
Type species: A. clavatus Desm.
Teleomorph unknown.

a. Section CLAVATI (automatically generated)
A. clavatus group Thom & Church 1926.

Only one section, defined as the subgenus.
Type species: A. clavatus Desm.
Teleomorph unknown.

5. Subgenus NIDULANTES subgen. nov

Stipites leves, hyalini vel pigmentati. Vesiculae subglobosae ad modice clavatae, in dimidio superiore fertilia. Metulae praesentes. Conidiorum capitula columnaria vel radiantia, varie colorata. Species typica A.

nidulellus Samson & W. Gams (anamorphe Emericellae nidulantis (Eidam) Vuill.].

Conidiophore stipes smooth-walled, hyaline or pigmented. Vesicles subglobose to somewhat clavate, fertile in the upper half. Metulae present. Conidial heads columnar or radiate, variously pigmented. Hülle cells often present.
Type species: A. nidulellus Samson & W.Gams [anam. of Emericella nidulans (Eidam) Vuill.].
Teleomorphs: Emericella Berk. & Br. (Nidulantes), Fennellia Wiley & Simmons (Flavipedes).

a. Section NIDULANTES automatically generated
A. nidulans group Thom & Church 1926.

Conidiophore stipes brown, smooth-walled, commonly less than 250 um long. Vesicles hemispherical to flask-shaped. Metulae present, covering the upper half of the vesicle. Conidial heads typically columnar, conidial masses in green shades. Hülle cells typically abundantly produced, globose to irregularly ovate or pyriform.

Type species: A. nidulellus Samson & W.Gams (anam. of Emericella nidulans (Eidam) Vuill.)
Teleomorph: Emericella Berk. & Br..

b. Section VERSICOLORES sect. nov.
A. versicolor group Thom & Church 1926

Stipites leves, hyalini vel dilute brunnei, plerumque >300 µm alti. Vesiculae ovatae ad ellipsoideae. Metulae praesentes, dimidium vel tres quartos vesiculae obtegentes. Conidiorum capitula radiantia vel laxe columnaria, conidiorum massa plerumque viridis. Cellulae crassitunicatae ("hülle" dictae) plerumque copiosae, globosae. Species typica A. versicolor (Vuill.) Tirab.

Conidiophore stipes smooth-walled, hyaline or pale brown, mostly >300 µm long. Vesicles ovate to ellipsoidal. Metulae present, covering the upper half to three quarters of the vesicle. Conidial heads radiating or loosely columnar, conidial masses usually in some shades of green. Hülle cells usually abundant, globose.
Type species: A. versicolor (Vuill.) Tirab.
Teleomorph unknown.

c. Section USTI sect. nov.
A. ustus group Thom & Raper 1945.

Stipites brunnei, leves. Vesiculae hemisphaercae. Metulae praesentes. Conidiorum capitula radiantia vel late columnaria, conidiorum massa ochracea, olivacea vel umbrina. Species typica A. ustus (Bain.) Thom & Church.

Conidiophore stipes brown, smooth-walled. Vesicles hemispherical. Metulae present. Conidial heads radiate or broadly columnar, conidial masses drab, olivaceous or dull brown.
Type species: A. ustus (Bain.) Thom & Church
Teleomorph unknown.

d. Section TERREI sect. nov.
A. terreus group Thom & Raper 1945.

Stipites leves, hyalini. Vesiculae hemisphaericae. Metulae praesentes. Conidiorum capitula compacte columnaria. Conidiorum massa ochracea vel cinnamomea vel aurantio-brunnea. Cellulae crassitunicatae ("hülle" dictae) absentes, sed cellulae inflatae, globsae vel ovales, inspissatae, in hyphis submersis formatae. Species typica A. terreus Thom.

Conidiophore stipes smooth-walled, hyaline. Vesicles hemispherical. Metulae present. Conidial heads compactly columnar, conidial masses buff, cinnamon, to orange-brown. Hülle cells absent, but globose to ovate, relatively heavy-walled hyaline cells formed from submerged hyphae.
Type species: A. terreus Thom.
Teleomorph unknown.

e. Section FLAVIPEDES sect. nov.
A. flavipes group Thom & Church 1926.

Stipites leves vel modice incrustati, hyalini vel brunnei. Vesiculae subglobosae vel ovales vel elongatae. Metulae praesentes, plerumque totam vesiculam obtegentes. Conidiorum capitula late vel irregulariter columnaria, conidiorum massa alba vel avellanea vel vinaceobubalina. Cellulae crassitunicatae ("hülle" dictae) vel alia elementa inspissata praesentia. Species typica A. flavipes (Bain. & Sart.) Thom & Church.

Conidiophore stipes smooth-walled or slightly encrusted, hyaline or brown. Vesicles subglobose to ovate or elongate. Metulae present, usually covering the entire vesicle. Conidial heads broadly or irregular columnar, conidial masses white, avellaneous to vinaceous buff. Hülle cells or heavy-walled hyphal elements present.
Type species: A. flavipes (Bain. & Sart.) Thom & Church.
Teleomorph: Fennellia Wiley & Simmons.

6. Subgenus CIRCUMDATI subgen. nov

Stipites pigmentati vel hyalini, leves vel asperulati. Vesiculae perfecte globosae, totae fertiles. Metulae saepe adsunt. Conidiorum capitula plerumque radiantia, conidiorum massa varie colorata. Species typica A. alutaceus Berk. & Curt.

Conidiophore stipes pigmented or hyaline, smooth-walled or roughened. Vesicles perfectly globose, fertile over the entire surface. Metulae often present. Conidial heads typically radiate, conidial masses variously pigmented.
Type species: A. alutaceus Berk. & Curt.
Teleomorphs: Petromyces Malloch & Cain, Chaetosartorya Subram., Hemisartorya Rai & Chowdheri.

a. Section WENTII sect. nov.
A. wentii group Thom & Raper 1945.

Stipites leves vel asperulati, hyalini vel dilute pigmentati. Vesiculae globosae ad pyriformes vel hemisphaericae, totae fertiles. Metulae praesentes vel absentes. Conidiorum capitula globosa vel laxe columnaria, conidiorum massa flavo-brunnea vel sordide ochracea. Species typica *A. wentii* Wehmer.

Conidiophore stipes smooth-walled to roughened, hyaline or lightly pigmented. Vesicles globose to pyriform or hemispherical, fertile over the entire surface. Metulae present or absent. Conidial heads globose to loosely columnar, conidial masses yellow-brown to dull buff.
Type species: *A. wentii* Wehmer.
Teleomorph unknown.

b. Section *FLAVI* sect. nov.
A. flavus-oryzae group Thom & Church 1926; *A. flavus* group Raper & Fennell 1965.

Stipites hyalini, plerumque verrucosi. Vesiculae clavatae, lageniformes, globosae vel subgloboae, in fere tota superficie fertiles. Metulae praesentes vel absentes. Conidiorum capitula radiantia, conidiorum massa flavoviridis vel intense olivaceo-brunnea. Species typica *A. flavus* Link : Fr.

Conidiophore stipes hyaline, usually warted. Vesicles clavate, flask-shaped, globose or subglobose, fertile over most of their surface. Metulae present or absent. Conidial heads radiate, conidial masses yellow-green to deep olive-brown.
Type species: *A. flavus* Link : Fr..
Teleomorph unknown.

c. Section *NIGRI* sect. nov.
A. niger group Thom & Church 1926

Stipites leves, hyalini vel sub vesicula fusci. Vesiculae globosae vel fere globosae, nonnumquam fuscae. Metulae praesentes vel absentes, saepe fuscae. Conidiorum capitula typice radiantia (in paucis speciebus columnis divergentibus disposita), conidiorum massa nigra. Species typica *A. niger* van Tieghem.

Conidiophore stipes smooth and hyaline or pigmented below the vesicle. Vesicles globose or nearly so, sometimes dark brown. Metulae present or absent, often pigmented. Conidial heads typically radiate (in some species in divergent columns), conidial masses in shades of black.
Type species: *A. niger* van Tiegh.
Teleomorph unknown ("*Saitoa* Rajendran & Muthappa" not being an acceptable teleomorph).

d. Section *CIRCUMDATI* (automatically generated)
A. ochraceus group Thom & Church 1926.

Conidiophore stipes usually roughened and often pigmented. Vesicles mostly globose, occasionally elongate. Metulae present. Conidial heads at first globose, later splitting into several divergent columns, conidial masses in yellow, buff or ochraceous shades.
Type species: *A. alutaceus* Berk. & Curt. (= *A. ochraceus* Wilhelm)
Teleomorph *Petromyces* Malloch & Cain

e. Section CANDIDI sect. nov.
A. candidus group Thom & Raper 1945.

Stipites leves, hyalini. Vesiculae globosae vel subglobosae, totae fertiles. Metulae praesentes. Conidiorum capitula globosa, deinde nonnumquam in complures columnas divisa, conidiorum massa alba vel cremea. Species typica A. candidus Link : Fr.

Conidiophore stipes smooth-walled, hyaline. Vesicles globose or subglobose, fertile over the entire surface. Metulae present. Conidial heads globose, later sometimes splitting into several columns, conidial masses white or cream-coloured.
Type species: A. candidus Link : Fr.
Teleomorph unknown.

f. Section CREMEI sect. nov.
A. cremeus group Raper & Fennell, 1965.

Stipites plerumque hyalini, leves. Vesiculae magnae, globosae, totae fertiles. Metulae praesentes vel absentes, nonnumquam ambo modi in una vesicula. Conidiorum capitula laxe radiantia, conidiorum massa bubalino-brunnea vel dilute viridis vel coeruleo-viridis. Species typica A. cremeoflavus Samson & W.Gams [anamorphe Chaetosartoryae cremeae (Kwon & Fennell) Subram.].

Conidiophore stipes mostly hyaline, smooth-walled. Vesicles large, globose, fertile over the entire surface. Metulae present or absent, sometimes on the same vesicle. Conidial heads loosely radiate, conidial masses buff-brown, pale yellow-green or blue-green.
Type species: A. cremeoflavus Samson & W.Gams [Chaetosartorya cremea (Kwon & Fennell) Subram. anam.].
Teleomorph: Chaetosartorya Subram.

g. Section SPARSI sect. nov.
A. sparsus group Raper & Fennell 1965.

Duae formae conidiophororum: magna massam pigmentatam conidiorum ferentia, et reducta brevia. Stipites prioris formae plerumque brunnei et incrustati. Vesiculae globosae, totae fertiles. Metulae plerumque praesentes. Conidiorum capitula globosa, radiantia vel irregulariter divisa, conidiorum massa dilute grisea vel viridis vel olivaceo-bubalina. Species typica A. sparsus Raper & Thom.

Two types of conidiophores: large heads with light grey, green, or olive-buff conidial masses, and reduced heads near the agar surface. Stipes of the larger conidiophores usually in brown shades and encrusted. Vesicles globose, fertile over the entire surface. Metulae usually present. Conidial heads globose, radiate to irregularly split, conidial masses light grey through shades of green to olive-buff.
Type species: A. sparsus Raper & Thom.
Teleomorph ?Hemisartorya Rai & Chowdheri.

REFERENCES

MALLOCH, D. (1985). Taxonomy of the Trichocomaceae. In "Filamentous Microorganisms, biomedical aspects. T. Arai ed. Tokyo: Japan Scient. Soc. Press. pp. 37-45.

PITT, J. I. (1979). The genus Penicillium and its teleomorphic states Eupenicillium and Talaromyces. London: Academic Press.

RAPER, K.B. & FENNELL, D. I. (1965). The genus Aspergillus. Baltimore: Williams & Wilkins Co.

SAMSON, R.A. (1979). A compilation of the Aspergilli described since 1965. Stud. Mycol. 18: 38 pp.

THOM, C. & CHURCH, M. B. (1926). The Aspergilli. Baltimore: Williams & Wilkins Co.

THOM, C. & RAPER, K.B. (1945). A manual of the Aspergilli. Baltimore: Williams & Wilkins Co.

Chapter 2:

TERMINOLOGY AND METHODOLOGY

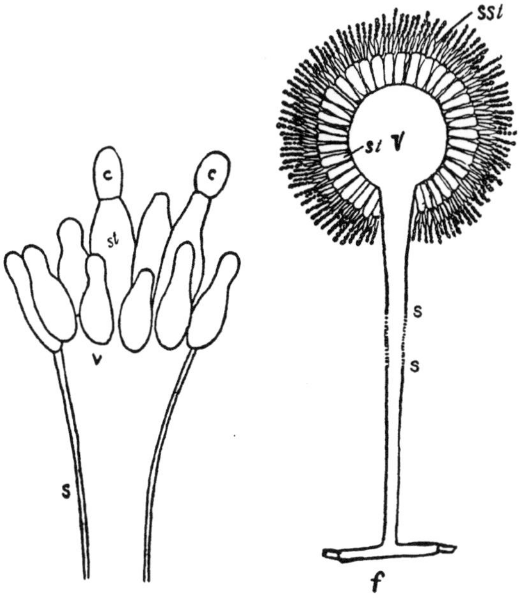

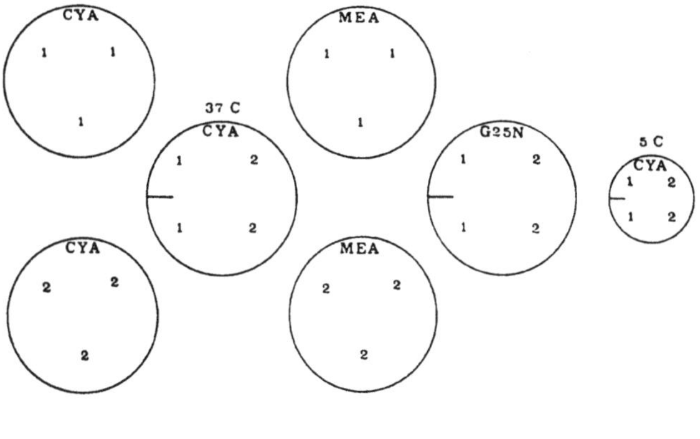

THE PENICILLUS

Carlos RAMÍREZ

Laboratory of General and Applied Mycology
Instituto "Jaime Ferrán" de Microbíologia
28006 Madrid, Spain

INTRODUCTION

The asexual reproductive, brush-like structure of the genus Penicillium, known as the penicillus is the characteristic fruiting structure present in species of this genus. It provides the basis for the delimitation of the genus, permitting distinction between this and other hyphomycete genera, and is also the most satisfactory basis for interspecific taxonomy.

The term conidiophore is understood to cover the entire system of differentiated sporulating structures, including the conidiogenous cells or phialides. It consists of a stipe supporting the brush-like structure or branched part known as the penicillus.

The penicillus may consist of a single whorl of phialides, or may be twice to several times ramified, with the branching system symmetrical or asymmetrical. The penultimate cells below the phialides are known as metulae. All cells between metulae and stipe are referred to as branches by some authors (Samson et al., 1976); others use the term rami (Smith, 1946; Pitt, 1979), or where more than one stage of branching is present, the term ramuli is used as well (Pitt, 1979). The terms monoverticillate (Raper and Thom, 1949) are used for conidiophores bearing only a cluster of phialides.

Structures producing a terminal verticil of metulae and phialides are termed biverticillate (Thom, 1910; Biourge, 1973; Zaleski, 1927; Raper and Thom, 149; Pitt 1979) or one-stage- branched (Samson et.al., 1976). Conidiophores developing lower branches at one or two levels below the verticil of metulae are termed two- and three-stage-branched (Samson et al., 1976), corresponding to the terms terverticillate and quaterverticillate of Pitt (1979), respectively.

Ramigenous conidiophores (Thom, 1930; Samson et al., 1976) include the Subsection Divaricata of Raper and Thom (1949) and

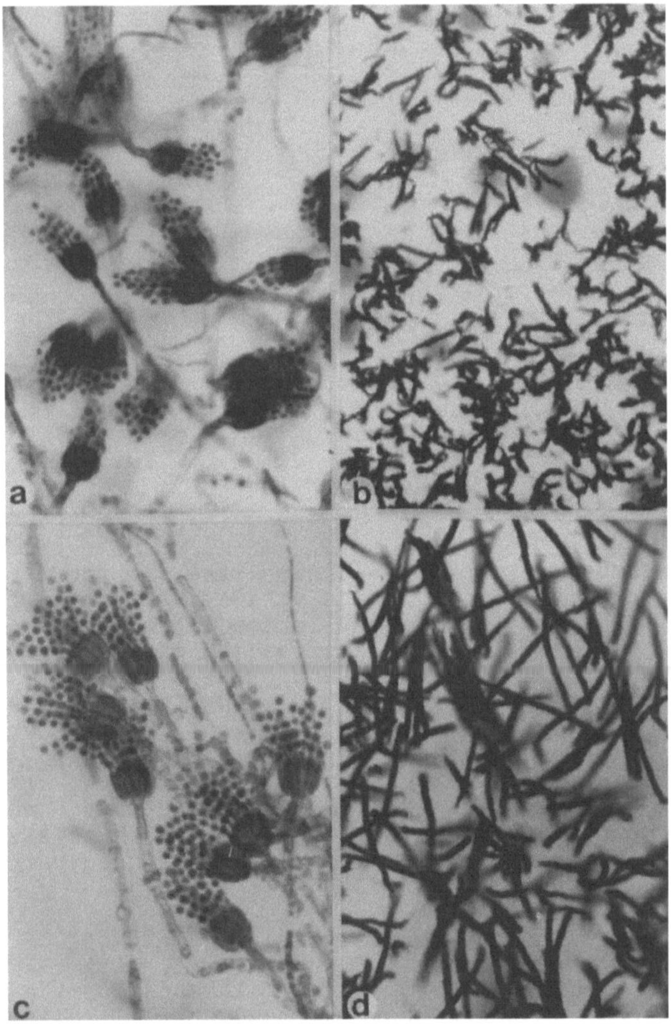

Fig. 1. a-b. Eupenicillium pinetorum Stolk IJFM 19.053, penicilli (x 300) and habit (x 70) respectively; c-d. P. frequentans Maire, IJFM 5955, penicilli (x 300) and habit (x 70) respectively.

consist of a terminal penicillus and a number of lower conidiophores arising from different levels of the stipe. The term branch-like metulae is occasionally used for metulae occurring singly or in irregular groups at a lower level of the stipe.

A number of types of penicillus may be produced by species of the genus Penicillium, determined by the number, shape and configuration of the constitutive elements. Five types are recognized by Pitt (1979) and used as a basis for subgeneric classification.

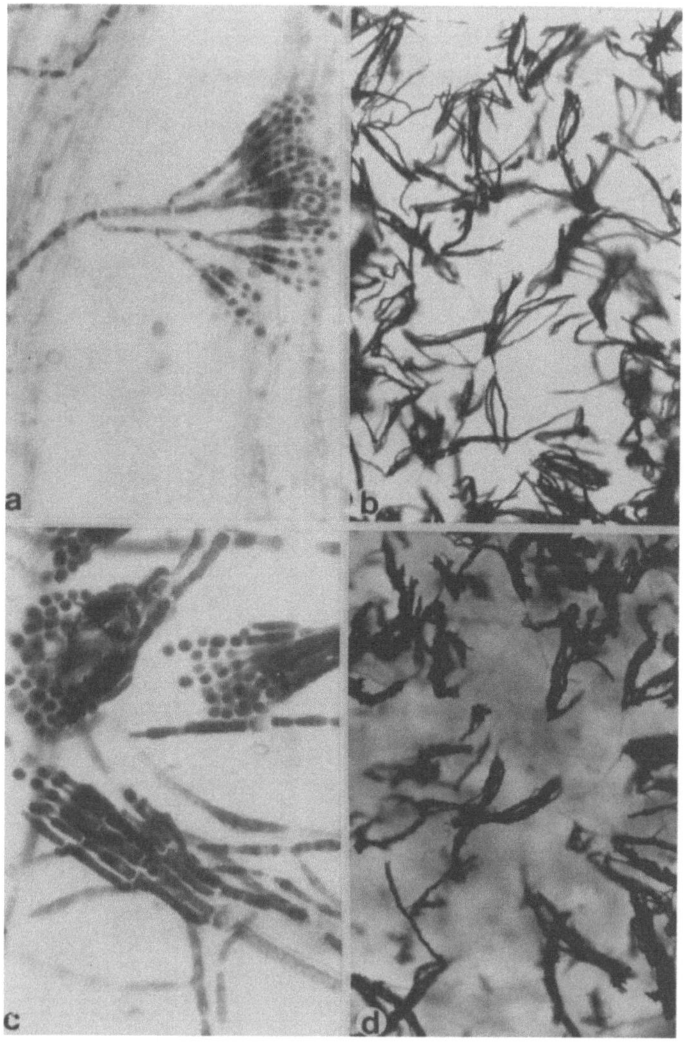

Fig.2. a-b.P. chrysogenum Thom, IJFM 1423, penicilli (x 300) and habit (x 70) respectively; c-d. P. roqueforti Thom, IJFM 19.068, penicilli (x 300) and habit (x 70), respectively.

Patterns of branching are important usually from a configuration viewpoint, rather than because of their morphology. However, some species may characteristically produce stipes metulae or branches, with apical swellings or vesiculation. Wall texture of these elements also has taxonomic significance.

Phialides are conidiogenous cells, i.e., the reproductive functional apparatus in Penicillium species. Two distinct but intergrading shapes may be observed: lageniform or ampulliform or bottle-shaped, which is the type developed in most species

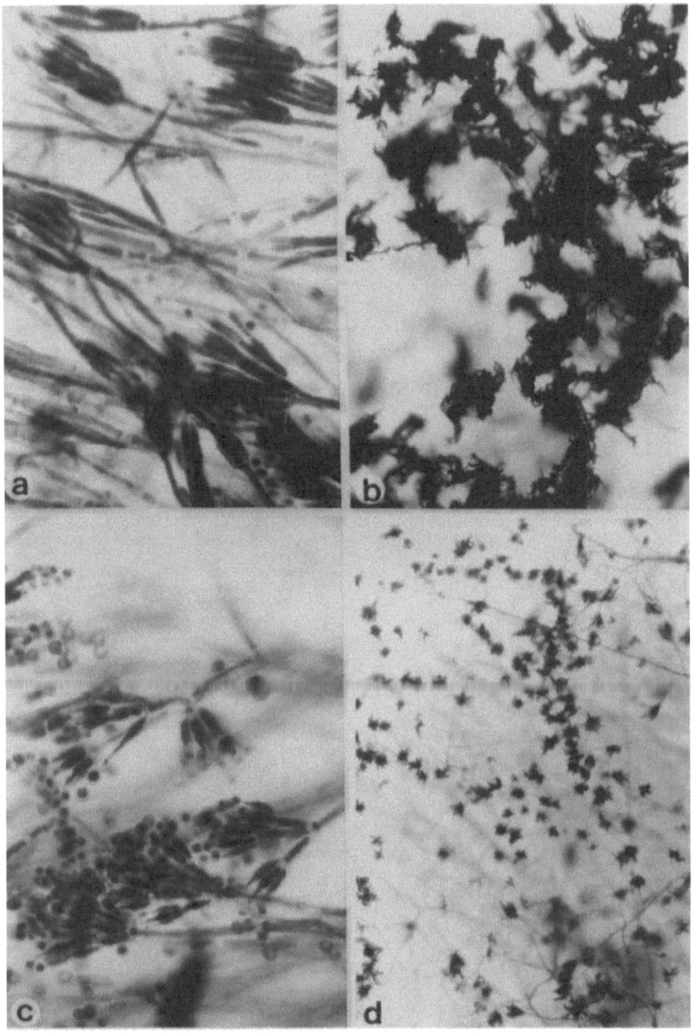

Fig. 3. a-b. P. funiculosum Thom IJFM 19.100 penicilli (x 300) and habit (x 70) respectively; c-d. P. verruculosum Peyronel, IJFM 19.104, penicilli (x 300) and habit (x 70) respectively.

of Penicillium. They may taper at the apex to a more or less conspicuous neck or collulum (Zaleski, 1927), which from inconspicuous, or short and varies, to very long and slender, as in P. janthinellum Biourge, where the phialide tapers abruptly producing a conspicuously long neck of fairly uniform diameter. The second type is acerose or needle-shaped, nearly cylindrical, narrowing at the apex to a smaller conoidal collulum terminating in a minute apical pore.

Conidia are highly developed, single-celled structures, with walls usually heavy, more or less characteristically ornamented. Furthermore, conidia may remain connected in chains of varying length and aggregation, or may fall apart quickly. The chains of conidia may diverge widely, become a tangled mass, be closely aggregated into columnar masses, or stand almost rigidly parallel. This general arrangement of conidial chains is usually fairly constant for each species, and can, in my opinion and that of Raper and Thom (1949), be a useful character in the identification of species. The accompanying photographic examples are presented in support of this opinion (Figs. 1-3).

In summary, I believe that *Penicillium* species can be distinguished by the pattern of the conidiophores, the shape of phialides and conidia, the conidial arrangement, and finally, by the fine ornamentation of conidia as viewed with the scanning electron microscope (SEM).

REFERENCES

BIOURGE, P. 1923. Les moissisures du groupe *Penicillium* Link. Cellule 33: 7-331.

PITT, J.I. 1979. "The genus *Penicillium* and its teleomorphic states *Eupenicillium* and *Talaromyces*" London: Academic Press.

RAMÍREZ, C. 1982. "Manual and Atlas of the Penicillia" Amsterdam: Elsevier Biomedical Press,

RAPER, K.B. and THOM. C 1949. "A Manual of the Penicillia". Baltimore: Williams & Wilkins.

SAMSON, R.A., STOLK, A.C. and HADLOK, R. 1976 Revision of the subsection Fasciculate of *Penicillium* and some allied species. Stud. Mycol., Baarn 11: 1-47.

SMITH, G., 1946 "An Introduction to Industrial Mycology". London: Edward Arnold.

THOM, C. 1910. Cultural Studies of species of *Penicillium*. Bull. Bur. Anim. Ind. US Dep. Agric. 118: 1-109.

ZALESKI, K. 1927. Über die in Polen gefundenen Arten der Gruppe *Penicillium* Link. I, II, and II, - Bull. Acad. pol. Sci. Math. Nat., Ser. B; 1927: 417-563.

DESCRIPTIVE TERMINOLOGY OF THE CONIDIOGENOUS STRUCTURES IN ASPERGILLUS AND PENICILLIUM

David W. MINTER, David L. HAWKSWORTH,
Agnes H.S. ONIONS and Zofia KOZAKIEWICZ

Commonwealth Mycological Institute
Kew, Surrey TW9 3AF, United Kingdom

SUMMARY

The special descriptive terms used for conidiogenous structures in Aspergillus and Penicillium are examined. Terms such as 'metula', ramulus', 'sterigma', 'vesicle', 'stipe' and 'foot cell' are considered misleading, confusing and unnecessary. They should be discarded and replaced by the more accurate descriptive terms currently applied throughout the rest of the Hyphomycetes. A simplified general system recommended for adoption in these genera is illustrated diagrammatically.

INTRODUCTION

Many terms currently employed to describe conidiogenous structures in Aspergillus and Penicillium are used only for fungi of these two genera. Structures very similar or identical to those in Aspergillus and Penicillium are, however, well-known in other Hyphomycetes: conidiogenous structures with elements recalling those of Aspergillus are, for example, to be found in Custingophora, Gliocephalis, Goidanichiella, Gonatobotryum, Haplobasidion and Harziella; and of Penicillium in, for example, Clonostachys, Gliocladium, Harpographium, Phialocephala, Thysanophora and Verticicladiella. For illustrations of these genera see Carmichael et al. (1980).

For such fungi researchers have found no need for the special terms used in Aspergillus and Penicillium. Since these terms are in any case often ill-defined or used in other senses in different groups of fungi, they are a source of confusion to specialists and non-specialists alike, and tend to make research on these two genera isolated from that of other Hyphomycetes. Reform is therefore clearly desirable. From the user's standpoint, the system of terms applied should be as simple as possible and of a descriptive nature, with none or a minimum of terms peculiar to any genus. Terms which convey some meaning in a modern language and can therefore be translated into other languages are preferable to those which are obscure and derived from a dead language.

Kozakiewicz (1978) noted the considerable volume of literature published in recent years on the form of the Aspergillus conidial head relative to its development and function during sporulation, and it follows from this emphasis in recent research that it is also extremely important to ensure that any terminology adopted can enable and encourage researchers to describe aspects of anatomy, development and function independently of each other for species of these two genera. This paper accordingly considers the structures involved and the terms being used to describe them, and it proposes reforms in line with current practice in other hyphomycete genera.

ASPERGILLUS

Traditional terms

The two principal forms taken by species of Aspergillus are shown, labelled with terms used by recent authors, in Fig. 1. Some of these terms i.e. conidium, conidiogenous cell, conidiophore and phialide, are used generally in the Hyphomycetes, and the comparatively few problems which they present are not unique to Aspergillus. The other terms i.e. sterigma, metula, vesicle, stipe and foot cell, are largely confined to use in Aspergillus (some also to Penicillium), or present serious problems if used more widely. These terms will now be reviewed briefly.

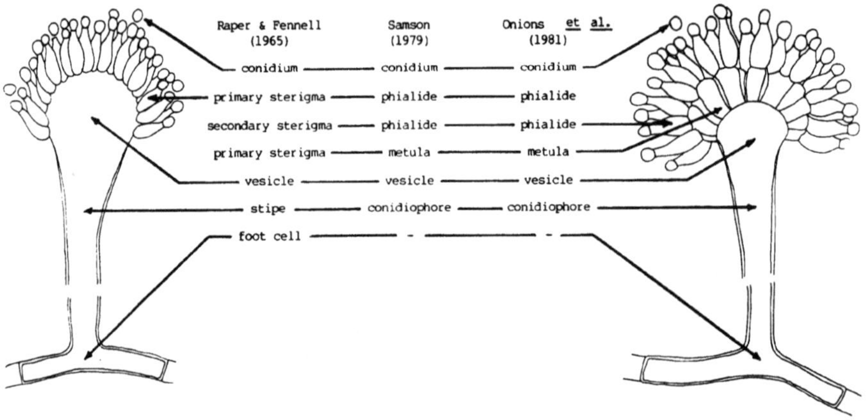

Fig. 1. Terms used by some recent authors to describe the different conidiogenous structures in Aspergillus.

Conidium, conidiogenous cell and conidiophore

From its use in the early days of hyphomycete taxonomy to denote certain specialized types of anamorphic propagules (Vuillemin, 1910), the conidium has gradually become an almost universal term for any anamorphic spore (Hawksworth et al.,

1983), with the first Kananaskis Conference (Kendrick, 1971), Hughes (1953) and Cole and Samson (1979) marking significant stages in this evolution. Not all mycologists agree that the present situation is ideal, however, and there is a growing need for a thorough reappraisal of the universal use of this term, and of the use of terms which are derived from conidium (conidiogenous cell, conidiophore, conidioma). Most of the controversy centres on whether certain large and complex multi-septate spores, e.g. those of many aquatic Hyphomycetes, should be called conidia (Descals, 1985).

As, however, the spores produced by both *Aspergillus* and *Penicillium*, have always been defined as conidia, there seems no immediate problem in applying the term "conidium" to the spores of these genera. There are, however, some difficulties with the terms "conidiogenous cell" and "conidiophore", but these difficulties, discussed below, relate to the general application of these terms and are not unique to *Aspergillus* and *Penicillium*.

Phialide

The term "phialide" has recently been reviewed (Minter et al., 1983 b). Although there are no reasons why the term should not be applied in *Aspergillus* and *Penicillium*, it has been clearly demonstrated that this has been used in such a wide range of senses in the past that it now conveys little or no information. Minter et al., (1983 b) concluded that it was preferable to use the term "conidiogenous cell" accompanied by a detailed description of shape, development and function.

Sterigma

The term sterigma was widely used for conidiogenous cells in the nineteenth century, and has continued into the twentieth century in some works on *Aspergillus* to describe phialides. In these works a distinction has been made between primary and secondary sterigmata, the primaries giving rise secondary sterigmata, the secondary then producing only conidia. In *Penicillium* an analogous use of the term sterigmata has been made by some authors, although many species do not produces swollen apices. In these two genera therefore the term sterigma has been used instead of phialide or conidiogenous cell, but in a modified manner. In the rest of the Hyphomycetes, however, such usage has been generally abandoned for over half a century (Kendrick, 1971) and the term is currently restricted to the spore-bearing protrusions of basidia (Hawksworth et al., 1983).

Metula

The term metula was introduced by Westling (1911). It clearly derives from the diminutive of *meta*, the Latin noun for a pyramidical or conical object. Metula has been used principally in *Aspergillus* and *Penicillium*, where it represents the cell proximally adjacent to the phialide. No need has been found for this term in the other genera similar to

Aspergillus and *Penicillium* noted above (Carmichael et al., 1980; Ellis, 1971, 1976). Furthermore, use of the term within these two genera is confused with that of primary sterigma. Species of *Aspergillus* with metulae have sometimes been described as having biseriate, and those without, uniseriate conidial heads. This usage of biseriate and uniseriate is confusing since these terms are used much more extensively in the Ascomycetes to describe the arrangement of ascospores in asci, and it is not recommended here.

Vesicle

The term vesicle has a long history of use in *Aspergillus* for the distinctive swollen apex of the conidiophore. Unfortunately, the term is also used in other fields of mycology in different senses, most significantly in *Pythium* for an evanescent extra-sporangial structure in which zoospores are differentiated. In their monograph of *Gonatobotryum* and similar Hyphomycete genera, which have swollen conidiophore heads resembling those of *Aspergillus*, Walker and Minter (1981) described this feature without recourse to specialized terms. "Vesicle" has the further disadvantage that it suggests that the swollen apex is a separate structure from the main supporting hypha, which it is not: in most species of *Aspergillus* the supporting hypha and the swollen apex are all parts of a single cell.

Stipe and foot cell

The terms stipe and foot cell are also problematic. Stipe has been used in *Aspergillus* for the part of the conidiophore which is roughly cylindrical and located below the swollen apex. In *Aspergillus* both of these parts are usually only different regions of a single cell. It therefore seems inappropriate to use the term stipe in this context, particularly in view of its extensive use outside *Aspergillus* to describe massive multi-hyphal supporting structures which are a part of basidiomata or ascomata (Ahti, 1982; Hawksworth et al., 1983). Similarly, foot cell is a misnomer, since it too is often only a part of the same usually single-celled stucture, although in some cases it may be delimited by a septum.

From the foregoing review it follows that there is little justification for retaining the terms sterigma, metula, vesicle, stipe and foot cell for use in *Aspergillus* studies; these obsolete and confusing terms should now be abandoned. Practical use of the term phialide is also limited, and the recommendation made by Minter et al. (1983 b) to substitute the term conidiogenous cell and detailed description of development, form and fuction is adopted here.

ALTERNATIVE TERMS

The principal uses of the terms under discussion are to identify without ambiguity which part of the fungus is under consideration and to enable the mycologist then to make clear statements about the morphological characteristics of that part and how it develops or functions (Kozakiewicz, 1978). It follows that there is a need to distinguish concepts of anatomy, development and funtion. An ideal solution would be to have a terminology based on anatomy, with anatomically equivalent parts of different fungi having the same name.

a. Terminology based on anatomy

In a terminology based on anatomy, there must be some agreement about what parts of different fungi are anatomically equivalent. In the case of Aspergillus, the swollen apex and the cells distal to that apex make a useful starting point. In Fig. 2, the two principal forms of growth of Aspergillus are illustrated; in the simpler, one layer of cells is distal to the swollen apex whereas in the more complex there are two layers of cells.

Comparing these two forms, the following questions arise. Are both layers of distal cells in the more complex form anatomically equivalent to the one distal cell layer of the simpler form (Fig. 2A)? If not, which of the distal cell layers are anatomically equivalent? Are the distal cells adjacent to the swollen apex in both forms anatomically equivalent (Fig. 2B)? If so, what does the terminal cell in the more complex form represent? Is it, for example, anatomically equivalent to the first conidium of the simpler form (Fig. 2C)? Although such questions are now being critically addressed (Kozakiewicz, 1978), it is apparent that many of them cannot yet be answered completely for all groups in Aspergillus.

The terminology currently in use in the Hyphomycetes precludes a simple answer to such questions. "Conidiogenous cell", for example, is a single term central to present hyphomycete systems, but it embraces a mixture of concepts: "conidiogenous" relates to function (and indeed development), while "cell" is a distinctly anatomical term. As a result there is no agreement as to whether the conidiogenous cell is a distinct anatomical unit which should be comparable throughout the Hyphomycetes (as the ascus is throughout the Ascomycotina), or whether it simply identifies in functional terms a cell which is actually producing anamorphic propagules. Many researchers accordingly find it hard to conveive the possibility of, for example, conidiogenous cells which have lost their function of producing conidia, or of conidia which do not function as propagules. Such phenomena, however, undoubtedly exist within the Hyphomycetes and may even, arguably, be common (Hughes, 1981; Minter, 1985).

With the term conidiophore there is a similar confusion as to whether the word relates to the functional aspect of the structure ("conidiophores lift conidia above the substratum"), or to the structure itself as an anatomical entity ("conidiophores are certain defined cells adjacent to the conidiogenous cells or conidia irrespective of whether they lift those conidia up or not"). This problem was considered at the first Kananaskis Conference (Kendrick, 1971), but not satisfactorily resolved. In Aspergillus the same issue takes the form of a question of whether the term "conidiophore" should include the conidiogenous cells or be limited to the part of the structure which ends in the swollen apex. These unresolved problems lead to the conclusion that, however ideal it might be to base our terminology on anatomy, this is not at present possible.

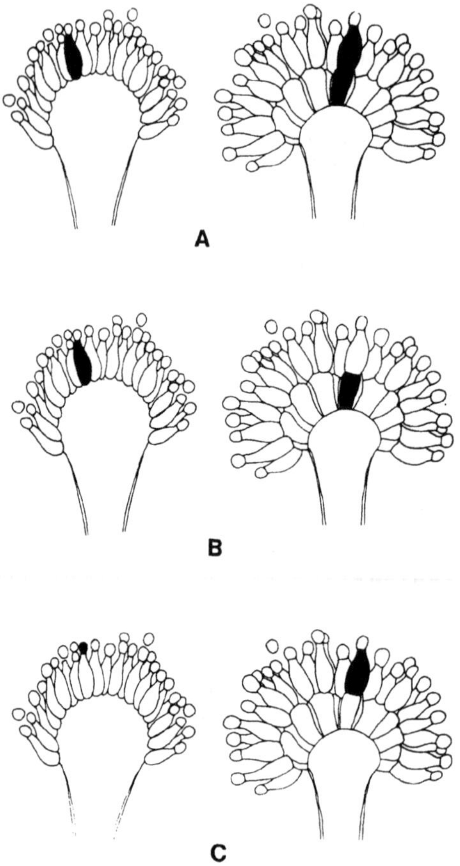

Fig. 2. Three examples of how different (shaded) parts of the conidiogenous structures of different species of Aspergillus could be anatomical equivalents. 2A: conidiogenous cell on the left conidiophore hypothetically equivalent to conidiogenous cell plus supporting cell on the right conidiophore. 2B: conidiogenous cell on the left conidiophore hyptheti-cally equivalent to supporting cell on the right conidiophore. 2C: first conidium on the left conidiophore hypothetically equivalent to conidiogenous cell on the right conidiophore.

b. Terminology based on development

While a terminology describing developmental features is an essential tool in looking at Hyphomycetes, it is hard to imagine how such a terminology could be used to identify effectively and without ambiguity different parts of a fungus; indeed, the part of the fungus under discussion must be identified before one can begin to describe its development. Developmental terms such as "holoblastic", "enteroblastic", "sympodial" and "percurrent" are therefore not suitable for naming the different parts of Aspergillus or other Hyphomycetes.

However, in Aspergillus and Penicillium, as elsewhere in the deuteromycetes, there is a need for researchers to be far more careful over how they describe the features of conidial development. Minter et al. (1982, 1983a) and Minter (1984) have stressed the importance of describing separately each stage of the process (conidial initiation, maturation, delimitation and secession, and proliferation and regeneration of the conidiogenous cell). In the two genera under discussion, perhaps the most significant aspect to which these proposals relate is the question of how their conidiogenous cells are modified to produce more than one conidium. It is not clear whether these fungi use a system of replacement wall building apices, or the wall building ring mode of conidial production, or whether they can change facultatively from one mode to the other. Further research is needed to elucidate this fundamental point.

c. Terminology based on function

The most satisfactory scheme, given our present state of knowledge of the Hyphomycetes, is to base our terminology on function (Fig. 3). Pending a more widespread review of the terminology, it is recommended that the term conidiogenous cell is used in a functional sense to identify all cells actually producing conidia. The cell bearing the conidiogenous cell or cells, when present, can then be referred to as a cell supporting the conidiogenous cell or, conveniently, when no ambiguity is likely, as the supporting cell. The conidium may be defined as a specialized non-motile asexual propagule, usually caducous (i.e. seceding readily), not developed by cytoplasmic cleavage; the conidiophore is the hypha on which the conidiogenous cells are borne. The different parts of the conidiophore can then be referred to as the swollen apex. the median part and the basal part.

This system offers several major advantages over the old terms. Firstly, it follows exactly the definitions of conidium and its derivatives in the current edition of Ainsworth and Bisby's Dictionary of the Fungi (Hawksworth et al., 1983). Secondly, it brings the terminology for Aspergillus into line with that of the rest of the Hyphomycetes. Thirdly, it uses effectively only one term (conidium) derived from a dead language. Conidiogenous cell and conidiophore derive from conidium, and the remaining terms, supporting cell, swollen apex, median part and basal part, can easily be translated into any language. Fourthly, the terms biseriate and uniseriate are no longer necessary, since the swollen apex of the conidiophore can now be described as bearing conidiogenous cells directly or on supporting cells. Lastly these terms can be used to distinguish anatomical, functional and developmental aspects of the fungus. For example, in Fig. 3, it is possible to assert that "the conidiogenous cells on the left conidiophore are anatomically equivalent to the supporting cells on the right conidiophore", or "the swollen apex on the left conidiophore is an organ anatomically equivalent to the supporting cells on the right conidiophore". Only when such assertions can be made unambiguously will be ideas they can contain be able to be tested.

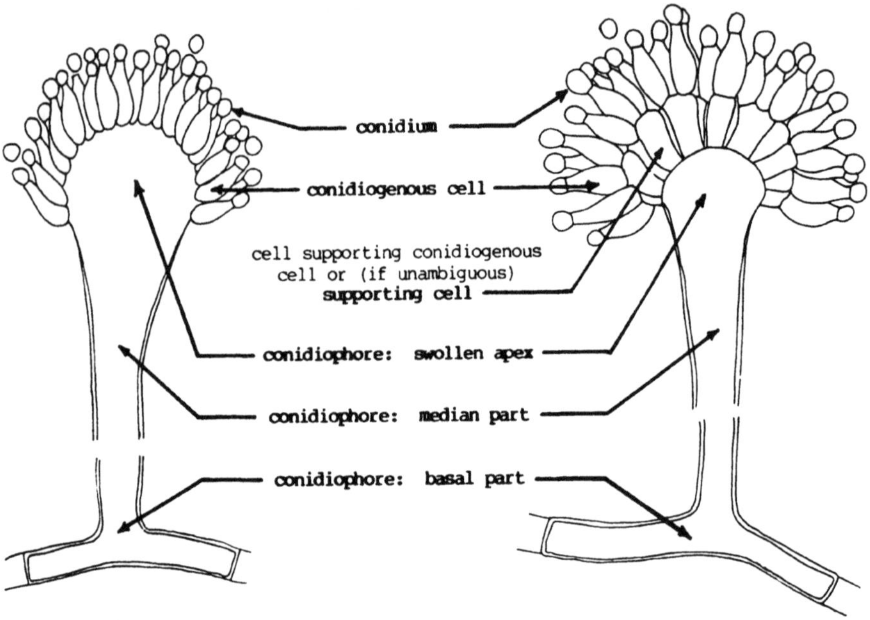

Fig. 3. Terms recommended to describe the different conidiogenous structures in *Aspergillus*.

PENICILLIUM

Different terms currently used to describe the conidiogenous structures in *Penicillium* are indicated in Fig. 4. It will be evident from the arguments used for *Aspergillus* that term such as sterigma, metula, stipe and phialide are equally inappropriate in *Penicillium*. It might also be noted that coremium (not shown in Fig. 4) is now generally considered synonymous with synnema (Hawksworth et al., 1983) although a careful review of the usage of this term is still required. Conidium, conidiogenous cell, conidiophore and supporting cell (Fig. 5) may, however, be applied without dificulty to *Penicillium* if defined in a functional sense, as proposed for *Aspergillus* above, and the observations on developmental terminology in *Aspergillus* also apply to *Penicillium*.

In comparison with *Aspergillus*, *Penicillium* has no median part or basal part to the conidiophore and in most subgenera no specialized swollen apex. In these respects descriptions of *Penicillium* are often simpler than those of *Aspergillus*. *Penicillium*, however, may have a complex series of branches on the conidiophore, and it is these complex branching systems which present the *Penicillium* taxonomist with the greatest terminological problems. Critical studies of the development of conidial heads of species of this genus have yet to be published, but are currently in progress in our laboratory. Preliminary studies suggest that the development is irregular.

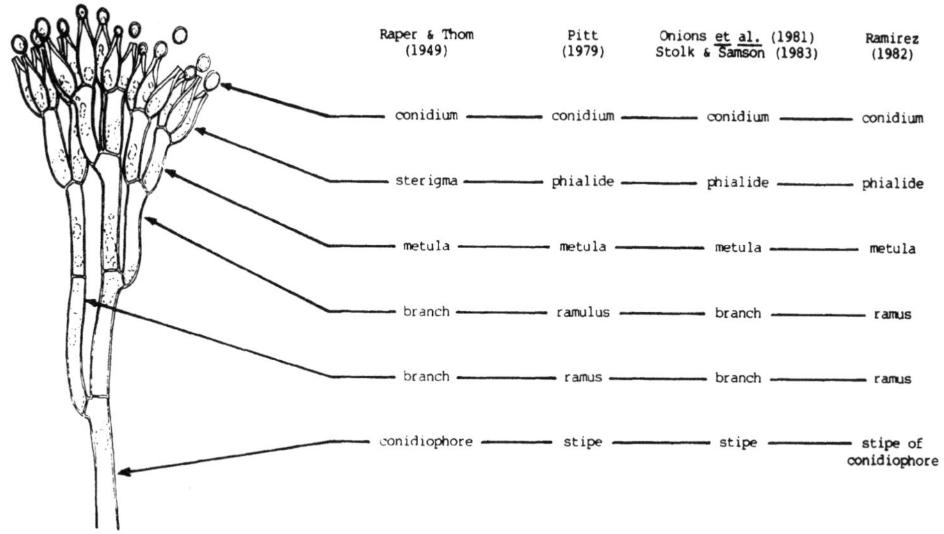

Fig. 4. Terms used by some recent authors to describe the different conidiogenous structures in *Penicillium* (the arrows are referred to in the text).

In the simplest examples, conidiogenous cells are produced singly, in sequence, at the apex of an unbranched conidiophore, eventually forming a whorl (often also called a verticil). Most species are, however, more complex than this: after producing a similar apical whorl of conidiogenous cells, they produce proximally one to several more or less defined whorls of branches of the conidiophore, in varying degrees of complexity, and these branches in turn give rise at their apices to more whorls of conidiogenous cells. In the most complex species, the branches themselves may branch again, producing further subsidiary whorls of branches, each bearing a whorl of conidiogenous cells at its apex. The order in which these subsidiary branches and conidiogenous cells are produced is still not known.

Among the terms currently used to describe these complex branching systems (Fig. 4) are metula, ramus and ramulus, while the structure as whole is often referred to as the *penicillus*. The term metula has already been rejected, and it is hard both to justify usage of the term penicillus in a restricted sense to mean the branched head of the conidiophore (Thom, 1910, and later authors including Westling, 1911, and Zaleski, 1927) and to see how it differs in the broad sense usage from the much more widely used term conidiophore. We are left with ramus and ramulus.

Ramus and ramulus

The term *ramus* has been used to denote all branches of the *Penicillium* conidiophore below the branches caused by supporting cells (Fig. 4). Branches caused by supporting cells have been excluded from past definitions of ramus because they have traditionally been considered different from those lower down the conidiophore. The exact nature of this difference has not, however, been adequately demonstrated. It therefore seems more consistent to regard all branches of the *Penicillium* conidiophore as being of the same fundamental nature. Ramulus has been used to denote branches immediately proximal to the branches caused by supporting cells. Some authors have accordingly used ramus in a restricted sense for only those branches proximal to the ramuli. The distinction between ramus and ramulus (like that between ramus and the branches caused by supporting cells) has not been adequately argued, and it is recommended here that the term ramulus is abandoned.

In the past there has been a confusion between anatomical and functional definitions of the term ramus. For some authors ramus is defined as an anatomical entity, i.e. as a cell proximal to supporting cells. Other authors use the term in a functional sense, i.e. as causing a branch in the conidiophore. Confusion between these two usages can result when considering conidiophores where the cell proximal to supporting cells (Fig. 4, dark arrow) is not the same as the one causing the branching (Fig. 4, light arrow). In view of the difficulties already outlined in using terms based on anatomy, it is recommended that the functional definition be generally adopted. Ramus should therefore be used in translation in the language any given author using: in English it may be rendered as *branch*, and the most proximal cell of the branch may be called a branch cell (Fig. 5).

Counting whorls

The greatest confusion in *Penicillium* terminology arises from the lack of a standarized way of counting whorls. This results partly from the variability of the fungi themselves. All *Penicillium* species with more complex branching systems also exhibit some conidiophores with less complex branching and, conversely, many species with simpler systems often produce a few slightly more complex conidiophores. The predominant branching type is therefore used for counting whorls. Confusion has also arisen because of a lack of agreement on what constitutes a whorl. Raper and Thom (1949), Pitt (1979), Ramirez (1982) and Onions et al. (1981) have counted whorls of both conidiogenous cells and branches (so that, for example, the most simple species has one whorl, and a species with one branch below the conidiogenous cells has two whorls), whereas Samson et al. (1976) have counted only the whorls of branches (so that the most simple species have no whorls).

For consistency we recommend that the practice of counting whorls of conidiogenous cells and branches separately is adopted, listing both, and beginning at the top of the conidiophore and working proximally. Some additional effort is involved in using this system, but the confusions of previous workers could thereby be avoided in the future.

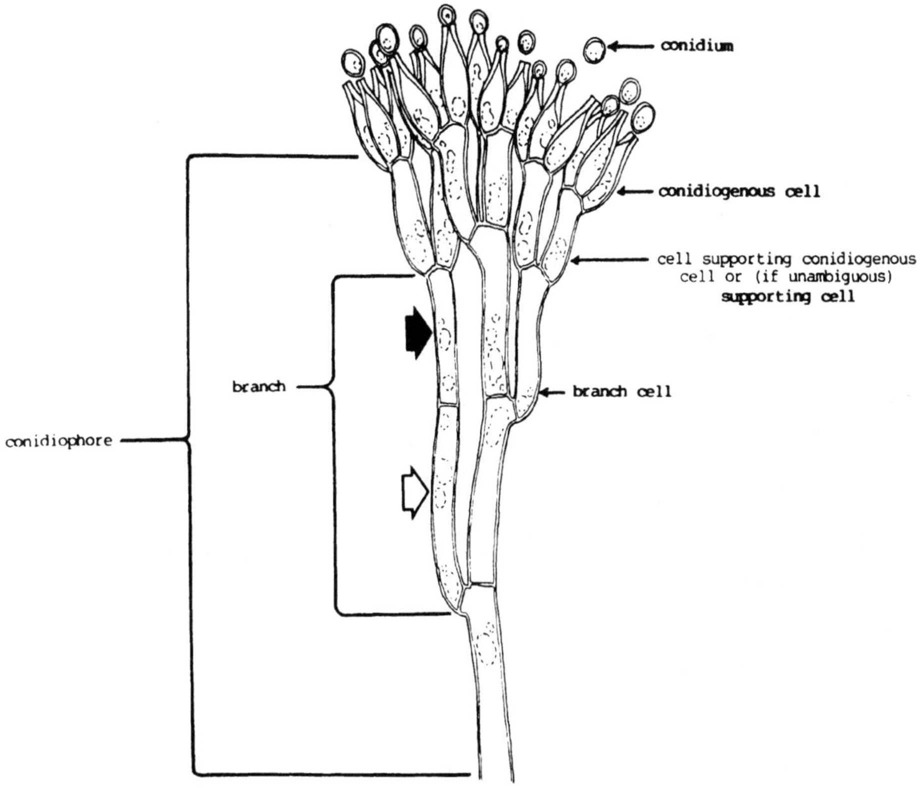

Fig. 5. Terms recommended to describe the different conidiogenous structures in Penicillium.

Acknowledgements

The conclusions presented in this paper are largely based on the results of a round-table discussion involving many of the mycologists at the CMI and designed to establish a generally acceptable simplified system for these genera. Their comments and stimulating discussions are gratefully acknowledged. This study was undertaken in connexion with Science and Engineering Research Council (SERC) contract no. SO/17/84.

REFERENCES

AHTI, T., 1982 - The morphological interpretation of cladoniiform thalli in lichens. - Lichenologist 14: 105-113.

CARMICHAEL, J.W., KENDRICK, W.B., CONNERS, I.L. and SIGLER, L., 1980."Genera of Hyphomycetes". Edmonton: University of Alberta Press.

COLE, G. and SAMSON, R.A., 1979."Patterns of development in conidial fungi". London, San Francisco and Melbourne: Pitman.
DESCALS, E., 1985 - Conidia as modified hyphae. - Proc. Indian Acad. Sci. (Plant Sci.) 94: 209-227.
ELLIS, M.B., 1971."Dematiaceous Hyphomycetes". Kew: Commonwealth Mycological Institute.
ELLIS, M.B., 1976. "More Dematiaceous Hyphomycetes". Kew: Commonwealth Mycological Institute.
HAWKSWORTH, D.L., SUTTON, B.C. and AINSWORTH, G.C., 1983 - "Ainsworth and Bisby's Dictionary of the Fungi". Kew: Commonwealth Mycological Institute.
HUGHES, S.J., 1953 - Conidiophores, conidia and classification. - Can. J. Bot. 31: 577-659.
HUGHES, S.J., 1981 - Mucronate hyphopodia of Meliolaceae are phialides. - Can. J. Bot. 59: 1514-1517.
KENDRICK, W.B., 1971 - Taxonomy of Fungi Imperfecti. University of Toronto Press, Toronto and Buffalo.
KOZAKIEWICZ, Z., 1978 - Phialide and conidium development in the Aspergilli. - Trans. Br. mycol. Soc. 70: 175-186.
MINTER, D.W., 1984 - New concepts in the interpretation of conidiogenesis in deuteromycetes. - Microbiological Sciences 1: 86-89.
MINTER, D.W., 1985 - A re-appraisal of the relationships between Arthrinium and other hyphomycetes. - Proc. Indian Acad. Sci. (Plant Sci.) 94: 281-308.
MINTER, D.W., KIRK, P.M. and SUTTON, B.C., 1982 - Holoblastic phialides. - Trans. Br. mycol. Soc. 79: 75-93.
MINTER, D.W., KIRK, P.M and SUTTON, B.C., 1983a. Thallic phialides. - Trans. Br. mycol. Soc. 80: 39-66.
MINTER, D.W., SUTTON, B.C. and BRADY, B.L., 1983b. What are phialides anyway? - Trans. Br. mycol. Soc. 81: 109-120.
ONIONS, A.H.S., ALLSOPP, D. and EGGINS, H.O.W., 1981."Smith's Introduction to industrial mycology". London: Edward Arnold.
PITT, J.I., 1979. "The genus Penicillium and its teleomorphic states Eupenicillium and Talaromyces". London: Academic Press
RAMIREZ, C., 1982. "Manual and atlas of the Penicillia". Amsterdam: Elsevier Biomedical Press.
RAPER, K.B. and THOM, C., 1949. "A manual of the Penicillia". Baltimore: Williams and Wilkins.
SAMSON, R.A., STOLK, A.C. and HADLOK, R., 1976 - Revision of the subsection Fasciculata of Penicillium and some allied species. - Stud. Mycol., Baarn 11: 1-47.
THOM, C. 1910. Cultural studies of species of Penicillium Bull. Bur. Anim. Ind. U.S. Dep. Agric. 82: 1-39.
VUILLEMIN, P., 1910 - Materiaux pour une classification rationelle des Fungi Imperfecti. - C.r. hebd. Seanc. Acad. Sci., Paris 90: 882-884.
WALKER, J.C. and MINTER, D.W., 1981 - Taxonomy of Nematogonium Gonatobotrys, Gonatobotryum and Gonatorrhodiella. - Trans. Br. mycol. Soc. 77: 299-319.
WESTLING, R., 1911 - Über die grünen Species der Gattung Penicillium. - Ark. Bot. 11: 1-156.
ZALESKI, K. 1927. Über die in Polen gefundenen Arten der Gruppe Penicillium Link. I, II, and II, - Bull. Acad. pol. Sci. Math. Nat., Ser. B; 1927: 417-563.

DIALOGUE FOLLOWING PROF. HAWKSWORTH'S PRESENTATION OF THE PAPER BY D.W. MINTER et al.

PITT: Although Dr. Minter and his co-workers seem uncertain what a phialide actually is, I think everybody else in this room has a very good idea: a phialide is the type of a conidigenous cell Aspergillus and Penicillium produce. If you use the term conidiogenous cell, after saying that phialide has been used in several senses, you must realize that conidiogenous cell is also a very broad term. At least with phialide, we know precisely what we are talking about. There is no ambiguity whatever in the term phialide in these genera.

If we use the term metula, it does not need to be translated into different languages, the way "supporting cell" has to be. If we use metula as a term for the cell, which supports a phialide, which has been its only usage, there is absolutely no ambiguity.

HAWKSWORTH: Although in these genera everybody knows what these terms mean, this is not necessarily so throughout the Hyphomycetes and Coelomycetes. It seemed unnecessary to us to have a separate system of anatomical terminology in these two genera. Why create a barrier in communication between Penicillium and Aspergillus taxonomists and those working with other Deuteromycetes.

PITT: I don't think there is a barrier at all. It could be a barrier if we start to use "conidiogenous cell" - we have to explain what we mean. With the word "phialide" we all know exactly what we mean. If you can bring other Hyphomycete taxonomists into line with us, all the better.

MALLOCH: I agree!

SEIFERT: I've been using the word "metula" for other Hyphomycetes. In too many cases in Hyphomycetes, we have no adequate terminology to describe conidiophores or branching, and anything we can adopt from Penicillium and Aspergillus, such as the word "metula" will be useful.

SAMSON: I'm really against the term "supporting cell". It is not specific and if you look at the most simple structure it means that the stipe becomes the supporting cell, if you go along with the terminology you propose.

HAWKSWORTH: In that case we would say that the supporting cell is absent. The conidiophore bears the conidiogenous cell.

SAMSON: If we replace the word "vesicle" with "swollen apex", the word "stipe" with "median part" and begin to use the word "basal part", we are not being any more precise about what we are describing. In fact, these definitions are much less precise.

GAMS: The papers by Dr. Minter and co-workers about conidium ontogeny have been received with a very mixed reaction. Dr. Minter goes to a lot of trouble to show, how confused the term "phialide" should be and if we go back to Vuillemin, the term is confused. However, since S.J. Hughes in 1953 introduced the ontogenetic terms of conidiogenesis, everybody has been clear about what the term "phialide" means. Although there is a certain amount of variation in the morphology of the phialide, the ontogenetic processes are remarkably uniform. Therefore, this is an indispensable, concise term for this type of conidiogenous cell. Why should we replace it with something vague? I am strongly opposed to replacing a shorthand expression with someting less precise.

Another point is the use of the word "conidiophore". Since Kananaskis I we have used the term conidiophore to include the conidiogenous cell and any branching structures supporting them. If we follow this, we then need a separate term to describe what we now call the stipe. There has never been any implication in my mind that a stipe must be an aggregate structure. In some of these fungi, the stipe is so long that it comprises almost the whole conidiophore. Why should we call this then the "median part", as you propose for _Aspergillus_?

HAWKSWORTH: Doesn't it worry you that these terms are used in different ways for other fungi?

PITT: When I worked at CMI on _Penicillium_ taxonomy in 1974, I asked Dr. M. B. Ellis what he would call the structure holding up the penicillus? He said:" No question, that is a stipe". That is good enough for me.

RAMIREZ: The word phialide was orginally used to mean "flask-shaped".

ONIONS: I was present at our discussion at CMI on terminology with two _Penicillium_ and _Aspergillus_ taxonomists, and 13 or so general taxonomists. There was a strong feeling that we should adopt these generalized terms to be consistent with the rest of Hyphomycete taxonomy. We also thought that terms should be used which could be translated into any language.

KLICH: For me the Latin names are a point of reference which greatly assist me translating works in other languages.

SEIFERT: I checked the terms "stipe" and "vesicle" in both the Concise Oxford Dictionary and some biological glossaries. The term stipe is defined simply as a "stalk" or "stem", and is used throughout botany for such structures as the stem of a fern frond, the support of a carpel in flowering plants, and for various structures in the fungi. Vesicle is used in biology and geology for "small bladder, cell or hollow structure". Although vesicles in _Aspergillus_ and _Pythium_ are completely different structures, they are both vesicles according to the definition of the word.

SAMSON: We also must take into consideration that we will have to use our terminology for teaching purposes.

WILLIAMS: I'd like to make a plea for Latin. Latin provides us with simple, straight-forward terms, which need only be learnt once and which are universal.

ONIONS: I think that it is the forward-looking attitude to use this new terminology, but I will have some difficulties adapting to it.

GOCHENAUR: I personally have no problems with the terms phialide and metula. The undergraduate students that I teach, however, isolate fungi from soil and must study a broad range of isolates. They have difficulty with the proliferation of specialized terms that we use for particular genera. They appreciate a simplified terminology using words like "conidiogenous cell", with a minimum of specialized terms. Special terms are very difficult for students, who are not mycologists.

SEIFERT: Do you think that it is any easier for a student to use the Minter system of describing the seven different processes involved in conidium ontogeny every time he wants to say "phialide"?

GOCHENAUR: No, but they would like to look at this and say it is a conidiogenous cell. That, they can understand.

HAWKSWORTH: CMI has run into this problem in teaching courses abroad, especially for students from less developed countries, where we are trying to keep the amount of specialized terminology to a minimum. The term phialide has been used in lots of different senses, so that it now has almost nothing to do with shape.

MALLOCH: Are you teaching this system in the third world? Are we going to have "supporting cell" in Swahili? At least if they are using the Latin terms we know what they are describing in almost any language.

SEIFERT: In your paper you say that the metulae of a biseriate _Aspergillus_, for example, is possibly anatomically equivalent to the phialide of an uniseriate. This seems like a very peculiar comparison to draw. What exactly does this mean?

HAWKSWORTH: This is related to the way these things develop.

SEIFERT: But why is that anatomical equivalence?

ONIONS: We were trying to differentiate between anatomical, developmental and functional terms.

KOZAKIEWICZ: Looking at the development of an _Aspergillus_ with both light microscopy and the SEM, in a head which is to become biseriate, the initial becomes hour-glass shaped. Sometimes phialides do not produce conidia, but elongate and develop a septum; the top become a phialide and produces conidia, the bottom is a metula. These two cells are derived from the same origin.

PITT: But if you look at a microscopic preparation of a freshly sporulating _Aspergillus_, the metulae all finish development simultaneously, then two or three phialides develop simultaneously from each metula. This is a very well organized system; metulae and phialides each have a specific function, and are not equivalent.

(At this point the discussion following the presentation was stopped. At the roundtable discussion the matter of terminology was picked up again which resulted in the following discussion).

HENNEBERT: As we distinguish a hierarchy of names in nomenclature, we also recognize informal ranks of descriptive terms. It is true that the name of a specific organ, such as phialide, is not the same as a "class" of organs such as conidiogenous cell. These are two different categories of words and we cannot use one for the other. This would mean a loss of precision. If you want a chair and just ask for furniture you might get a table.

HAWSKWORTH: The problem with phialide is that it has been used for a shape.

HENNEBERT: I would like to see a discussion on the differences between primary branch and secondary branches, or any similar terminology. I prefer the Latin names ramus and ramulus, because they have a hierarchy. A ramus is bigger than a ramulus. The word "branch" is applicable to any level of branching. It is less precise.

GAMS: I do not like the concept of rami and ramuli. I would prefer to use just branch. However, we should not discuss this without considering the terminology of branching patterns. How are we going to define branching patterns in Penicillium?

SAMSON: We (Samson et al., 1976) introduced a new terminology to indicate the stages of branching, excluding the whorls of phialides. To me a whorl (or verticil) of metulae or branches is more than two.

PITT: My point of view was that in 1979 I continued using terminology that had been introduced a long time ago. Dierckx and Biourge used the name Biverticillium subgenerically for species which clearly had whorls of phialides and then a whorl of metulae. If Biverticillium is biverticillate, then clearly a terminal whorl of phialides is monoverticillate and if you have a branch underneath the metulae, you have three stage-branching, for which I coined the term terverticillate. I start counting branch points from the conidium down the largest side branch. I must include the phialides; otherwise I would need to explain why I leave out that particular branch point.

GAMS: In Penicillium expansum there is a solitary branch at the lowest level of the penicillus. Here you do not have a proper verticil, because there is only one branch. How can you call this terverticillate?

PITT: I coined the term terverticillate to avoid having to say "three stage branched both down the main axis and the side branch". I defined the word to mean that, regardless of whether there was one or more rami at the bottom. I'm always using verticil in its strictest sense.

SAMSON: All right. If you define it as terverticillate and you call it terverticillate and I call it two-staged-branched, then we both know what we're talking about and there is no problem.

GAMS: What about the term "fractional penicillus"?

PITT: I don't use the term fractional penicillus. Some people use it to describe a reduced penicillus with only two or three phialides on a stipe. This is simply a monoverticillate penicillus. The term fractional can also refer to a penicillus with metulae and phialides arising at a single level. Here I would simply describe the penicillus that way or use the term "concurrent metulae and phialides". The term fractional to me is confusing, and best avoided.

SAMSON: One other point we should discuss: the terms "uni- and biseriate" as applied to the conidial heads of *Aspergillus*.

GAMS: The terms are self-explanatory and I do not consider them a burden at all. It is equally simple to say metulae present or absent.

MALLOCH: Or you can also say metulate and ametulate.

PITT: I use metulate and phialidic. But I'm coming to the point of view that uniseriate and biseriate are equally satisfactory.

SAMSON: One final thing, does anyone have a good term for the head of an *Aspergillus*?

SEIFERT: How about capitulum? This is used for terminal conidial heads of synnematous Hyphomycetes.

PITT: That sounds like a very useful term. I've been looking for such a term for years.

THE NAMING OF CHEMICAL VARIANTS IN PENICILLIUM AND ASPERGILLUS

John I. PITT and David L. HAWKSWORTH

CSIRO, Division of Food Research, North Ryde
NSW 2113, Australia and Commonwealth Mycological
Institute, Kew, Surrey, TW9 3AF, United Kingdom

SUMMARY

It is felt inappropriate to name variants of Penicillium and Aspergillus species distinguished solely by secondary metabolite profiles within the formal systematics of the genera while their taxonomic significance is uncertain. In cases where reference to such variants is necessary, it is proposed that the tested and widely used "Chemotype" terminology now used to communicate such variation in the lichen-forming fungi also be adopted in Penicillium and Aspergillus. Each chemotype is given a Roman numeral, Chemotype I including the metabolite profile of the type of the species name. To avoid ambiguity, the differentiating compounds can be mentioned in parenthesis, e.g. "Chemotype II (citrinin)". The Chemotype terminology avoids the necessity of creating new Latinized names and asigning ranks to them while the situation is unclear, and does not preclude the use of formal scientific names in the rank of 'species, ' variety' or 'form' when the taxonomic importance of such variation has been assessed.

INTRODUCTION

The recognition of strains within morphologically defined species which appear to consistently produce characteristic and constant profiles of secondary metabolites has led to these being given Roman numerals within accepted species names in Penicillium (e.g. Ciegler et al. 1973, 1981; Frisvad 1981, Frisvad and Filtenborg, 1983). There has been some discussion amongst Penicillium taxonomists as to how such strains can be indicated unambiguously while their taxonomic significance is uncertain. Various possibilities have been discussed to avoid having to create formal scientific names.

However, this problem is not unique to genera such as Penicillium and Aspergillus. Indeed, considerable attention has been paid to this question in regard to phanerogams and more especially to the lichen-forming fungi. A number of special terms have been proposed to accomodate this type of variation over the last 35 years, including "strain" (Lamb, 1951) and "chemovar" (Tétényi, 1958), but in the lichen-forming fungi the term "Chemotype"(Santessson, 1968) has become generally accepted for this as the term does not imply any taxonomic status or genetic basis and provides a satisfactory and now well-tested solution to this problem. An account of the history of the use of terms to differentiate chemical races is provided by Hawksworth (1976).

We can see no justification for adopting a separate system for *Penicillium* and *Aspergillus* and recommend that "Chemotype" be used for chemically characterized strains of morphologically indistinguishable fungi, where these strains have either (1) no taxonomic importance; or (2) where the taxonomic significance of the variation is still uncertain. This system also has the advantage that it is normally used with Roman numerals and so is consistent with the numbering system adopted by recent workers in *Penicillium*.

The systematics of the lichen-forming fungi went through a period from about 1965-1975 when a number of authors distinguished almost any chemically characterized strain as a separate species and instituted formal Latinized binominals for them. However, this practise is not now generally accepted, and while chemical features are often an essential aid to identification or in the definition of species concept, they are not now normally used as the sole basis for species separations.

It should also be pointed out that what appear to be clear differences between isolates by thin-layer chromatography (TLC), may not be so clear when examined by more sophisticated techniques. For example, two isolates may produce an identical range of biosynthetically related compounds but concentrate different ones so that on TLC they appear to be quite distinct when the difference is really a quantitative one, i.e. a failure to detect products present only in trace amount. This type of variation is refered as "chemosyndromic" (Culberson and Culberson, 1976) and the possibility that it could be occurring in some *Penicillium* and *Aspergillus* species cannot be excluded at this stage.

In view of the extensive discussion on the use of secondary metabolite characters that has taken place in the lichen-forming fungi since their first use in 1866, it seems rather a misuse of time to enact the same debate in the non-lichenized fungi. We therefore strongly recommend that those mycologists involved in the use of secondary metabolite data in taxonomy consider the situation that has evolved in their use in the lichen-forming taxa (see discussions in Brodo, 1978; Hale, 1983; Hawksworth, 1976; Hawksworth and Hill, 1984; Sheard, 1978). This is especially so as many of the compounds are closely related, produced by similar biosynthetic routes, and in some cases are even identical to ones present in *Penicillium* and *Aspergillus*.

In using the "Chemotype" system of nomenclature, it is sometimes helpful to indicate the key products produced in parenthesis, e.g. Chemotype I (cyclopiazonic acid), Chemotype

II (citrinin). Chemotype I should always include the race represented by the nomenclatural type of the name. Some authors omit Roman numerals and name the compounds (e.g. Swinscow and Krog, 1979), but we do not recommend this for general use as this approach can be unnecessarily cumbersome.

It is important to stress that the chemotype system of naming does not preclude the use of formal taxonomic ranks and scientific names when the taxonomic value of the variation is clear. Such recognition may be at ranks other than species, of which "variety" is often found to be the most appropriate in the lichen-forming fungi (e.g. Stevens, 1983), although "form" can also be used for some situations (e.g. Gilbert et al., 1981)

REFERENCES

BRODO, I.M. 1978. Changing concepts regarding chemical diversity in lichens. Lichenologist 10: 1-11.

CIEGLER, A., FENNELL, D.I. SANSING, G.A., DETROY, R.W. and BENNETT, G.A. 1973. Mycotoxin producing strains of Penicillium viridicatum: classification into subgroups. Appl. Microbiol. 26: 271-278.

CIEGLER, A., LEE, L.S. and DUNN, J.J. 1981. Production of naphthoquinone mycotoxins and taxonomy of Penicillium viridicatum. Appl. environ. Microbiol. 42: 446-449.

CULBERSON, W.L. and CULBERSON, C.F. 1976. Chemosyndromic variation in lichens. Syst. Bot. 1: 325-339.

FRISVAD, J.C. 1981. Physiological criteria and mycotoxin production as aids in identification of common asymmetric Penicillia. Appl. environ. Microbiol. 41: 568-579.

FRISVAD, J.C. and FILTENBORG, O. 1983. Classification of terverticillate Penicillia based on profiles of mycotoxins and other secondary metabolites. Appl. environ. Microbiol. 46: 1301-1310.

GILBERT, O. L., HENDERSON, A., and JAMES, P. W. 1981. Citrine-green taxa in the genus Candelariella. Lichenologist 13: 249-251.

HALE, M.E. 1983. "The Biology of Lichens". 3rd edn. London: Edward Arnold.

HAWKSWORTH, D.L. 1976. Lichen chemotaxonomy. In "Lichenology: Progress and Problems". D.H. Brown, R.H. Bailey and D.L. Hawksworth, eds. London: Academic Press. pp. 139-184.

HAWKSWORTH, D.L. and HILL, D.L. 1984. "The Lichen-forming Fungi".Glasgow: Blackie.

LAMB, I. M. 1951. Biochemistry in the taxonomy of lichens. Nature Lond. 168: 38.

SANTESSON, R. 1968. Lavar. Some aspects in lichen taxonomy. Svensk Natur. 1968: 176-184.

SHEARD, J.W. 1978. The taxonomy of the Ramalina siliquosa species aggregate (lichenised Ascomycetes). Can. J. Bot. 56: 915-938.

STEVENS, N. G. 1983. Tropical-subtropical Ramalinae in the Ramalina farinacea complex. Lichenologist 15: 213-229.

SWINSCOW, T. D. and KROG, H. 1979. The fructicose species of Usnea subgenus Usnea in East Africa. Lichenologist 11: 207-252.

TÉTÉNYI, P. 1958. Proposition à propos de la nomenclature des races chimiques. Taxon 7: 40-41.

MEDIA AND INCUBATION CONDITIONS FOR PENICILLIUM AND ASPERGILLUS TAXONOMY

John I. PITT

CSIRO, Division of Food Research
North Ryde, N.S.W. 2113 Australia

SUMMARY

This paper reviews the various media and incubation conditions which have been used in the past, and are in use now, for the taxonomy of Penicillum and Aspergillus species. Recommendations for standardisation of media and incubation conditions are put forward.

INTRODUCTION

About 80 or 90 years ago, taxonomists began to grow microorganisms in pure culture, and to use morphological observations and cultural characteristics in taxonomic judgments. Among fungal taxonomists, students of Penicillium and Aspergillus were pioneers in this area, for three reasons: first, these fungi grew readily on almost any kind of artificial medium; second, their fruiting structures were often ephemeral and fresh growth was needed for taxonomic evaluations; and third, because growth on agar media allowed the observations of a variety of characters, growth habits and pigments for example, which could not be seen on natural habitats.

Some of the pioneers recognised clearly the need for standardisation of the cultural conditions they used and, naturally enough, used the kinds of substrates which readily came to hand, brewed from natural ingredients. Thom (1910), in his pioneering monograph on Penicillium, used 15% gelatine, and potato or bean agar for all of his observations. Westling (1911) made all his descriptions from colonies on prune gelatine. Taking this approach to extremes, Biourge (1923) used no fewer than 13 media for his observations, recording data on all 13, which now makes interpretation of his descriptions slow and often difficult.

Czapek agar

Although Raulin recognised the advantages of defined media in biochemical tests over a century ago, Thom and Church (1926) were apparently the first Aspergillus and Penicillium taxonomists to systematically use a defined medium in a taxonomic study. They used a medium adapted by Dox (1910) from a formulation by Czapek, which is now sometimes known as Czapek-Dox agar, but more usually as Czapek agar (Cz). Minor variations on the formulation have been suggested. Czapek and Dox used KH_2PO_4, but Thom and Church (1926) recommended the neutral pH provided by the monobasic salt, K_2HPO_4, and this has been used by most authors since. Dox (1910) used 3 g $NANO_3$ as the nitrogen source; Thom and Church (1926) used 2 g, but most later workers have reverted to 3 g. The usual formula for Czapek agar is given in Table 1.

Table 1. Czapek agar and some variations.

	Cz	Steep	CYA	C20S	CY20S	G25N
$NaNO_3$	3.0	3.0	3.0	3.0	3.0	3.0
K_2HPO_4	1.0	1.0	1.0	1.0	1.0	1.0
$MgSO_4.7H_2O$	0.5	0.5	0.5	0.5	0.5	0.5
KCl	0.5	0.5	0.5	0.5	0.5	0.5
$FeSO_4.7H_2O$	0.01	0.01	0.01	0.01	0.01	0.01
Yeast extract			5		5	5
Steep liquor		10				
Sucrose	30	30	30	200	200	
Glycerol						330[a]
Water, ml	1000	1000	1000	1000	1000	1000

[a]Glycerol 25% (w/w), i.e. 250 g in 750 g water. Customary agar levels in all formulations

When used for the identification of Penicillium species, Cz has the disadvantage that variations in water and agar quality can add or subtract trace elements or chelating agents, which in turn can influence colony characters, including sporulation, in Penicillium species (Smith, 1949). Smith recommended the addition of trace elements, zinc and copper, to overcome this problem. My experience with Smith's modification to Cz has been variable, with the added metals sometimes being inhibitory and producing less satisfactory results than the standard formulation. A second disadvantage of Cz is that as nitrate is the sole carbon source, growth of many species is slow, because a great deal of energy is used to synthesise amino acids and other complex nitrogenous compounds.

Raper and Thom (1949), in their study of *Penicillium*, provided a solution to both problems at a single stroke, by adding a small quantity of corn steep liquor, a byproduct of corn starch manufacture, to Cz. Corn steep liquor provides both trace minerals and organic nitrogen, producing faster and more uniform growth of fungi, especially Penicillia. Corn steep liquor has its own drawbacks. Readily available in Peoria, Illinois, where Raper and Thom worked, and even in Sydney, Australia, it is not obtainable at all in many parts of the world.

Czapek yeast extract agar

After trials of alternatives, Pitt (1973) recommended the substitution of corn steep liquor by yeast extract, now readily available as a dried product from commercial media manufacturers. Yeast extract has the advantages of corn steep liquor in assisting rapid and uniform fungal growth, while at the same being readily available and quite stable. Moreover, because it is always made from cells of *Saccharomyces cerevisiae*, it is of reasonably standard composition. The formula for Czapek Yeast Extract Agar (CYA; Pitt, 1973) is given in Table 1.

I have used CYA now for 15 years in laboratories on three continents, and have obtained very satisfactory results. For almost all species I have studied, colonies on CYA have a very similar appearance to those on Cz agar, but grow rather more rapidly. This enables taxonomic determinations, for both *Penicillium* and *Aspergillus*, to be completed in 7 days. This is a big advantage. I used CYA as the principal medium in my *Penicillium* monograph (Pitt, 1979) as did Ramirez (1982), although he failed to acknowledge its source, and gave quite unnecessarily complicated directions for its preparation.

I wish to recommend CYA to this workshop as a standard medium, for use in the taxonomy of both *Penicillium* and *Aspergillus*.

In Pitt (1979), the preparation of CYA and other Cz based media was simplified by using a 100-fold concentrate (Czapek concentrate) of the inorganic salts, apart from K_2HPO_4, as a stock solution. This is added at the rate of 10ml/litre of medium. Apart from a slight precipitate of iron, which appears to have no ill effects, this solution keeps indefinitely without sterilisation.

Malt extract agar

Beer wort has been used in media manufacture for many years, but suffers from the same disadvantages as corn steep liquor: availability depends on locality, and composition is quite variable. Thom and Church (1926) again were pioneers. They introduced the use of malt extract in media for *Aspergillus*, as a source of complex nitrogen and carbohydrate.

Unlike Cz, which has enjoyed a virtually standard formulation for the past 50 years, malt extract agars vary widely in composition. Thom and Raper (1945) introduced a formulation based on that of Blakeslee (1915), containing both glucose and peptone as well as malt extract (Table 2). Pitt (1979) and Ramirez (1982) used the same formulation in their Penicillium monographs. However, a variety of other media based on malt extract are in current use. The formulae of some are shown in Table 2.

Table 2. Some current malt extract agar-formulations

Ingredient	Oxoid	Difco malt agar	Difco malt extract agar	Raper & Pitt	Stolk & Samson
Malt extract	30	30	12	20	20-40
Peptone	5		0.78	1	
Glucose			2.75	20	
Maltose			12.75		
pH	5.4	5.5	4.7	5.5	

If CYA (or Cz) is an effective medium for the identification of Penicillium and Aspergillus species, is a medium based on malt extract also necessary? Pitt (1973) pointed out that a strong case for use of two basic media, one with an inorganic nitrogen source, and one with complex nitrogen. Particularly in Penicillium, development patterns on CYA and agars made with malt extract are different, and in both Penicillium and Aspergillus, development of teleomorphs and sclerotia varies from one medium to the other. Much more taxonomic information in thus gained by using both media.

All modern taxonomists working with Penicillium and Aspergillus have used a "malt extract agar". But on which one should we standardise? A case can be made for a total reevaluation of the basis for media formulation for these genera, and some recent studies in my laboratory suggest that it is much needed. However, modern taxonomy is based on specified formulations, and until or unless a complete reappraisal is carried out, a formulation in common use should be standard. There are two obvious choices (Table 2): the medium used by Thom and Raper (1945), Raper and Thom (1949), Raper and Fennell (1965), Pitt (1979) and Ramirez (1982), or malt extract alone as used by Samson (1974) for Paecilomyces, Samson et al. (1976, 1984) and Stolk and Samson (1984).

In my view, the choice is clear. The formulations used at the Centraalbureau are simply too weak, too lacking in energy sources, to be suitable for Aspergillus species (see for example the colony photographs of A. niger, A. ochraceus and A. flavus in Samson et al., 1984). Colony development is too sparse to provide the maximum taxonomic information. I wish to recommend that the Malt Extract Agar introduced by Thom and Raper (1945) be adopted as standard for both Penicillium and Aspergillus taxonomy and, indeed, for other genera which form compact colonies and require a concentrated energy source. I also wish to recommend that the acronym "MEA", in use for a long time for this medium, be adopted as standard for it, and its use for other formulations be discouraged to avoid confusion.

Czapek agar with 20% sucrose

One other medium has been used by students of Aspergillus for many years in Eurotium taxonomy, and that is Czapek agar with 20% sucrose. Thom and Raper (1941) credit Elizabeth Dale (1909) with first demonstrating the value of high sucrose media in Eurotium taxonomy. Thom and Raper (1941) made Cz with 3%, 5%, 20% and 40% sucrose, and concluded that 20% sucrose was the most satisfactory for nearly all known species of Eurotium.

Over the past 25 years I have formulated a wide variety of experimental modifications of this medium, and have concluded that the formulation of Thom and Raper (1941) is the optimum one. The water activity, 0.98, is low enough to promote healthy growth of Eurotium species, although still well above the optimum for growth of any species. Less sugar, i.e. higher water activity, decreases growth and prolongs ascospore maturation. More sugar, 25-40%, reduces water activity and promotes rampant growth of most species. At the same time colonies are often less colourful, and hence there is a loss of taxonomic information. One modification is worthwhile, however. I have found that the addition of 0.5% yeast extract is beneficial, for the same reasons that it is in Cz, and I would like to recommend Czapek Yeast Extract Agar with 20% Sucrose (CY20S) as the standard medium for identification of Eurotium species.

Other media

Other media have been recommended for use with Penicillium and Aspergillus from time to time. Stolk and Scott (1967), and in later papers, have used oatmeal agar for cleistothecial development and maturation in Eupenicillium, with impressive results (although the formulation of this medium was not published in any of their papers). However as the production of mature ascospores of many Eupenicillium species remains touched with witchcraft, I see no purpose in attempting to standardise such media at this time. At the Commonwealth Mycological Institute, a medium known as Malt Czapek is used as a maintenance medium (Booth, 1971; Smith and Onions, 1983). It is apparently very suitable for this purpose, but is unlikely to have any particular value in taxonomy.

I introduced 25% glycerol nitrate agar (G25N; Table 1; Pitt, 1973) for a specific purpose, to broaden the base of information used in the identification of *Penicillium* species. Unlike other taxonomic media, G25N has a substantially reduced water activity (ca. 0.93). While not entirely independent of inherent growth rates seen on CYA or MEA, which have water activities near 0.995, colony diameters on G25N do provide a whole new dimension in *Penicillium* taxonomy. Pitt (1973; 1979) reported that colony diameters after 7 days on G25N at 25° showed a strong relationship with taxonomic position within the genus (Pitt, 1979, Fig. 13). G25N is of assistance to the less expert mycologist, especially in helping him to decide into which subgenus an unknown isolate should be placed. G25N is especially helpful in the area of species with biverticillate penicilli, and provides an important assurance to the inexpert that an isolate has been correctly placed in Subgen. *Furcatum* or Subgen. *Biverticillium*. It is not my purpose here to recommend G25N as a standard medium, but merely to point out its value in *Penicillium* taxonomy and my belief that it can be of value in other genera also.

Incubation temperatures

Temperature control in early taxonomic studies consisted of placing cultures on the laboratory bench at room temperature, and this happens to be very satisfactory for most fungi. As temperature controlled rooms became available, taxonomists began to specify incubation temperatures. General agreement is that temperatures near 25° are the most suitable for *Penicillium* and *Aspergillus* species. The only point worth noting is that *Penicillium* cultures should not be incubated above 27°. Some species have upper growth temperature limits near 30° (Mislivec and Tuite, 1970) and, in our experience, colonies may become atypical at 28° or above.

Abe (1956), Udagawa (1959) and in later papers Pitt (1973; 1979) have recommended incubation also at 30° for *Penicillium* species. Because of its relationship to human body temperature, 37° has been used for a long time in bacteriology, and its already a standard incubator temperature. This temperature is very suitable for species differentiation in the genera of interest here. Fifty six percent of the 150 *Penicillium* species accepted by Pitt (1979) grew at least occasionally at 37° In a recent more limited study, 12 of 21 common *Aspergillus* species (57%) were found to grow at 37° on CYA (J.I. Pitt and A.D. Hocking, unpublished). Based on all these considerations, I recommend that when an elevated temperature is considered to be desirable in *Penicillium* and *Aspergillus* taxonomy, 37° be the temperature used.

Pitt (1973) introduced the low temperature growth in *Penicillium* identification, again with the idea of broadening the taxonomic base. The temperature chosen was 5°, because this is the temperature of household refrigerators, and so was readily obtainable. This temperature is of value in *Penicillium* taxonomy, as 52% of the species accepted by Pitt (1979) germinated at least sometimes at 5° on CYA in 7 days. Few Aspergilli will germinate under these conditions, however. It is recommended that, where a low temperature is considered desirable for Penicillium taxonomy, 5°C is the temperature of choice.

Incubation time

It should never be forgotten that taxonomic works, especially on industrially important genera like Penicillium and Aspergillus, are written not for taxonomists, but for people who often need rapid fungal identifications. Any system which speeds this process must be beneficial. To conclude this paper, may I ask a simple question? Determinative taxonomy of all Penicillium and Aspergillus species which do not produce teleomorphs can be completed after 7 days incubation on currently available media. Why, then, are taxonomic papers still being produced which describe colonies after incubation for 10 days, 2 weeks or more?

REFERENCES

ABE, S. 1956. Studies on the classification of the Penicillia. J. gen. appl. Microbiol., Tokyo 2: 1-344.

BIOURGE, P. 1923. Les moissisures du groupe Penicillium Link. Cellule 33: 7-331.

BLAKESLEE, A.F. 1915. Lindner's roll tube method of separation cultures. Phytopathology 5: 68-69.

BOOTH, C. 1971. Introduction to general methods. In "Methods in Microbiology. Vol. 4". C. Booth, ed. London: Academic Press. pp. 1-47.

DALE, E. 1909. On the morphology and cytology of Aspergillus repens de Bary. Annls mycol. 7: 215-225.

DOX, A.W. 1910. The intracellular enzyms of Penicillium and Aspergillus. Bull. Bur. Anim. Ind. US Dep. Agric. 120: 1-70.

MISLEVEC, P.B. and TUITE, J. 1970. Temperature and relative humidity requirements of species of Penicillium isolated from yellow dent corn kernels. Mycologia 62: 75-88.

PITT, J.I. 1973. An appraisal of identification methods for Penicillium species: novel taxonomic criteria based on temperature and water relations. Mycologia 65: 1137-1157.

PITT, J.I. 1979. "The Genus Penicillium and its Teleomorphic States Eupenicillium and Talaromyces". London: Academic Press.

RAMÍREZ, C. 1982. "Manual and Atlas of the Penicillia". Amsterdam: Elsevier Biomedical.

RAPER, K.B. and FENNELL,D.L. 1965. "The Genus Aspergillus". Baltimore: Williams and Wilkins.

RAPER, K.B. and THOM, C. 1949. "A Manual of the Penicillia". Baltimore: Williams and Wilkins.

SAMSON, R.A. 1974. Paecilomyces and some allied Hyphomycetes. Stud. Mycol., Baarn 6: 1-119.

SAMSON, R.A. STOLK, A.C. and HADLOK, R. 1976. Revision of the subsection Fasciculate of Penicillium and some allied species. Stud. Mycol., Baarn 11: 1-47.

SAMSON, R.A., HOEKSTRA, E.S. and Van OORSCHOT, C.A.N. 1984. "Introduction to food-borne fungi". 2nd edn. Baarn Netherlands: Centraalbureau voor Schimmelcultures. 248 pp.

SMITH, D. and ONIONS, A.H.S. 1983. "The preservation and maintenance of living fungi". Kew, Surrey: Commonwealth Mycological Institute.

SMITH, G. 1949. The effect of adding trace elements to Czapek-Dox culture medium. Trans. Br. mycol. Soc. 32: 280-283.
STOLK, A.C. and SCOTT, D.B. 1967. Studies on the genus Eupenicillium Ludwig. I. Taxonomy and nomenclature of Penicillia in relation to their sclerotioid ascocarpic states. Persoonia 4: 391-405.
STOLK, A.C. and SAMSON, R.A. 1983. The ascomycetes genus Eupenicillium and related Penicillium anamorphs. Stud. Mycol. Baarn 23: 1-149.
THOM, C. 1910. Cultural studies of species of Penicillium. Bull. Bur. Anim. Ind. US Dep. Agric. 118: 1-109.
THOM, C. and CHURCH, M.B. 1926." The Aspergilli". Baltimore: Williams and Wilkins.
THOM, C. and RAPER, K.B. 1941. The Aspergillus glaucus group. US Dep. Agric. Misc. Pub. 426: 1-46.
THOM, C. and RAPER, K.B. 1945. "A Manual of the Aspergilli". Baltimore: Willkiams and Wilkins.
UDAGAWA, S. 19549. Taxonomic studies of fungi on stored rice grains. III. Penicillium group (Penicillia and related genera). J. agric. Sci. Tokyo 5: 5-21.
WESTLING, R. 1911. Über die grünen Spezies der Gattung Penicillium. Ark Bot. 11: 1-156.

DIALOGUE FOLLOWING DR. PITT'S PAPER

GAMS: What you have been talking about is macromorphology, obtained on different media. For micromorphology we find better development of the penicillus on 2% malt agar than on Czapek. The penicilli are very regular and well-formed, although growth is less luxurious than on Czapek.

SAMSON: We are not using malt agar for our colony descriptions, but only for micromorphology. I also agree that for identification of these genera malt agar with yeast extract (MEA) is a useful medium.

PITT: For micromorphology I do not get such good results with Czapek, but with Czapek yeast extract agar the penicilli are fine. Sometimes I go to MEA to verify what I've seen on CYA.

GOCHENAUR: Those of us who work with soils find a significant number of isolates which will not grow on Czapek agar. The reason is not that they can't utilize nitrate, but that they can't utilize sucrose, and therefore have no carbon source. Do you often encounter this sort of isolate?

PITT: Very rarely. The isolates which Raper and Thom mentioned as not growing on Czapek agar and which they tended to place in separate species, I find grow quite well on CYA, and in many cases can be placed in other recognized species.

GOCHENAUR: I agree that we need standardized media. I have found Bacto-Yeast Nitrogen Base Agar without amino acids and without ammonium sulphate, which supplies all the trace elements and all the known vitamins, useful. This gives a very a reproducible medium to which can be added an organic or inorganic nitrogen source and glucose or whatever carbon source you want. That would be reproducible all around the world.

PITT: The reason I would not recommend Bacto-Yeast nitrogen base as a standard is that it is made by only a single company, and is therefore difficult to obtain in many places. It is also expensive.

GOCHENAUR: For describing new species, I think it would be very useful if we had a medium which is absolutely standardized.

PITT: My point is that taxonomic descriptions will be used by people all around the world, and therefore should be described reasonably from cheap, readily available media.

(continuation of discussion at later point)

WICKLOW: Why do you use G25N and not, say G35N? Was it because the medium would not solidify beyond that point, or did you find that there was some taxonomic cut off point?

PITT: When I started work on Penicillium in Peoria in 1968, I sought to broaden the taxonomic base by looking for physiological tests which were unaffected by single genes, unlike colours and so on. Because of my food-mycological background, I decided to try low water activity media. I looked at sucrose, but found high concentrations were difficult to use in practise. Sodium chloride worked better but sometimes with inocula more than three to four weeks old, erratic growth resulted. I therefore turned to glycerol, and found that 25% gave reasonably rapid growth but allowed a useful differentiation within the genus.

KLICH: If we are working towards standardized media, at least one of them should be fully defined. Is anyone working towards that?

PITT: It is not easy. Czapek is a defined medium, but there are two problems with it. One is that sometimes sporulation is poor. Depending on the quality of the water or of the agar, the medium isn't quite right. The other problem is that development on it is on the slow side. What we need is a fully defined medium which provides rapid, reproducible results, but I think it will take time to develop this.

(Continuation of discussion at later round table)

SAMSON: We sometimes get peculiar colony colours with Czapek agar in various different countries. Colonies may be yellow rather than green.

PITT: I have seen this effect in Great Britain also, but I don't know what causes it.

FRISVAD: I believe that copper in the water supply may be the problem. Copper is necessary for melanin production. A minimum amount of copper is required for production of the green colour. When we use an inhibitor of melanin production we get a similar effect - the colonies are usually brown or yellow.

BRIDGE: At CMI we have used Czapek as our defined medium, but we find that the quality of water and agar are absolutely critical. Czapek has an inorganic nitrogen source, but organic nitrogen may contaminate the medium from various sources, including the water supply, the agar and the autoclave.

PITT: The reason I work with CYA is that it avoids the problems of variation in water and agar quality. The trace ingredients which might be variable in a defined medium are supplied in abundance by the yeast extract.

WILLIAMS: We grew *Penicillium chrysogenum* on CYA with tap or distilled water and three different batches of yeast extract, and saw quite different colony characters. Colonies varied from velutinous to floccose; sporulating to non-sporulating. We, therefore, normally use tap water in medium making or add a simple mineral supplement.

PITT: I have recently found that pH is an important factor in reproducing results with any of these media.

BEUCHAT: Can you comment on temperature control of ascospore production?

SAMSON: For *Eupenicillium* we find 25-27°C works well. For *Talaromyces* and *Neosartorya*, 30°C is best.

PITT: For *Byssochlamys*, 30°C appears to be critical for ascospore production. Neither 25° or 37°C is satisfactory.

BEUCHAT: For production of ascospores in *Neosartorya fischeri* we have tried a variety of formulations and now we use water agar with sodium acetate. Very little mycelium is produced, but ascospore production is very good.

RAMIREZ: Dr. Pitt has suggested that identifications be done after seven days, but I can determine microscopic morphology after three days using slide cultures.

SAMSON: Perhaps our previous papers have given the wrong impression, but we often carry out our identifications within seven days, or even in four to five days with some common species.

GOCHENAUR: At what temperature do you incubate your cultures?

SAMSON: At 25°C.

GOCHENAUR: Do you have incubators at 25°C or do you use room temperature.

SAMSON: No, we have incubators. Our room temperatures can drop as low as 10°to 15°C and are during the day around 20°C.

GOCHENAUR: In the United States our room temperatures are around 25°C, so it is very difficult to stabilize incubators at 25°C.

PITT: I believe 22-27°C is satisfactory range for incubating Penicillia.

ONIONS: What is the effect of light? I think that light can be important. We grow our cultures in the dark.

SAMSON: Except for synnematous species which may require light to form the synnemata, I think light is not important. We carried out a large experiment incubating cultures in diffuse daylight, and observed no differences from cultures incubated in the dark.

PITT: Our cultures are always incubated in the dark, but if after one week sporulation is poor, we may place them on a window sill for a further two days or so. Then light often appears to assist sporulation.

PATERSON: Light affects metabolite production. Some metabolites are degraded by light.

NIRENBERG: Is CO_2 concentration important with these fungi? *Emericella nidulans* grows more quickly in Petri dishes sealed with parafilm.

PITT: Normally in *Aspergillus* and *Penicillium* free access to oxygen is essential for the promotion of sporulation.

MALLOCH: *Sclerocleista citrispora* is called the "bashful *Aspergillus*" because it only reproduces sexually in the dark.

EVALUATION OF SOME MORPHOLOGICAL FEATURES OF ASYMMETRIC PENICILLIA ON DIFFERENT MEDIA

Olga FASSATIOVÁ

Department of Cryptogamic Botany
Faculty of Sciences, Charles University
Prague, Czechoslovakia

SUMMARY

Isolates of 10 species of Sect. Asymmetrica of the genus Penicillium were cultivated on three nutrient agar media: Czapek, Potato-carrot agar and synthetic nutrient agar(SNA). Morphological features of the conidiophores during their development were followed directly in petri dishes. In all isolates studied, the beginning of conidiophore development has common characteristics and its final structure on all media used corresponds to the taxonomical classification of each species. Variability of the penicillus within a species frame corresponds to various stages of its individual development. The structure of the conidiophores could be best followed on Potato-carrot agar. When investigating the formation of the conidiophore in Penicillium soppii on this medium an abnormal form of the primary conidiogenous cell was observed, from which several chains of conidia were simultaneously produced, in a manner similar to a polyblastic phialide.

INTRODUCTION

Data concerning species concepts in the genus Penicillium are abundant (Raper and Thom, 1949; Samson et al. 1976; Pitt, 1979; Ramírez, 1982). The large role of Penicillia in nature, especially their destructive and toxinogenic effects, contributes to critical appraisals of their diagnosis.

It is generally accepted that for differentiation of species in Penicillium, morphological features are of first importance. However, the appearance of colony is highly variable in a single species, especially in older isolates that are regularly transferred. The structure of the conidiophore and character of the penicillus exhibit a variability in some species of Sect. Asymmetrica also. Such species cannot be identified accurately without considering other detailed morphological and physiological properties.

In this communication the morphology of conidiophore formation is investigated and a simple method is used to evaluate constancy of some features of branching. Among other things, it is assumed that the development of vegetative and conidial stages in Penicillium under natural conditions is little understood at present. This fact has recently been stressed by Ramírez (1983).

To compare the development of the conidiophore and the penicillus, two media containing reduced nutrients were used in addition to Czapek-agar. The ontogenic development of conidiophores was followed by direct observation on Petri dishes at a low magnification. For this purpose 10 species represented by 1- to 2 isolates, were selected from biverticillate and terverticillate in Sect. Asymmetrica. Details of the conidiophores are illustrated by drawings.

A great deal of attention has recently been devoted to the morphology of the conidiogenous cells and the formation of conidia. On the other hand, the morphology of conidiophore formation has been much less studied.

MATERIALS AND METHODS

Isolates from the Sect. Asymmetrica: P. verrucosum var. verrucosum Samson et al. CCF 1197 (= IMI 91920); CCF 1228 P. verrucosum var. cyclopium Samson et al. CCF 1241 (= ATCC 8507); CCF 1229 P. expansum Link CCF 1443, CCF 1234 (= ATCC 1117); P. griseofulvum Dierckx CCF 1742, CCF 1743; P. canescens Sopp CCF 192; P. nalgiovense Laxa CCF 1728 (=ATCC 10472); P. soppii Zaleski CCF 1864; P. steckii Zaleski CCF 640; P. paxilli Bainier CCF 1914; and P. camemberti Thom CCF 861, CCF 1899.

Agar media used: Czapek agar (Cz) (Raper and Thom, 1949); Potato carrot agar (PCA; Booth, 1971); Synthetic nutrient agar (SNA; Nirenberg, 1981): 1.0 g KH_2PO_4, 1.0 g KNO_3, 0.5 g $MgSO_4$, 0.5 g KCl, 0.2 g glucose, 0.2 g saccharose, 1,0 l water and 20 g agar. Incubation was at 24° C.

Marginal and submarginal zones of the colony were observed in the light microscope (Lu-Zeiss) at an enlargement of 150-300 x. Control preparations were made in lactophenol.

RESULTS

In comparison with Cz, colonies on PCA has less abundant mycelium. Macroscopically, the layer of conidiophores and conidia was predominant, and was often coloured identically with that on Cz. Colonies on SNA formed minimal amounts of aerial mycelium that could hardly be observed by the unaided eye. The layer of conidiophores and conidia on the latter medium could only be seen in certain species and only in the middle of the colony, whose diameter was smaller than that of colonies on Cz and PCA. Characteristics of colonies of species observed on PCA will not been presented here as their significance for species taxonomy appears to be negligible. Features of the colony on Cz cannot be evaluated in this study because only small numbers of isolates have been used.

In all species studied, the beginning of ontogenic development of the conidiophores was morphologically identical on all media used. Faster maturation was observed on Cz and PCA than on SNA. During development of the colony a whole range of conidiophores was observed, beginning with a single phialide and ending with a complete penicillus. Primary, i.e. first conidiogenous cells, usually not typical phialides, grow from vegetative hyphae and produce chains of conidia (Fig. 2a). From the primary conidiogenous cell either an additional one is formed (Fig. 2b) or several new conidiogenous cells, with a typical phialide shape, grow from it forming a whorl (Fig. 4a). The initial penicillus is always monoverticillate. A further cell is soon formed laterally from the primary conidiogenous cell below this verticillus, giving rise at its apex to a new verticil of phialides (Fig. 3a). Thus the initial (primary) conidiogenous cell and the new lateral cell assume the function of metulae. The initial conidiophore is always divaricate in all species followed. Then the stipe is elongated and 1-2 lateral branches are formed.

In species of Subsect. Divaricata, the beginning of development of the conidiophore proceeds according to the above mentioned scheme; however, the penicillus remains divaricate. In P. canescens and P. nalgiovense the conidiophore was clearly asymmetrical (Fig. 4b), but P. paxilli, P. soppii and P. steckii symmetrical verticils of metulae were produced at the apex of the stipe (Figs. 2f, 1c).

In the Subsect. Velutina, Lanata and Fasciculata, the development of the conidiophore proceeded in a similar way, however, both metulae and branches were parallel to the main axis of the conidiophores during the final development phase. The irregularity of conidophores characteristic of P. camemberti could be observed during its ontogenic development. In this species phialides bud into others (Fig. 5a-d). In P. griseofulvum lateral conidiogenous cells or metulae were produced on irregular levels divaricately (Fig. 3a).

The development of the conidiophores could be best observed on PCA. The time required for the production of a large number of complete conidiophores was 5-8 days. As on this medium substantially less mycelium is formed than on Cz, growth did not prevent the penetration of transmitted light.

In P. soppii on PCA an interesting abnormity was observed during the first days of development. The first conidiogenous cell ('primary phialide') swelled apically and several chains of conidia were simultaneously formed from it (Fig. 1a,c,e). Sometimes even an elongated sterile hyphae was simultaneously formed with chains of conidia (Fig. 1b,e). In other cases several other conidiogenous cells proliferated from it. They were of elongated, finger-like shape and each of them produced a chain of conidia (Fig. 1d). The unusual development of the conidiogenous cell which could be designated as a polyblastic one, and deserves more detailed observation in the electron microscope.

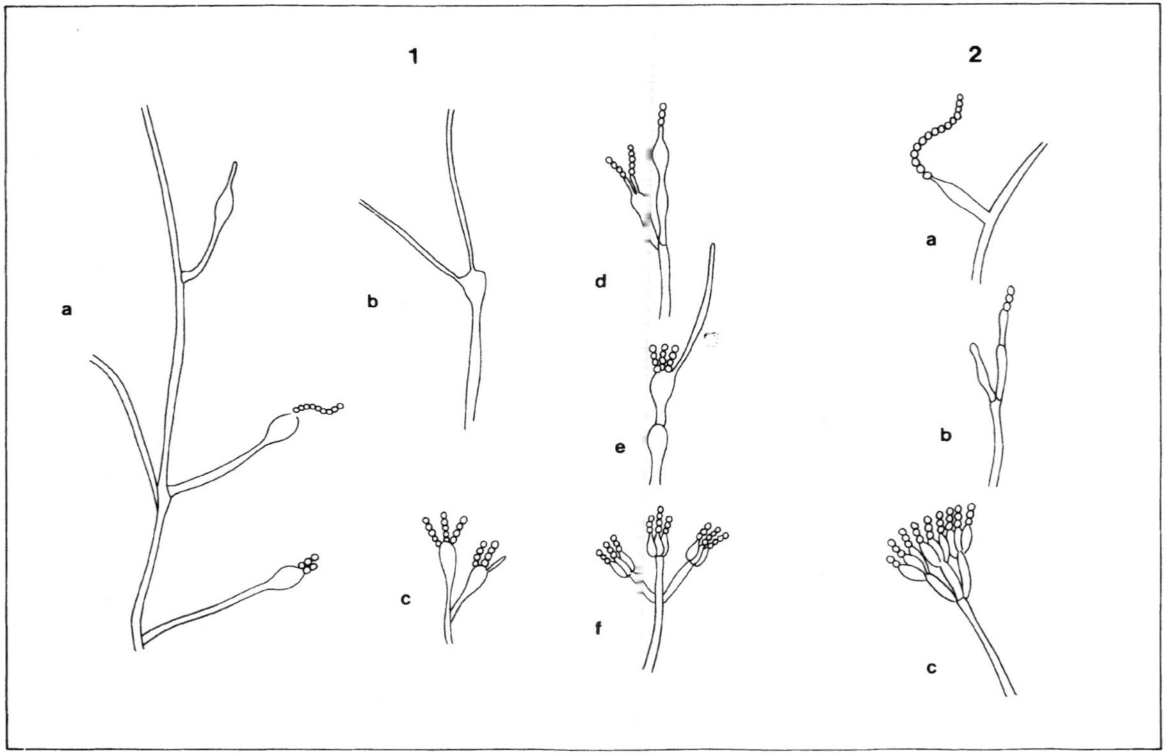

Fig. 1. P. soppii (a, b, c, d, e, - conidiophores on PCA with abnormal phialides; f- normal conidiophores on SNA); Fig. 2. P. paxilli (a, b - the conidiogenous cells in the first stage of development on PCA; c - the typical conidiophore on SNA). Fig.3. P. griseofulvum (a - the conidiophore in the first stage of development on PCA; b - the typical conidiophore on SNA)

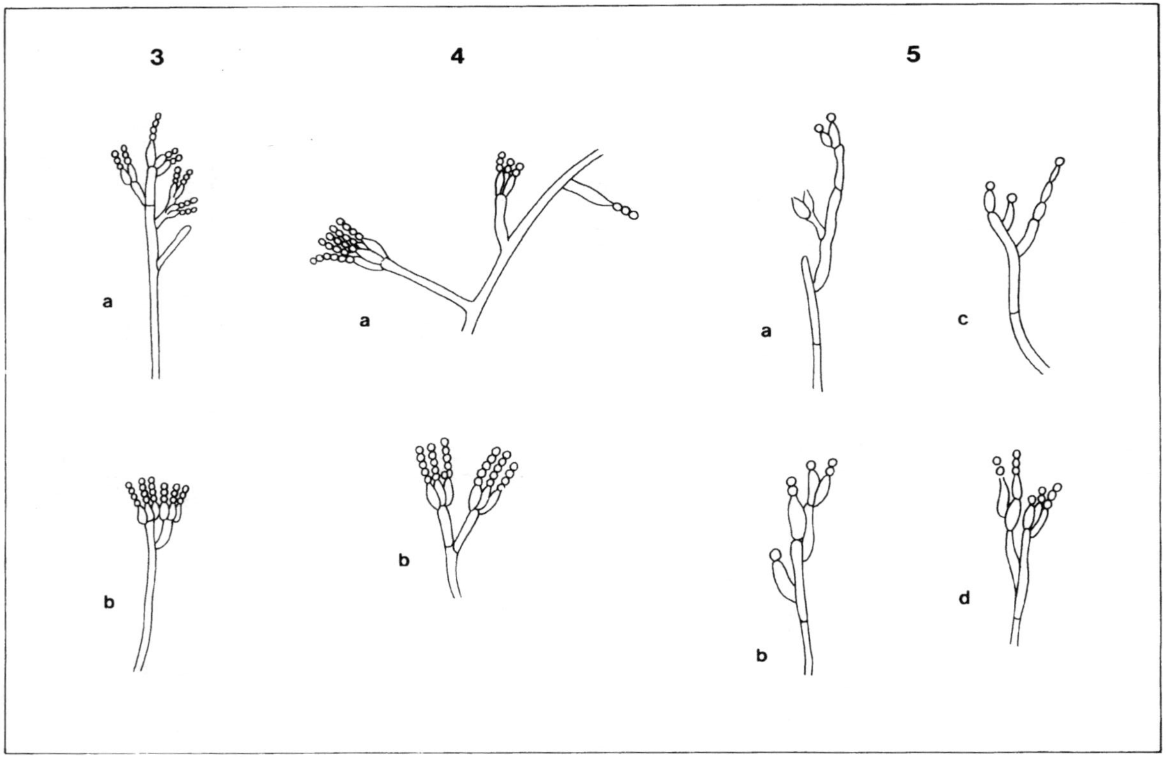

Fig. 3. P. griseofulvum (a - the conidiophore in the first stage of development on PCA; b - the typical conidiophore on SNA); Fig. 4. P. nalgiovense (a - the conidiophore in the first stage of development on PCA; b - the typical conidiophore on SNA); Fig. 5. P. camemberti (a, b, c, d - various stages of the development of the conidiophores on PCA).

DISCUSSION

In addition to Cz, PCA, SNA was used to attain more natural coniditions for the growth of the conidiophore. Nirenberg (1981) used SNA for the isolation of Fusaria and induction of macroconidia formation and could identify them by means of direct observation on Petri dishes. The method of direct observation of viable cultures at low magnification is naturally not sufficient for evaluation of all microscopic features of conidiophore in Penicillia. It served only to determine stability of certain features. Further details of the conidiophores, i.e. size and shape of phialides and cell wall texture can be defined only in preparations at higher magnifications. It was also found that various types of conidiophores in asymmetrical Penicillia present on slides in lactophenol are apparently different developmental stages during formation of complete conidiophores.

As far as the ontogenic development of the conidiophore is concerned, the work of Zachariah and Metitiri (1971) is relevant. They followed the development of phialides in wild and mutant strains of P. claviforme and in mutant strains observed chains of phialides budding from each other. They noticed the acropetal origin of some cells, whereas during formation of conidia in the Penicillium the basipetal direction is apparently involved. This conversion was explained as being due to mutagenic treatment. In the strains studied in this paper the formation of new phialides from the primary one by budding was rather frequent. Multiple budding was observed only in P. camemberti. The strains used were not changed by mutagenic treatment.

The unusual method of formation of conidia in several chains from the apex of the primary conidiogenic cells in P. soppii could be explained by assuming that its apical part contains several areas capable of budding, as the primary conidiogenous cell can, after a certain time, produces a whorl of secondary phialides at its apex.

REFERENCES

BOOTH, C. 1971. "Methods in Microbiology. Vol. 4". London: Academic Press.

NIRENBERG, H.I. 1981. A simplified method for identifying Fusarium spp. occuring on wheat. Can. J. Bot. 59: 1599-1609.

PITT, J.I. 1979. "The genus Penicillium and its teleomorphic states Eupenicillium and Talaromyces". London: Academic Press.

RAMÍREZ, C. 1983. Species concepts in Penicillium based on morphological characters. Abstracts Third. Inter. Mycol. Congress Tokyo, p. 247.

RAPER, K.B. and THOM, Ch. 1949. "A Manual of the Penicillia". Baltimore: Williams and Wilkins.

SAMSON, R.A., STOLK, A.C. and HADLOK, R. 1976. Revision of the subsection Fasciculata of Penicillium and some allied species. Stud. Mycol., Baarn 11: 1-47.

ZACHARIAH, K. and METITIRI, P.O. 1971. The organization of the penicillius of Penicillium claviforme Bainier. In, "Taxonomy of Fungi Imperfecti" B. Kendrick. ed. Toronto; University of Toronto Press, p. 120-131.

DIALOGUE FOLLOWING DR. FASSATIOVÁ'S PAPER

GAMS: I wonder if the very weak media you use are causing suboptimal development. Perhaps because there are insufficient nutrients to develop a proper cell wall. Perhaps the wall is more elastic than usual and this may explain some of the phenomena you have seen.

NIRENBERG: Sometimes weak media are useful for following development and observing the roughness of the walls. When using a weak medium, however, it is important to consider the quality of the agar and the length of time for autoclaving. Autoclaving for too long results in PH changes.

PATERSON: Is the development of penicillus structure similar on these weak media to that on the media normally used?

FASSATIOVÁ: Yes, the structures are similar, although they may take longer to develop, and the branching density may be less.

BRIDGING THE GAP:

TAXONOMISTS AND FOOD MYCOLOGISTS

Larry R. BEUCHAT

Department of Food Science, University of Georgia
Agricultural Experiment Station
Georgia 30212 U.S.A.

SUMMARY

The selection of certain groups, or species of Penicillium and Aspergillus for research using gene manipulation to develop strains capable of producing increased quantities of primary or secondary metabolites for use in food processing would undoubtedly be made easier with knowledge of innate metabolic growth and morphological characteristics of the fungi. These characteristics, of course, are among the major factors considered when classifying Penicillium and Aspergillus. Knowledge of taxonomic schemes would also be beneficial to industrial food mycologists responsible for producing high-quality foods requiring metabolic by-products or fermentation actvities of these genera. Examples of these include certain cheeses, soy sauces and fermented beverages. The objective of this paper will be to identify more clearly potential benefits which can be derived from increased mutual understanding of food mycology and taxonomy as they relate to food processing and preservation.

The food mycologist is a rare breed of scientist who finds his way to the profession through various and sundry routes. His academic training may be in food science, plant pathology, veterinary medicine, microbiology, botany, or one of several other recognized curricula offered by institutions of higher learning. In the U.S.A. at least, there is no university which offers a degree with a major in food mycology. Likewise, I am not aware of a university that offers a degree in Penicillium and Aspergillus taxonomy. And there are only a few exceptional individuals worlwide who can justifiably claim they are food mycologists as well as experts in taxonomy. I am not one of these individuals, yet I can envision the benefit and value to be derived from research programs designed and executed to capitalize on knowledge integrated from both areas of expertise. I would like to share some thoughts concerning these areas of study and directions which might be taken concurrently and cooperatively by Penicillium and Aspergillus taxonomists and food mycologists to achieve these objectives.

Advances in molecular biology, cell and tissue culture, and related fields "will likely have an impact on agriculture comparable to that of the discovery of the laws of inheritance in the late 1800's." So states a recent report by the U.S. National Research Council. It is my firm belief and expectation that through the so-called biotechnology revolution, increased use and new uses of Penicillia and Aspergilli to benefit mankind not only hold great promise but will also revolutionize some areas of food technology and manufacture beyond anything we might presently imagine. I have read that human genes have been transferred to Petunia plants and that these plants are producing human hormones. It is not unreasonable, then, to think that genes from an apple tree could be transferred to a Penicillium or Aspergillus species which would then produce large quantities of ethyl-2-methylbutyrate, a compound largely responsible for apple aroma and flavour. The compound could then be extracted and used by the food technologists. An ever greater accomplishment would be to induce Penicillia or Aspergilli to produce apples. However, the probability of succeeding in this and other similar endeavours will depend upon food mycologists and technologists increasing their understanding of biochemical pathways and growth characteristics unique to various species and strains of Penicillium and Aspergillus. In turn, the taxonomist, who knows these traits so well, will benefit greatly by better understanding the needs of the food mycologists. To illustrate mutual gains to be made by both groups of scientists, the following examples are offered.

The use of Penicillia and Aspergilli directly in manufacturing processes has been practiced for centuries. Many indigenous fermented foods, especially in the Orient, depend in no small way upon the enzymatic activities of Aspergillus oryzae as an intimate part of the manufacturing process. Prime examples of products of fermentation in which A. oryzae plays an integral part are soy sauce, miso and sake. Penicillium species also are used as a direct and necessary part of food manufacture. Anyone who has consumed Roquefort or Camembert cheeses can attest to the fact that Penicillia do indeed have pronounced effects on flavour, aroma, texture and appearance of these fermented dairy products. Still, certain problems associated with otherwise beneficial attributes of Aspergilli and Penicillia in fermented foods exist. Aside from desirable as well as undesirable biochemical activities of these moulds, the time required to produce high-quality products is unreasonably long, in the order of months or even years. By understanding biochemical capabilities and potential for change through genetic engineering or some other mechanism, it seems reasonable to believe that the manufacturing time for these cheeses could be substantially shortened. But it will require close interaction and an exchange of information and ideas between the food mycologists and the taxonomist if this is to be accomplished.

In addition to using living, growing cultures of Penicillia and Aspergilli as part of the manufacturing process, primary and secondary by-products of these fungi are being used and have great potential for increased use in the food and beverage industry. Some examples are as follows.

The potential for enzyme production by *Penicillium* and *Aspergillus* species for use in the food industry has not been fully exploited. Proteases, peptidases and amylases from Aspergilli, especially, are currently being used on an industrial scale but should experience increased use with advances in immobilized enzyme technology. Super strains of *Aspergillus* capable of producing large quantities of highly active enzymes are being developed through cooperative efforts of food scientists, biochemical engineers and mycologists. Selection of strains for this purposes is influenced greatly by their taxonomic position.

While interest in using fungi as sources of protein for food and feed has subsided somewhat in recent years, there nevertheless exists a tremendous potential for production of relatively high-quality protein from Penicillia and Aspergilli. With the advent of genetic engineering, I suspect we will see a resurgence or research activity in this area. Taxonomists will be very helpful in charting the direction of such research.

There is tremendous research activity in industry laboratories to develop sweeteners for use in foods and beverages. These include starch hydrolysates, hydrogenated glucose syrup, sugar alcohols such as xylitol, sorbitol, mannitol and lactitol, and glucopyranosyl derivatives of sorbitol and mannitol. Aspartame, a dipeptide of L-phenylalanine as the methyl ester and L-aspartic acid has intense sweetness as does acesulfame K. Each sweetener is characterized by unique properties. Usually, its sweetness value is several fold greater than sucrose and it also may have low or no caloric value. Special solubility, stability or other functional properties may also be attributed to these non-conventional sweeteners. The use of a yeast (*Sporobolomyces odorus*) to produce aspartame already shows promise for commercial application. Since several genera of fungi, including *Penicillium* and *Aspergillus*, can be induced to accumulate certain polyalcohols, e.g., mannitol and arabitol, it would be reasonable to conduct extensive studies to explore the feasibility of using these genera for large scale production of sweeteners or humectants for the food industry.

Another area in which the food mycologist could benefit by knowledge of taxonomy is that of oil content and composition of Penicillia and Aspergilli. It is well known that fatty acid profiles of fungi grown at elevated temperatures differ from profiles of the same fungi grown at reduced temperatures. In this age of nutrition-conscious consumerism, at least in developed countries, the demand for foods containing oils with a high proportion of unsaturated fatty acids is evident. Foods containing low or no cholesterol are highly desirable among some groups of consumers.

Thus the use of oils in food products which did not originate from animals has received increased attention. The oil content of some fungi can reach 60% or more on a dry weight basis. Is it not possible, through knowledge of biochemical capabilities and growth characteristics (areas of great familarity to the taxonomist), to select and manipulate Penicillia and Aspergilli for the sole purpose of producing high-quality oils for use in the food industry? Certainly, there is an increasing need for such oils as the expense of producing oils from plants and especially animals continues to increase at a dramatic rate.

Large quantities of organic acids derived from fungal fermentation are used in the food industry. These include citric and gluconic acids, both commercially produced using Aspergillus niger. As high as a 10% solution of citric acid has been obtained by Japanese researchers. One has to wonder just how much higher this concentration can go, but through genetic approaches involving recombination and hydridization perhaps this level can be increased to greater than 10%. The selection and manipulation of strains for this purpose will require input from scientists knowledgeable about all phases of metabolism and growth habits, and thus would necessarily include input from taxonomists.

The addition of flavouring agents to foods and beverages is practiced on a world-wide basis. These agents may have had their origins from plants or animals, or they may have been synthesized in the laboratory. With a few notable exceptions, e.g., nucleotides from some yeasts, extracts from mushroom mycelia and other fleshy fungi, and distillates from certain fermented beverages, flavouring compounds have not been extensively produced on an industrial scale in which fungi have been used as the manufacturing agent. All of you are familiar with the wonderfully wide and mysterious array of fragrances that can be evolved from cultures of Penicillium and Aspergillus. The sensory nature and intensity of these fragances will vary substantially depending, of course, upon the species and isolate at hand as well as its age, the composition and pH of the growth medium, availability of oxygen, growth temperature and a multitude of other variables.

The food industry is constantly searching for superior sources of flavouring agents to enhance existing and newly developed products. The blending of several dozen, indeed, hundreds of chemical components to yield desirable flavour notes upon incorporation into foods is a common practice. Subtle flavour or aroma notes variously described as earthy, peppery, green or flowery, for example, are quite desirable in certain foods. It is these fragrances that are not uncommon to many strains of Penicillium and Aspergillus. Unfortunately, meaningful communication seldom occurs between technologists responsible for developing flavour profiles in foods and biochemical engineers responsible for designing processes for manufacturing flavouring agents. Mutual understanding of the desires of the technologists with regard to flavour and aroma in product development and the few mycologists who are dedicated to harnessing the ability of Penicillia and Aspergilli as well as other moulds for the purpose of deriving useful ingredients for the food industry is almost non-existent. It seems to me that great value could come to both parties by increased communication. The selection of a species or strain of *Penicillium* and *Aspergillus* based on its ability to produce a specific aromatic or flavour compound followed by intensive research to optimize production, extraction and purification for use in the food industry would be a rational approach to success. Only through greater understanding by the food technologists, food mycologists and taxonomist of needs, possibilities and limitations of utilizing Penicillia and Aspergilli as organisms for synthesizing desirable flavouring compounds can mutual benefit be realized.

In addition to utilizing valuable by-products from Penicillia and Aspergilli as sensory modifiers in foods, we should also consider the detrimental metabolites of these genera, viz, mycotoxins. If, in the course of routine examination of raw ingredients or processed foods, the presence of high populations of a species of *Penicillium* or *Aspergillus* is detected, does this provide a basis for concluding that the ingredient or food is not wholesome? The answer is no. Such a judgment cannot be made unless species identification is first carried out, followed by chemical and/or biological analyses for the suspect mycotoxin. Certainly, knowledge of taxonomic principles is necessary to accomplish this task. Thus the quality assurance supervisor could benefit greatly be an increased skill in *Penicillium* and *Aspergillus* identification techniques.

Accepting the premise that mutual benefit can be derived by increased understanding of *Penicillium* and *Aspergillus* taxonomy by the food mycologist and, conversely, increased understanding of food science and technology by the *Penicllium* and *Aspergillus* taxonomist, then how do we go about bridging this gap? Obviously, workshops like the one in which we are participating this week can be of great value. But really we must nurture the development of a broader understanding of both food mycology and *Penicillium* and *Aspergillus* taxonomy in courses taught in our institutions of higher learning. If there is a course in food microbiology taught in the United States which devotes more than 10 minutes to discussing taxonomy of any mould genus, I am not aware of it. On the other hand, courses designed to identify moulds are often taught under the auspices of Departments of Microbiology, Botany or Plant Pathology, and the student rarely, if ever, receives a lecture on the value of *Penicillium* and *Aspergillus* to the food industry. There is a reason for this, of course, and that is that the food microbiology professor too often knows little or nothing about taxonomy of moulds in general and the taxonomy professor often knows little about applications of *Penicillium* and *Aspergillus* in the modern food industry. There are always exceptions, however, as evidenced in part by those of you attending this workshop. The point to be made is that we must make a conscientious effort to cross each other's borders. Furthermore, we both must remain abreast with advances in biotechnologies so that developments which might strengthen productivity on a collaborative basis can be quickly recognized.

DIALOGUE FOLLOWING DR. BEUCHAT'S PAPER

SAMSON: Have you any suggestions as to how we can reach people in industry who still use old names which are now taxonomically incorrect?

BEUCHAT: These people are not normally exposed to current taxonomic information. One approach in the USA is that we have university extension services which can provide an excellent vehicle for disseminating new information and new names, in this case, that are currently recognized and widely used. They organize workshops and give lectures to people from food industry. Even professionals tend to use incorrect names when talking to industrial people because they are mostly interested in communicating. Perhaps it is time we stop taking this approach and started using the correct names all the time. Eventually, everyone would learn them.

PITT: One approach is to write comprehensive manuals directed at this audience, using correct nomenclature. We're engaged in such a project at the moment.

KLICH: A slower but useful process which I use in distributing cultures, is to use both old and new names until people get used to seeing the new name.

HAWKSWORTH: The International Commission on the Taxonomy of Fungi (ICTF) of IUMS plans to publish a series of bulletins from time to time in Microbiological Sciences providing information on name changes in industrially and microbiologically important fungi.

Chapter 3:

ADVANCES IN PENICILLIUM TAXONOMY

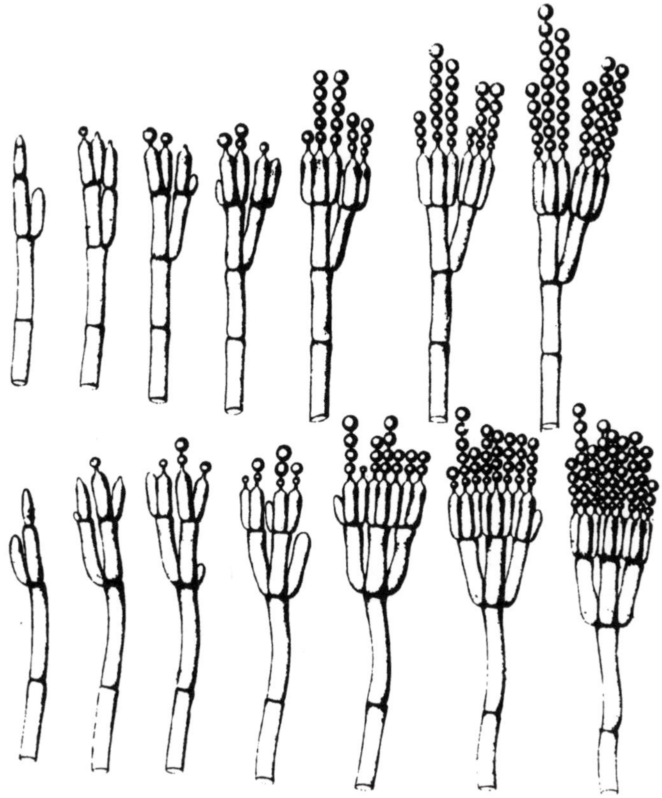

THE CLOSELY RELATED SPECIES OF SUBGENUS PENICILLIUM - A PHYLOGENIC EXPLORATION

Anthony P. WILLIAMS and John I. PITT and Ailsa D. HOCKING

Leatherhead Food Research Association, Leatherhead Surrey KT22 7RY UK and CSIRO, Division of Food Research, North Ryde, NSW 2113 Australia

SUMMARY

In this paper we are attempting to describe the interrelatedness of the terverticillate species of subgenus Penicillium. A model which describes the observable species interfaces is presented and it is postulated that intermediate species may be common ancestors of two or more modern species.

INTRODUCTION

Mycologists who attempt to identify the terverticillate species in subgenus Penicillium frequently encounter isolates which fail to key property for one or more of several reasons. These isolates show characters of more than one species, yet are apparently healthy fungi with no evidence of cultural deterioration. In the preface to his chapter on subgenus Penicillium, Pitt (1979) describes many of the interfaces between species encountered when large numbers of isolates are examined. However, his model is two-dimensional and other interfaces are sometimes observed (e.g. between P. roqueforti and P. aurantiogriseum; and P. crustosum and P. aurantiogriseum). We are attempting here to provide one possible explanation for the diversity of interspecies relationships which is frequently observed.

Perhaps of all the fungi, those in subgenus Penicillium have taken the most advantage of man's agricultural advancement. They colonise his cereal crops before and after harvest; several species cause extensive rots to various of his cultivated fruits; others damage Allium species and two have been utilised for the enhancement of flavour in cheeses. However, the genome of these fungi appears to be unstable: isolates in culture tend to degenerate rapidly or to change morphological appearance between consecutive subcultures. This may well be the result of the rapid adaptation of a relatively few ancient species to take advantage of the many new nutritional niches provided by man during the few millennia of his agricultural activity.

If a group of fungi is evolving rapidly it may be presumed that many of the common ancestors of more modern species still exist and are active. These may, indeed, be represented by the intermediate species which we often isolate and these intermediates may give us a valuable insight into the interrelations and the evolutionary pattern within subgenus Penicillium.

It is necessary, first, to ask what constitutes an advanced species. In evolutionary terms the successful fungus will be either the one which manages to disseminate the most spores from its colony or that which manages to take advantage of a particular available niche.

In his chapter entitled "Polyglotta", Pitt (1979) described several morphological and physiological characteristics which might represent an advanced species of Penicillium. Several of these, along with the possible evolutionary consequences, are listed in Table 1. It should be noted that successful colonisation of a special niche may result in an arrest of advancement, so that not all highly successful Penicillia would necessarily be morphologically advanced.

Table 1. Possible signs of evolutionary advancement in subgenus Penicillium

Primitive character	Advanced character	Possible consequence
Much mycelium per spore	Little mycelium per spore	Metabolic energy used for spore production
Light sporulation per phialide	Heavy sporulation per phialide	Metabolic energy used for spore production
Scattered single phialides	Complex three or more branched penicillus	Many more spores per conidial apparatus
Smooth stipes	Rough stipes	Reduction of arthropod predation
Diverse habitats	Special niches	Reduced competition
Floccose colonies	Velvety colonies	Metabolic energy used for spore production; spores more readily exposed to dispersal by air movement.
Velvety colonies	Fasciculate to coremial colonies	
Weak conidial pigmentation	Strong conidial pigmentation	Protection of genome from ultraviolet light.
Hygrophilic	Xerotolerant	Ability to grow on some stored foods.

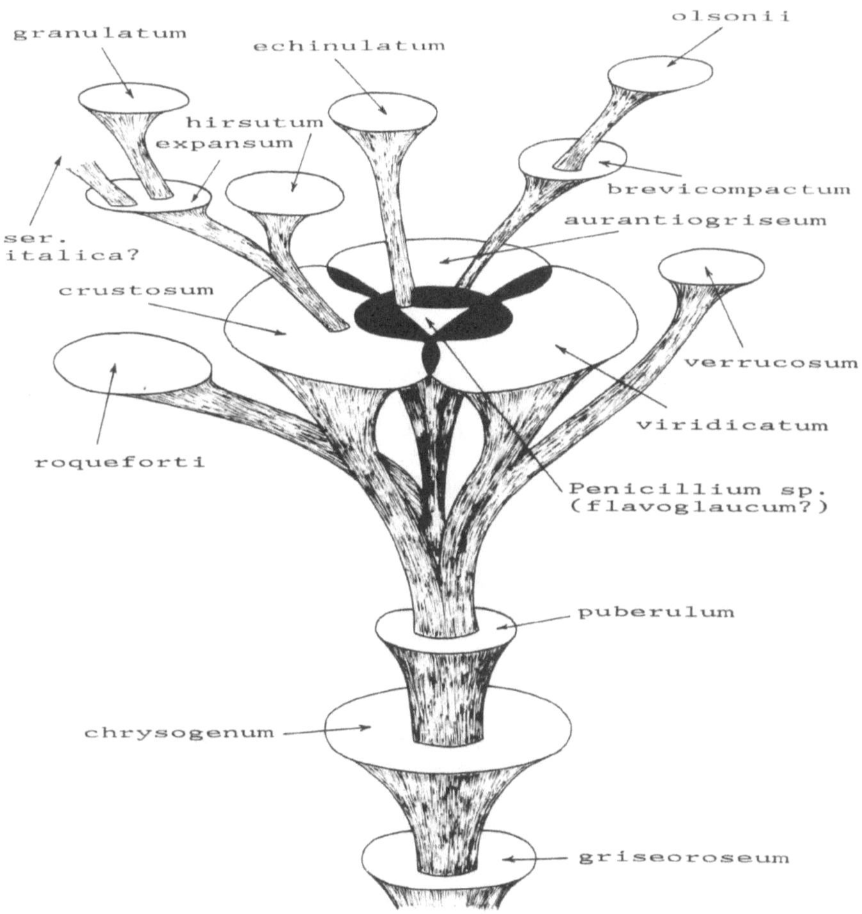

Fig. 1. A "nested trumpet" diagram to explain the possible species interrelationships in the evolution of the subgenus Penicillium.

Fig. 1 is a pictorial representation of the possible interrelationships of many of the major species of subgenus Penicillium. A few species have been omitted because their position is unclear. These include those in series Italica, some species of which may be close to P. expansum; P. camemberti, which is probably a domesticated species with its wild type somewhere in the middle of the P. crustosum/ P. aurantiogriseum/ P. viridicatum complex; P. griseofulvum; P. olivicolor and P. atramentosum.

Several authors (Raper and Thom, 1949; Frisvad, 1981; Anne, this publication) have suggested that P. citrinum and P. chrysogenum are quite closely related and hence that the terverticillate species (subgenus Penicillium) may have arisen from an advanced species in subgenus Furcatum such as P. citrinum.

In Fig. 1, the ancestor in subgenus *Furcatum* would be just below the foot of the page and the species in subgenus *Penicillium* then ascend the page in a sequence of increasing complexity. Wherever possible, the most distinctive and advanced species have been placed away from the ascending central core.

The following list contains all of the species which we have attempted to place and gives a brief summary of our reasons for those placements. The nomenclature of Pitt (1979) is used.

P. griseoroseum

This was placed by Pitt (1979) in subgenus *Furcatum*. Colonies are floccose, stipes smooth and penicilli often possess a subterminal ramus. *P. griseoroseum* is very closely related to *P. chrysogenum*.

P. chrysogenum

This species produces velvety colonies and smooth stipes. Penicilli are terverticillate but less compact than those of most other species in the subgenus.

P. puberulum

Colonies of this species are at most weakly fasciculate and stipes are at most finely roughened. The penicillus is terverticillate and similar to *P. aurantiogriseum* and *P. viridicatum*. Conidia are greenish grey. *P. puberulum* is probably a direct ancestor of *P. aurantiogriseum*.

P. aurantiogriseum

P. aurantiogriseum is a large and variable species which interfaces with *P. viridicatum*, *P. crustosum*, *P. roqueforti*, *P. puberulum* and possibly *P. brevicompactum*. Conidia are blue to greyish blue and stipes are smooth to strongly roughened. The colonies are fasciculate and variable in size.

P. viridicatum

This is similar in all major respects to *P. aurantiogriseum* but the conidia are bright yellow green. Interfaces exist with *P. aurantiogriseum*, *P. crustosum* and *P. verrucosum*.

P. crustosum

P. crustosum is a distinctive species, not very fasciculate, but with highly roughened stipes. Conidia are produced in huge numbers on malt extract agar. *P. crustosum* consistently causes a limited rot in apples and has interfaces with *P. roqueforti*, *P. aurantiogriseum*, *P. viridicatum* and *P. expansum*.

P. roqueforti

This is a partially domesticated species which is used to

ripen blue cheeses. Colonies are velvety but the stipes are very rough or tuberculate, possibly as a protection from predation by cheese mites. *P. roqueforti* intergrades with *P. crustosum* (with which it shares a very similar penicillus) and with *P. aurantiogriseum*.

P. verrucosum

This distinctive yellow green species which produces ochratoxin A. Stipes vary from finely to strongly roughened, while penicilli may be either rather simple with 2-3 stage branching or rather complex with 3-4 stage branching. This species has its closest affinity with *P. viridicatum* but, from the variability of its penicillus morphology, probably diverged from *P. viridicatum* a long time ago.

P. expansum

A species which seems to be related to several others in subgenus *Penicillium*. It has found a highly successful niche due to its ability to cause a destructive rot of pomaceous fruit. It may be derived from *P. crustosum*, which is capable of causing a similar but less extensive rot. Stipes are almost invariably smooth and the fasciculate or coremial habit is strongly marked. Apart from similarities to *P. crustosum*, *P. expansum* also shows distinct interfaces with *P. granulatum* and *P. hirsutum*.

P. granulatum

This is an advanced species by all the criteria considered here. Penicilli are complex, terverticillate to quaterverticillate; conidia are highly pigmented, stipes are conspicuously roughened while colonies are compact and microcoremial. Like *P. expansum*, to which it seems related, *P. granulatum* usually produces orange-brown pigments in culture.

P. hirsutum

This variable species has either smooth stipes and rough conidia or rough stipes and smooth conida. It is usually strongly fasciculate and is isolated either from barley or rotting bulbs. Similarities can be seen to *P. expansum* although in all probability the two species diverged a long time ago.

P. echinulatum

Although *P. echinulatum* is now considered a distinct species, it was originally thought to be a variety of *P. aurantiogriseum*. Neat terverticillate penicilli with rough stipes, fasciculate growth and dark green conidia suggest that this is a fairly advanced species with origins somewhere between *P. aurantiogriseum*, *P. viridicatum* and *P. crustosum*.

P. brevicompactum

By most criteria, *P. brevicompactum* is not a particularly advanced species, exhibiting smooth stipes and velvety growth. However the compact penicillus with many metulae and phialides permits the production of high numbers of conidia from each conidial apparatus. Pitt (1979) described isolates which intergrade with *P. aurantiogriseum* and *P. puberulum* and the species has accordingly been placed centrally in the evolutionary scheme.

P. olsonii

The multiramulate heads and larger deep colonies distinguish this species clearly from *P. brevicompactum*. However there is an unmistakable similarity in the penicillus structure. Conidia are again produced in large numbers. *P. olsonii* has been isolated from a wide variety of foods but is of relatively infrequent occurrence.

Penicillium sp

A rather common group of Penicillia exists with dark green conidia, rough stipes, velvety to fasciculate colonies and penicilli typical of *P. aurantiogriseum* and *P. viridicatum*. Pitt (1979) enlarged his species concepts for *P. viridicatum* and *P. crustosum* to accomodate these isolates, which correspond to *P. palitans sensu* Raper & Thom.

Samson et al. (1976) named them *P. verrucosum* var. *melanochlorum*, while Frisvad (1981) and Frisvad and Filtenborg (1983) called them *P. melanochlorum* and *P. mali* respectively. Some of these isolates fail to key in the Pitt (1979) scheme, so Williams and Pitt (this publication) have slighty enlarged the species concept for *P. aurantiogriseum* in order to accomodate them. However, experience now suggests that these fungi almost certainly represent a recognisable species, of which the earliest description is probably that of *P. flavoglaucum* Biourge. We intend to confirm the identity and typification of this species, which appears to have interfaces at least in *P. puberulum*, *P. aurantiogriseum*, *P. viridicatum*, *P. crustosum* and *P. echinulatum*.

The evolutionary scheme shown in Fig. 1 is only one possible explanation of the phylogenic relationships of members of subgenus *Penicillium*. It does, however, have certain advantages. For example, it simplifies making comparisons of the species concepts of various authors. Raper and Thom (1949) included four species (*P. chrysogenum*, *P. meleagrinum*, *P. cyaneofulvum* and *P. notatum*) in their *P. chrysogenum* series. These are now considered to be synonymous. In Fig. 1, *P. chrysogenum* in their sense would occupy the back of the disc, under the blue fasciculate species; *P. meleagrinum* would occupy the front of the disc, under the green fasciculate species and *P. notatum* would be placed somewhere between *P. chrysogenum* and *P. griseoroseum* (= *P. cyaneofulvum*). Similarly *P. aurantiogriseum sensu* Pitt can be divided into Raper and Thom's *P. cyclopium* (adjacent to *P. viridicatum*) and *P. martensii* (adjacent to *P. crustosum*).

Samson et al. (1976) placed several of the fasciculate species into one large species (*P. verrucosum*) with several varieties. These can be seen on the large central disc of Fig. 1, where *P. aurantiogriseum* and *P. crustosum* were called *P. verrucosum* var. *cyclopium*; *P. viridicatum* was called *P. verru-*

cosum var verrucosum and the unnamed central species was called P. verrucosum var. melanochlorum. Stolk & Samson (1985; this publication) have restored these varieties to species status. Frisvad and Filtenborg (1983), and Frisvad (1985; this publication) have divided a number of the species of subgenus Penicillium into chemotypes (compare Pitt and Hawksworth, this publication) on the basis of secondary metabolite production and substrate utilisation. Again it is easy to locate many of those chemotypes in Fig. 1, and indeed it may well be possible to confirm or correct placements in the proposed evolutionary sequence by reference to common metabolic profiles.

Wicklow and Filtenborg (pers. commun) have described some isolates of subgenus Penicillium from the underground food stores of kangaroo rats. These fungi have colonised a niche which has been undisturbed for up to 9 million years. One species is quite separate from any yet described in subgenus Penicillium, having velvety colonies like P. chrysogenum, green conidia and rough stipes. It is however, possible to place this species in Fig. 1. as lying somewhere beteen P. chrysogenum and one of the fasciculate species, or perhaps even as an ancestor of P. roqueforti. Unfortunately we have no way of knowing when this fungus became separated from its siblings and colonised its present niche.

Models such as we have described in this paper help to illustrate that the taxonomies currently in use differ only by degree. Time and further research will tell us a lot more about the interrelations of these economically important fungi. We hope that this paper will be a useful contribution to the understanding of species relationships of subgenus Penicillium.

REFERENCES

FRISVAD, J.C. 1981. Physiological criteria and mycotoxin production as aids in classification of common asymmetric Penicillia. Appl. environ. Microbiol. 41: 568-579.

FRISVAD, J.C. and FILTENBORG, O. 1983. Classfication of terverticilate Penicillia based on profiles of mycotoxins and other secondary metabolites. Appl. environ. Microbiol. 46: 1301-1310.

PITT, J.I. 1979. "The Genus Penicillium and its Teleomorphic States Eupenicillium and Talaromyces". London: Academic Press.

RAPER, K.B. and THOM. C. 1949. "Manual of the Penicillia". Baltimore: Willias and Wilkins.

SAMSON, R.A., STOLK, A.C. and HADLOK, R. 1976. Revision of the subjection Fasciculata of Penicillium and some allied species. Stud. mycol., Baarn 11: 1-47.

DIALOGUE FOLLOWING MR. WILLIAMS' PAPER

SAMSON: I like your evolutionary concept and the diagram. I think P. brevicompactum does not originate from P. palitans but rather from P. chrysogenum.

WILLIAMS: You are probably correct. *P. brevicompactum* definitely comes out of the middle, but it's difficult to indicate exactly where.

SAMSON: Concerning *P. verrucosum* and *P. viridicatum*, Pitt (1979) placed these two species in separate series. Your diagram indicates that they are more closely related. Have you then revised that concept?

WILLIAMS: I see them as being more closely related, but *P. verrucosum* comes out lower down and runs more or less parallel to *P. viridicatum*.

SAMSON: Where would you put *P. verrucosum* var. *melanochlorum* (= *P. mali*)?

WILLIAMS: In the centre; they would key out as *P. palitans* in Raper and Thom.

SAMSON: I would put *P. camemberti* very close to *P. aurantiogriseum*, because I believe it is a domesticated form of the latter species.

A REVISED KEY TO PENICILLIUM SUBGENUS PENICILLIUM

Anthony P. WILLIAMS and John I. PITT

Leatherhead Food Research Association, Randalls Road Leatherhead, Surrey, KT22 7RY, UK and CSIRO Division of Food Research, North Ryde, NSW 2113 Australia

SUMMARY

The standard description of Penicillium aurantiogriseum Dierckx has been expanded slightly to include isolates with slow growth rate, grey-green conidia and/or smooth-walled stipes. An amended dichotomous key is provided.

INTRODUCTION

During recent studies on the mycoflora of mould-spoiled foods in the United Kingdom, more than three hundred isolates belonging to subgenus Penicillium have been examined. The majority of these isolates were identified successfully using the methods of Pitt (1979). However, of more than sixty isolates of P. aurantiogriseum Dierckx, approximately half failed to key properly for one of the following reasons:

A. colony diameters on CYA after 7 days at 25 °C less than 30 mm (occasionally as small as 18 mm);
B. stipe texture smooth on CYA and occasionally also on MEA;
C. blue element in colony colour on CYA reduced to Methuen colour 26E3-4, (Kornerup and Wanscher, 1978).

The following amended key is proposed as an alternative to that of Pitt (1979; 320-322). Penicillium aurantiogriseum is keyed out in various places to take into account these variations. The standard description of the species (Pitt, 1979; 9 353-4) has accordingly been slightly modified.

1. Conidia borne as cylinders, with at least a proportion remaining so at maturity.......................... Sect. Cylindrosporum Ser. Italica 5

Conidia borne as ellipsoids or spheroids and remaining so at maturity.......................................2

2 (1b). Colonies on G25N not exceeding 10 mm diam; conidia en masse brown; penicilli often irregular.......... Sect. Inordinate, Ser. Arenicola, P. arenicola (111)

Colonies on G25N exceeding 10 mm diam; conidia en masse green or olive; penicilli characteristically in regular verticils 3

3 (2b). Penicilli often with 3 or more appressed rami; stipes occasionally exceeding 1 mm long.................... Sect. Coronatum, Ser. Olsonii, P. olsonii (110)

Penicilli with no more than two rami; stipes not exceeding 1 mm long....................................4

4 (3b). At most one ramus, borne subterminally; penicilli often biverticillate ....spp. from Subgen. Furcatum Chap. 9

One to two rami characteristically produced, borne terminally ....Sect. Penicillium -- 8

Sect. CYLINDROSPORUM. Ser. Italica (spp. 106-109)

5 (1a) Conidia en masse olive.......... ...P. digitatum (107)

Conidia en masse green................................6

6 (5b). Colonies on CYA at 25°C exceeding 50 mm diam......... ....P. resticulosum (108)

Colonies on CYA at 25°C not exceeding 50 mm diam.....7

7 (6b). Mycelium salmon; conidial walls spinulose........... ....P. fennelliae (109)

Mycelium white; conidia smooth-walled.P. italicum (106)

Sect. PENICILLIUM

8 (4b) Colonies floccose; conidia en masse persistently white or pale grey green; isolated from cheese or cheese factory........Ser. Camemberti, P. camemberti (99)

Colonies with surface texture velutinous or fasciculate, conidia en masse blue, green, grey or rarely olive; source inconsequential 9

9 (8b). Colonies on CYA at 25°C not exceeding 30 mm diam... ....Ser. Urticicola. 23

Colonies on CYA at 25°C exceeding 30 mm diam........10

10 (9b). Stipes smooth-walled............ Ser. Expansa 11

Stipes commonly rough-walled.... Ser. Viridicata 16

Ser. Expansa (spp. 90-92)

11(10a). Conidia *en masse* greyish blue, blue or blue green 12

Conidia *en masse* green..............................14

12 (11a). On CYA at 25°C exudate, soluble pigment, and/or reverse colouration yellow; sometimes growth at 37°C, velutinous or floccose on MEA....................13

On CYA at 25°C, exudate, soluble pigment, and/or reverse colouration pale, orange, reddish brown or brown; no growth at 37°C fasciculation evident on MEA ....*P. aurantiogriseum* (98)

13 (12a) Colonies velutinous to lightly floccose; conidiogenesis moderate to heavy; penicilli predominantly terverticillate...............*P. chrysogenum* (91)

Colonies floccose; conidiogenesis sparse; penicilli commonly biverticillate ....*P. griseoroseum* (66)

14 (11b). On CYA at 25°C exudate, soluble pigment and reverse colouration red; conidia *en masse* dark green. *P. atramentosum* (92)

On CYA at 25°C exudate, soluble pigment and reverse colouration orange or brown; conidia *en masse* grey green 15

15 (14b). Exudate on CYA at 25°C near maroon; mycelium on MEA often deep yellow; penicilli sometimes quaterverticillate.........................*P. hirsutum* (96)

Exudate on CYA at 25°C orange brown; mycelium on MEA white or inconspicuous; penicilli not quaterverticillate ....*P. expansum* (90)

Ser. Viridicata (spp. 93-97)

16 (10b). Conidia *en masse* greyish blue *P. aurantiogriseum*(98)

Conidia *en masse* green or greyish green 17

17 (16b) Conidial wall rugose or spinose...................18

Conidia smooth-walled............................19

18 (17a). Exudate on CYA at 25°C pale yellow to pale brown; soluble pigment absent; mycelium on MEA inconspicuous.........................*P. echinulatum* (97)

Exudate on CYA at 25°C near maroon; soluble pigment deep yellow to orange brown; mycelium on MEA conspicuous, often deep yellow ....*P. hirsutum* (96)

19 (17b). Colonies on CYA and MEA at 25°C exceeding 40 mm diam; reverse commonly deep green...... *P. roqueforti* (95)

Colonies on CYA and MEA at 25°C not exceeding 40 mm diam; reverse pale or brown......................20

20 (19b) Exudate on CYA at 25°C clear to orange or brown; mycelium on MEA white or inconspicuous; penicilli terverticillate.................................21

Exudate on CYA 25°C near maroon; mycelium on MEA often deep yellow; penicilli sometimes quaterverticillate ....*P. hirsutum* (96)

21 (20a). Colonies on CYA and MEA at 25°C exceeding 35 mm and 30 mm respectively; conidia *en masse* dull green; colonies on MEA often producing masses of detached conidia when jarred...............*P. crustosum* (94)

Colonies on CYA and MEA at 25°C not exceeding 35 mm and 30 mm respectively; conidia on MEA adhering to the colonies......................................22

22 (21b) Conidia *en masse* bright yellow green to dull green ....*P. viridicatum* (93)

Conidia *en masse* greenish grey ....P. *aurantiogriseum* (96)

Ser. *Urticicola* (spp. 100-105)

23 (9a). Conidia *en masse* olive......... *P. olivicolor* (102)

Conidia *en masse* green, blue or grey..............24

24 (23b). Phialides commonly 4.5-6µm long.*P. griseofulvum* (100)

Phialides exceeding 6 µm long 25

25 (24b). Stipe walls smooth to finely roughened...........26

Stipe walls commonly rugose......................29

26 (25a). Penicilli broad, often 40 µm or more across due to numerous apically enlarged metulae ....*P. brevicompactum* (103)

Penicilli narrow, commonly 30 µm or less across with metulae cylindroidal 27

27 (26b) Colonies on CYA low, dense and velutinous; exudate absent or limited and clear; soluble pigment absent ....*P. puberulum* (101)

Colonies on CYA deeply floccose, definitely fasciculate or minutely coremial; exudate clear to orange or brown; soluble pigment, if present, orange to brown...28

28 (27b). Colonies on CYA deep, usually coremial; conidia *en masse* dull green; stipes smooth on CYA..*P. expansum* (90)

Colonies on CYA moderately deep, fasciculate at least at the margins; conidia *en masse* greyish blue to greyish green; stipes on CYA usually finely roughened, sometimes smooth .....*P. aurantiogriseum* (98)

29 (25b). Conidia *en masse* greyish blue, rarely greyish green; on CYA at 25°C colony margins fasciculate, but not microcoremial or coremial...*P. aurantiogriseum* (98)

Conidia *en masse* bright yellow green or dull green; on CYA at 25°C colony margins velutinous, fasciculate or coremial..................................30

30 (29b). Colonies on CYA and MEA at 25°C exceeding 28 mm and 25 mm diam respectively ...... *P. viridicatum* (93)

Colonies on CYA and MEA at 25°C not exceeding 28 mm and 25 mm diam respectively 31

31 (30b). On CYA at 25°C, colony texture velutinous to fasciculate, mycelium white, soluble pigment absent; conidia usually spheroidal ......*P. verrucosum* (105)

On CYA at 25°C, colony texture fasciculate to coremial, mycelium white to yellow, yellow to brown soluble pigment typically produced; conidia ellipsoidal to subspheroidal.........*P. granulatum* (104)

STANDARD DESCRIPTION

98. *Penicillium aurantiogriseum* Dierckx

Annls Soc. Scient. Brux. 25: 88, 1901.

CYA, 25°C, 7 days: Colonies typically 25-35 mm diam, occasionally as small as 18 mm, conspicuously radially sulcate, moderately deep, with surface texture smooth to granular with the unaided eye, but under low magnifications seen to consist predominantly of fascicles of immature conidiophores; mycelium white, usually inconspicuous, but occasionally dominating the colony appearance in floccose isolates; conidiogenesis typically moderate to heavy, Greyish Turquoise to Dull Green (M. 24-25E3, or even as green as 26E3-4) or between Glaucous Blue-Green and Glaucous Grey (R. 94-109), sometimes in paler or brighter shades at the margins or centrally brown; exudate typically conspicuous at least near the margins, and rarely completely absent, clear or pale brown; soluble pigment produced by some isolates, brown to reddish brown; reverse varying from isolate to isolate, pale, light to brilliant orange, or reddish to violet brown.

MEA, 25°C, 7 days: Colonies 20-35 mm diam, plane or rarely radially sulcate, low or sometimes centrally umbonate, relatively sparse with surface texture finely granular to the unaided eye; marginal hyphae usually entirely subsurface, bearing fascicles of conidiophores; mycelium usually subsurface, but in uncommon isolates conspicuous at colony centres or over much of the colony surface and then bright yellow; conidiogenesis typically moderate to heavy, Dull Green (M. 25-26D-E3-4) or near Glaucous Blue-Green (R.94), athough in uncommon isolates limited or even lacking; exudate absent; soluble pigment sometimes produced, yellow brown reddish brown; reverse pale, orange, or reddish brown.

G25N, 25°C, 7 days: Colonies 15-24 mm diam, usually radially sulcate, moderately deep, dense, with a granular surface texture; conidiogenesis absent to heavy, deep grey green, often with bluish marginal areas, or centrally brown; exudate and soluble pigment absent; reverse pale, yellow or brown.

5°C, CYA, 7 days: Colonies 2-5 mm diam, often centrally floccose or umbonate, of white mycelium, reverse pale or deep orange brown.

37°C, CYA, 7 days: No growth

Conidiophores borne singly or in fascicles, mostly from subsurface hyphae, with stipes commonly 200-400 x 3.0-4.0 µm, although sometimes longer or of indeterminate length because of fasciculation, with walls usually finely to conspicuously roughened but in some isolates smooth, bearing terminal terverticillate or less commonly biverticillate penicilli; rami usually borne singly, 15-25 x 3.0-3.5 µm; metulae in verticils of 3-4, measuring 10-12(-15) x 2.8-3.5 µm; phialides 5-8 per metula, slender, ampulliform, typically 9-10 x 2.5-2.8 µm, with undistinguished collula; conidia subspheroidal to ellipsoidal, commonly 3.5-4.0 x 2.5-3.2 µm, smooth walled, usually borne in long quite well defined columns.

REFERENCES

PITT, J.I. 1979. "The Genus *Penicillium* and its Teleomorphic States *Eupenicillium* and *Talaromyces*. London: Academic Press.

REVISION OF RECENTLY DESCRIBED PENICILLIUM TAXA

Carlos RAMÍREZ

Laboratory of General and Applied Mycology
Instituto "Jaime Ferrán" de Microbiologia
28006 Madrid, Spain

SUMMARY

A list of recently described Penicillium taxa, ofwhich 13 are recognized as new taxa, illustrations of 9 of them and a Latin diagnosis to validate P. nordicum Dragoni & Cantoni are presented. The 13 recognized taxa are: P. chalybeum Pitt & Hocking, P. corynephorum Pitt & Hocking, P. flavido-stipitatum Ramírez & Gonzalez, P. jugoslavicum Ramirez & Muntanola-Cvetkovic, P. lapatayae Ramírez, P. mclennaniae Yip, P. mali Gorlenko & Novobranova, P. nordicum Dragoni & Cantoni ex Ramirez, P. patens Pitt & Hocking, P. radiato-lobatum Lorinczi, P. resinae Qi Tong, P. rubefasciens Quintanilla and P. turrispainensis Ramírez.

INTRODUCTION

The purpose of this compilation is to present short descriptions with our own camera lucida drawings of new Penicillium taxa that appeared before 1981, mostly from relatively obscure publications, but not included in the "Manual and Atlas of the Penicillia" (Ramírez, 1982) and those that appeared after that date, so that those interested in identifying these species will not have to search the literature for descriptions of them. However, according to substantial broadening of species concepts in Penicillium proposed by Stolk & Samson (1983), some of these newly described taxa have here been reduced to synonymy.

Alphabetical list of new Penicillium taxa.

1. P. castellae Quintanilla IJFM 7304
2. P. cieglerii Quintanilla IJFM 7673
3. P. caerulescens Quintanilla IJFM 7673
4. P. chalybeum Pitt & Hocking
5. P. coalescens Quintanilla IJFM 19.021

6. *P. corynephorum* Pitt & Hocking
7. *P. flavido-stipitatum* Ramírez & Gonzalez IJFM 7824
8. *P. jugoslavicum* Ramírez & Muntanola-Cvetovic IJFM 7785
9. *P. lapatayae* Ramírez IJFM 19.012
10. *P. mali* Gorlenko & Novobranova CBS 500.73
11. *P. mariaecrucis* Quintanilla IJFM 7300
12. *P. mclennaniae* Yip IMI 253.561
13. *P. mediolanensis* Dragoni & Cantoni ATCC 44200
14. *P. nordicum* Dragoni & Cantoni ATCC 44219
15. *P. michaelis* Quintanilla IJFM 7757
16. *P. patens* Pitt & Hocking
17. *P. radiatolobatum* Lorinczi CBS 340.79
18. *P. resinae* Qi Tong
19. *P. rubefaciens* Quintanilla IJFM 7755
20. *P. sajarovii* Quintanilla IJFM 7674
21. *P. severskii* Shechovtsov IJFM 19.000
22. *P. smithii* Quintanilla IJFM 7675
23. *P. turrispainensis* Ramírez IJFM 19.011
24. *P. vaccaeorum* Quintanilla IJFM 7756

Descriptions of the new Penicillium species

1. *Penicillium castellae* Quintanilla - Avan. Nutr. Mej. Anim. 23: 333-343. 1982.

From the original description, and my observations with the camera lucida and SEM, *P. castellae* may be regarded as a sclerotial strain of *P. raciborskii* Zaleski.

2. *Penicillium cieglerii* Quintanilla - Avan. Nutr. Mej. Anim. 23: 333-343. 1982.

From the original description and my observations with the camera lucida and SEM, *P. cieglerii* may be regarded as a sclerotial strain of *P. novae-caledoniae* Smith. It also produces the characteristic strawberry pink colour on the colony reverse.

3. *P. caerulescens* Quintanilla- Mycopathologia 82: 101-104. 1983.

From the original description and my observations with the camera lucida and SEM, *P. caerulescens* may be regarded as a sclerotial strain of *P. raciborskii* Zaleski. It produces the dark colour on the colony reverse as reported by Ramírez (1982) for *P. fagi* = *P. raceborskii* fide Stolk and Samson, 1983).

4. *Penicillium coalescens* Quintanilla - Mycopathologia 84: 115-120. 1983

From the original description, and my observations with the camera lucida and SEM, *P. coalescens* has to be considered as synonymous with *P. dendriticum* Pitt.

5. *Penicillium flavido-stipitatum* Ramírez & Gonzalez - Mycopathologia 88: 3-7. 1984 -- Fig. 1.

The correct placement of this species appears to be in the *P. canescens* series (Ramírez, 1982), but it can easily be separated from all species of the series by its characteristically yellow-stained stipes and by its spirally ornamented conidia.

6. *Penicillium jugoslavicum* Ramírez & Muntanola-Cvetovic - Mycopathologia 88: 79-81. 1984. -- Fig. 2.

P. jugoslavicum is very closely related to species of the *P. frequentans* series (Ramírez, 1982), especially to *P. frequentans* and *P. purpurescens*. Differences in the fine ornamentation of conidia as viewed with the SEM, the colony appearance and the arrangement of conidial chains, warranted, in my opinion, its description as a new species.

7. *Penicillium lapatayae* Ramírez, 1985 - Mycopathologia (in press). Fig. 3.

P. lapatayae appears to be most satisfactorily assignable to the *P. thomii* series (Ramírez, 1982), within the *P. turbatum* subseries, but colony appearance and the fine ornamentation of conidial walls as viewed with the SEM are different from *P. turbatum* and the other species included in the subseries.

8. *Penicillium mali* Gorlenko & Novobranova, 1983 - Mikologia et Fitopatologia 17: 464-467.-- Fig. 4.

Note: due to wrong bibliographic information, a description of *P. mali* was published in Ramírez (1982: 389), one year before its publication by Gorlenko and Novobranova, and under the single authorship of Dr. T.I. Novobranova.

9. *Penicillium mariaecrusis* Quintanilla - Avan. Nutr. Mej. Anim. 23: 333-343. 1982.

From the original description and my observations with the camera lucida and SEM, *P. mariaecrusis* seems quite similar to *P. soppii* Zaleski, but with some marked differences in gross colony appearance.

10. *Penicillium mclennaniae* Yip -- Trans. Br. mycol. Soc. 77: 202-203. 1981 -- Fig. 5.

P. mclennaniae seems to be best placed in the subgenus Furcatum, section Furcatum and series Oxalica (Pitt, 1979). However, it has no particular affinity with any species in the series.

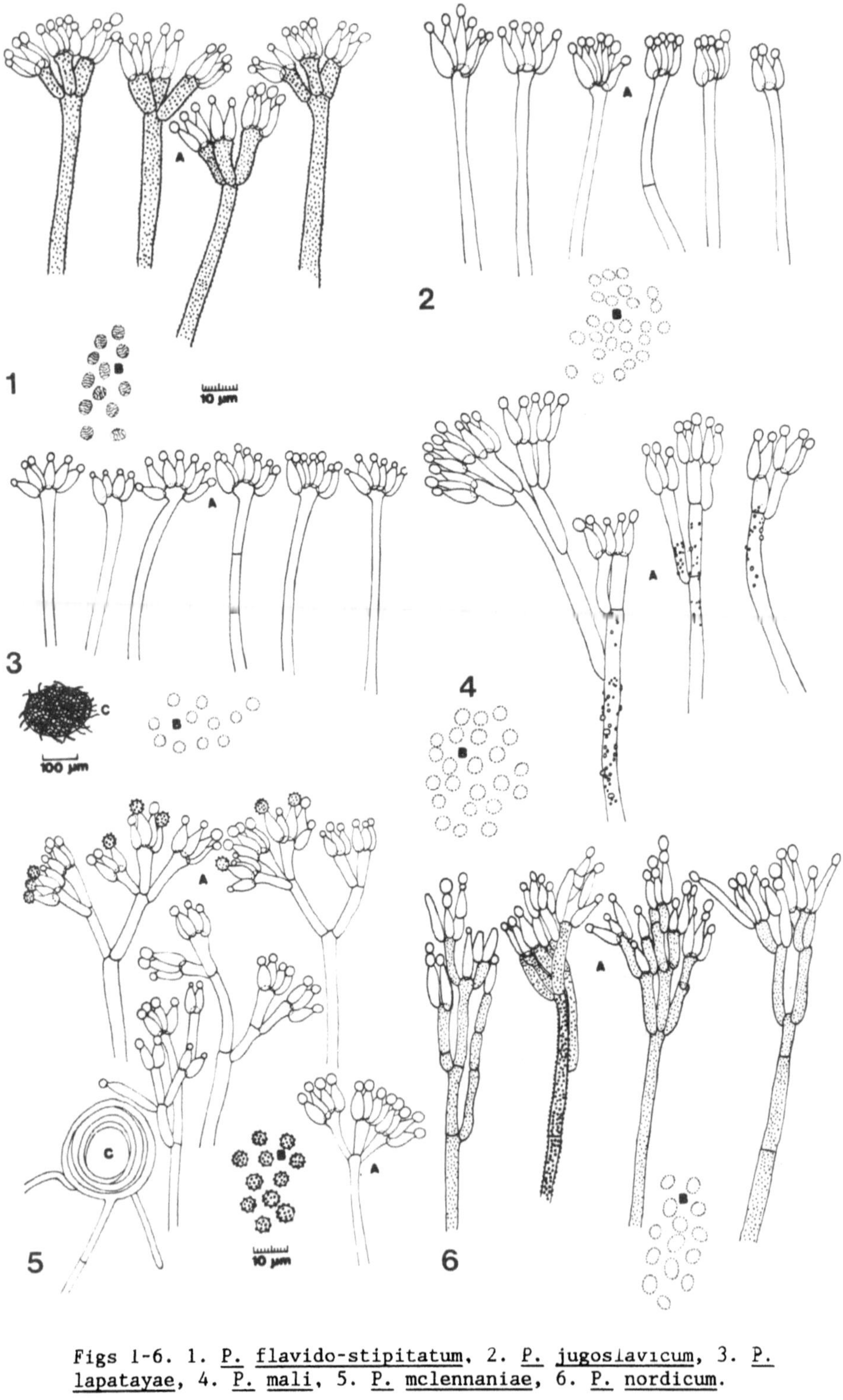

Figs 1-6. 1. *P. flavido-stipitatum*, 2. *P. jugoslavicum*, 3. *P. lapatayae*, 4. *P. mali*, 5. *P. mclennaniae*, 6. *P. nordicum*.

11. <u>Penicillium</u> <u>mediolanensis</u> Dragoni & Cantoni - Ind. Aliment. 155: 281-284. 1979.

The species was invalidly published, without a Latin diagnosis,(Art. 36). From my observations with the camera lucida and SEM, <u>P.</u> <u>mediolanensis</u> shows a great resemblance to <u>P.</u> <u>verrucosum</u> var. <u>cyclopium</u>, and has to be regarded as synonymous with it.

12. <u>Penicillium</u> <u>nordicum</u> Dragoni & Cantoni - Ind. Aliment. 155: 281-284. 1979. Fig. 6.

<u>P.</u> <u>nordicum</u> was described without a Latin diagnosis. In order to validate the species a Latin diagnosis is given below:

Coloniae in agaro Czapekii satis celeriter crescentes, post duas hebdomadas ad 25°, 60 mm in diametro assecutae sunt, lanosae, cum fasciculati margine, denso texto basali coacto mycelio, albae; exsudatio sine colore; odore mucido; reverso albido cum fusco centro. Coloniae in agaro hordeaceo lente crescentes, post duas hebdomadas ad 25°, 30 mm in diametro assecutae sunt, planae, albae, pulverulentae; reverso flavido; odor vehemens mucidus. neque ascomata neque sclerotia visa sunt. Penicilli ramosi, densi, asymmetrici, compacti; rami pleraque 18-35 µm in longitudine per 3 µm in diametro; metulae pleraque 10-28 µm in longitudine per 3 µm in diametro; phialides paper numero, 2-4 per verticillium, plerumque 8-20 µm in longitudine per 2.5-3 µm in diametro; conidia subglobosa vel ovoida, levia, 4-6 µm in diametro maximo.

Typus cultura ATCC 44219 ex longanonibus Italiae a Dragoni et Cantoni isolata est in anno 1978.

13. <u>Penicillium</u> <u>michaelis</u> Quintanilla - Mycopathologia 80: 73-82. 1982

From the original description and my own observations with the camera lucida and SEM, <u>P.</u> <u>michaelis</u> closely resembles <u>P.</u> <u>stoloniferum</u> Thom in important features that it has to be regarded as synonymous with it.

14. <u>Penicillium</u> <u>radiato-lobatum</u> Lorinczi - Publ. Soc. Nat. Rom. Pent. Stin. Sol. 10 B: 433-445. 1972. -- Fig. 7

The species diagnosis is based on the type strain CBS 340.79 isolated from alluvial soil from the valley of Popi, Roumanian locality of Cluj. <u>Penicillium</u> <u>radiato-lobatum</u> seems to be best placed within the <u>P.</u> <u>nigricans</u> series (Ramírez, 1982) or the series <u>Canescentia</u> (Pitt, 1979), because of its divaricate penicilli and spinose conidia.

15. <u>Penicillium</u> <u>rubefasciens</u> Quintanilla - Mycopathologia 80: 73-82. 1982

<u>P.</u> <u>rubefasciens</u> may be best placed into Pitt's sect. <u>Furcatum</u> (Pitt, 1979), either in series <u>Citrina</u> between <u>P.</u>

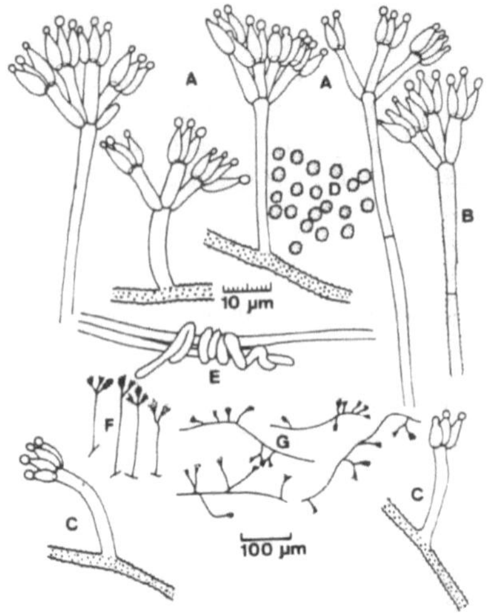

7

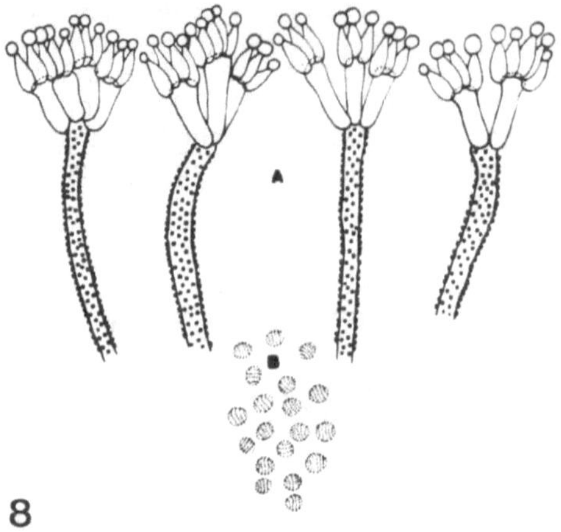

8

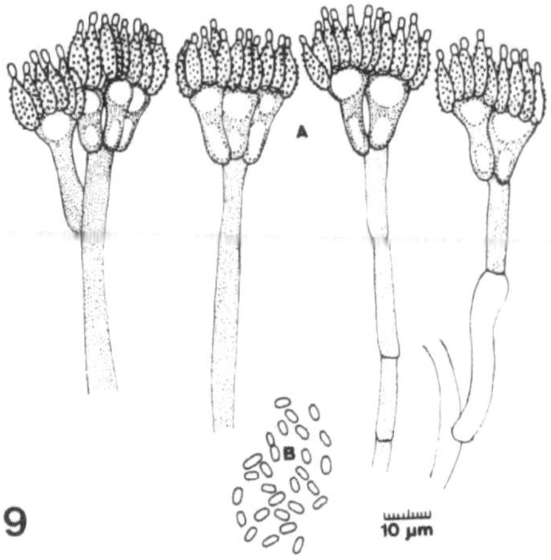

9

Figs. 7-9. 7. P. radiolobatum, 8. P. rubefasciens, 9. P.turrispainensis.

novaezeelandiae van Beyma and P. miczynskii Zaleski, or in the series Oxalica, close to P. madritii Smith. P. rubefaciens is one of the few species of Penicillium showing coloured reddish stipes. This character, together with the deep red diffusing pigment and the fine ornamentation of conidia (Fig. 8) allows maintenance of P. rubefasciens as a separate species.

16. Penicillium sajarovii Quintanilla - Avan. Nutr. Mej. Anim. 22: 539-541. 1981.

From the original description and my observations with the camera lucida and SEM, P. sajarovii may be regarded as synonymous with P. canescens Sopp as the two species have identical penicilli and identically ornamented conidia.

17. Penicillium severskii Schechovtzov - Mikrobiologia 43: 122-123. 1981.

From the original description, and my own observations with the camera lucida and SEM I believe that P. severskii cannot be separated from P. raciborskii Zaleski.

18. Penicillium smithii Quintanilla - Avan. Nutr. Mej. Anim. 23: 333-343. 1982.

From the original description and my observations with the camera lucida and SEM, I believe that P. smithii may represent a deficient mutant of P. raciborskii because of the very scanty growth on Cz. However, the penicilli and conidia are identical.

19. Penicillium turrispainensis Ramírez - Mycopathologia (in press).-- Fig.9

P. turrispainensis appears to be most satisfactorily assignable to the group of Penicillia showing rugulose phialides and cylindrical conidia, for which the genus Geosmithia Pitt was created, i.e., the P. pallidum series (Ramirez, 1982). It shows sufficient differences from all the species included in the series to warrant its recognition as a new taxon.

20. Penicillium vaccaeorum Quintanilla - Mycopathologia 80: 73-82. 1980

From the original description and my observations with the camera lucida and SEM, P. vaccaeorum cannot be separated from P. citreonigrum Dierckx, and should be regarded as a synonym of it.

CONCLUSION

From the 24 species of Penicillium recently described as new by several authors, eleven are regarded here, as synonyms, as follows:

P. castellae Quintanilla = P. raciborskii Zaleski

P. cieglerii Quintanilla = P. novae-caledoniae Smith

P. caerulescens Quintanilla = P. raciborskii

P. coalescens Quintanilla = P. dentriticum Pitt

P. mariae-crusis Quintanilla = P. soppii Zaleski

P. mediolanensis Dragoni & Cantoni = P. cyclopium Westling

P. michaelis Quintanilla = P stoloniferum Thom

P. sajarovii Quintanilla = P. canescens Sopp

P. severskii Shechovtsov = *P. raciborskii*

P. vaccaeorum Quintanilla = *P. citreonigrum* Dierckx

P. smithii Quintanilla = *P. raciborskii*

THE GENUS COREMIUM AND THE SYNNEMATOUS PENICILLIA

Keith A. SEIFERT and Robert A. SAMSON

Centraalbureau voor Schimmelcultures
P.O. Box 273, 3740 AG BAARN, Netherlands

SUMMARY

Species attributed to Coremium, a genus long considered a synonym of Penicillium, are revised based on the study of type specimens. Three new combinations in Penicillium are proposed: P. vulpinum (Cooke & Massee) Seifert & Samson for the species previously known as P. claviforme, P. coprophilum (Berk. & Curt.) Seifert & Samson for P. concentricum, and P. glandicola (Oud.) Seifert & Samson for P. granulatum Bain. Species of Coremium excluded from Penicillium, or of doubtful identity, are listed.

INTRODUCTION

The genus Coremium was described by Link (1809) with C. glaucum Link as the type species. The genus was considered a synonym of Penicillium by Fries (1832), but the name was commonly used for the synnematous species of Penicillium for many years, until it was abandoned after the first Penicillium monograph of Thom (1930). Unfortunately, the approximately 50 species assigned to Coremium have never been systematically revised, although the identities of some species were established by Thom (1930), Samson (1974) and Pitt (1979).

In this study as many type specimens of species attributed to Coremium were examined as possible, with the aim of stabilizing the nomenclature of the synnematous species. Not surprisingly, some new combinations are necessary. The experience of trying to identify Penicillium species from dried herbarium specimens, and the concurrent examination of freshly collected specimens, leads us to emphasize the importance of being able to recognize species of Penicillium as they occur in nature. Modern taxonomy of Penicillium has been modeled around the large number of isolates of species occuring in foodstuffs and in soil. This emphasis has distorted the taxonomy of species which occur on decaying vegetation or on such substrates as dung, with the result that such species usually have to be cultivated before they can be identified.

As we hope to show, with some species culturing is not necessary for identification, particularly when synnematous species are involved. The descriptions given in this paper are based only on specimens on the natural substrate.

Two other synnematous genera usually considered synonyms of Penicillium, Pritzeliella Henn. and Floccaria Grev. were also studied. The synonymy of Pritzeliella with Penicillium was confirmed and its type species, P. caerulea Henn. identified. No authentic material of Floccaria was found in K or E. The only described species, F. glauca Grev., is discussed under Coremium glaucum in the check list.

PART 1: THE COPROPHILOUS SYNNEMATOUS PENICILLIA

A number of species described in Coremium, and the type of Floccaria, F. glauca, were originally found on dung, and an examination of the type specimens confirms that at least two synnematous Penicillia occur on this substrate. Penicillium claviforme Bain. has been known as a frequent inhabitant of rabbit dung in some parts of the world, particularly in western Europe. The second coprophilous species was recently described as Penicillium concentricum Samson et al. (1976). At the time this species was described, the authors were not aware how commonly the fungus is encountered on dung in the wild. Pitt (1979) considered the type culture of P. concentricum an abnormal strain of P. italicum; again this author was unaware of the abundance of the species in nature. These two coprophilous species of Penicillium are described below based on specimens rather than cultures, and following this the relationship between the two species and P. italicum is clarified. Older epithets for both P. claviforme and P. concentricum have been discovered, and the appropriate new combinations are therefore proposed.

Penicillium vulpinum (Cooke & Massee) Seifert & Samson, comb. nov., Fig. 1.

= Coremium vulpinum Cooke & Massee -- Grevillea 16: 81. 1888 (basionym).
= Penicillium claviforme Bain. -- Bull. Soc. mycol. Fr. 21: 127. 1905 = Coremium claviforme (Bain.) Peck -- Bull. N.Y. St. Mus. 131: 16. 1909.
= Coremium silvaticum Wehmer -- Ber. dt. bot. Ges. 31: 373. 1914. = Penicillium silvaticum (Wehmer) Biourge -- Cellule 33: 105. 1923. = Penicillium silvaticum (Wehmer) Gaumann -- Vergl. Morphol. Pilze, p. 177. 1926 (superfluous combination).

Possible synonym -- Stilbum coprogenum Ellis & Everhart -- J. Mycol. 4: 116. 1888

Synnemata solitary, scattered or gregarious, cylindrical with an ellipsoidal capitulum, unbranched, white becoming brown or reddish brown from the base up, smooth to slightly lanose, 1000-2200 µm tall, 75-150 µm wide. Hyphae of stipe parallel, 2-3.5 µm, with smooth walls. Conidiophore branching

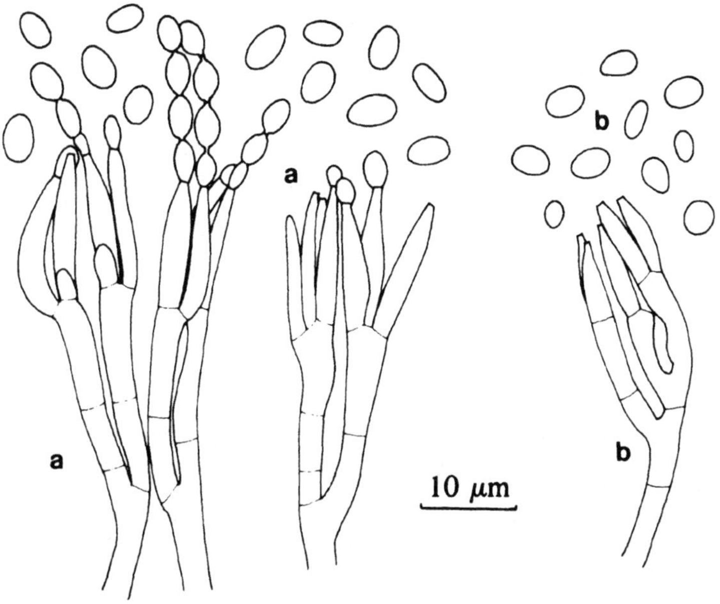

Fig. 1. Penicillium vulpinum. a. conidiophores and conidia from holotype of Coremium vulpinum. b. conidiophore and conidia from Stilbum villosum herb. Fuckel 1894 no. 2604.

modified 1-stage penicillate, usually with one or two basal dichotomous branches without septa giving rise to metulae 7-13 x 2.5-3 µm. Phialides 8-14 x 2-3 µm, single, in pairs, or usually in terminal whorls of 3-5. Conidia green or grey-green in mass, subglobose to ellipsoidal, 3-5.5 x (2-)3-3.5 µm, with smooth walls.

HABITAT: Dung, usually that of rabbit.

MATERIAL EXAMINED:

Coremium vulpinum, holotype, on dung, herb. M.C. Cooke (K). Stilbum coprogenum, holotype, on dung of ? raccoon, USA, N.J., Newfield, Sept. 1887, herb. J. B. Ellis (NY). As S. villosum, Herb. Fuckel 1894, no. 2604, on dung of fox, Lange near Hallgarten, autumn (G).

Penicillium coprophilum (Berk. & Curt.) Seifert & Samson, comb. nov., Fig. 2.

= Coremium coprophilum Berk. & Curt. -- J. Linn. Soc. Lond., Bot. 10: 363. 1869.
= Stilbum humanum P. Karst. -- Revue mycol. 1888: 75.
= Pritzeliella caerulea Henn. -- Hedwigia Beiblatt 42: 88. 1903.
= Penicillium concentricum Samson, Stolk & Hadlok -- Stud. Mycol. 11: 17. 1976.

Synnemata solitary, gregarious or sometimes crowded, cylindrical-capitate, subulate-capitate to somewhat spathulate, unbranched, sometimes proliferating apically to give 1 or 2 new synnemata, white, becoming light brown from the base up, occasionally reddish brown at base, smooth, 375-1800 µm tall, 25-150 (-300) µm wide. Hyphae of stipe parallel, 2.5-5 µm wide, with smooth walls. Conidiophore branching 2-stage penicillate, metulae 9-14 x 3-5 µm. Phialides in whorls of 3-5, 8-12 x 2-3.5 µm. Conidia globose to ellipsoidal, green in mass, 3-4 x 2.5-3.5 µm.

HABITAT: Dung, commonly that of rabbit or deer. Also isolated from soil and maize by Samson et al.(1976).

MATERIAL EXAMINED:

Coremium coprophilum, holotype, on rabbit dung, leg. M.C. Cooke (K). Penicillium coprophilum, on rabbit dung, leg. G.S. de Hoog, Netherlands, De Vuursche, 30 Nov. 1983 (CBS-H 3464). On rabbit dung, leg. K.A. Seifert, Netherlands, Baarn, Groeneveld, 8-9 Sept. 1984 (CBS-H 3677). Pritzeliella caerulea, holotype, on dung, leg. E. Pritzel no. 12, Australia, N. Queensland, Oberer Barruz River, Aug. 1902 (S). Stilbum humanum, holotype, on dung of dog or human, leg. P.A. Karsten, Finland, Tavastia australis, Tammela, Mustiala, 4 Oct. 1869, herb. P. A. Karsten no. 2475 (H).

Samson et al. (1976) commented on the similarities between these two species (as P. claviforme and P. concentricum). Cultures of the two are easily distinguished, because those of P. vulpinum produce long positively phototropic synnemata and very few mononematous conidiophores, while P. coprophilum forms loose synnemata in crowded concentric rings with an abundance of mononematous conidiophores. Collections made from nature are not so easy to distinguish, however, especially since the radiating columns of conidia commonly seen in cultures of P. vulpinum form only in closed Petri dishes. However, the peculiar conidiophore branching of P. vulpinum is usually not difficult to detect. The reddish colour at the base of the synnemata of P. vulpinum may be more distinct than in P. coprophilum, but the latter species often has discolouration at the basal part of the stipe. The capitulum of P. vulpinum is usually conspicuously ellipsoidal, contrasting to the looser, more divergent capitulum of P. coprophilum. Conidiophores of P. coprophilum tend to have tighter whorls of phialides than conidiophores of specimens from the natural substrate.

Pitt (1979) considered P. coprophilum (as P. concentricum) a synonym of P. italicum. Unfortunately, Pitt studied only the type culture of P. concentricum, and concluded that it was a slightly deviating isolate of P. italicum. P. coprophilum is an extremely common fungus on rabbit dung in the Netherlands, and the type isolate of P. coprophilum does not deviate from isolates from the collections cited above. The conidia of P. italicum are usually more cylindrical and longer than those of P. coprophilum, although degenerated strains of

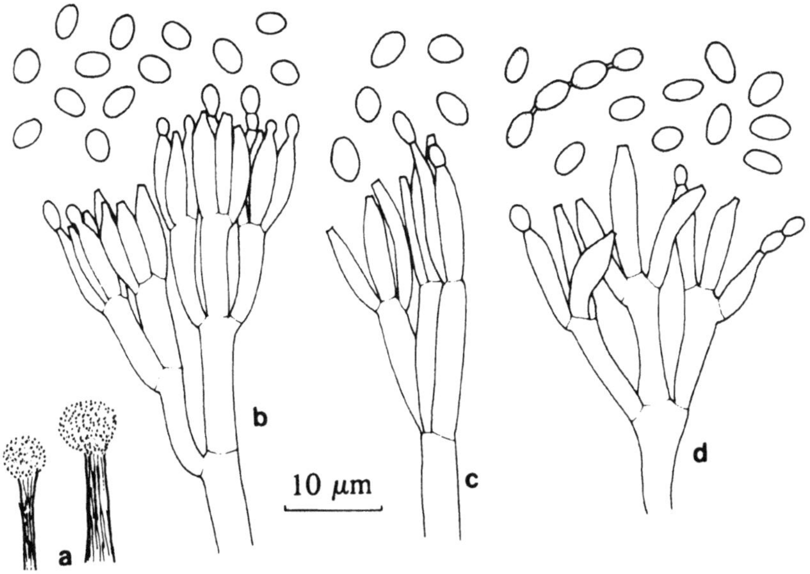

Fig. 2. Penicillium coprophilum. a. habit, about 25x. b. conidiophores and conidia from culture freshly isolated from dung, and c. conidiophores and conidia from original specimen, CBS-H 3464. d. conidiophore and conidia from holotype of Coremium coprophilum.

P. italicum do tend to have somewhat ellipsoidal conidia. Pitt (1979) reported that the type strain of P. concentricum caused small lesions on citrus fruit. We have been unable to reproduce this result with cultures of P. coprophilum freshly isolated from dung. Although the relationship between P. coprophilum and P. italicum is close, the widely different ecological niches, the comparative morphological uniformity of the populations in these niches, and the differences in the shape and size of the conidia seem to warrant their separation.

Stilbum coprogenum is here included as a possible synonym of Penicillium vulpinum. The synnemata and conidia on the type specimen of S. coprogenum seem typical of P. vulpinum, but no conidiophores or phialides were observed.

PART 2: COREMIUM GLANDICOLA, AN EARLIER NAME FOR PENICILLIUM GRANULTATUM

Penicillium glandicola (Oud.) Seifert & Samson, comb. nov., Fig. 3.

= Coremium glandicola Oud.-- Ned. kruidk. Arch. 3, ser II, 4: 918. 1903 (basionym).
= Penicillium granulatum Bain. -- Bull. Soc. mycol. Fr. 21: 126. 1905.

For further synonyms see Samson et al. (1976).

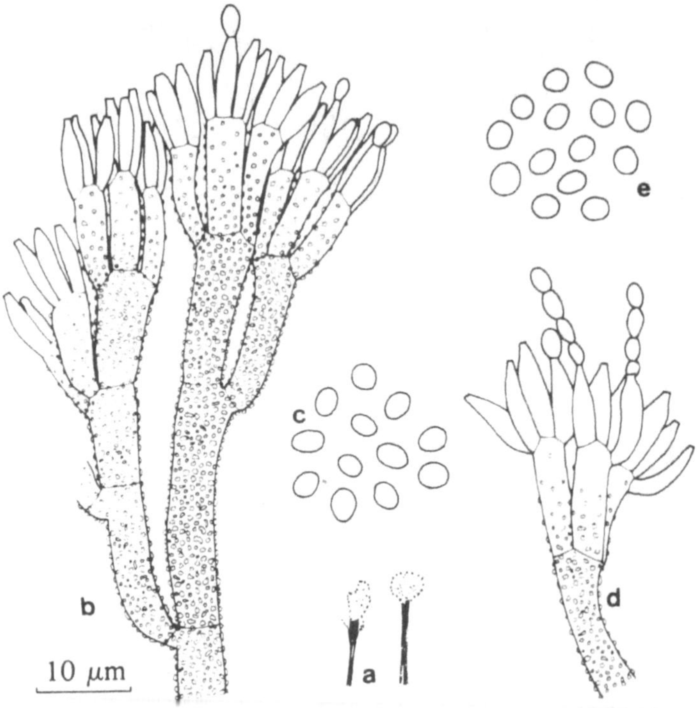

Fig. 3. Penicillium glandicola. a. habit from Fautrey no. 96. b. conidiophore and c. conidia from holotype of Coremium glandicola. d. conidiophore and e. conidia from Fautrey no. 96.

Synnemata crowded to caespitose, forming greyish green to cream-coloured colonies 2-10 mm across, cylindrical capitate, 300-500 μm tall, with a whitish to cream-coloured stipe 30-50 μm wide, and a grey-green spore mass. Hyphae of stipe parallel, verrucose, 3-5.5 μm wide. Conidiophore branching 1-2 stage penicillate, with verrucose rami 12-18 x 3.5-5.5 μm, and metulae 10-13 x 3-5 um which are verrucose at the base but tend to become smooth-walled near the apex, in whorls of 3-5 on the rami. Phialides in compact whorls of 3-6 on metulae, 7-10 x 2-3 μm, with smooth walls. Conidia yellow-green in mass, subglobose to ellipsoidal 2.5-4 x 2.5-3 μm, in chains which easily fall apart.

HABITAT: acorns of Quercus.

MATERIAL EXAMINED:

Coremium glandicola, holotype, on cotyledons of Quercus robur, leg. J. Rick, Netherlands, Valkenburg, July 1901, herb. Oudemans (L). as Coremium glaucum, two specimens on acorns of Quercus, leg. Fautrey, July 1893, no. 96 and no. 260 (UPS).

The verrucose hyphae, rami and metulae of this fungus are distinctive, and there is no question that the species is identical with Penicillium granulatum, despite the somewhat less ellipsoidal conidia. Penicillium verrucosum was also considered, but the species is not known to produce such well defined synnemata. Pitt (1979) speculated that C. glandicola might be Penicillium vulpinum (as P. claviforme), but did not examine authentic material.

PART 3: CHECKLIST OF SPECIES ATTRIBUTED TO COREMIUM

Coremium Link -- Mag. Ges. naturf. Freunde, Berlin 3:19. 1809.

C. album (Cost.) Sacc. & Trav. -- Syll. Fung. 19: 428. 1901 = Synpenicillium album Cost. -- Bull. Soc. mycol. Fr. 4: 67. 1888 = Penicillium costantini Bain. -- Bull. Soc. mycol. Fr. 22: 205. 1906 = Scopulariopsis costantini (Bain.) Dale -- Annls mycol. 12: 57. 1914. Costantin's illustration is obviously Doratomyces putredinus (Corda) Morton & Smith, as noted by Morton & Smith (1963).

C. alphitopus Secr. -- Mycol. Suisse 3: 539. 1833 = Penicillium expansum Link fide Thom (1930). Secretan's names are invalid according to Article 23 of the ICBN, because he did not employ a consistent binomial nomenclature.

C. arbuscula H. Fischer -- Ber. dt bot. Ges. 27: 502. 1909, see also Centbl. Bakter. Abt. II, 26: 57. 1910 = Penicillium arbuscula (H. Fischer) Biourge -- Cellule 33: 100. 1923. Type not in B. The diagnosis describes a soil isolate with highly branched synnemata 2 cm tall with dry red spores. Its identity is unknown.

C. aureum (Hedgecock) Sacc. -- Syll. Fung 22: 1443. 1913 = Graphium aureum Hedgecock -- Rep. Mo. Bot. Gdn 17: 94. 1906. Examination of authentic and cotype material (FP 46667, 46645, 46659, in K, UBC) demonstrates that this is a species of Phialographium Upadhyay & Kendrick.

C. bicolor Fr. apud Liljebl. -- Sv. Fl. 3: 678 = Penicillium bicolor (Fr.:Fr.) Fr. -- Syst. mycol. 3: 408. 1832. Thom (1930) suggested that this might be Penicillium duclauxii Delacroix. There is no material in UPS.

C. borzianum Sacc. apud Borzi -- Jb. wiss. Bot. 16: 450. 1885. Type not in PAD. Diagnosis not seen.

C. breve Petch -- Trans. Br. mycol. Soc. 11: 259. 1926. See Coremium gracile below. = Paecilomyces farinosus (Holm) Brown & Smith fide Samson (1974, K!)

C. briardi (Vuill.) Sacc. & D. Sacc -- Syll. Fung. 18: 643. 1906 = Penicillium briardi Vuill. -- Bull. Soc. mycol. Fr. 20: 218. 1904. Possibly Metarhizium anisopliae (Metsch.) Sorokin fide Pitt (1979). The illustration accompanying the diagnosis supports this interpretation.

C. candidum (Pers.) Nees -- Syst. Pilze, p. 39, 1837 = Monilia candida Pers. -- Syn. meth. fung. p. 692. 1801 = Peni-

cillium candidum β *coremium* Fr.:Fr. --Syst. mycol 3: 409. 1832 = *Penicillium candidum* var. *coremioides* Sacc. -- Syll. Fung. 4: 80. 1886. The only specimen marked *Monilia candida* in herb. Persoon is an *Aspergillus* species, probably *A. glaucus*. A second specimen in the same folder, marked "Penicillium candidum Lk, Coremium candidum Pers., herb. Pers. ded. Nees" (L 910.258-933) is *Penicillium expansum* Link.

C. candidum (Corda) Lindau -- Rabenh. Krypt.-Fl. 1, Pilze 9: 332. 1910 = *Stysanus candida* Corda = *Penicillium candidum* (Corda) Biourge -- Cellule 33: 101. 1923. See *Coremium stysanoides* below.

C. cinereo-album (Bonorden) Sacc. -- Syll. Fung. 10: 687. 1892 = *Scleromitra cinereo-alba* Bonorden -- Handb. allgem. Mykol. p. 141, 1851. = *Penicillium cinereo-album* (Bonorden) Biourge --Cellule 33: 101, 1923. Bonorden kept no herbarium. Pitt (1979) regarded the original illustration as suggestive of *P. vulpinum* (as *P. claviforme*) but considered the species unidentifiable.

C. cinereum Rivolta -- Parass. veget. p. 500. 1873. Type not in TO, presumed to be lost. The diagnosis and illustration are suggestive of a species of *Doratomyces*.

C. citrinum Pers. -- Mycol. Eur. I: 43. 1822 = *Monilia penicillus* Pers. -- Syn. meth. fung. p. 692. 1801 = *Penicillium citrinum* (Pers.) Biourge -- Cellule 33: 102. 1923. Type not in herb Persoon (L) under either *Coremium* or *Monilia*. Nomen dubium.

C. claviforme (Bain.) Peck, see *P. vulpinum*, part 1 above.

C. coccineum Schw. -- Proc. Am. Phil. Soc. Phila. 4: 282. 1831. The type (ad lignum cariosum, George, leg. Dr. Torrey, PH, isotype in K) appears to represent an immature fruiting body of a member of the Aphyllophorales; the developing hymenium was presumably mistaken for gregarious synnemata. Nomen dubium.

C. compressum (Berk. & Br.) Petch -- Ann. R. Bot. Gard. Peradeniya 10: 176. 1927. = *Stilbum compressum* Berk. & Br. -- J. Linn. Soc. Lond., Bot. 14: 97, 1875. The holotype of *Stilbum compressum* (no. 742, on fruit of *Citrus decumana*, Ceylon, Peradeniya, Aug. 1868, herb. Berk., K) consists of determinate synnemata with poorly developed 1-stage branched conidiophores, and cylindrical conidia 4.5-7 x 2.5-3 um. Although the synnematous habit suggests *P. italicum* Wehmer, conidia of this length have not been recorded for this species. The conidiophore branching is somewhat reminiscent of *P. digitatum* Pers., a species not known to produce synnemata. Given these anomalies, it seems best to consider *Stilbum compressum* a nomen dubium.

C. concentricum Schw. -- Proc. Am. Phil. Soc. Phila. 4: 282. 1831 = *Sporocybe concentrica* (Schw.) Sacc. -- Syll. Fung. 4: 608. 1886. The type (on leaves of *Laurus aestivalis*, Bethlehem & Salem, PH) contains immature or decapitated conidiomata of what may be a Mycosphaerella anamorph. Nomen dubium.

C. coprophilum Berk. & Curt., see P. coprophilum , part 1 above.

C. cordyceps Henn. -- Hedwigia 43: 396. 1904. The holotype (on Lostustidice, leg. E. Ule. no. 5801, Brasil, Amazonas, Rio Jurua, Jurua-Miry, Sept. 1901, B) is a species of Hymenostilbe Petch, and will be considered in a future publication.

C. elasticae Koord. -- Bot. Untersuch. Java p. 239. 1907 = Penicillium elasticae (Koord.) Biourge -- Cellule 33: 102. 1923. The illustration accompanying the diagnosis portrays nodding synnemata with chains of conidia. This undoubtedly represents the hyphophores of a lichen of the family Asterothyriaceae, such as those illustrated by Vězda Vězda (1979).

C. fimetarium Schw. -- Proc. Am. Phil. Soc. Phila. 4: 283. 1831. The type (Bethlehem, in fimo vaccino, PH) contains no synnematous fungi. Nomen dubium.

C. flavo-virens Schw. -- Proc. Am. Phil. Soc. Phila. 4: 282. 1831. Examination of the holotype (in foliis dejectis, Bethlehem, herb. Schw., PH) reveals that it is a mononematous species of Penicillium similar to P. variabile Sopp, P. rugulosum Thom or P. verruculosum Peyronel. The specimen is in poor condition, and it is not possible to decide unequivocably which species it actually represents. Best considered a nomen dubium.

C. flavum Rivolta -- Parass. veget. p. 500. 1873. Type not in TO, presumed to be lost. The diagnosis and illustration suggest Sphaeronaemella fimicola Marchal.

C. fragariastri Schw. -- Proc. Am. Phil. Soc. Phila. 4: 282. 1931 = Graphiothecium fragariastri (Schw.) Sacc. -- Syll. Fung. 4: 624. 1886. Type not in PH or K. Identity unknown.

C. fruticulosum (Cumino) Sacc. -- Syll. Fung. 22: 1444. 1913 = Mucor fruticulosum Cumino -- Act. Ac. Taur. p 267, 1805. Type not seen. Not recognizable from illustration.

C. fuscum Rivolta -- Parass. veget. p. 500. 1873. Type not in TO, presumed to be lost. The diagnosis and illustration are suggestive of a species of Doratomyces.

C. glandicola Oud., see part 2 above.

C. glaucum Link --Mag. Ges. naturf. Freunde, Berlin 3: 19. 1809. This has been considered a synonym of Penicillium expansum Link since Raper and Thom (1949). A specimen in herb. Persoon marked "Coremium glaucum Lk., herb. Pers. mis. Link" (L 910.258-933) may be authentic for the name, and appears to be identical with P. expansum. Floccaria glauca Grev. (Scottish Crypt. Flora, vol. 6, tab. 301, 1827) is a facultative synonym. No mention of Coremium glaucum is made in the text accompanying the plate of F.

glauca, but in the index to the work, the following appears "Floccaria glauca, Grev. est Coremium glaucum". This correction indicates that Greville considered his species the same as Link's, and in the absence of a type in K or E, this synonymy is best followed. See also C. vulgare var. Linkii.

C. glaucum var. fimicola Marchal -- Bull. Soc. R. bot. Belg. 34: 143. 1895 = Penicillium fimicola (Marchal) Biourge -- Cellule 33: 103. 1923. The illustration accompanying the diagnosis looks more like a myxomycete than a Penicillium. See also C. vulgare var. fimicola.

C. gracile MacBride -- Mycologia 18: 131. May-June 1926. The description is not detailed enough to allow the species to be identified.

C. gracile Petch -- Trans. Br. mycol. Soc. 11: 260. Dec. 1926 = Spicaria gracilis (Petch) Petch -- Trans. Br. mycol. Soc. 16: 241. 1932. A later homonym of C. gracile MacBride, and a synonym of Paecilomyces farinosus (Holm) Brown & Smith fide Samson (1974, K!).

C. hiemale Bonorden -- Bot. Ztg 19: 203. 1861 = Penicillium hiemale (Bon.) Biourge -- Cellule 33: 103. 1923. Bonorden kept no herbarium. Described as having white conidia, this species is unlikely to belong to Penicillium fide Pitt (1979).

C. incanum P. Karst. -- Hyphom. fenn. p. 463. 1892. Type not seen. Described as occurring on larvae of an insect, this is probably a species of Paecilomyces.

C. leucocephalum Wallroth, a lapsus calami for Cephalotrichum leucocephalum Wallroth by Saccardo, Syll. Fung. 4: 611. 1886.

C. leucopus Pers. -- Mycol. Eur. I: 42. 1822 = Penicillium leucopus (Pers.) Biourge -- C. r. Seanc. Soc. Biol. 82: 877. 1919. Authentic specimens of C. leucopus (L 910.258-923, L 910.258-920) examined are Penicillium expansum Link.

C. luteolum S. Camara apud Serafin -- Agronomia lusit. 17: 304. 1955. Possibly Briosia ampelophaga Cavara fide Sutton (1973).

C. necans Oud. -- Ned. kruidk. Arch. 3, ser II, 4: 918. 1903 = Penicillium necans (Oud.) Biourge -- Cellule 33: 104. 1923. A synonym of Paecilomyces farinosus (Holm) Brown & Smith fide Samson (1974, L!).

C. nigrescens (Jungh.) Penzig -- Syll. Fung. 4: 583. 1886 = Penicillium nigrescens Jungh. -- Verh. batav. Genoot. Kunst. Wet. 17: 5. 1839. Unidentifiable from illustration fide Pitt (1979). Type not in L. Original publication not seen.

C. niveum Corda -- Icones fung. II: 17. 1838 = Penicillium niveum (Corda) Biourge -- Cellule 33: 104. 1923. Type not in PRM. The illustration accompanying the diagnosis looks like a species of Sphaeridium Fres.

C. pulcherrimum Petch -- Trans. Br. mycol. Soc. 11: 259. 1926 = Spicaria (Isaria) pulchella (Petch) Petch -- Annls Cryptog. exot. 6: 234. 1933. Samson (1974) examined a sterile authentic collection in PC and suggested this species might be a synonym of Paecilomyces amoeneroseus (Henn.) Samson. Examination of the holotype (K) confirms this synonymy.

C. rhopalinum Batista & Peres -- Univ. Recif. Inst. Micol. Publ. 266: 8. 1960. Type not seen. Described as a leaf parasite with mucoid conidia, this is not likely to be a Penicillium.

C. silvaticum C. Wehmer, see P. vulpinum, part 1 above.

C. stysanoides Sacc. -- Syll. Fung. 4: 582. 1886 = Stysanus candidus Corda -- Icones fung. 1: 22. 1837. Type not in PRM. Nomen dubium. See also Coremium candidum (Corda) Lindau.

C. subiculatum Schw. -- Proc. Am. Phil. Soc. Phila. 4: 282. 1831. Type not in PH. Nomen dubium.

C. swantonii A. L. Smith -- Trans. Br. mycol. Soc. 6: 156. 1918 = Spicaria (Isaria) swantonii (A. L. Smith) Petch -- The Naturalist 1931: 103. A synonym of Paecilomyces farinosus (Holm) Brown & Smith fide Samson (1974, K!).

C. syphiliticum Hallier -- Flora, Jena 51: 295. 1868. The original publication places the species in both Penicillium and Coremium. Considered by its author as a form of Coniothecium syphiliticum Hallier, the identity of the species is unknown.

C. triostei H. C. Greene -- Trans. Wisc. Acad. Sci. Arts Letters 41: 127, 1952. Type not seen. The original diagnosis suggests a synnematous species of Ramularia.

C. tuberculosum Gasparr. --Rend. Acc. Sci. Napoli 1: 134. 1852. Original publication not seen.

C. vulgare Corda -- Prachtflora p. 53. 1839. Considered a synonym of Penicillium expansum Link fide Thom (1930) or P. vulpinum (as P. claviforme) fide Raper & Thom (1949). Type not in PRM. Corda's illustration includes specimens from a number of different substrates and no doubt includes several different species of synnematous Penicillium. Best considered a nomen dubium.

C. vulgare var. linkii Sacc. -- Syll. Fung. 4: 582. 1886. See C. glaucum Link.

C. vulgare var. fimicola (March.) Sacc. -- Syll. Fung. 4: 582. 1886. See C. glaucum var. fimicolum.

C. vulpinum Cooke, see P. vulpinum, part 1 above.

Acknowledgements

K.A. Seifert thanks the Natural Science and Engineering Research Council of Canada for the award of a post graduate scholarship. We are also grateful to the curators of the herbaria mentioned in the text for the loan of specimens.

REFERENCES

FRIES, E. 1832. "Systema mycologicum vol. 3." Gryphiswaldae.

LINK, H. F. 1809. Observationes in Ordines plantarum naturales, Dissertatio 1, Mag. Ges. naturf. Freunde, Berlin 3: 1-42.

PITT, J. I. 1979. "The genus *Penicillium* and its teleomorphic states, *Eupenicillium* and *Talaromyces*." Academic Press, London.

RAPER, K. B. and THOM, C. 1949. "A Manual of the Penicillia." Williams & Wilkins, Baltimore.

SAMSON, R. A. 1974. *Paecilomyces* and allied hyphomycetes, Stud. Mycol. 6: 1-119.

SAMSON, R. A., STOLK, A. C. and HADLOK, R. 1976. Revision of the subsection *Fasciculata* of *Penicillium* and some related species, Stud. Mycol. 11: 1-47.

THOM, C. 1930. "The Penicillia". Williams & Wilkins, Baltimore.

VĚZDA, A. 1979. Flechtensystematische Studien XI. Beiträge zur Kenntnis der Familie Asterothyriaceae, Folia geobot. phytotax., Praha 14: 43-94.

DIALOGUE FOLLOWING DR. SEIFERT'S PRESENTATION

PITT: Given the information put forward earlier by Dr. Frisvad, and the information you present here, I am now happy to recognize *P. concentricum* as a species distinct from *P. italicum*. Although I am satisfied that the old names you have found appear to be based on good specimens, in general I am reluctant to take up old names based on dried herbarium material unless there is absolute certainty about the application of the name. If there is any uncertainty, taxonomy is much better served by leaving the name aside.

SEIFERT: I agree. However, sometimes, even though the specimen may be poor, the substratum may be quite specialized, and we have to ask, "What else can this be?"

SAMSON: This might be a bigger problem in *Aspergillus*, where older type specimens may be quite readily recognizable.

WICKLOW: Have you ever grown cultures of noncoprophilous species on dung?

SAMSON: We've grown *P. italicum* on dung and it grows well, but produces no synnemata.

INHERENT PROBLEMS IN PENICILLIUM TAXONOMY

John I. Pitt

CSIRO, Division of Food Research
North Ryde, NSW 2113 Australia

SUMMARY

Inherent variability in Penicillium isolates constitues a serious problem for the taxonomist. Some aspects of the variability are discussed, especially as its effects new physiological approaches to taxonomy. A personal view of the future of Penicillium taxonomy is also included.

I believe that substantial benefits can accure from the use of physiological and biochemical tests in Penicillium and Aspergillus taxonomy. The principal ones which come to mind are to confirm or disprove traditional, morphological taxonomic relationship, to aid taxonomic judgement in identifying difficult isolates and to build up information on a single isolate or a cluster of closely related isolates within a species. However, I do not see these approaches as a panacea for all our taxonomic problems or as providing taxonomic systems in their own right. In this short paper I wish to sound some notes of caution. There are problems involved in the use of such tests, which I feel are not always fully appreciated, and I wish to draw attention to some of them.

It seems to me that a basic problem in this whole field is the inherent variability present in Hyphomycetes, more than in many other organisms, due to the lack of genetic feedback in these haploid, asexual fungi. I have talked and written about this in various places over the years, but I think it is worth saying once more: a conidium of Penicillium and Aspergillus germinates and grows into a colony. One or more conidia from that colony germinate, grow and form conidia in their turn, and so on -- generation after generation. In the absence of true sexuality, i.e. a mating process between different individuals, there is no direct genetic control over the development of different lines, radiating out from a single conidium, unchecked except for natural selection, for endless generations. Inherent variability is the result, and this inherent variability is a taxonomic quagmire.

It is easy to take a few isolates, run them through a battery of tests, find a correlation, and publish the data as useful taxonomic system. Finding correlation among small groups of Hyphomycete isolates is easy, but when large numbers are studied, inherent variability causes these first correlations to be obscured or negated. One could cite many examples of this.

In my view, the assessment of inherent variability in a taxonomic system requires the examination of hundreds of isolates. A colleague once remarked to me that in her opinion a person should work on the taxonomy of large genus for 5 years before contemplating the first publication. This is a rather extreme view. However, before my first publication on Penicillium (Pitt, 1973), I had studied more than 500 isolates from throughout the genus, most of them authenticated by Raper and Thom (1949), over a period of 5 years,and I must say I am in agreement with her. I am not saying that new ideas on systems which may be of taxonomic value should not be published promptly, but I am saying that such ideas are just that, untill a great deal of work has been done, especially to assess isolate to isolate variability in currently accepted species, preferably throughout the genus concerned.

A second type of variability in Penicillium and other Hyphomycete genera is variability in expression of a single genome. Some isolates look the same in culture every time they are examined, while others are subtly different. In my laboratory, if we encounter an isolate which does not fit a known species, we grow it again on the standard media using 7 days old inocula - and we will sometimes even examined it a third time. In many cases the morphological and physiological characters of such an isolate will subtly change and it will now fit a known species. I am sure we have all had this experience. What happens to the biochemical properties of such an isolate? Are metabolic and biochemical tests from any isolate always reproducible when done repeatedly?

A further problem is that Hyphomycetes are eukaryotes, able to control the expesssion of genes and regulate enzymic pathways. So that the fact, that a particular enzyme or metabolite is not found in particular test, is no proof that such an enzyme or metabolite cannot be made by that isolate. This is a well known problem with reverse colours in Penicillium. Some times ago I recall having an unusually variable CYA plate of P. bilaii, a species which normally produces a conspicuously deep brown soluble pigment. The plate was inoculated with my usual 3-point mass inoculum from a needle point of conidia mixed with weak agar. One colony had abundant soluble pigment, one a little and one none at all. Clearly this was due to gene regulation, and this is a problem in studying eukaryotic genera of this kind. Negative information is no information.

In his recent studies Frisvad (1981, 1984; Frisvad and Filtenborg, 1983) has within several species, produced groups which differ by the addition or subtraction of a single metabolite. Are such slightly different groups worth separating? Or are we not just looking at variability within a species? We do not know nearly enough about variation within a single genome. So decide what such variability is, i.e. how valid any particular test is, I believe one needs to make 100 single spore isolates from a single colony and grow and examine them all.

A third kind of variability in Penicillium and Aspergillus is cultural deterioration - a problem of which morphologists are well aware. From the past 60 to 80 years experience, it is known that some Penicillium isolates are remarkably stable in culture, while others rapidly deteriorate. What happens to their biochemistry? I suggest that great care is needed in assessing the biochemistry of older isolates - and many cultures are old.

A question which needs to be addressed by anyone doing taxonomic work on a large genus such as Penicillium relates to the kind of taxonomy they wish to do. There are always going to be two kinds of taxonomy: theoretical and practical. Perhaps this is often much the same thing as phylogentic and determinative taxonomy, but not always. Is the taxonomist interested in working out the ultimate varaibility in part of all the genus or is he interested in providing a practical classification?

If all the variation seen in Penicillium, even in morphology and colour, were accurately catalogued, there would be hundreds and hundreds of clearly distinguished types. If then we add in biochemical variation, I believe we are looking at 1000 to 2000 morphologically and biochemically different variants. We could call these species and the temptation is high to do this when using modern techniques. A taxonomy with 1000 or more species of Penicillium simply is not practical. The biochemical taxonomists must be just as prepared to lump together slightly different isolates, just as the morphological taxonomists have been. I have now examined in detail somewhere near 10,000 Penicillium isolates, and I am still encountering new combinations of characters which I simply have not come across before.

This is the real problem in Penicillium taxonomy: inherent variability. It seems to go on and on, ad infinitum. Any successful taxonomy must be able to take this variability into account. What do I see as the future of Penicillium taxonomy? I had always hoped that "The Genus Penicillium" (Pitt, 1979) would be a springboard for renewed interest in the taxonomy of this genus. This has clearly been so. But I am uneasy at the direction inwhich some newer research is going.

Let me say first what I had hoped would happen. I hoped that others would look at some of the more obvious deficiencies and weaknesses in my work and attempts to remedy them, using my work as a basis. For example, is P. ochrochloron Biourge really distinct from P. janthinellum Biourge? Can we split P. janthinellum Biourge? Do P. paxilli Bain. and P. miczynskii Zaleski, as I have defined them, each include two subtly different, distinguishable species? Can P. glabrum (Wehmer) Westling be adequately separated from P. spinulosum Thom? Where does P. claviforme Bain. really fit? I believe, that studies of secondary metabolites and biochemical tests can be invaluable at this level.

In this context I am grateful to Stolk and Samson (1983) for pointing out that my interpretation of P. raciborskii Zaleski is probably incorrect, and to those at this workshop who provided new information which indicates that my reduction of P. concentricum Samson et al. and P. hordei Stolk, based on limited data, was incorrect.

Work by Amelia Stolk, De B. Scott, R. A. Samson and myself in the 1970's was all based on K. B. Raper's work, which in turn was based on C. Thom, P. Biourge and others. I was extremely careful to maintain continuity with Raper's work, so that Penicillium taxonomy could be seen to be evolving and that changes in concepts were minimised. Continuity in taxonomy must exists, both because types are fixed in botanical taxonomy, and must be accounted for; and because users of taxonomies - the people we do taxonomy for - must be able to see evolution, or they will ignore and chaos will result.

Unfortunately, and this point I alluded to earlier, I see a current trend to pull names out of the air without regard for typification or to base test on cultures which are not clearly authenticated, both of which mean that some taxonomist has to sort it all out later. And I see a tendency to tear down a reasonably ordered system, to move parts of the genus from place to place without adequate explanation. I am inherently concerned whether my work stands the test of time. My aim was to provide a new working basis in Penicillium, to stimulate interest in what to me is both a fascinating and frustrating genus. But I am concerned that some who are leaving apart or starting afresh will not rebuild on firm foundation, or not build at all. The results may be taxonomic chaos into the next century.

REFERENCES

FRISVAD, J. C. 1981. Physiological criteria and mycotoxin production as aids in identification of common asymmetric Penicillia. appl. environ. Microbiol. 41: 568-579.

FRISVAD; J. C. 1984. Expresssions of secondary metabolism as fundamental characters in Penicillium taxonomy. In "Toxigenic Fungi - their Toxins and Health Hazard", H. Kurata and Y Ueno, eds. Amsterdam: Elsevier, pp.98-106.

FRISVAD, J.C. and FILTENBORG, O. 1983. Classification of terverticillate Penicillia based on profiles of mycotoxins and other secondary metabolites. Appl. environ. Microbiol. 46: 1301-1310.

Pitt, J.I. 1973. An appraisal of identification methods for Penicillium species: novel taxonomic criteria based on temperature and water relations. Mycologia 65: 1137-1157.

Pitt, J.I. 1979. "The Genus Penicillium and its teleomorphic States Eupenicillium and Talaromyces". London: Academic Press.

Raper, K.B. and Thom, C. 1949. "A Manual of the Penicillia". Baltimore: Williams and Wilkins.

Stolk, A.C. and Samson, R.A. 1983. The Ascomycete genus Eupenicillium and related Penicillium anamorphs. Stud. Mycol. Baarn 23: 1-149.

DIALOGUE FOLLOWING DR. PITT'S PAPER

SAMSON: I agree with Dr. Pitt that one of the main reasons for this workshop was to get together and learn to understand each others concepts.

WICKLOW: I don't have the same fears that Dr. Pitt has. I think it should be possible to structure the organization of the data, so that the information that a particular person needs can be extracted.

SAMSON: But let's not mix up these systems. A person who wants to identify a Penicillium should not have to do all these biochemical tests.

PITT: At this point, Dr. Samson and I seem to agree fairly well on morphological species concepts within the subgenus Penicillium. There is a whole infrastructure of biochemical information building up which we must accomodate. Can we as a group decide whether we should split these biochemically recognizable groups out as species, or hold with our morphological species and recognize these things as varieties, or some other name at a lower level?

KLICH: How about chemovar or chemotype?

HAWKSWORTH: Chemotype.

(This discussion led to the creation of the paper by Pitt and Hawksworth; see chapter 2).

GOCHENAUR: Strain degeneration can be a problem. I have done some very detailed nutritional studies, and was pleased to see that wild type strains are identical with degenerated strains of the same species. Morphological characters may degenerate, but the biochemical pathways appear to remain unaffected.

FRISVAD: As a further example, we recently looked at 150 isolates of Penicillium chrysogenum from various sources. They all produced the same profile of secondary metabolites, without exception.

SAMSON: But if you had an isolate with that profile together with one extra metabolite, you would call it a different species.

FRISVAD: No, I wouldn't.

WICKLOW: Actually, Dr. Frisvad gives it a Roman numeral and puts it on hold, with a red flag that says, this might be something different.

PITT: Dr. Frisvad, we are generally very impressed with your work, but in the long run we have to correlate your data with morphology. You do your metabolite profiles first. The test will come when you do your work in the other direction - identify your isolates on morphology first, and then do the metabolite profiles and see if the two fit together.

FRISVAD: Usually I do the morphology first.

PATERSON: Strain variation can be accounted for by numerical taxonomy as long as it is confined to a few characters. You mention the problem of degenerate type isolates: we have found it possible to obtain biochemical data from dried types.

MALLOCH: I have observed the tendency amongst Penicillium taxonomists to use a strictly anatomical species concept for anamorphs, even to the point where some of these have different teleomorphs. On the other hand, we have Dr. Frisvad's work which has a strictly biological species concept. I wonder if these two different systems will ever be reconciled, or if there will always be tension between these two camps. Is it possible, in your view, that a biological system can ever be practical?

PITT: I think we're pretty close in Penicillium taxonomy. The fundamental characters we are using in Penicillium really are phylogenetically based. I am hopeful that metabolite profiles and biochemistry will fit this all very nicely.

FRISVAD: Although you criticize my use of secondary metabolites, all morphological taxonomists use colour for taxonomic purposes, as Dr. Pitt did in Monascus (Hawksworth and Pitt, Aust. J. Bot. 31: 51-61, 1983). Colours are secondary metabolites.

PITT: That's correct. We use colours in Penicillium too. The use of secondary metabolites doesn't inherently trouble me provided that it is considered and integrated into morphological taxonomy. In this way we can achieve a taxonomy with 100 or 200 Penicillium species. If we take account of individual differences in metabolite patterns we might come out with a thousand or more species.

CHRISTENSEN: I would hate to see a thousand species of Penicillium, I hope we can keep to one or two hundred, but at the same time we must not lose sight of the fact that there could be 2000 biological entities and we must not lose the biochemical and ecological information represented by these variations. We need some kind of system that will recognize the biological reality.

KLICH: A point not brought out yet is, that we have to realize that taxonomy is dynamic and that there is no such thing as the ultimate taxonomy. It's a man-made system that will continue to develop both as we acquire new knowledge and as we require more detailed information.

SAMSON: All monographers are aware that there are many isolates that they can't fit into their systems.

RAMIREZ: I have a Linnaean concept of species, so morphology is the primary criterion, and the rest is only addenda.

MALLOCH: I think we have to consider the idea that these fungi may evolve physiologically while retaining the same method of dispersal. The morphological characters we use in taxonomy are merely a reflection of the dispersal, and may not be representative of the biological speciation that is occurring.

PITT: If we try to account for all the variation we see in *Penicillium* we might end up with 2000 species, which we do not need.

MALLOCH: You may not need them, but you may have to accept them.

A NEW TAXONOMIC SCHEME FOR PENICILLIUM ANAMORPHS

Amelia C. STOLK and Robert A. SAMSON

Centraalbureau voor Schimmelcultures
P. O. Box 273, 3740 AG BAARN, The Netherlands

SUMMARY

The genus *Penicillium* is divided into ten sections: *Torulomyces* (Delitsch) Stolk & Samson, *Aspergilloides* Dierckx, *Eladia* (G. Smith) Stolk & Samson, *Divaricatum* Raper & Thom ex Pitt, *Inordinate* Pitt, *Ramosum* Stolk & Samson, *Penicillium*, *Coremigenum* (Biourge) Pitt, *Biverticillium* (Dierckx) Stolk & Samson, and *Geosmithia* (Pitt) Stolk & Samson. A key to the section and short descriptions of each section are given. Sections which are subdivided in series are keyed out.

INTRODUCTION

At the Centraalbureau voor Schimmelcultures taxonomic studies on the Penicillia and Aspergili have been carried out since the beginning of the Century. Mycologists such as Wehmer, Blochwitz, Westling and Zaleski sent their strains including many type isolates for deposition in the collection. F. van Beyma thoe Kingma was particularly interested in these two genera and he achieved recognition for the publication of several new species, such as *P. sclerotiorum*, *P. euglaucum* and *P. novae-zeelandiea*, which are still recognized taxa. Van Beyma examined cultures on several media and made notes of the morphological features as well as camera lucida drawings. These notes and drawings are still kept in the CBS archives and represent a valuable taxonomic resource. In many cases these notes and drawings were very useful when we studied type isolates which had deteriorated after frequent transfers.

After the Second World War, the senior author began as a young student at the CBS. Prof. J. Westerdijk assigned her to a special request from K. B. Raper and C. Thom to prepare all available *Penicillium* cultures for their monographic study. After a careful check of the purity of each culture and morphological comparison with its protologue, the cultures were sent to Raper and Thom, who returned them after their study in 1949, accompanied with their comments. In the 1950's and 1960's many *Penicillium* and *Aspergillus* isolates were studied, but because the work in the collection was of primary importance, only a few publications were prepared. De B. Scott, who collected many sclerotial strains from South Africa, solicited taxonomic help during a visit at the CBS in 1966. The senior author also had collected many similar isolates over the years and with De B. Scott a monographic study of *Eupenicillium* was initiated and subsequently published (Stolk and Scott, 1967; Scott and Stolk 1967, Stolk, 1968). It is worth noting that the discovery of many of these teleomorphic states was not only a result of patient cultivation of the sclerotial *Penicillium* cultures, but also due to the fact that CBS keeps old cultures for two to three years after transferring them. In these old cultures the ascospores of, for example in *E. lapidosum* were discovered. When in 1970 the junior author began at CBS, the study of the teleomorph genus *Talaromyces* was commenced and published (Stolk and Samson, 1971, 1972).

In the 1970's the number of *Penicillium* and *Aspergillus* strains sent to CBS for identification increased considerably. Identification of Penicillia at that time was carried out following the Raper and Thom's keys, but the need was felt to reinvestigate particularly the fasciculate Penicillia, because of their common occurrence and great variability. Prof. R. Hadlok, who had isolated thousands of Penicillia from meat and other food products in Giessen (FRG), joined our study and subsequently we published our account of the *Fasciculata* (Samson et al., 1976).

Following this study and those of some other *Penicillium* groups (Samson et al., 1977 a and b), the senior author started a revision of Eupenicillium, then retired, but continued her work at home. In our paper (Stolk and Samson, 1983) we presented the results of this revision and proposed teleomorphic species which are linked to what were formerly considered strictly anamorphic taxa. In this paper we are presenting our taxonomic scheme for all *Penicillium* anamorphs. A more detailed account will be published elsewhere.

Generic delimitation

The genus *Penicillium* is closely related to *Aspergillus* *Paecilomyces* and *Merimbla*. *Penicillium* species with simple conidiophores, such as *P. glabrum* and some species of the section *Eladia* superficially resemble those of uniseriate Aspergilli e.g. *A. restrictus* G. Smith. Moreover both genera are similar in phialide shape and both produce conidia in dry long chains. However, in *Penicillium* phialides are produced successively on the apices of the stipes and/or metulae, whereas in *Aspergillus* they are borne simultaneously at the apex of the stipes. In addition, the stipes of *Penicillium* differ from those of *Aspergillus* in being usually thin-walled and septate. The typical structure of the conidiophore base ("footcell") in *Aspergillus* is usually absent in *Penicillium*, though it may rarely be present in a few species (e.g. *P. varians* G. Smith).

The generic distinction between *Penicillium* and *Paecilomyces* is more difficult. Both genera develop branched penicilli, but those of *Paecilomyces* are more irregularly and verticillately branched. The phialides of *Paecilomyces* typically have much longer necks than those of *Penicillium*. However, the phialides of *Penicillium* section Divaricatum may be somewhat intermediate. In addition, *Paecilomyces* differs from *Penicillium* in producing more divergent conidial chains and no green conidial pigments.

Pitt (1979) erected the genus *Merimbla* to accomodate *P. ingelheimense* v. Beyma (= *P. avellaneum* Thom and Turesson), a species which does not belong in *Penicillium*, because of the shape of the robust conidiophores, the phialides and the conidial colour. *Merimbla* superficially resembles *Aspergillus*, but metulae and phialides are borne successively and stipes lack the typical conidiphore base structure. The genus resembles *Paecilomyes* more closely, the main difference being the shorter phialidic necks. *M. ingelheimense* is closely related to *P. lineatum* Pitt (= *P. striatum* Raper & Fennell). Both species produce a *Hamigera* teleomorph (Stolk & Samson, 1971). Their anamorphs differ, however, in structure. The conidiophores of *P. ingelheimense* are robust, whereas those of *P. lineatum* are slender. *P. lineatum* does not have green colours and does not belong in *Penicillium*. The phialides of *P. lineatum* have relatively long necks and resemble those of *Paecilomyces*. A renewed study of the relationship between *Merimbla* and *Paecilomyces* as well as *P. lineatum* should be undertaken.

Pitt (1979) erected the genus *Geosmithia* to accomodate species of the *P. pallidum*-series. In our opinion a separate genus is inappropriate because the morphological differences are small. If the genus *Geosmithia* was accepted then other genera for species belonging to the sections *Biverticillium*, *Inordinate*, *Digitata* and others, which show slight morphological differences, must also be established. For a further discussion see also under Section *Geosmithia*.

Species concept

As we pointed out in our *Eupenicillium* monograph (Stolk and Samson, 1983: 12) we are applying a rather broad species concept, based mainly on the combination of conidiophore branching and phialide shape. An additional character is the shape and ornamentation of the conidia. Colony diameter is not considered a primary character, but is sometimes applied to distinguish species and varieties. In our taxonomic decisions the characters of telomorphs were given primary importance, resulting in species concepts which sometimes deviate from those of other authors.

In contrast with our earlier proposals to regard morphologically similar taxa of *Penicillium verrucosum* complex as varieties, we believe that for practical reasons these varieties can better be regarded as species. In addition, the profiles of their secondary metabolites (see Frisvad, 1983) support this view.

Table 1. Proposed taxonomic scheme for *Penicillium* anamorphs

A. Section *Torulomyces* (Delitsch) Stolk & Samson, comb. nov.

B. Section *Aspergilloides* (Dierckx) Pitt

C. Section *Eladia* (G. Smith) Stolk & Samson, comb. nov.

1. Series *Canescentia* Raper & Thom ex Pitt
2. Series *Inflata* Stolk & Samson, ser. nov.

D. Section *Divaricatum* Raper & Thom ex Pitt

1. Series *Atroveneta* Stolk & Samson, ser. nov.
2. Series *Purpurea* Stolk & Samson, ser.nov.
3. Series *Oxalica* Raper & Thom ex Pitt
4. Series *Janthinella* Thom ex Pitt
5. Series *Ochrosalmonea* Stolk & Samson, ser.nov.
6. Series *Dupontii* Stolk & Samson, ser. nov.

E. Section *Inordinate* Pitt

F. Section *Ramosum* Stolk & Samson, sect. nov.

1. Series *Lanosa* Stolk & Samson, ser. nov.
2. Series *Chrysogena* Raper & Thom ex Stolk & Samson
3. Series *Urticicola* Fassatiová

G. Section *Penicillium*

1. Series *Digitata* Raper & Thom ex Stolk & Samson
2. Series *Gladioli* Stolk & Samson
3. Series *Italica* Raper & Thom ex Pitt
4. Series *Expansa* Raper & Thom ex Fassatiová
5. Series *Olsonii* Pitt

H. Section *Coremigenum* (Biourge) Pitt

I. Section *Biverticillium* (Dierckx) Stolk & Samson, comb. nov.

1. Series *Miniolutea* Pitt
2. Series *Verruculosa* Stolk & Samson, sect. nov.

J. Section *Geosmithia* (Pitt)Stolk & Samson, comb. nov.

KEY TO THE SECTIONS

1 Conidiophores consisting of a short stipe and a single, small phialide ... Section *Torulomyces*

Conidiophores with distinct, relatively long stipe and phialides in whorls ... 2

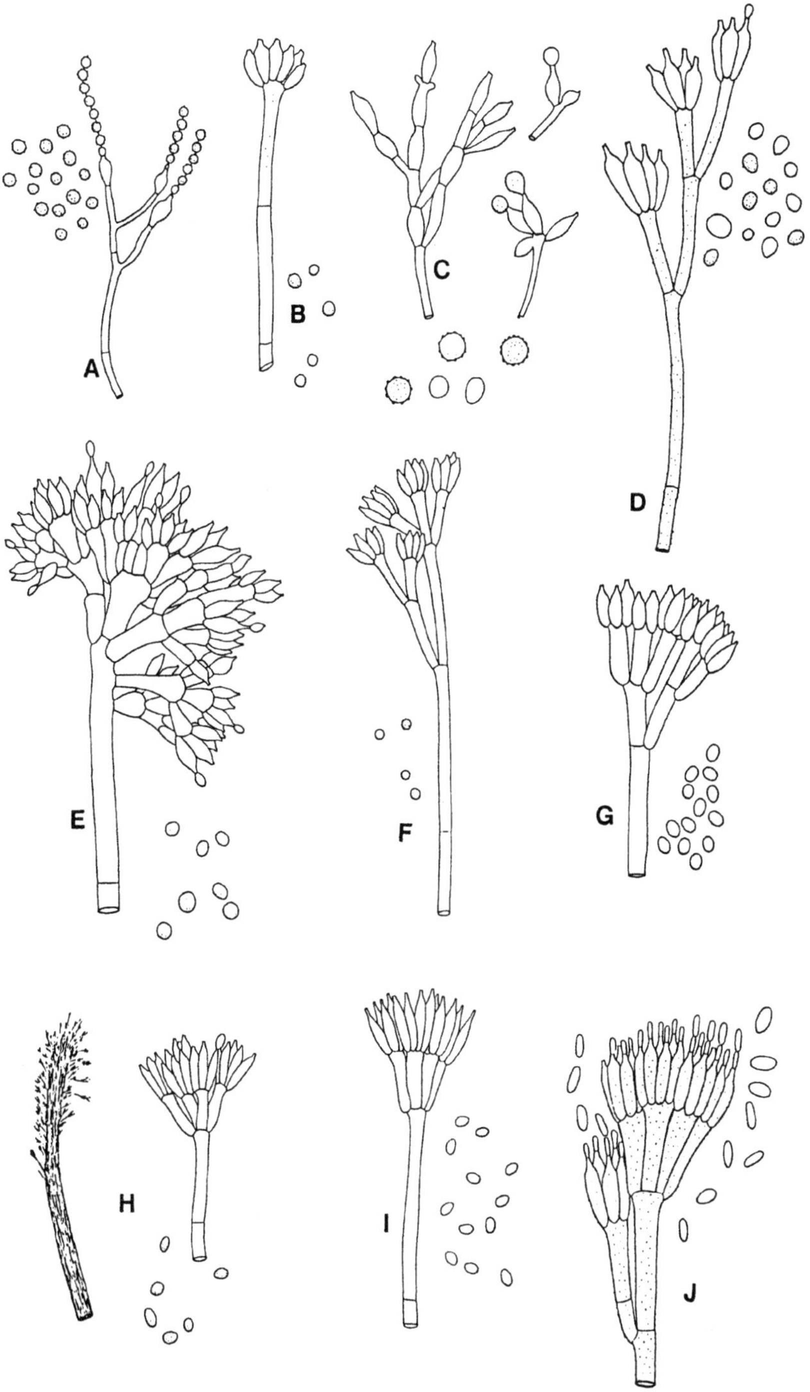

Fig. 1. Conidiophore types of the various sections. A. TORULOMYCES (P. lagena), B. ASPERGILLOIDES (P. glabrum), C. ELADIA (P. sacculum), D. DIVARICATUM (P. simplicissimum), E. INORDINATE (P. arenicola), F. RAMOSUM (P. lanosum), G. PENICILLIUM (P. expansum), H. COREMIGENUM (P. duclauxii), I. BIVERTICILLIUM (P. minioluteum), J. GEOSMITHIA (P. emersonii).

2 Conidiophores strictly simple or simple to one-stage-branched with 2-3 apical metulae 3

Typical conidiophores, with more developed branching, number of metulae usually larger 4

3 Phialides relatively short, with a broadly cylindrical, sometimes slightly swollen base, tapering abruptly at the apex into a very short, occasionally more pronounced, definitely narrowed neck; conidia globose to subglobose, typically roughened and pigmented, sometimes appearing smooth-walled and hyaline Section *Eladia*

Phialides not short, showing a cylindrical, occasionally broadly cylindrical base, narrowing at the apex into a short, but distinct neck; conidia globose, subglobose or ellipsoidal, smooth or roughened Section *Aspergilloides*

4 Conidiophores rarely simple, mostly one-stage-branched, rarely with a lower branch 5

Conidiophores one- to three-stage branched, rarely irregularly branched 8

5 Phialides lanceolate (acerose); conidiophores usually one-stage-branched, occasionally two-stage-branched, typically with symmetrical verticils of metulae Section *Biverticillium*

Phialides and penicilli different 6

6 Phialides slender, with a cylindrical base, tapering abruptly into a narrow neck, which is typically, relatively long, sometimes short and less conspicuous; conidiophores variable, simple to one-(two)-stage-branched, occasionally slightly irregular Section *Divaricatum*

Phialides relatively short, with a broadly cylindrical or slightly swollen base and a short, rarely more pronounced, definitely narrowed neck 7

7 Metulae in most species divergent; conidia globose to subglobose, definitely roughened, sometimes appearing smooth-walled, rarely ellipsoidal and striate; in well-developed penicilli conidia in divergent, occasionally parallel columns, often tangled in reduced structures Section *Eladia*

Metulae in adpressed to slightly divergent verticils of 2-6; conidia globose, subglobose or ellipsoidal, in parallel to tangled chains, or columns Section *Ramosum*

8 Conidiophores robust, complex, compact; irregularly one-three-stage-branched. sometimes with metulae or branches arising on

the apical part of the stipe; phialides with a somewhat swollen base and prominent, narrow neck Section Inordinate

Conidiophores and phialides different 9

9 Phialides cylindrical with a short, sometimes inconspicuous, narrowed neck; conidia cylindrical to strongly ellipsoidal; conidiophores mononematous, with all elements adpressed and typically rough-walled; cultures not green Section Geosmithia

Phialides and conidia different; cultures green 10

10 Conidiophores strongly synnematous, forming true synnemata with well-developed stalks, differing in length; conidiophores either borne over nearly the entire surface of the stalks or only produced on their apex and forming a broadened capitulum; phialides with a long, cylindrical base and a short neck Section Coremigenum

Conidiophores mononematous and/or loosely synnematous 11

11 Conidiophores typically with all elements adpressed; phialides with a broad or narrow, cylindrical base and slightly narrowed at the apex, often with a relatively wide neck Section Penicillium

Conidiophores relatively loose in structure; branches, when present, often arising well below the terminal verticil of metulae, usually more or less divergent; phialides with a broadly cylindrical, occasionally slightly swollen base and a short or inconspicuous neck Section Ramosum

A. SECTION TORULOMYCES (Delitsch) Stolk & Samson, comb. nov.

Genus Torulomyces Delitsch - Systematik der Schimmelpilze, Neudamm: 91. 1943 (Basionym).

Conidiophores mononematous, simple. Phialides single, relatively small, consisting of a swollen base and a small, abruptly narrowed neck. Conidia in very long, simple chains. Sclerotia sometimes present.

Teleomorph: Eupenicillium

Type species: P. lagena (Delitsch) Stolk & Samson

P. lagena (Delitsch) Stolk & Samson (teleomorph: E. limoneum Gochenaur & Zlattner)

In our Eupenicillium monograph (Stolk and Samson, 1983) we transferred Torulomyces lagena Delitisch to the genus Penicillium. The reasons for this transfer are discussed on p. 103 of that paper. P. lagena differs from all known species of Penicillium in producing single phialides rather than whorls. For this reason we propose to erect the new section Torulomyces.

B. SECTION ASPERGILLOIDES Pitt emend Stolk & Samson

Subgenus Aspergilloides Dierckx - Anns Soc. Scient. Brux. 25: 85. 1901
Subgenus Monoverticillium Biourge - Cellule 33: 265. 1923
Section Monoverticillata Raper & Thom - Manual Penicillia: 126. 1949 (nom. inval.)

Conidiophores mononematous, simple to one-stage-branched. Stipes long and often vesiculate, in some species short. Phialides in whorls, flask-shaped, consisting of a broad cylindrical base and an abrupty narrowed neck. Conidia globose, subglobose or ellipsoidal, with walls smooth or rugulose, occasionally striate. Conidial chains in columns. Sclerotia in some species present. Teleomorph Eupenicillium.

Type species: P. aurantiobrunneum Dierckx = P. glabrum (Wehmer) Westling

P. catenatum Scott (teleomorph: E. catenatum Scott)
P. corylophilum Dierckx
P. cyaneum (Bain. & Sart.) Biourge = P. dierckxii Biourge (teleomorph: E. cinnamopurpureum Scott & Stolk)
P. decumbens (teleomorph: E. meridianum Scott)
P. donkii Stolk
P. glabrum (Wehmer)Westling = P. frequentans Westling
P. lapidosum Raper & Fennell (teleomorphs: E. lapidosum Scott & Stolk and E. terrenum Scott)
P. lividum Westling var. lividum
P. lividum Westling var. thomii (Maire) Stolk & Samson*
P. papuanum Udagawa & Horie (teleomorph: E. parvum (Raper & Fennell) Stolk & Scott)
P. sclerotiorum van Beyma
P. spinulosum Thom = P. purpurescens (Sopp) Biourge

In the section Aspergilloides we have placed both species with strictly simple conidiophores as well as species with simple and branched conidial structures. The number of metulae is small, up to 3. As far as the phialides are concerned the section is heterogenous, since different phialide types occur. The species are only grouped together because of their simple and little-branched conidiophores.

In our Eupenicillium paper we placed P. thomii in synonymy with P. lividum, because the neotype cultures produced identical conidiophores and phialides. In this paper we regard P. thomii as a variety of P. lividum. There are many intergrading strains between the varieties and they can only be distinguished by their conidial ornamentation. P. donkii, placed in synonymy with P. dierckxii (= P. cyaneum) in 1983 is regarded as a separate species, because of its more rapid growth.

* Basionym: P. thomii Maire -- Bull. Soc. Hist. nat. Afr. N. 8: 18. 1917.

P. catenatum, represented only by its type strain, is tentatively placed in the section Aspergilloides. It produces only a few, very small, fractional conidiophores on hay decoction agar, clearly representing strongly reduced, atypical conidial structures. Since the teleomorphs of P. catenatum and P. tularense are identical, we considered the two species synonyms in our Eupenicillium monograph despite the different conidiophores. However, as demonstrated by Pitt (1979) the two species differ in rate of growth at 37 C (P. catenatum 15-22 mm, P. tularense 0 mm). Since we do not know with certainty the structure of a typical P. catenatum and because of the differences in temperature relations, we prefer, at present, to keep the two species separated. Consequently P. catenatum is placed in the section Aspergilloides and P. tularense in the section Ramosum. Thus E. catenatum belongs in Eupenicillium section Lapidosa and E. tularense in section Eupenicillium.

C. SECTION ELADIA (G. Smith) Stolk & Samson, comb. nov.

Genus Eladia G. Smith - Trans. Br. mycol. Soc. 44: 47. 1961 (basionym)
Subgenus Furcatum Pitt - Genus Penicillium: 233. 1980

Conidiophores mononematous, one-stage-branched, in some strains ramigenous or reduced and simple. Phialides in whorls, borne terminally, occasionally subterminally on metulae or stipes; flask-shaped, relatively short, consisting of a broadl cylindrical occasionally slightly or more definitely swollen base and a short, sometimes prominent, abruptly narrowed neck. Conidia globose to subglobose, thick-walled, rugulose or echinulate, rarely ellipsoidal and striate, sometimes appearing smooth-walled, especially in species with small conidia; hyaline or pigmented, often brown; sometimes in young cultures at first ellipsoidal, hyaline and smooth-walled. Conidial chains often with more or less distinct connectives between the conidia. Sclerotia present in some species. Conidial areas greyish blue-green, greyish or olivaceus.

Teleomorph: Eupenicillium or absent

Type species: P. sacculum Dale

1. Series Canescentia Raper & Thom ex Pitt
2. Series Inflata Stolk & Samson, ser. nov.

Key to the series

1. Conidiophores, typically Aspergillus-like, with stipes apically enlarged, bearing 2-10 radiately arranged metulae, occasionally ramigenous; conidia globose to subglobose, striate, in divergent chains — Series Inflata

1. Conidiophores not Aspergillus-like. Conidia globose and roughened, rarely ellipsoidal and longitudinally striate, typically in columns, in reduced strains often in tangled chains — Series Canescentia

Thom (1930) and Raper and Thom (1949) regarded *Penicillium sacculum* Dale as a *Scopulariopsis*. Smith (1961) introduced the genus *Eladia* to accomodate it, since he did not believe it was a *Scopulariopsis*. In our opinion, the species should be regarded as a reduced *Penicillium*. The characteristic swollen phialides with abruptly narrowed neck, the globose, roughened conidia and the conidial chains, though short (2-7 conidia) are also found in other species of *Penicillium* (e.g. *P. janczewskii*). In fresh isolates of *P. sacculum* phialides occur in verticils of 2-4, but they are also irregularly borne on the stipes. In old cultures phialides are arranged more irregularly and they produce chains of only 2-3 conidia. These short conidial chains do not warrant recognition of *Eladia* as genus. We regard *Eladia* a section of *Penicillium*, in which *P. sacculum* as well as a number of related species are placed.

The section *Eladia* is divided into 2 series: *Canescentia* and *Inflata*, which may be separated by the structure of their conidiophores and by their conidia.

In our paper on *Eupenicillium*, we placed *P. lassenii* (teleomorph *E. lassenii*) in synonymy with *P. citreonigrum* (teleomorph *E. euglaucum*). Because of the structure of the conidiophores we prefer at present to regard *P. lassenii* a separate species. *P. citreonigrum* produces 0-2 (3) metulae, whereas *P. lassenii* is characterized by verticils of 2-4 (5). Moreover, the ascospores of *E. lassenii* are slightly thinner than those of *E. euglaucum* and the ridges are wide.

C.1. Series *Canescentia* Raper & Thom ex Pitt

Series *Canescentia* Raper & Thom ex Pitt
Raper & Thom - Manual Penicillia: 315. 1949 (nom. inval.)
Pitt: - Genus *Penicillium*: 251. 1979
Series *Citrina* Raper & Thom ex Pitt
Raper & Thom - Manual Penicillia: 338. 1949 (nom. inval.)
Pitt - Genus *Penicillium*: 290. 1979
Series *Citreonigra* Pitt - Genus *Penicillium*: 218. 1980
Series *Restricta* Raper & Thom ex Pitt
Raper & Thom - Manual Penicillia: 222. 1949 (nom. inval.)
Pitt - Genus *Penicillium*: 205. 1979
Series *P. adametzi* Raper & Thom - Manual Penicillia: 227. 1949 (nom. inval.)
Series *P. nigricans* Raper & Thom - Manual Penicillia: 303. 1949 (nom. inval.)

Conidiophores typically one-stage-branched, with a terminal verticil of 2-5 divergent rarely adpressed metulae, occasionally ramigenous with the stipes bearing single, divergent branches, but sometimes reduced and simple. Phialides relatively short with a broadly cylindrical to slightly or more definitely swollen base, and a short, occasionally more pronounced narrowed neck. Conidia globose to subglobose, rarely pear-shaped or ellipsoidal (*P. striatisporum*) typically rugulose to echinulate, but sometimes appearing smooth-walled or striate. Conidial chains commonly in divergent, occasionally in parallel columns, in reduced strains tangled. Sclerotia present in some species .

smooth or roughened, when ellipsoidal or fusiform occasionally striate. Sclerotia present in some species.

Teleomorphs: Eupenicillium, Talaromyces

Type species: P. janthinellum Biourge = P. simplicissimum (Oudem.) Thom

1. Series Atroveneta Stolk & Samson ser. nov.
2. Series Purpurea Stolk & Samson, ser. nov.
3. Series Janthinella Thom ex Pitt
4. Series Oxalica Raper & Thom ex Pitt
5. Series Ochrosalmonea Stolk & Samson ser. nov.
6. Series Dupontii Stolk & Samson, ser. nov.

Key to the series

1. Thermophilic, optimum temperature 45°C, chlamydospores usually present; teleomorph: Talaromyces — Series Dupontii

   Not thermophilic, chlamydospores absent — 2

2. Phialide necks narrow, prominent — 3

   Phialide necks slightly or definitely narrowed, short, occasionally prolonged, not pronounced — 5

3. Conidia globose to subglobose, thick-walled, rugulose or echinulate, often pigmented; typically with pigmented stipes — Series Atroveneta

   Conidia and stipes different — 4

4. Conidiophores simple to one-stage-branched, metulae divergent; conidia with transverse striations, brown; teleomorph: Talaromyces or absent — Series Purpurea

   Conidiophores variable in structure, simple to one-(two)stage-branched; conidia in various shapes, smooth or roughened, in a few species with transverse striations; teleomorph: Eupenicillium or absent — Series Janthinella

5. Conidia subglobose, pear-shaped or broadly ellipsoidal, smooth-walled, no striations, in tangled chains — Series Ochrosalmonea

   Conidia cylindrical to strongly or broadly ellipsoidal, smooth-walled, sometimes with transverse striations; in long, parallel chains, typically forming conspicuous columns — Series Oxalica

The section Divaricatum is characterized by simple to one-stage-branched or irregularly two-stage-branched conidiophores and phialides with a prominent narrow neck, which may be long in some species. The section is divided into six

Teleomorph: Eupenicillium or absent

Type species: P. canescens Sopp

P. canescens Sopp = P. viridobrunneum Sopp = P. raciborskii Zaleski
P. citreonigrum Dierckx (teleomorph: E. euglaucum van Beyma) Zaleski (sclerotial)
P. citrinum Thom
P. fuscum (Sopp) Biourge (teleomorph: E. pinetorum Stolk)
P. janczewskii Zaleski = P. nigricans Thom
P. jensenii Zaleski (sclerotial)
P. lassenii Paden (teleomorph: E. lassenii Paden)
P. megasporum Orpurt & Fennell
P. melinii Thom
P. sacculum Dale
P. senticosum Scott (teleomorph: E. senticosum Scott)
P. striatisporum Stolk

C. 2. Series Inflata Stolk & Samson, ser.nov.

Series in sectione Eladia; conidiophora Aspergillo adsimile; conidia striata in catenis divergentibus connexa.
Species typica: P. inflatum Stolk & Malla

Conidiophores Aspergillus-like, one-stage-branched, with a verticil of 2-10 divergent, sometimes radiately arranged metulae, occasionally ramigenous with the stipe bearing a few single branches, sometimes reduced and small, simple or bearing only a few metulae. Stipes thin-walled, varying in length. Stipes and metulae enlarged at the apex. Aspergillus conidiophore base structure ("footcell") absent. Phialides with a broad, sometimes slightly swollen base and a short, abruptly narrowed neck. Conidia globose to subglobose, brownish, striate. Conidial chains divergent, in well-developed penicilli radiately arranged. Sclerotia and teleomorph absent.

Type species: P. inflatum Stolk & Malla

P. inflatum Stolk & Malla

D. SECTION DIVARICATUM Raper & Thom ex Pitt

Raper & Thom - Manual Penicillia: 255. 1949 (nom. inval.)
Pitt - Genus Penicillium: 238. 1980

Conidiophores mononematous, variable, ranging from simple to one-(two)-stage-branched, but predominantly one-stage-branched, sometimes slightly irregular with phialides and metulae in one cluster. Phialides in whorls, flask-shaped, consisting of a cylindrical, rarely slightly swollen base and at the apex tapering abruptly, occasionally more slowly into a relatively narrow neck, which is typically prominent, but occasionally short. Conidia in various shapes, with walls series, which are separated by the structure of the conidiophores and the conidia. Most teleomorphs belonging to species of this section are Eupenicillia. P. purpureus and P. dupontii, however, have Talaromyces teleomorphs.

The series Purpurea and Atroveneta are somewhat intermediate between the section Eladia and Divaricatum. Both are placed in Divaricatum because of the (prominent) long narrow phialide necks. The series Ochrosalmonea is placed in Divaricatum because of the structure of their conidiophores, which resemble those of the series Janthinella. The phialides have relatively narrow necks, but these are shorter than those of Janthinella. The series Dupontii contains only the thermophilic species P. dupontii. The conidiophores are consistent with those of this section. The phialide necks are distinct but not very long.

The series Oxalica is placed in the section Divaricatum, because of the variable structure of the conidiophores, resembling those of P. ochrosalmoneum, P. ochrochloron and other species of this section, and because of the shape and ornamentation of the conidia. The phialides differ from most other species of this section, since they do not show the characteristic, long narrow necks. Their phialide necks are usually short, and slightly or definitely narrowed. Phialides of this type, though smaller also occur in the series Ochrosalmonea. The conidia are cylindrical to ellipsoidal and in some strains have transverse striations, as observed in P. ochrochloron, though they often appear smooth-walled.

D. 1. Series Atroveneta Stolk & Samson, ser. nov.

Series in sectione Divaricatum; phialides cum collibus angustatis sed distinctis; conidia (sub)globosa; conidiophores stiptibus pigmentatis.
Species typica: P. atrovenetum G. Smith

Conidiophores predominantly one-stage-branched, but ranging from simple to one-stage-branched, rarely two-stage-branched with one branch below the verticil of metulae; walls of all elements slightly or more definitely pigmented, sometimes hyaline, stipes rough-walled or smooth. Metulae 2-5, usually divergent. Phialides with a cylindrical or slightly swollen base, tapering abruptly into a prominent, relatively narrow neck. Conidia globose to subglobose, rarely broadly ellipsoidal, rugulose or echinulate, olivaceous or brown in mass. Conidial chains parallel, in loose columns or tangled. Sclerotia rarely present.

Teleomorph: Eupenicillium

Type species: P. atrovenetum G. Smith

P. atrovenetum G. Smith
P. velutinum van Beyma (teleomorph: E. stolkiae Scott)

D. 2. Series Purpurea Stolk & Samson, ser.nov.

Species in sectione Divaricatum; phialides collibus angustatis sed distinctis; conidia transverse striate. Teleomorphosis: Talaromyces.

Species typica: P. purpureum Stolk & Samson

Conidiophores simple to one-stage-branched, with 2-4 metulae. Phialides, usually showing a slightly or strongly swollen base and a relatively long, narrow neck. Conidia subglobose to ellipsoidal, or apiculate, hyaline and smooth, when young, becoming olive-brown, thick-walled when mature with conspicuous transverse sometimes spiral bands. Conidial chains divergent or tangled.

Type species: P. purpureum Stolk & Samson

P. daleae Zaleski
P. purpureum Stolk & Samson - (teleomorph: Talaromyces purpureus (Müller & Pacha-Aue) Stolk & Samson)

The teleomorph Talaromyces purpureus differs from the other species of this genus and is accommodated by us (Stolk and Samson, 1972: 56) in a separate section Purpurea.

D. 3. Series Janthinella Thom ex Pitt

Thom - Penicillia: 338. 1930
Pitt - Genus Penicillium: 239. 1979

Conidiophores simple to two-stage-branched, but predominantly one-stage-branched, sometimes irregularly branched. Phialides with a cylindrical, rarely slightly swollen base and a relatively long, narrow neck. Conidia subglobose, ellipsoidal or fusiform, with walls smooth, roughened or striate. Conidial chains loosely parallel, in loose columns or tangled. Sclerotia present in some species.

Teleomorphs: Eupenicillium, Talaromyces

Type species: P. simplicissimum (Oudem.) Thom

P. humuli van Beyma
P. inusitatum Scott (teleomorph: E. inusitatum Scott)
P. ochrochloron Biourge var. ochrochloron (sclerotial)
P. ochrochloron Biourge var. paraherquei (Abe)Stolk & Samson, comb. nov.*
P. simplicissimum (Oudem.) Thom (teleomorph: E. javanicum (van Beyma) Stolk & Scott)

*Basionym: P. paraherquei Abe ex G. Smith -- Trans. Br. mycol. Soc. 46: 335. 1963

D. 4. Series Oxalica Raper & Thom ex Pitt

Raper & Thom - Manual Penicillia: 376. 1949 (nom. inval.)
Pitt - Genus Penicillium: 273. 1979

Conidiophores mononematous, simple to one- or two-stage-branched, with all elements closely adpressed and smooth-walled. Phialides large with a long and wide cylindrical base and a short, definitely narrow neck. Conidia cylindrical to ellipsoidal, transversally striate, but often appearing smooth-walled. Conidial chains in long parallel columns. Sclerotia and teleomorph absent.

Type species: P. oxalicum Currie & Thom

P. fennelliae Stolk
P. oxalicum Currie & Thom

D. 5. Series Ochrosalmonea Stolk & Samson, ser. nov.

Series in sectione Divaricatum; phialides collibus brevis, sed longis, indistinctis; conidia (subglobosa), piriformis vel late ellipisoidea.
Species typica: P. ochrosalmoneum Udagawa

Conidiophores variable, irregular, simple to one-or two-stage-branched, occasionally developing 2 or 3 whorls of phialides one above the other, or with whorls, consisting of both metulae and phialides. Branches, (when present) somewhat adpressed, or divergent. Metulae, when present, in verticils of 2-4. Phialides slender, with a relatively, long cylindrical base and tapering slowly or more abruptly into a relatively narrow neck. Conidia subglobose to ellipsoidal, smooth-walled. Conidial chains tangled. Sclerotia sometimes present.

Teleomorph: Eupenicillium

Type species: P. ochrosalmoneum Udagawa

P. abidjanum Stolk (teleomorph: E. abidjanum Stolk)
P. ochrosalmoneum Udagawa (teleomorph: E. ochrosalmoneum Scott & Stolk)

D. 5. Series Dupontii Stolk & Samson, ser.nov.

Species in sectione Divaricatum; phialides collibus longis sed angustis, chlamydosporae adsunt. Species thermophilae.Teleomorphosis: Talaromyces.
Species typica: P. dupontii Griffon & Maublanc

Conidiophores simple to one-stage-branched, with a verticil of 2-4(-5) divergent or somewhat adpressed metulae, occasionally with a lower branch or metula. Stipes short, arising from aerial hyphae. Phialides with a cylindrical base and a short to relatively long, narrowed neck. Conidia ellipsoidal, smooth-walled. Conidial chains divergent. Chlamydospores usually present. Thermophilic, optimum temperature 45°C.

Teleomorph: Talaromyces

Type species: P. dupontii Griffon & Maublanc

dupontii Griffon & Maublanc (teleomorph: Talaromyces thermophilus Stolk)

The teleomorph Talaromyces thermophilus differs from the other species of this genus and is accommodated by us (Stolk and Samson, 1972: 53) in a separate section Thermophila.

E. SECTION INORDINATE Pitt

Section Inordinate Pitt - Genus Penicillium: 394. 1979

Conidiophores complex, usually compact, irregularly one- to three-stage-branched, sometimes also producing branches below the terminal penicillus on the subapical portion of the stipe. Phialides in whorls, flask-shaped, with a somewhat swollen base and a prominent, narrow neck. Conidia globose to subglobose, sometimes ranging from broadly ellipsoidal to globose, thick-walled, rugulose, yellowish, in divergent chains, with dinstinct connectives between the conidia. Conidial areas brown. Sclerotia and teleomorph absent.

Type species: P. arenicola Chalabuda

P. arenicola Chalabuda

The section Inordinate contains only one species: P. arenicola Chalabuda. This brown species is not a typical Penicillium. It produces large, irregular and complicated, up to three-stage-branched penicilli, developing on long, thick-walled, brown stipes. The phialides are characterized by relatively long necks, reminiscent of the section Divaricatum. They occur in apical verticils, but also on the subapical portion of the stipe and in this respect they resemble the species of Merimbla. P. arenicola differs from the section Divaricatum by its complicated penicilli and brown stipes. In our opinion the section Inordinate represents an intermediate species between Penicillium and Merimbla. In accordance with Pitt (1979) it is placed for the time being in Penicillium, because of its many resemblances with other species of this section.

F. SECTION RAMOSUM Stolk & Samson, sect. nov.

Conidiophora metulis apppressis vel subdivergentibus; rami divergentibus; conidia (sub)globosa vel ellipsoidea in catenis parallelis vel implexis connexa.
Species typica: P. lanosum Westling

Conidiophores mononematous (in series Urticicola synnematous), strictly one-stage-branched or ranging from one- to two-, three- or more-stage-branched, depending on the species. Branches typically, more or less divergent, rarely some of them a little adpressed, usually arising well below the terminal verticil of metulae. Metulae in whorls of 2-6, adpressed to slightly divergent. Phialides in clusters, flask-shaped, consisting of a cylindrical to broadly cylindrical, rarely slightly swollen base and a short, but distinct, sometimes inconspicuous neck. Macrophialides occasionally present. Sclerotia produced in some species.

Teleomorph: Eupenicillium or absent

Type species: P. lanosum Westling

1. Series Lanosa Stolk & Samson ser. nov.
2. Series Chrysogena Raper & Thom ex Stolk & Samson
3. Series Urticicola Fassatiová

The section Ramosum differs from the section Penicillium in the looser structure of the conidiophores and the slightly smaller phialides. In the section Ramosum the branches, when present, are commonly divergent and usually arise well below the terminal verticil of metulae, whereas those in the section Penicillium are, when typical, strongly adpressed. The phialides of Ramosum resemble those of the section Eladia. The sections may be separated by the different structure of their conidiophores. In some species macrophialides may occur also.

We have divided the section into three series: Lanosa, Chrysogena and Urticicola. The series Urticicola contains only one species, P. griseofulvum, which does not seem closely related to any of the other species of Penicillium. The series fits best in the section Ramosum despite the divergent branches and very small phialides. P. griseofulvum differs from the other species of this section in the complicated, synnematous conidiophores, and the small phialides with very short or inconspicuous necks.

Key to the series

1.	Conidiophores mononematous, one-, two-(three-)stage-branched with branches more or less divergent	2
	Conidiophores mono- and/or synnematous	3
2.	Conidia in parallel to tangled chains	Series Lanosa
	Conidia in well-defined columns	Series Chrysogena

3. Synnemata small, loose; conidiophores complicated with up to 5 or more divergent branches; phialides very small, with a broadly cylindrical to slightly swollen base and a very short or inconspicuous neck   Series Urticicola

Synnemata well developed, various in structure; phialides often small at first, regularly after a number of transfers becoming larger   Section Coremigenum

F. 1. Series Lanosa Stolk & Samson, ser. nov

Series in sectione Ramosum; conidiophora mononematous, conidia in catenis parallelis vel implexis connexa.
Species typica: P. lanosum Westling

Conidiophores mononematous strictly one-stage-branched or one- to two-(three-)stage-branched, with walls smooth or roughened. Branches, single, more or less divergent arising well below the terminal verticil of metulae, sometimes lacking. Phialides with a cylindrical, often broadly cylindrical base, narrowed at the apex into a small, but distinct neck. Conidia in parallel to tangled chains, sometimes at first in loose columns, globose to broadly ellipsoidal. Sclerotia present in some species.

Teleomorph: Eupenicillium or absent.

Type species: P. lanosum Westling

P. coralligerum Nicot & Pionnat (sclerotial)
P. lanosum Westling
P. manginii Duche & Heim (sclerotial)
P. novae-zeelandiae van Beyma (sclerotial)
P. osmophilum Stolk & Veenbaas-Rijks (teleomorph: E. osmophilum Stolk & Veenbaas-Rijks)
P. soppii Zaleski (teleomorph: E. shearii Stolk & Scott)
P. tularense Paden (teleomorph: E. catenatum Scott)

F. 2. Series Chrysogena Raper & Thom ex Stolk & Samson

Series Chrysogena Raper & Thom - Manual Penicillia: 355. 1949 (nom. inval.)

Series in sectione Ramosum; conidiophora mononematous; conidia in columnas aggregata.
Species typica: P. chrysogenum Thom

Conidiophores mononematous, (one-), two- to three-stage-branched, with all elements smooth-walled. Branches single, occasionally two per stage, usually divergent, rarely slightly adpressed. Phialides usually with a wide, very short, but occasionally slightly longer neck. Conidia subglobose to ellipsoidal, smooth-walled. Conidial chains forming columns. Sclerotia and teleomorph absent.

Type species: P. chrysogenum Thom

P. chrysogenum Thom

F. 3. Series Urticicola Fassatiová

Series Urticicola Fassatiová - Acta Univ. Carol., Biol. 12: 324. 1977
Series Urticae Raper & Thom - Manual Penicillia: 531. 1949 (nom. inval.)

Conidiophores mono- and synnematous, variable in pattern, two- five- or more-stage-branched. Branches single, divergent; in terminal penicilli sometimes slightly adpressed. Metulae short, numbering 2-4. Phialides very small, slightly swollen, with a very short or inconspicuous neck; in crowded whorls. Conidia subglobose to ellipsoidal. Conidial chains parallel, producing somewhat divergent columns or becoming tangled. Sclerotia and teleomorph absent.

Type species: P. urticae Bain. = P. griseofulvum Dierckx

P. griseofulvum Dierckx

G. SECTION PENICILLIUM

Subsection Fasciculata Raper & Thom - Manual Penicillia: 467. 1949 (nom. inval.)
Subsection Lanata Raper & Thom - Manual Penicillia: 419. 1949 (nom. inval.)
Subsection Velutina Raper & Thom pro parte - Manual Penicillia: 336. 1949 (nom. inval.)
Subsection Funiculosa Raper & Thom pro parte - Manual Penicillia: 445. 1949 (nom. inval.)

Conidiophores mono- and/or loosely synnematous, in a few species sometimes definitely synnematous, especially in marginal areas of the colony; one- to two-, sometimes to three-stage-branched; all elements adpressed. Phialides in clusters, flask-shaped, consisting of a broad or narrow cylindrical base and a short, sometimes longer, relatively wide neck. Sclerotia present in some species.

Teleomorph: Eupenicillium or absent

Type species: P. expansum Link

1. Series Digitata Raper & Thom ex Stolk & Samson
2. Series Gladiolii Raper & Thom ex Stolk & Samson
3. Series Italica Raper & Thom ex Pitt
4. Series Expansa Raper & Thom ex Fassatiová
5. Series Olsonii Pitt

Key to the series

1. Conidia cylindrical to strongly or broadly ellipsoidal 2

Conidia globose, subglobose or occasionally broadly ellipsoidal 4

2. Conidiophores poorly developed, irregularly simple to one-(two-)stage branched, consisting only of a few elements, cultures olivaceous Series Digitata

Conidiophores well-developed, one- to two-stage branched; cultures blue green or dull green 3

3. Conidiophores mono- and/or synnematous, sometimes well-developed synnemata at the colony margins Series Italica

Conidiophores strictly mononematous; conidia in conspicuous long, parallel columns series Oxalica (sect. Divaricatum)

4. Conidiophores mononematous with short, complex and compact one- to two-(three-)stage-branched penicilli; conidia subglobose, pear-shaped to broadly ellipsoidal, in tangled or divergent chains series Olsonii

Conidiophores mononematous, and/or sometimes loosely synnematous; penicilli not short and complex; conidia not in divergent chains 5

5. Conidiophores usually mononematous, in a few species synnematous with distinct, sometimes feather-like synnemata, slender; branches well below the terminal verticil of metulae; phialides relatively narrow; sometimes macrophialides present Series Gladiolii

Conidiophores mononematous, or monematous and/or slightly and loosely synnematous in marginal areas, not slender; phialides when typical large with a wide neck Series Expansa

The section Penicillium is characterized by one- to two- or three-stage-branched conidiophores with adpressed elements and relatively large, cylindrical phialides, usually with distinct, wide necks. It differs from section Ramosum in the shape of the phialides and the adpressed structure of the penicillia. The section is divided into five series: the series Digitata, Gladiolii, Italica, Expansa and Olsonii.

As mentioned above, the varieties of P. verrucosum proposed by Samson et al. (1976) are here regarded as distinct species.

G. 1. Series Digitata Raper & Thom ex Stolk & Samson

Series Digitata Raper & Thom - Manual Penicillia: 385. 1949 (nom. inval.)

Series in sectione Penicillium; conidiophora male evoluta, signis paucis; culturae olivacae; conidia cylindrica ad ellipsoidea.
Species typica: P. digitatum (Pers.: Fr.) Sacc.

Conidiophores mononematous, irregular in pattern, ranging from simple to irregularly one-, occasionally two-stage-branched, varying greatly in complexity, with identity of metulae and branches often poorly established, usually consisting only of a few elements. Phialides large. Conidia cylindrical to ellipsoidal, large, up to 4-8 um in length, smooth-walled. Conidial chains tangled. Conidial areas in dull olivaceous shades. Sclerotia and teleomorph absent.

Type species: P. digitatum (Pers.: Fr.) Sacc.

P. digitatum (Pers.: Fr.) Sacc.

G. 2. Series Gladiolii Raper & Thom ex Stolk & Samson

Series Gladiolii Raper & Thom - Manual Penicillia: 471. 1949 (nom. inval.)

Series in sectione Penicillium, conidiophora mononematica vel synnematica, penicilli appressi, collum phialidis angustatum.
Species typica: P. gladioli Mc Culloch & Thom

Conidiophores mono- and/or synnematous, often with one of these features predominating, depending on the species, slender, one-two- to three-stage-branched, with all elements appressed, in a few species aggregated into synnemata. Branches often arising relatively low on the stipe. Branches and metulae with walls smooth or strongly roughened. Phialides slender with a relatively narrow base and a small narrowed neck. Conidia globose or ellipsoidal, with walls usually smooth, only in P. hordei (occasionally in P. turbatum) roughened. Conidial chains parallel to tangled, sometimes in loose, divergent or adpressed columns. Sclerotia present in some species.

Teleomorph: Eupenicillium or absent

Type species: P. gladioli Mc Culloch & Thom

P. gladioli Mc Culloch & Thom (sclerotial)
P. glandicola (Oud.) Seifert & Samson = P. granulatum Bain.
P. hordei Stolk
P. nilense Pitt (teleomorph: E. crustaceum Ludwig)

P. sinaicum Udagawa & Ueda (teleomorph: E. sinaicum Udagawa & Ueda)
P. turbatum Westling (teleomorph: E. baarnense (van Beyma) Stolk & Scott)

G. 3. Series Italica Raper & Thom ex Pitt

Series Italica Raper & Thom - Manual Penicillia: 563. 1949 (nom. inval.)

Conidiophores sometimes mononematous, more commonly loosely to definitely synnematous, especially in marginal areas, one- to two- stage-branched, with walls smooth. Synnemata typically with short to long stalks up to 1 mm or more, in length and terminal conidial heads. Stipes sometimes slightly sinuous. Phialides parallel, relatively large, with a broad, cylindrical base and a short, but distinct wide neck. Phialide necks merge often almost imperceptibly into the chains of conidia. Conidia cylindrical to ellipsoidal, smooth-walled. Colony colour pale grey-green white or avellaneous. Teleomorph absent.

Type species: P. italicum Wehmer

P. italicum Wehmer

G. 4. Series Expansa Raper & Thom ex Fassatiova

Series Expansa Raper & Thom ex Fassatiová
Raper & Thom - Manual Penicillia: 508. 1949
Fassatiova- Acta Univ. Carol., Biol. 324. 1977
Series Viridicata Raper & Thom ex Pitt
Raper & Thom- Manual Penicillia: 481. 1949 (nom. inval.)
Pitt - Genus Penicillium: 334. 1979
Series Camemberti Raper & Thom ex Pitt
Raper & Thom - Manual Penicillia: 421. 1949 (nom. inval.)
Pitt - Genus Penicillium: 358. 1979
Series P. roqueforti Raper & Thom - Manual Penicillia: 392: 1949 (nom. inval.)
Series P. ochraceum Raper & Thom - Manual Penicillia: 475. 1949 (nom. inval.)
Series P. cyclopium Raper & Thom - Manual Penicillia: 490. 1949 (nom. inval.)
Series P. ochracea Raper & Thom ex Fassatiová - Acta Univ. Carol., Biol. 12: 324. 1977

Cultures usually granular. Conidiophores mono- and/or synnematous, often with one of these features predominating, depending on the species; usually two-, to three-stage-branched, occasionally reduced and one-stage-branched, commonly all elements adpressed, sometimes aggregated into small fascicles or synnemata, especially in marginal areas of the colonies. Phialides with a relatively large and broad base and a wide narrowed neck. Conidia ellipsoidal to subglobose, smooth-walled. Conidial chains parallel, forming loose columns or becoming tangled. Sclerotia sometimes present, teleomorph absent.

Type species: P. expansum Link

P. album G. Smith
P. aurantiogriseum Dierckx (= P. cyclopium Westling)
P. camemberti Thom
P. commune Thom
P. crustosum Thom
P. echinulatum Raper & Thom ex Fassatiová
P. expansum Link
P. hirsutum Dierckx
P. melanochlorum(Samson et al.)Frisvad
P. olivicolor Pitt
P. roqueforti Thom
P. verrucosum Dierckx

G. 5. Series Olsonii Pitt

Series Olsonii Pitt - Genus Penicillium: 392. 1979
Series Brevicompactum Raper & Thom - Manual Penicillia: 404. 1949 (nom. inval.)
Series Herquei Raper & Thom - Manual Penicillia: 658. 1949 (nom. inval.)

Conidiophores mononematous,usually with a long stipe and a relatively short, compact penicillus; one- to two-, or three-stage-branched. Branches 1-6, adpressed; metulae and phialides numerous and closely crowded. Phialides with a cylindrical base and a small, narrowed neck, smaller than those in the other series of the section Penicillium. Conidia pear-shaped, but ranging from subglobose to broadly ellipsoidal, with walls finely roughened, sometimes appearing smooth. Conidial chains divergent. Sclerotia and teleomorph absent.

Type species:P. olsonii Bain. & Sart

P. olsonii Bain. & Sart.
P. brevicompactum Dierckx
P. herquei Bain. & Sart.

Since no type material of P. olsonii is known to exist, Pitt (1979) designated IMI 192502 (= CBS 232.60) as neotype. The conidiophores of CBS 232.60 are large, complex and compact, one- to three-stage-branched consisting of closely adpressed branches, secondary branches and metulae. The conidial chains are divergent. P. brevicompactum is placed in the section Olsonii because of its strong resemblance with P. olsonii. The two species differ only in the number of branches. In P. brevicompactum branches are usually borne singly, though occasionally 2-3 may be present, whereas typically in P. olsonii, verticils of up to 6 branches occur. However, in reduced strains of P. olsonii smaller verticils of branches are formed. Thus the difference between P. olsonii and P. brevicompactum is very small and in fact the two species merge. The one-stage-branched species, P. herquei is placed in the series Olsonii because of its relatively short, compact penicilli and its long divergent conidial chains. The series

Olsonii is placed in the section Penicillium because of the structure of its conidiophores. Moreover, the phialides approximate those of section Penicillium, but are often slightly smaller.

H. SECTION COREMIGENUM (Biourge) Pitt

Section Coremigenum (Biourge) Pitt - Genus Penicillium: 403. 1979
Subsection Coremigena Biourge - Cellule 33: 31. 1923

Conidiophores strongly synnematous, forming true synnemata, often in concentric zones, occasionally a few mononematous conidiophores present. Synnemata various; composed of a stalk and a fertile upper portion, with strongly interwoven penicilli or with conidiophores retaining their individuality, in other species Isaria-like bearing single conidiophores over nearly the entire length of the stalk. Conidiophores various in structure, one-stage-branched, one- to two-, rarely three-stage-branched or irregularly branched with penicilli interwoven and phialides forming a hymenium-like structure. Phialides in clusters, flask-shaped, when typical, consisting of a long, cylindrical base and a short narrowed neck, commonly parallel, in fresh isolates of a few species at first sometimes small. Sclerotia and teleomorph absent.

Type species: P. duclauxii Delacroix

- P. clavigerum Demelius
- P. coprophilum (Berk & Curt.) Seifert & Samson (= P. concentricum Samson et al.)
- P. duclauxii Delacroix
- P. isariiforme Stolk & Meyer
- P. vulpinum (Cooke & Massee) Seifert & Samson (= P. claviforme Bain.)

The section Coremigenum is characterized by well-developed synnemata and long, parallelly arranged phialides. The structure and length of the synnemata differ markedly in the different species. After a number of transfers the synnemata show a tendency to become smaller. In P. vulpinum (= P. claviforme) they are club-shaped, composed of a distinct sterile stalk and a blue-green capitulum. The penicilli are irregularly branched and closely interwoven, and their parallel arranged phialides form a hymenium-like layer. P. clavigerum, P. duclauxii and P. isariiforme produce Isaria-like synnemata and their stalks bear conidiophores over nearly their entire length, though more concentrated in terminal and subterminal portions. P. duclauxii differs from other species in producing strictly one-stage-branched conidiophores, which form short tangled chains. In contrast, the conidiophores of P. clavigerum and P. isariiforme are one-to two-stage-branched, though sometimes more irregular and their conida form columns. The synnemata of P. coprophilum are very small and consist of a short stipe and a green conidial head with the conidiophores retaining their individuality. In fresh isolates the phialides are relatively small and the branches adpressed to slightly divergent. Such structures are suggestive of those of the section Ramosum. However, after a number of transfers the phialides become larger, the branches more adpressed, thus becoming more characteristic of Coremigenum. A number of species typically producing well-defined synnemata have been placed in others sections. P. italicum produces well-developed synnemata, sometimes with stalks partly submerged in the agar. It is placed in section Penicillium, because of its large phialides with a wide neck.

The synnematous species P. hordei and P. glandicola differ from species in section Coremigenum in producing looser, sometimes featherly synnemata. They are classified in the series Gladiolii (sect. Penicillium) because of the structure of the conidiophores and phialides. P. griseofulvum producing small loose synnemata, is placed in the section Ramosum, because of its loose, partly divergent conidiophores and its very small phialides. P. dendriticum and P. pseudostromaticum belong in the section Biverticillium, because of their lanceolate phialides.

I. SECTION BIVERTICILLIUM (Dierckx)Stolk & Samson, comb. nov

Subgenus Biverticillium Dierckx apud Biourge - Cellule 33: 31, 1923 (Basionym).
Section Biverticillatum-Symmetrica Thom -- Penicillia. 436, 1930

Conidiophores mono and/or synnematous, usually with one of these features predominating, depending on the species; one- to occasionally two-stage branched with a verticil of numerous symmetrical adpressed rarely divergent metulae. The lengths of metulae and phialides are approximate equal. Phialides lanceolate. Conidia usually ellipsoidal or fusiform. Sclerotia present in a few species.

Teleomorph: Talaromyces, Trichocoma or absent

Type species: P. minioluteum Dierckx

1. Series Miniolutea Pitt
2. Series Verruculosa Stolk & Samson

Key to the series

1. Phialides with a swollen base and sharply tapered neck, conidiphores usually one-stage-branched, teleomorph absent ... Series Verruculosa

   Phialides mostly typically lanceolate (acerose), long with a narrow base and tapering slowly to a thin neck. Conidiophores one-stage-branched, sometimes two to three stage-branched. Talaromyces or Trichocoma teleomorph sometimes present ... Series Mineolutea

Dierckx described the subgenus Biverticillium, which he typified with P. minioluteum Dierckx. However, in agreement with our present scheme, we prefer to regard Biverticillium as a section of Penicillium. The section is characterized by lanceolate (acerose) phialide and one-stage-branched symmetrical penicilli. In some species lower branches may also occur. Unfortunately we have not yet completed the study of this group and consequently the proposed division of this section is provisional.

I. 1. Series Miniolutea Pitt

Series Miniolutea Pitt - Genus Penicillium: 419. 1979, p.p.
Series Islandicum Pitt - Genus Penicillium: 445. 1979.

Conidiophores one-stage-branched, sometimes two to three staged branched. Phialides lanceolate, usually long with a narrow base and tapering slowly into a thin conidium bearing tube. Sclerotia in some species present.

Teleomorph: Talaromyces, Trichocoma or absent.

Type species: P. mineoluteum Dierxck

P. aurantiacum Miller et al.
P. brunneum Udagawa
P. dendriticum Pitt
P. diversum Raper & Thom
P. erythromellis Hocking
P. funiculosum Thom
P. islandicum Sopp
P. lignorum Stolk
P. minioluteum Dierckx
P. piceum Raper & Fennell
P. pinophilum Hedgecock
P. primulinum Pitt
P. purpurogenum Stoll
P. pseudostromaticum Hodges et al.
P. rugulosum Thom
P. tardum Thom
P. varians G. Smith
P. variabile Sopp

Anamorphs of Talaromyces

P. assiutense Samson & Abdel-Fattah (teleomorph: T. assiutensis Samson & Abdel-Fattah)
P. dangeardii Pitt (teleomorph: T. flavus (Klöcker) Stolk & Samson)
P. emmonsii Pitt (teleomorph: T. stipitatus (Thom) C.R. Benjamin)
P. galapagense Samson & Mahoney (teleomorph: T. galapagensis Samson & Mahoney)
P. intermedium Stolk & Samson (teleomorph: T. intermedius (Apinis) Stolk & Samson)
P. kloeckeri Pitt (teleomorph: T. wortmannii (Klöcker)C. R. Benjamin)

P. lehmanii Pitt (teleomorph: T. trachyspermus (Shear)Stolk & Samson)
P. luteum Zukal (teleomorph: T. luteus (Zukal) C.R. Benjamin)
P. mimosinum Hocking (teleomorph: T. mimosium Hocking)
P. ohiensis Pitt (teleomorph: T. ohiense Huang & Schmitt)
P. panasenkoii Pitt (teleomorph: T. panasenkoii Pitt)
P. sphaerum Pitt (teleomorph: T. rotundus (Raper & Fennell) C. R. Benjamin)
P. spirrillum Pitt (teleomorph: T. helicus (Raper & Fennell) C.R. Benjamin)
P. udagawae Stolk & Samson (teleomorph: T. udagawae Stolk & Samson)

Anamorph of Trichocoma paradoxa Jungh.

P. trichocomae*(teleomorph: Trichocoma paradoxa Jungh.)

I.2. Series Verruculosa Stolk & Samson, ser. nov

Series in sectione Biverticillium; conidiophora univerticillata, metulis divergentibus, phialides inflatae ad basem, collibus angustis. Teleomorphosis abest.

Species typica: P. verruculosum Peyronel

Conidiophores one-stage-branched, with (2-)4-10, divergent metulae. Phialides with a swollen base and a sharply tapered, thin conidium-bearing tube. Sclerotia rarely present.

Teleomorph: absent

Type species: P. verruculosum Peyronel

P. loliense Pitt
P. resinae Qi Tong
P. verruculosum Peyronel

J. SECTION GEOSMITHIA (Pitt) Stolk & Samson, comb. nov

Geosmithia Pitt - Can. J. Bot 57: 2021-2030, 1979 (basionym)

P. pallidum-series Raper & Thom - Manual Penicillia: 458, 1949

Conidiophores one- to three-stage-branched, with all elements adpressed and roughened. Phialides cylindrical, without or with a short, abruptly narrowed neck, smooth or roughened. Conidia cylindrical or strongly ellipsoidal, smooth-walled, in long chains. Colonies developing not true green colours.

Teleomorph: Talaromyces or absent

Type species: P. lavendulum Raper & Fennell

---

* new species, will be described elsewhere.

P. argillaceum Stolk & al.
P. cylindrosporum G. Smith
P. emersonii Stolk (teleomorph: T. emersonii Stolk)
P. lavendulum Raper & Fennell
P. namyslowskii Zaleski
P. putterillii Thom
P. swiftii (teleomorph: T. bacillisporus (Swift) C. R. Benjamin)

Pitt (1979) proposed the new genus Geosmithia to include the species described by Raper & Thom (1949) in the P. pallidum series of Penicillium. He separated Geosmithia from Penicillium by following characters: formation of cylindrical conidia, rough-walled cylindrical phialides without neck and colony colours which are not green. However, cylindrical conidia occur in Penicillium species as well (Section Cylindrosporum Pitt). Phialide necks in Geosmithia, range from inconspicuous (e.g. P. pallidum, P. lavendulum) to distinct, but short (e.g. P. emersonii). Moreover the structures of the penicilli agrees very well with those of Penicillium. Consequently we regard Geosmithia as only a section of Penicillium.

Acknowledgements

The authors are grateful to Dr. J. A. Stalpers and Mr. K. A. Seifert for their help with the manuscript.

REFERENCES

DALE, E. 1912. Notes on three new species of Penicillium, P. echinatum, P. flexuosum and P. sacculum. Ann. Mycol. 24: 137.

FRISVAD, J. 1984. Expressions of secondary metabolism as fundamental charcters in Penicillium taxonomy. In "Toxigenic fungi - their toxins and health hazards". H. Kurata and Y. Ueno, eds. Amsterdam. Elsevier.

PITT, J. I. 1979. The genus Penicillium and its teleomorphic states Eupenicillium and Talaromyces. London: Academic press.

RAPER, K. B. & THOM, C. 1949. A manual of Penicillia. Baltimore: Williams & Wilkins.

SAMSON, R.A., STOLK A.C. & HADLOK. R. 1976. Revision of subsection Fasciculata of Penicillium and some allied species. Stud. Mycol. 11: 1-47.

SAMSON, R.A., ECKHARDT, C. & ORTH R. 1977 a. The taxonomy of Penicillium species from fermented cheeses. Antonie van Leeuwenhoek 43: 341-350.

SAMSON, R.A., HADLOK, R. & STOLK A.C. 1977 b. A taxonomic study of the Penicillium chrysogenum-series. Antonie van Leeuwenhoek 43: 169-175.

SCOTT, De B. & STOLK, A. C. 1967. Studies on the genus Eupenicillium Ludwig I. Perfect states of some Penicillia. Antonie van Leeuwenhoek33: 297-314.

SMITH, G. 1961. Some new and intersting species of micro-fungi. Trans. Br. mycol.Soc. 44: 42-50.

STOLK, A: C. 1968. Studies on the genus Eupenicillium

Ludwig III. Four new species of Eupenicillium. Antonie van Leeuwenhoek 34: 37-53.

STOLK, A.C. & SAMSON, R.A. 1971. Studies on Talaromyces and related genera I. Hamigera, gen. nov. and Byssochlamys. Persoonia 6. 341-357.

STOLK, A.C. & SAMSON, R.A. 1972. Studies on Talaromyces and related genera II. The genus Talaromyces. Stud. Mycol. Baarn 2, 1-40

STOLK, A.C. & SAMSON, R.A. 1983. The ascomycete genus Eupenicillium and related Penicillium anamorphs. Stud. Mycol., Baarn 23: 1-149.

STOLK, A. C. & SCOTT, De B. 1967. Studies on the genus Eupenicillium Ludwig I. Taxonomy and nomenclature of Penicillia in relation to their sclerotial ascocarpic states. Persoonia 4: 391-405.

THOM, C. 1910. The Penicillia. Baltimore: Williams & Wilkins.

DIALOGUE FOLLOWING DR. SAMSON'S PRESENTATION OF THE PAPER BY STOLK AND SAMSON

MALLOCH: One of the things that struck me as an ascomycete taxonomist was seeing two Talaromyces anamorphs classified with Eupenicillium anamorphs. Does this mean that your system is purely anatomical or is it an indication that these two species may not belong in Talaromyces?

SAMSON: That's difficult. Some Talaromyces also have Paecilomyces anamorphs, so there could be three anamorph types in this genus. The genus Talaromyces is certainly heterogeneous in this respect.

GAMS: There are two approaches to taxonomy. One is a classification based on your taxonomic insight into the relationships of the organisms. The second is to make it convenient for identification. I would prefer to see anamorphs associated with Eupenicillium in a different subgenus from those associated with Talaromyces.

SAMSON: We have monographs of these teleomorph genera, so you can use these if you've got the teleomorphs. What we have done here is to integrate the anamorphs in a single scheme.

MALLOCH: So this is a purely anatomical system and not one reflecting biological reality.

GAMS: I am disturbed by some of your lumping. Although I acknowledge the existence of intermediates, we must learn to cope with these interfaces. It does not necessarily mean you have to lump.

SAMSON: I believe that recognizing interfaces is difficult even for an expert.

CHRISTENSEN: Nature may recognize interfaces because sometimes they may be ecologically specific.

SAMSON: Maybe in soil, but there are so many isolates from foods that the interfaces are much more difficult to define.

GAMS: If you lump too much you lose information.

SAMSON: I agree.

GAMS: You would say if you have two different anamorphs with indistinguishable teleomorphs that these must be the same species. I wouldn't always agree.

UDAGAWA: We have often seen single terminal phialides on unbranched conidiophores in cultures of *Eupenicillium* and even *Talaromyces*. Is this what you are calling section *Torulomyces*?

SAMSON: No, not neceassarily. You also sometimes see reduced penicilli in *Talaromyces*, especially in *T. intermedius*, but I would not place these in section *Torulomyces*, because we believe that the normal penicillus is biverticillate.

PITT: The question of *Penicillium palitans* and *P. verrucosum* var. *melanochlorum*. How do you distinguish your variety from *P. palitans*, which is a well typified name which would have priority if it was recognized as different from *P. viridicatum*?

SAMSON: The notes on the type isolate of *P. palitans* made by Dr. van Beyma and Miss Stolk when the strain originally arrived at CBS do not fit the description by Raper and Thom. At that time we felt that *P. palitans* might be a *nomen dubium*.

Chapter 4:

ADVANCES IN ASPERGILLUS TAXONOMY

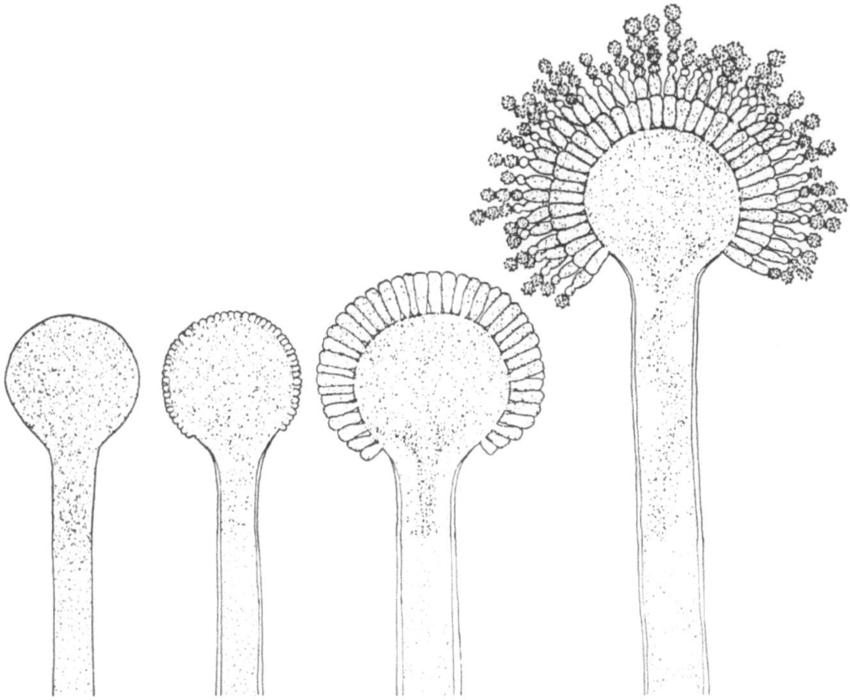

ASPERGILLUS: AN OVERVIEW

Martha CHRISTENSEN and Dorothy E. TUTHILL

University of Wyoming, Laramie, Wyoming, USA

SUMMARY

Since 1965, approximately 80 species in Aspergillus have been described as new and thousands of publications have contributed to our knowledge of biochemical and ultrastructural features. The group concept, developed in the monographs of 1926, 1945 and 1965 has been a tremendous aid in species identification. Sixteen of Raper and Fennell's 18 groups have been validated by the contributions of the last 20 years. Only the A. ornatus and A. wentii groups appear to be heterogenous. An analysis of the Aspergilli listed in 62 recent soil surveys has revealed highest absolute and relative numbers of species in the subtropics, particularly in cultivated soils, saline soils and subtropical deciduous forest soils. In temperate regions, highest numbers of species have been reported from desert soils and certain grassland soils. Unusual species in the soils of western US intermountain basins may have evolved following early migration, then isolation and specification through the Pleistocene.

INTRODUCTION

Three marvelous books on Aspergillus have been published since Link's definition of the genus in 1809, each in its time providing a compehensive review of the known species and ordering them in a reasonable, phenetic scheme. Thom and Church (1926) organized 69 species into 13 groups with just a few species, that did not fit well into the designated groups. That study was stimulated by the need for a reliable taxonomic treatment among industrialists and pathologists. Practicality continued to be a feature in the succeeding monographs. Thom and Raper (1945) introduced Czapek agar as a standard culture medium and provided concise accounts for 77 species in 14 groups. Raper and Fennell's monograph (1965) contains detailed descriptions of 132 species in 18 groups, with all taxa grown on at least two standard media. This monograph was more than a reworking, in that it contained descriptions for 28 new species, many of them isolated from nature by Raper, Fennell, or Kwon-Chung.

Since 1965, approximately 670 publications per year have added to our knowledge of Aspergillus. A high proportion of those studies have treated biochemical aspects of single species or groups of species, e.g. mycotoxin production, enzymatic capacities, and induction of biosyntheses; even a casual perusal of titles impresses the classical taxonomist with 1) a curiosity about the degree of concordance between biochemical and morphological features; and 2) a realization of the absolute requirement for a single, reliable taxonomic frame of reference if we are to assimilate new knowledgement.

The book on Aspergillus edited by Smith and Pateman (1977) contains a wealth of information on all aspects of the biology of Aspergillus species and can only be viewed as an indispensable guide to future investigation. In particular, Dorothy Fennell in the chapter on taxonomy carefully reviews the contributions of molecular biology to Aspergillus taxonomy and discusses its potential as a taxonomic tool. Her optimism is evident in the comment, "Despite the present limitations, it appears possible that, in far fewer that the 247 years that Aspergillus has been known, chemotaxonomic identification of Aspergilli will be a reality and ..... through a combination of conventional and molecular approaches...phylogenetic relationships..... will have been mapped". But she mentions, too, that for work-a-day mycologists and non-mycologists, morphology of the asexual state will continue to be the useful criterion in species identification.

Samson (1979) compiled the names of all Aspergillus species described since 1965, provided brief descriptions of the 32 new species which he accepted, and reviewed and keyed the genera to which ascomycetous states have been assigned. Our own scrutiny of his list has revaled that 64 species in Aspergillus were discovered and described between 1965 and 1979. The total number of validly described species from 1965 through 1984, as listed in the Index of Fungi, is approximately 80. The tabulation in Figure 1 does not include species in Raperia and Stilbothamnium. An increase in varieties as well as species has brought the total numer of taxa in Aspergillus to approximately 276. Thus, in the 20 years since Raper and Fennell (1965) numbers of species and total taxa have increased 61% and 84%, respectively.

Comparative studies and descriptions of new species in Aspergillus have been contributed by numerous workers worldwide since 1965, (e.g. Blaser, 1976; Christensen, 1981, 1982; Horie, 1980; Murakami, 1971; Samson, 1979). The new species have been soil and litter isolates from, primarily, India, Europe and the Soviet Union, Egypt and Syria, tropical Africa, Japan, and intermountain basins in western North America. Domsch et al. (1980) have succinctly characterized the genus Aspergillus and have provided keys to 26 commonly encountered species in eight groups. Their species accounts contain an abundance of biochemical and ecological data. Ascomycete states, now known for about 65 species assigned to 10 or 11 genera, have been studied in detail since 1955, primarily by mycologists in Canada, Germany, India and Japan (e.g., Blaser, 1976; Horie, 1980; Malloch and Cain, 1972; Rajendran and Muthappa, 1980; Subramanian, 1972; Tewari, 1985). An excellent synopsis of Aspergillus teleomorphs can be found in Fennell (1977).

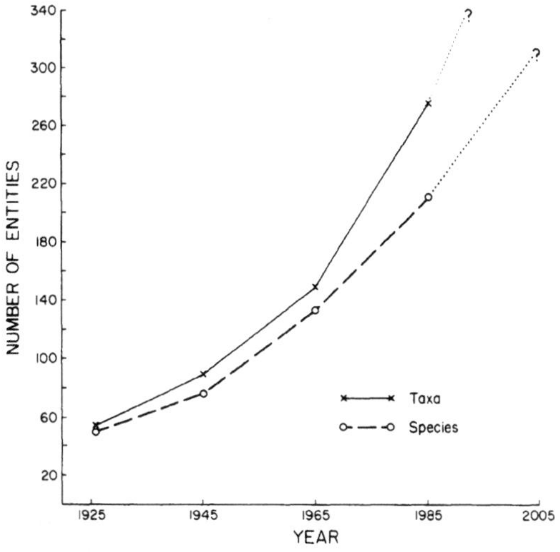

Fig. 1. Increment in number of species in Aspergillus and species plus varieties, beginning with Thom and Church (1926).

DEFINITION OF THE GROUPS AND SPECIES IDENTIFICATION

The group concept in Aspergillus, developed by Thom and Church (1926), evolved from the perception that, when many representatives of many species are examined, patterns of species displaying correlated morphological and cultural features are discernible. Thom, Raper, and Fennell were early-day pheneticists in that they identified groups of species, but with a "scattering of . ... questionable" forms (Thom and Church, 1926). The groups in Aspergillus are evident, accepted, and are an extremely useful aid as "... convenient slots for fast characterization of strains". Further, the continuing discovery that there is a correspondence between anamorph group and teleomorph group indicates that conidial structures may also evolve and hence merit the attention of cladists!

The features that delineate the groups are in part morphological, in part cultural and gradually, a body of definitive physological features is accumulating. However, the precise combination of features that "works best" varies somewhat from group to group. In our experience, the essential features to ascertain first in identifying and Aspergillus are: spore colour, head shape, length and unusual features of conidiophores, orientation and dimensions of metulae, phialide density per metula, and size, shape, and wall texture of conidia.

Our further comments on delineation and arrangements of species and the application of chemotaxonomic and numerical techniques are organized below under 3 subheadings.

Clarity and current status of Aspergillus groups

Five of the six strictly uniseriate groups of Raper and Fennell (1965), i.e. the A. clavatus, A, glaucus, A. cervinus, A. restrictus and A. fumigatus groups (Fig. 2), appear to be

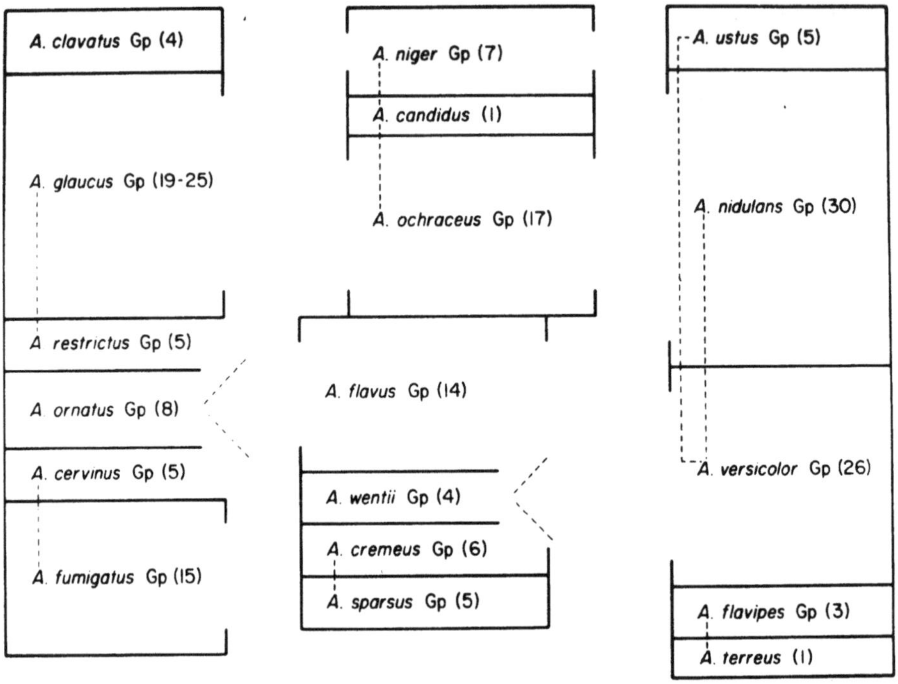

Fig. 2. The 18 groups of Aspergillus, with approximate numbers of species shown in parentheses and major related groups linked by dashed lines. The A. ornatus and A. wentii groups appear to be heterogenous.

well-defined in that each includes species that are similar morphologically. The A. glaucus group, recently revised by Blaser (1976), now contains about 25 species and is the largest of the strictly uniseriate groups. The existence of naturally occurring forms of A. fumigatus with tan - and buff-coloured conidia (Rai et al., 1974) may warrant expansion of our group concept. The A. ornatus group, described by Raper and Fennell (1965) as ".... an admittedly artificial grouping of species with uniseriate sterigmata and more or less clavate vesicles", clearly is heterogeneous in view of the recent reports that its species have teleomorphs in four different genera (Fennell, 1977; Samson, 1979).

The A. ochraceus, A. niger and A. candidus groups also appear to be natural assemblages (Hesseltine et al., 1972; Raper and Fennell, 1965) and, further, the 24-27 species in these groups somewhat resemble one another in vesicle size and metula-phialide configuration. Diminutive biseriate heads and elongating metulae in certain members in these groups are unifying features, but not in others, to our knowledge. The A. niger and A. ochraceus groups have been reviewed recently by Al-Musallam (1980) and Christensen (1982), respectively. Al-Musallam (1980) on the basis of quantitative features and multivariate analysis reduced six of the species in the A. niger-group to varietal status and added one new species.

The A. flavus group is reasonably well-defined morphologically, but there is an annoying overlap among the "domesticated species" (Christensen, 1981), i.e., A. flavus, A. oryzae, A. parasiticus, A. toxicarius and A. sojae. Murakami and his associates distinguish these five species by biochemical and ultrastructural features (e.g., Murakami, 1971; Murakami et al., 1982; Nasuno, 1974; Ushijima et al., 1982) and have recently included the absorption spectra of pigments formed on liquid Czapek medium (Yuasa et al., 1982). Christensen (1981) and Kozakiewicz (1982) also have reviewed the distinguishing cultural and morphological features for A. flavus, A. parasiticus and A. sojae. Tokunaga et al. (1973) and Kozakiewicz (1982) reported apparently basic differences in the pattern of development of metulae and phialides in the A. niger group and in A. flavus. Others have noted uniseriate diminutive heads in the A. flavus group, conidia mostly larger that those in the A. ochraceus group, and metulae with relatively few phialides (Christensen, 1981, 1982). With transfer of A. thomii to the A. flavus group, it will embrace about 15 species.

The A. wentii group, like the A. ornatus group, is heterogeneous. Because of its colour and robustness, Thom and Church (1926) compared A. wentii to A. ochraceus and A. tamarii and a recent analysis of mitochondrial DNA fragments again has linked it to A. tamarii and A. oryzae (Kozlowski and Stepien, 1982).

The A. cremeus, A. sparsus and A. ustus groups are well-defined and possibly allied. The five species in the A. ustus group, however, are strictly biseriate. Thirteen of the 16 species in these groups are known only from tropical or subtropical localities (Raper and Fennell, 1965; Samson, 1979).

In the four remaining strictly biseriate groups, the A. nidulans group is well-defined and is the largest group in Aspergillus, now having about 30 species, in contrast to six in 1945 (Christensen and States, 1982). Many of the recently described species have come from desert soils and it is likely that others are yet to be discovered. The A. versicolor group also is large (about 26 species) and according to Raper and Fennell (1965) ".... includes species that appear quite unlike ship to other species that obviously belong here". The group contains species that bridge toward A. nidulans, A. ustus and perhaps the A. flavipes group; at least 10 species appear to be solely or primarily tropical. Finally, the species in the A. flavipes and A. terreus groups, all strictly biseriate and with pink-tinged conidia, are similar to one another but differ markedly from members in the other 16 groups. The two core species are commonly encountered worldwide.

The potential contribution of biochemical data

Fennell (1977) described the detection and tabulation of biochemical characteristics for species in Aspergillus as a development which is bringing us to the "threshold of major advances" in Aspergillus taxonomy. Now, 8 years later, a tremendous expansion in knowledge of the biochemistry of Aspergillus continues and major advances are anxiously awaited. Since Fennell (1977) has thoroughly reviewed the topic and several biochemical/ultrastructural techniques will be discussed at this workshop, comments will be brief. The techniques which appear to be valuable as tools in recognition of Aspergilli are: scanning electron microscopy, organic acid and nigeran production, mycotoxin profiles, enzymes, long-chain fatty acids, pigment production and colour change with age, gel electrophoretic protein patterns, esterase and phosphatase in particular (Nealson and Garber, 1967), and pyrolysis gas liquid chromatography and chromatography mass spectrometry (Bobbitt and Nordin, 1978; Christensen, 1981; Fennell, 1977; Horie, 1980; Kozakiewicz, 1978; Lee et al., 1977; Seifert, 1982; Soderstrom and Frisvad, 1984; Stretton et al., 1976; Ushijima et al., 1982). Nuclear and mitochondrial DNA analysis will be helpful in ascertaining phylogenetic relatedness (Fennell, 1977; Kozlowski and Stepien, 1982).

Numerical taxonomy

With proper use, numerical phenetics and multivariate analysis can be powerful and convincing tools in biosystematics (Sheard, 1978), especially in groups that have relatively brief histories of natural selection. In *Aspergillus*, interspecific overlap and intraspecific breadth is evident in, for example, *A. flavus, A. oryzae, A. parasiticus, A. toxicarius* and *A. sojae*; *A. fumigatus*; and *A. niger* and its allies, i.e. in species that are principal forms in our manipulated landscapes. To date, two such groups have been subjected to analysis by numerical taxonomic methods (Al-Musallam, 1980; Christensen, 1981; Murakami, 1971).

Taxa in *Penicillium* and *Laccaria* have recently been examined by methods which objectively identify patterns of similarity. Soderstrom and Frisvad (1984) examined relationships among nine isolates of *Penicillium crustosum* by pyrolysis gas chromatography coupled with a pattern recognition analysis. The patterns in general agreed with prior groupings based on mycotoxin profiles and penicillus morphologies, and indicated to the investigators that there probably was not biological justification for treatment of the strains as representatives of a single species. Mueller (1985), in a phenetic analysis of the basidiomycete genus, *Laccaria*, identified ten distinct clusters by two different multivariate analysiss techniques; eight were given specific rank and two were treated as intraspecific taxa. Both Mueller (1985) and Sheard (1978) pointed out that development of reliable phenetic patterns requires the use of a large number of characters.

Clearly the beauty of numerical techniques is that they can objectively reveal similarities among individual units in the face of great morphological and chemical variation (Sheard, 1978). The methods, however, do not tell us precisely where to delineate taxa or what they call constellations. Systematists using numerical methods must assiduously practice good judgement, and must consider it obligatory to relate proposed new classification schemes to existing systems.

ECOLOGICAL AND BIOGEOGRAPHICAL CONSIDERATIONS

Most of the commonly encountered species in *Aspergillus* (all treated by Domsch et al., 1980) are both geographically widely distributed and capable of primary colonization of organic substrates (Ajello, 1980; Christensen, 1981, 1982; Domsch et al., 1980; Flannigan, 1969; Hering, 1965; Lacey, 1974). Those species are:

A. amstelodami	*A. melleus*	*A. ruber*
A. candidus	*A. nidulans*	*A. rugulosus*
A. carneus	*A. niger*	*A. sclerotiorum*
A. chevalieri	*A. niveus*	*A. sydowii*
A. clavatus	*A. ochraceus*	*A. tamarii*
A. japonicus	*A. restrictus*	*A. terreus*
A. flavipes	*A. oryzae*	*A. ustus*
A. flavus	*A. parasiticus*	*A. versicolor*
A. fumigatus	*A. repens*	*A. wentii*

An intensive survey of soil microfungi at 22 nonagricultural localities in Ohio, USA, yielded six Aspergilli among 180 identified species (Huang and Schmitt, 1975). All six are in the assemblage that we here are viewing as widely distributed and opportunistic. All of the eight Aspergilli listed by Christensen (1980) as prominent in two or more of 32 temperate region surveys in native communities also are in the widely distributed/opportunistic category. The five Aspergilli recorded from soils of montane grasslands in Ceylon among 62 species (Mueller-Dombois and Perera, 1971) are in this group; however, the Aspergilli in nearby Indian grasslands are more diverse (Dwivedi, 1966; Mishra, 1966).

If we accept 170-200 species in *Aspergillus*, i.e. 132 species from Raper and Fennell (1965) and half of those described since, and this includes 50 widely distributed species, then more than 120 species are either narrowly distributed geographically and ecologically, or have been missed because of inadequate sampling. As we will point out later, there is some evidence in support of the contention that a high proportion of the *Aspergillus* species are restricted in respect to geographic distribution, ecological specificity, or both.

The principal native habitats for those species commonly encountered in temperate regions are desert soils (*A. flavipes, A. fumigatus, A. niger, A. phoenicis, A. niveus, A. terreus, A. ustus*) and grassland soils (*A. alliaceus, A. candidus, A. flavipes, A. fumigatus, A. niveus, A. terreus*) (Christensen, 1980). Up to 20% of the species in US desert and grassland soils are Aspergilli, in contrast to 0-5% in forest soils (Table 1). Several species are known only from temperate region deserts (e.g. *A. leporis, A. spectabilis, A. bridgeri, A. bicolor, A. navahoensis*) (Christensen, 1981, 1982; Christensen et al., 1978; Christensen and States, 1982; Gochenaur, 1970; Ranzoni, 1968; Raper and Fennell, 1965; States, 1978).

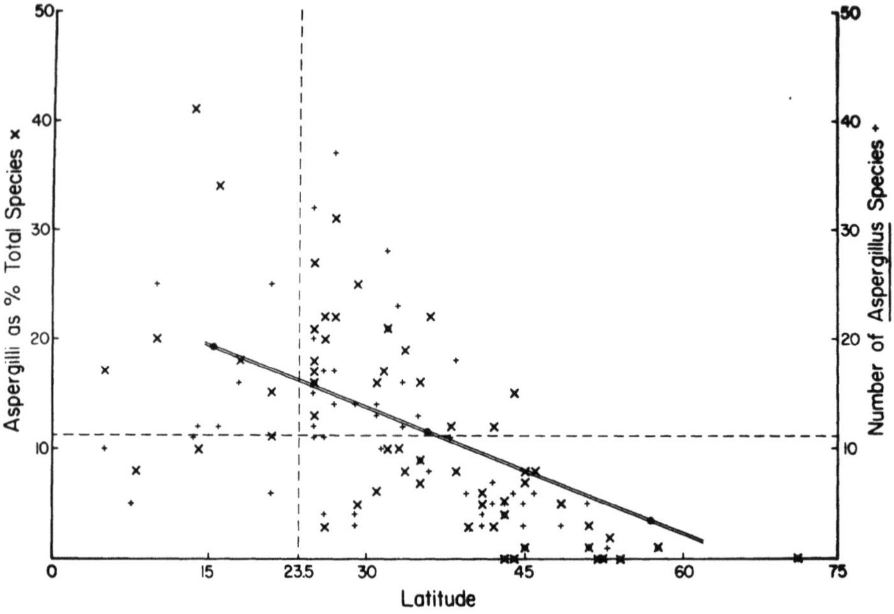

Fig. 3. Prominence of Aspergillus species in relation to latitude, based upon an analysis of 62 soil surveys at latitudes ranging from 3° to 71° (see text and table 1). The solid circles and double line indicate averages in aspergillii as a percent of total species through three succesive latidudinal segments, embracing 21, 21 and 20 surveys, repectively. The horizontal dotted line indicates overall average percent (11.4) and the small symbols (+) indicate absolute members of aspergilli in the 62 surveys.

Three closely related species in the A. cervinus group appear to be restricted to temperate northern hemisphere conifer forest soils and temperate southern hemisphere heath soils (Christensen et al., 1964). A. parvulus has been found only in the northern hemisphere, in the eastern USA, Britain and the Netherlands (Centraalbureau, 1978; Christensen et al., 1964; Commonwealth Mycological Institute, 1982; Gochenaur and Woodwell, 1974; Jeffreys et al., 1953), in the land area most recently split by continental drift. A. kanagawaensis and A. nutans, however, inhabit both hemispheres: A. kanagawaensis has been isolated from Japan, the USSR, Australia, and Wisconsin spodosols; A. nutans also has been isolated from Wisconsin spodosols, and additionally has been reported from a spodosol in Australia and from South Africa (Christensen et al., 1964; Raper and Fennell, 1965; Commonwealth Mycological Institute, 1982).

Many workers have observed that with progression toward the equator, Aspergilli seem to increase in prominence in soils both absolutely and relatively. Thus, Kamal and Bhargava (1972) obtained 37 Aspergillus species from soil collected in a teak forest at Gorakhpur, and Rai et al. (1971) listed 32 species from alkaline, dry soils in northeastern India. In the latter study, 21% of the 152 species isolated were Aspergilli. Those fugures contrast strikingly with six Aspergilli among 180 species (3%) from Ohio soils (Huang and Schmitt, 1975).

Further, it has been observed that Aspergilli often occur abundantly in both cropped and uncropped soils in tropical and subtropical regions.

In an attempt to verify that Aspergillus is predominantly a tropical and subtropical genus, we compared lists of species from 16 soil surveys, worldwide, at locations within 33° of the equator (entries 1-16, Table 1). Aspergillus species per survey ranged from 10 to 37 and total species ranged from 27 to 283 (average Aspergilli, 18.4; average total species, 103). The total number of Aspergilli in our tropical compilation was 76. Of that total, 36 species were recorded in single surveys and eight species were recorded in only two surveys. Approximately 58 species appear to be restricted in their occurrence to tropical or subtropical soils. Among those species are:

A. acanthosporus	A. deflectus	A. lucknowensis
A. aculeatus	A. dimorphicus	A. malodoratus
A. aeneus	A. diversus	A. ochraceoroseus
A. anthodesmis	A. egyptiacus	A. panamensis
A. apicalis	A. ellipticus	A. peyronelii
A. arenarius	A. elongatus	A. pseudodeflectus
A. biplanus	A. flaschentraegeri	A. pulvinus
A. chrysellus	A. funiculosus	A. puniceus
A. clavato-flavus	A. gorakhpurensis	A. purpureus
A. conjunctus	A. helicothrix	A. raperi
A. coremiiformis	A. insulicola	A. robustus
A. corrugatus	A. ivoriensis	A. silvaticus
A. cremeus	A. janus	A. xerophilus
A. cristatus	A. lanosus	A. zonatus
A. crystallinus	A. longivesica	Emericella desertorum

Refs: Al-Musallam (1980), Christensen (1981); Raper and Fennell (1965), Samson (1979), Samson and Mouchacca (1974).

More than 60 Aspergillii are known from Indian soils (Rai et al., 1974; Samson, 1979). Ogbonna and Pugh (1982) reported it to be the most frequently encountered genus in Nigerian soils, with 25 species and 22% of nearly 2000 isolates.

Table 1 indicates numbers of Aspergillus species present and Aspergilli as a percentage of total species in 62 soil surveys. Average number of microfungal species per survey was 96 and average number of Aspergilli was 10.1. Our four primary observations from examination of those data are:

1) Contrary to earlier surmise species in the genus Aspergillus appear to peak in abundance in the subtropics (Fig. 2). Indian flora is particularly rich: Average number of Aspergillus species in the Indian surveys is 19.8, i.e. 21.1% of the total species;

2) At single locations in both tropical and temperate regions, the number of species in cropped land commonly exceeds the number of species in uncultivated land (see entries 8, 11, 21, 22, 24, 25; 13, 14, 15, 30; 43-47 in Table 1.);

Table 1. Occurrence of Aspergillus species in soil in relation to country, vegetation and latitude (Complete citations available upon request).

Citation	Country	Vegetation	Latitude	Aspergilli No.	% Total Spp.
1. Varghese 1972	Malaysia	Forest	3-8°	10	17
17. Mueller-Dombois 1971	Ceylon	Montane grasslands	6-9	5	8
2. Ogbonna 1982	Nigeria	Savanna	10	25	20
3. Gochenaur 1970	Peru	Low desert	12-15	11	41
4. Gochenaur 1970	Peru	High desert	11-17	12	10
18. Nour 1956	Sudan	Mixed	16	12	34
5. Rao 1970	India	Mixed	18	16	18
19. Dutta 1965	India	Cereal crop	21	25	15
6. Rai 1971	India	"Usar"	22-28	32	21
20. Lee 1975	Hawaii, USA	Swamp	21	6	11
7. Moubasher 1970	Egypt	Mixed	22-32	17	17
21. Saksena 1963	India	Cereal crop	25	20	27
8. Dayal 1971	India	Legume crop	25	11	16
22. Dwivedi 1966	India	Grassland	25	15	18
9. Eicker 1974	S. Africa	Savanna	25	16	13
10. Gochenaur 1975	Bahamas	Tree crop	25	12	20
23. Wallace 1978	Florida, USA	Swamp	25-26	4	3
24. Dwivedi 1971	India	Sal forest	25-27	11	22
25. Mehrotra 1972	India	Cropland	26	17	31
26. Kanaujia 1977	India	Mixed	26-28	14	22
11. Kamal 1972	India	Teak forest	27	37	--
27. Ranzoni 1968	S.West. USA	Desert	28-34	13	6
12. Moustafa 1975	Kuwait	Saline	29	14	25
28. Eicker 1969	S. Africa	Tree crop	29	4	5
29. Eicker 1969	S. Africa	Savanna	29	3	5
13. Youssef 1974	Libya	Coastal	30-33	10	16
30. Borut 1960	Israel	Desert	31	14	17
14. Joffe 1966	Israel	Peanut crop	31-33	21	21
15. Joffe 1967	Israel	Citrus crop	32	28	10
31. Miller 1957	Georgia, USA	Mixed	31-35	16	10
16. Sizova 1967	Syria	Unknown	33	23	19
32. Al-Doory 1959	Iraq	Unknown	32-35	12	8
33. Horie 1977	Japan	Peanut crop	35	13	16
34. Warcup 1957	Australia	Cereal crop	35	9	7
35. England 1957	Oklahoma, USA	Grassland	35	9	9
36. Durrell 1960	Nevada, USA	Desert	36	9	22
37. States 1978	Arizona, USA	Native	38-39	18	8
38. Huang 1975	Ohio, USA	Native	39-40	6	3
39. McLennan 1954	Australia	Heathland	38	11	12
40. Scarborough 1970	Colorado, USA	Steppe	41	3	6
41. Gochenaur 1974	New York, USA	Forest	41	4	5
42. Christensen 1982	Wyoming, USA	Native	41-43	5	3
43. Wacha 1979	Iowa, USA	Cropland	42	7	12
44. Tuthill 1985	Iowa, USA	Grassland	43	0	0
45. Gochenaur 1967	Wisconsin, USA	Forest	43	6	4
46. Orpurt 1957	Wisconsin, USA	Grassland	43	4	4
47. Christensen 1962	Wisconsin, USA	Forest	43	5	5
48. Ruscoe 1973	New Zealand	Wet pasture	44	0	0
49. Clarke 1981	S. Dakota, USA	Grassland	44	6	15
50. Christensen 1969	Wisconsin, USA	Forest	44-46	1	1
51. Wicklow 1974	Oregon, USA	Forest	45	4	7
52. Montemartini 1975	Italy	Meadow	45	5	8

53. Luppi-Mosca 1960	Italy	Alpine grass	46	6	8
54. Singh 1976	Canada	Forest	48-49	3	5
55. Morrall 1974	Canada	Forest	49-57	0	0
56. Bissett 1979	Canada	Alpine	51	1	1
57. Brown 1958	Britain	Dune systems	51	5	3
58. Sewell 1959	Britain	Heathland	51-53	0	0
59. Apinis 1958	Britain	Grassland	53	1	2
60. Morall 1968	Canada	Forest	54	0	0
61. Soderstrom 1978	Sweden	Forest	56-61	1	1
62. Flannigan 1974	Alaska, USA	Tundra	71	0	0

3) Heathland soils in Australia contain more Aspergilli than do heatland soils in Britain (entries 39, 57, 58), but the Australian heaths also are at much lower latitudes;

4) Aspergilli account for the highest proportions of species in dry tropical soils (entries 2, 3, 6, 18), and in subtropical cropped lands (10, 14, 21, 25, 26), deciduous forest soils (24) and saline soils (12).

Although there is scant evidence for it now, it is predictable that as quantitative community surveys begin to replace floristic surveys in the tropics and subtropics, habitat specificity patterns will emerge there just as they have in temperate regions (Christensen, 1980).

The preceding introduction has been presented in part to prepare readers for a personal story and an intriguing related hypothesis. The senior author, a succession of graduate students, and about half-a-dozen mycology classes have been isolating and identifying fungi from Wyoming soils for 20 years. In the course of those surveys, we have encountered just five Aspergilli among approximately 183 identified species. Four of the Aspergili were new to science (A. leporis, A. spectabilis, A. bicolor, A. bridgeri) and all of them, along with A. navahoensis and A. campestris (also new, from Arizona and North Dakota, respectively), have come from the soils of isolated intermountain basins. The 183 total species from Wyoming soils is essentially equal to Huang and Schmitt's 180 species from native Ohio soils, but whereas all of the six Ohio Aspergilli were widely distributed, opportunistic species, four of the five Aspergilli from western US basin soils have been locally abundant, undescribed species. Numerous hay, silage, cubed feed, and stored grain samples conveyed to us in the last 15 years by Wyoming agricultural agencies ahve yielded 8 additional Aspergilli: A. candidus, A. clavatus, A. flavus, A. fumigatus, A. glaucus, A. niveus, A. parasiticus and A. terreus.

On the basis of these findings and the history of Quaternary climate in western North America (Mears, 1981), we are postulating the following biogeographical events:

1) Dispersal and establishment of Aspergillus species into mid-latitude regions during warming episodes in the late Tertiary or Pleistocene interglacial periods, or both;

2) Fragmentation and isolation of those mid-latitude populations during subsequent cold episodes, with local migration into intermountain basins and, perhaps, burial in accumulating loess deposits (Tuthill and Christensen, 1985);

3) Adaptation and speciation as a consequence of geographic isolation.

This scenario satisfactorily accounts both for the existence of undescribed species in non-tropical, undisturbed refugia, and for the continuance of "domesticated species" (Christensen, 1981) as primary colonizers in the degrading composts of our manipulated landscape.

In view of: 1) the high proportion of Aspergilli, perhaps 40% of the named species, that are known from single localities and single habitats, and 2) the likelihood that pockets far removed from a subtropical *Aspergillus* "center" may be harbouring recently-evolved species, we herewith strongly recommend Plant Hunting. Carefully conducted, quantitative surveys and examination of isolates fresh from nature probably is as essential to contemporary *Aspergillus* and *Penicillium* systematists as it was angiosperm systematists at the turn of the Century. Especially, it is imperative in our opinion that isolates newly wrested from nature serve as the major data base in our continuing attempt to conjure patterns in the conglomerate of taxa that comprise *Aspergillus*.

REFERENCES

AJELLO, L. 1980. Natural habitats of the fungi that cause pulmonary mycoses. *In* "Medical mycology", H.J. Preusser, ed. Stuttgart: Gustav Fischer. pp. 31-42.

AL-MUSAllAM, A. 1980. Revision of the black *Aspergillus* species. Ph-D Thesis, Utrecht.

BLASER, P. 1976. Taxonomische und physiologische Untersuchungen uber die Gattung *Eurotium* Link ex Fr. Sydowia 28: 1-49.

BOBBITT, T.F., NORDIN, J.H. 1978. Hyphal nigeran as a potential phylogenetic marker for *Aspergillus* and *Penicillium* species. Mycologia 70: 1201-1211.

CENTRAALBUREAU VOOR SCHIMMELCLTURES. 1978: "List of cultures. 29th Ed". Baarn: Centraalbureau voor Schimmelcultures. 388 pp.

CHRISTENSEN, M. 1980. Species diversity and dominance in fungal communities. *In*"The fungal community; its organization and role in the ecosystem", D.T. Wicklow and G.C. Carroll, eds. New York: Marcel Dekker. pp. 201-232.

CHRISTENSEN, M. 1981. A synoptic key and evaluation of species in the *Aspergillus flavus* group. Mycologia 72: 1056-1084.

CHRISTENSEN, M. 1982. The *Aspergillus ochraceus* group: two new species from western soils and a synoptic key. Mycologia 74: 210-225.

CHRISTENSEN, M., FENNELL, D.I. and BACKUS, M.P. 1964. *Aspergillus kanagawaensis* and related species in Wisconsin forest soils. Mycologia 56: 354-362.

CHRISTENSEN, M., RAPER, K.B. and STATES, J.S. 1978. Two new *Aspergillus nidulans* group members from Wyoming soils. Mycologia 70: 332-342.

CHRISTENSEN, M. and STATES, J.S. 1982. *Aspergillus nidulans* group: *Aspergillus navahoensis*, and a revised synoptic key. Mycologia 74: 226-235.

COMMONWEALTH MYCOLOGICAL INSTITUTE. 1982. "Catalogue of the culture collection of the Commonwealth Mycological Institute. 8th Ed". Kew, Surrey: Commonwealth Mycological Institute.

DOMSCH, K.H., GAMS, W. and ANDERSON, T.H. 1980. "Compendium of soil fungi". London: Academic Press.

DWIVEDI, R.S. 1966. Ecology of the soil fungi of some grasslands of Varanasi. II. Distribution of soil mycoflora, Tropical Ecology 7: 84-99.

FENNELL, D.T. 1977. *Aspergillus* taxonomy. *In* "Genetics and Physiology of *Aspergillus*", J.E. Smith and J.A. Pateman, eds. London: Academic Press. pp. 1-21.

FLANNIGAN, B. 1969. Microflora of dried barley grain. Trans. Br. mycol. Soc.. 53: 371-379.

GOCHENAUR, S.E. and WOODWELL, G.M. 1974. The soil microfungi of a chronically irradiated oak-pine forest. Ecology 55: 1004-1016.

HERING, T.F. 1965. Succession of fungi in the litter of a lake District oakwood. Trans. Br. mycol. Soc. 45: 391-408.

HESSELTINE, C.W., VANDEGRAFT, E.E., FENNELL, D.T., SMITH, M.L. and SHOTWELL, O.L. 1972. Aspergilli as ochratoxin producers. Mycologia 64: 539-550.

HORIE, Y. 1980. Ascospore ornamentation and its application to the taxonomic re-evaluation in *Emericella*. Trans. mycol. Soc. Japan 21: 483-493.

HUANG, L.H. and SCHMITT, J.A. 1975. Soil microfungi of central and southern Ohio. Mycotaxon 3: 55-80.

JEFFERYS, E.G., BRIAN, P.W., HEMMING, H.C.and LOWE, D. 1953. Antibiotic production by the microfungi of acid heath soils. J. gen.. Microbiol. 9: 314-341.

KAMAL and BHARGAVA, K.S. 1972. Studies on soil fungi from teak forests of Gorakhpur. II. A contribution to Indian Aspergilli. Mycopathol. Mycol. Appl. 47: 59-72.

KOZAKIEWICZ, Z. 1978. Phialide and conidium development in the Aspergilli. Trans. Br. mycol. Soc. 70: 175-186.

KOZAKIEWICZ, Z. 1982. The identity and typification of *Aspergillus parasiticus*. Mycotaxon 15: 293-305.

KOZLOWSKI, M and STEPIEN, P.P. 1982. Restriction enzyme analysis of mitochondrial DNA of members of the genus *Aspergillus* as an aid in taxonomy. J. gen. Microbiol. 128: 471-476.

LACEY, J. 1974. The microbiology of hay and straw. *In* "Aspergillosis and Farmer's Lung in Man and Animal", R. de Hallar and F. Suter, eds. Hans Huber. pp. 16-26.

LEE, J.D., DART, R.K. and STRETTON, R.J. 1977. Computerized classification of *Aspergillus*. Trans.Br. mycol. Soc. 69: 137-141.

MALLOCH, D. and CAIN, R.F. 1972. New species and combinations of cleistothecial Ascomycetes. Can. J. Bot. 50: 61-72.

MEARS, B.M Jr. 1981. Periglacial wedges and the late Pleistocene environment of Wyoming's intermontane basins. Quaternary Res. 15: 171-198.

MISHRA, R.R. 1966. Seasonal variation in fungal flora of

grasslands of Varanasi (India). Tropical Ecol. 7: 100-113.
MUELLER, G.M. 1985. Numerical taxonomic analysis on laccaria (Agaricales). Mycologia 77: 121-129.
MUELLER-DOMBOIS, D. and PERERA, M. 1971. Ecological differentiation and soil fungal distribution in the montane grasslands of Ceylon. Ceylon J. Sci. 9: 1-41
MURAKAMI, H. 1971. Classification of the koji mold. J. gen. appl. Microbiol., Tokyo 17: 281-309.
MURAKAMI, H., HAYASHI, K and USHIJIMA, S. 1982. Useful key characters separating three Aspergillus taxa: A. sojae, A. parasiticus, and A. toxicarius. J. gen. appl. Microbiol. 28: 55-60.
NASUNO, S. 1974. Further evidence on differentiation of Aspergillus sojae from Aspergillus oryzae by electrophoretic patterns of cellulase, pectin-lyase, and acid proteinase. Can. J. Microbiol. 20: 413-416.
NEALSON, K.H. and GARBER, E.D. 1967. An electrophoretic survey of esterases, phosphatases, and leucine aminopeptidases in mycelial extracts of species of Aspergillus. Mycologia 59: 330-336.
RAI, J.N., AGARWAL, S.C. and Tewari, J.P. 1971. Fungal microflora of 'Usar' soils of India. J. Indian bot. Soc. 50: 63-74.
RAI, J.N., TEWARI, J.P. and AGARWAL, S.C. 1974. The genus Aspergillus in India. In "Lower Plants of the Indian Subcontinent. Beih. Nova Hedwigia 47: 487-499.
RAJENDRAN, C. and MUTHAPPA, B.N. 1980. Saitoa, a new genus of Plectomycetes. Proc. Indian Acad. Sci. (Plant Sci.) 89: 185-191.
RANZONI, F.V. 1968. Fungi isolated in culture from soils of the Sonoran desert. Mycologia 60: 356-371.
RAPER, K.B. and FENNELL, D.I. 1965. "The genus Aspergillus" Baltimore: Williams and Wilkins.
SAMSON, .A. 1979. A compilation of the Aspergilli described since 1965. Stud. Mycol. 18: 1-38.
SAMSON, R.A. and MOUCHACCA, J. 1974. Some interesting species of Emericella and Aspergillus from Egyptian desert soil. Antonie van Leeuwenhoek 40: 121-131.
SEIFERT, R.M. 1982. Identification of some volatile constituents of Aspergillus clavatus. J. agric. Food Chem. 30: 786-790.
SHEARD, J.W. 1978. The taxonomy of the Ramalina siliquosa species aggregate (lichenized Ascomycetes). Can. J. Bot. 56: 915-938.
SMITH, J.E. and PATEMAN, J.A., Eds. 1977 "Genetics and physiology of Aspergillus". London: Academic Press. 552 pp.
SODERSTROM, B. and FRISVAD, J.C. 1984. Separation of closely related asymmetric Penicillia by pyrolysis gas chromatography and mycotoxin production. Mycologia 76: 408-419.
STATES, J. 1978. Soil fungi of cool-desert plant communities in northern Arizona and southern Utah. J. Arizona-Nevada Acad. Sci. 13: 13-17.
STRETTON, R.J., CAMPBELL, M. and BURNS, T.D. 1976. Pyrolysis gas chromatography as an aid to the identification of Aspergillus species. J. Chromatog. 129: 321-328.
SUBRAMANIAN, C.V. 1972. The perfect states of Aspergillus. Current Sci. 42: 755-761.
TEWARI, J.P. 1985. A new indeterminate stromatal type in

Petromyces. Mycologia 77: 114-120.
THOM, C. and CHURCH; M.B. 1926. "The Aspergilli" Baltimore: Williams & Wilkins.
THOM, C. and RAPER, K.B. 1945. "A Manual of the Aspergilli". Baltimore: Williams and Wilkins.
TOKUNAGA, M., TOKUNAGA, J. and HARADA, K. 1973. Scanning and transmission electron microscopy of sterigma and conidiospore formation in *Aspergillus* group. J. Electron Microscopy 22: 27-38.
TUTHILL, D.E. and CHRISTENSEN; M. 1985. *Aspergillus sepultus*, a new species in the *Aspergillus ochraceus* group. Mycologia 77 (in press).
USHIJIMA, S., HAYASHI, K. and MURAKAMI, H. 1982. The current taxonomic status of *Aspergillus sojae* used in shoyu fermentation. Agric. Biol. Chem. 46: 2365-2367.
YUASA, K., HAYASHI, K. and MIZUNUMA, T. 1982. A new criterion by which to distingish *Aspergillus sojae*, a koji-mold, from related taxa producing echinulate conidia. Agric. Biol. Chem. 46: 1683.1686.

DIALOGUE FOLLOWING DR. CHRISTENSEN'S PAPER

MALLOCH: Have you ever tried to make a distributional map of the different groups of Aspergilli? Are there geographical correlations?

CHRISTENSEN: It's a nice suggestion. I haven't done that. It would be fun to guess. Of the *A. sparsus* and *A. cremeus* groups, thirteen of seventeen species are of tropical origin. Groups dominating in the tropics tend to be different from those found predominantly in temperate zones.

MALLOCH: I'm thinking particularly of the Ascomycetes. The places they occur in the world might give us a clue to evolution in *Aspergillus* teleomorphs and in *Aspergillus*.

CHRISTENSEN: *Emericella nidulans* and related species are found mostly in deserts.

MALLOCH: How about the *A. ochraceus* group?

CHRISTENSEN: I have a feeling that they are also an ancient group.

THE THEORY AND PRACTICE OF DISTINGUISHING SPECIES OF THE ASPERGILLUS FLAVUS GROUP

Maren A. KLICH and John I. PITT

U.S. Department of Agriculture, Southern Regional Research Center, New Orleans, LA, 70179, USA and CSIRO, Division of Food Research, North Ryde NSW, 2113, Australia

SUMMARY

Sixty-two Aspergillus flavus group isolates from the SRRC and FRR culture collections were studied and identified independently by the authors. Strains were grown under standard conditions and observations recorded for a variety of morphological features at various stages of colony development. Aflatoxin production and response on a special A. flavus and A. parasiticus medium were also recorded. A discriminant analysis was performed to further study species relatedness. It was concluded that A. flavus, A. parasiticus, A. sojae, A. oryzae, and A. tamarii could be reliably separated from one another by morphological and microscopic criteria. The most salient distinguishing features were: 1) colony colour; 2) conidial size and roughening; 3) vesicle diameter; 4) conidiophore length; and 5) presence of metulae.

Aflatoxins were indicators of both A. flavus and A. parasiticus, while the presence of G aflatoxins was a distinguishing feature of A. parasiticus. Columnarity of conidial heads was found to be an inconsistent character even within isolate; therefore, A. flavus var. columnaris was not distinct from the species. Isolates previously identified as A. toxicarius were identified as either A. flavus or A. parasiticus.

INTRODUCTION

The Aspergillus flavus group contains several closely related species possessing a high degree of morphological variability within and between taxa. Some members of the group produce aflatoxins, the most potent naturally formed carcinogens. Other members of the group are used in food fermentation. The necessity for accurately distinguishing these species is obvious

Mycologists since Raper and Fennell (1965) have differed in opinion regarding the most important characteristics for both taxonomy of the group and for identification. Murakami, (1971)

Murakami et al. (1982) and Christensen (1981) agreed that roughness of conidial walls, colour of old cultures and conidial diameter are all important in identification. Although Murakami (1971), Murakami et al.(1982) and Raper and Fennell (1965) considered the presence of metulae to be an important key character, Christensen (1981) choose to deemphasize it as an identification character because it is so variable in most members of the group. Kozakiewicz (1982) and Christensen (1981) also considered conidiophore length, vesicle shape and vesicle diameter to be important. Kozakiewicz (1982) unlike the other authors mentioned above, found "no significant difference" in conidial roughening under light microscopy between *A. flavus* Link and *A. parasiticus* Speare. She did, however note distinct differences between these species under SEM.

The objectives of this study were to assess our own methodology in identifying species within the group, thereby pinpointing the features emphasized in our own routine identifications, and to utilize this information gathered from a number of isolates in discriminant analyses. In this way we hoped to provide criteria for the effective identification of unknown isolates.

Several of the taxa in the *A. flavus* group were not considered here because they were either quite rare of very distinctive morphologically. For a recent review of the entire group see Christensen (1981).

MATERIALS AND METHODS

Isolates of *A. flavus* Link, *A. flavus* var. *columnaris* Raper and Fennell, *A. oryzae* (Ahlb.) Cohn, *A. parasiticus* Speare, *A. sojae* Sakguchi and Yamada, *A. tamarii* Kita, and *A. toxicarius* Murakami were examined. The isolates selected for the study were from the collections at SRRC (USDA, New Orleans, LA, USA) and FRR (CSIRO, Food Research, North Ryde, NSW, Australia), and were chosen to represent a wide variety of source material, a wide range of initial isolation dates, and as much morphological variability as possible. Sixty-two cultures, labelled only with numbers, were exchanged and identified. Following discussions on the dispositions made by each of us, we carefully noted the characteristics on which the identifications were finally based.

During colony development, measurements were made based on 5, 10 and 20 day old cultures on Czapek (Cz) and Malt Extract agars (MEA) (Raper and Fennell, 1965), incubated at 25°C in diffuse light. Measurements included colony diameter, conidiophore length and roughening, vesicle diameter, conidial size and roughening, percent biseriate heads, shape of the mature conidial heads, and colour. Conidial and conidiophore ratings were based on a scale of one of three with one being smooth to slightly rough in appearance and three being very rough under 100x bright field illumination.

Aflatoxins B_1, B_2, G_1 and G_2 were measured in fourteen days old cultures on Czapek Yeast Extract agar (CYA; Pitt, 1979), and MEA according to the method of Filtenborg and Frisvad (1980). Cultures negative by this technique were extracted with 10 ml chloroform for one minute, filtered, evaporated to a small volume and spotted on Merck 5 TLC plates. Plates were developed in chloroform: acetone 90:10, and examined under UV light. Aflatoxin responses were rated on a scale of one of to three with three being the greatest response.

The intensity of the orange yellow pigmentation on the reverse of colonies on A. flavus & A. parasiticus-Agar (AFPA; Pitt et al., 1983) were rated on a scale of one to three with three being the most intense response.

To demonstrate some of the possible applications of computers in classification and identification of closely related species, the data for each species were analyzed using a discriminant analysis. This was performed using the Nearest Discriminant Analysis Procedure available from Statistical Analysis Systems Institute, Inc., SAS Circle, P.O. Box 8000, Cary NC 27511-8000, USA. A total of 43 characteristics (variables) were measured on the 62 isolates used in the initial analysis. Subsets of these variables were used in subsequent analyses.

RESULTS AND DISCUSSION

Taxonomic observations

Aspergillus flavus

The quantitative morphological results for the 28 A. flavus isolates are summarized in Table I. Conidia very rarely had a roughening rating higher than one. This seems to be a consistent character for the species. Murakami et al. (1982) referred to the conidia of A. flavus as being roughened but not quite echinulate. Although Kozakiewicz (1982) saw no difference in the conidia of A. flavus and A. parasiticus under the light microscope, we feel that the differences she saw under SEM are detectable under 100x magnification. Conidial diameters were generally within the same range as previous reports. There was no apparent relationship between conidia and size and the length of time the isolates had been in culture collections. However, among the isolates collected within the past three years, those from Australia and New Zealand tended to have smaller conidia than those collected in the United States. Four isolates had maximal conidiophore lengths of less than 400 µm, the lower end of the range for A. flavus reported by Christensen (1981) or Kozakiewicz (1982). Longer conidiophores were not necessarily associated with cultures kept for longer periods in culture collections. The strain examined (FRR 2801) of A. flavus neotype designated by Kozakiwicz (1982) had a maximal conidiophore length of less than 500 µm.

When examined after 5 and 10 days incubation, all but one A. flavus isolate had at least 20% biseriate heads on Cz. Colours of twenty day old cultures ranged from yellow green to olive green. Cultures examined either produced no afltoxins or only B aflatoxins, with one exception, SRRC 26 (NRRL 3161)

Table 1. Microscopic morphological features of isolates examined of the A. flavus group

Species	Roughness	Conidia Diameter (μm)	Maximum Conidiophore Length (μm)	Maximum Vesicle Diameter (μm)	Percent Biseriate Heads
A. flavus	1(2)	3.0-7.0(8.0)	340-1650	22-48	(10)30-100
A. oryzae	1	3.5-8.5	680-5900	24-53	<10-80
A. parasiticus	(1)2-3	(3.5)4.0-7.0	(170)250-470(920)	15-32	0(20)
A. sojae	(1)2-3	4.5-7.0(13)	490-730	17-35	0
A. tamarii	(1)2-3	3.0-8.0(9.5)	(317)470-1090	24-49	<10-100

which produced high levels of both B and G aflatoxins. All isolates tested gave a bright orange yellow resonse of at least 2 on AFPA.

Aspergillis flavus var. columnaris

Attempts to constistently differentiate A. flavus var. columnaris from A. flavus var.flavus were not successfull. Raper and Fennell (1965) distinguished it by consistently columnar heads on MEA, predominantly uniseriate heads, conidiophores generally less than 500 μm, and vesicles less than 40 um. Christensen (1982) found the maximum length of the conidiophores of this variety to be 660 μm, with metulae on 10-25% of the heads. The cultures we examined with consistently columnar heads on MEA had up to 70% biseriate heads. Other A. flavus isolates varied greatly in the percent columnar heads on MEA. Two cultures of the A. flavus neotype (FRR 2807) were examined. One had virtually 100% columnar heads on ME, the other about 80%. We believe, that the character of column formation by A. flavus is variable within and between isolates and, because of this inconsistency, consider the variety to be indistinguishable from the species.

Aspergillus parasiticus

Conidia of A. parasiticus isolates were quite distinct from those of A. flavus. A. parasiticus conidia were consistently roughened (rating of 2-3). Only one of the 10 isolates examined (SRRC 1008) had a rating of one on Cz. The maximum conidial size from our isolates (7 μm) was greater than the 5.5 um previously reported by Christensen (1981) and Raper and Fennell (1965). Conidia of cultures maintained in collections were not larger than those from new isolates. As in A. flavus, conidia of the isolates from New Zealand and Australia tended to be smaller than those from other locations. The maximum conidiophore length was generally between 250 μm and 470 μm, slightly lower than in previous reports. Vesicle diameters were consistently under 35 μm. This feature can be quite useful in separating A. flavus isolates with large vesicles from A. parasiticus. Predominantly uniseriate heads are characteristic of A. parasiticus. Only two of our isolates had any biseriate heads, with a maximum in one of 20% biseriate heads on 5 day old cultures on Cz. Even this isolate, when examined at 10 days had fewer than 20% biseriate heads. Colony colour at 20 days was generally dark green to olive green. All A. parasiticus isolates tested produced both B and G aflatoxins, and all but one had a bright orange yellow response of 3 on AFPA.

Aspergillus toxicarius

Murakami (1971) defined A. toxicarius as a biseriate species. He defined biseriate as "greater than several percent biseriate." A. toxicarius also occasionally produces sclerotia. Several isolates previously identified as A. toxicarius were examined. Two were separate strains of NRRL 2999, one received directly from NRRL (SRRC 77) and another indirectly, from a collection in South Africa (SRRC 255). SRRC 77 was a typical A. parasiticus, with strictly uniseriate heads. SRRC 255 possessed about 10% biseriate heads. The other A. toxicarius isolates fit fairly comfortably into either A. flavus or A. parasiticus. We feel that the occasional occurrence of biseriate heads and the presence of sclerotia do not warrant a separate species.

Three of our isolates had changed drastically in morphology since they were initially described as fresh isolates. Perhaps they change rapidly in morphology as they are transferred. This could only be determined by repeated observations of a group of fresh isolates over time. In the computer analyses (see below) several of the A. toxicarius isolates did seem to represent the border between A. flavus and A. parasiticus. However, they did not fit the description of A. toxicarius.

Aspergillus oryzae

The conidial roughening ratings for A. oryzae were quite similar to those for A. flavus. The conidia, however tended to be larger with a fairly wide variation in size within an isolate. The maximum conidiophore length measured was 5900 μm, and conidiophores were generally longer than those of A. flavus (Table 1). The presence of metulae was an inconsistent character in this species, but at least 10% of the heads were biseriate in all of our isolates. Twenty day old cultures were buff to citrine in colour. This species produced no aflatoxins. The colour rating on the reverse of AFPA plates was 1-2.

Aspergillus sojae

This species is distinct from A. oryzae in that it generally has rough conidia (rating of 2-3), shorter conidiophores, and smaller vesicles. Like A. parasiticus it produces heads with phialides only, but generally has longer conidiophores, and turns light brownish olive to buffy brown in age. Isolates examined produced no aflatoxins, and colour ratings on AFPA were never greater than two.

Aspergillus tamarii

This species generally has roughened conidia (rating 2-3), with thick walls. Conidia tend to be variable in size, but are fairly constant for any one isolate. The conidial size range of our isolates was broader than previously reported, and the maximum conidiophore lengths were shorter than previously reported (Table 1; Christensen, 1981; Raper and Fennell, 1965). Vesicle diameters were similar to those of *A. flavus* and *A. oryzae*. As in *A. oryzae*, the presence of metulae was a variable character in this species, and no aflatoxins were produced. All seven isolates tested produced a brown coloration on the reverse of AFPA.

Agreement in our identifications was greater than either of us had really expected, especially among the *A. sojae* and *A. oryzae* isolates. We found that we used relatively few characters in identification. Generally, colony colour, conidial size and surface texture, and presence of metulae were of primary importance. Other characters, notably gross colony morphology (best reflected in conidiophore length), and vesicle diameter were used secondarily. These characters were selected for use in the discriminant analysis (Groups 2 and below).

Discriminant Analysis

In discriminant analysis a mathematical function (distance function) is developed based on characters of the identified species. The nearest neighbour distance function compares two observations by computing their standardized distance from each other. In this type of analysis, each character carries equal weight. In our data set, for example, the computer program would consider conidiophore roughening to be of the same importance as conidial roughening or colony color. In the analysis, all isolates are identified or reidentified based on the distance function.

The characters used in our demonstration of discriminant analysis were varied to determine the influence they would have on reidentification of known species and identification of others. We used four groupings of characteristics (Table 2). The first included all of the characters. The second and third groups included characters we chose because we felt they were important in identification. The fourth set was developed using a stepwise discriminant analysis, a statistical procedure which selects a set of characters that discriminates best among species. (It should be noted that other sets of characters could, hypothetically, discriminate just as well.)

Selection of important characters using the stepwise discriminant analysis is a useful tool. In our data set, the analysis selected many of the same characters that we have considered important. Interestingly, conidial roughening was

Table 2. Character groupings used in discriminant analysis.

1	2	3	4
All characters	Colour (20-Cz)	Colour (20-Cz)	Colour (20-Cz)
	Conidia	Conidia	Conidia
	Roughening Cz	Roughening Cz	
	" MEA	" MEA	
	Mean diameter Cz	Mean diameter Cz	Mean diameter Cz
	" " MEA	" MEA	
	Max. diameter Cz		Max. diameter Cz
	" " MEA		" " MEA
			Range diameter Cz
	Vesicles	Vesicles	Vesicles
	Mean diameter Cz		
	" " MEA		Mean diameter MEA
	Max. diameter Cz	Max. diameter Cz	
	" " MEA	" " MEA	
			Min. diameter MEA
			Range diameter Cz
	Conidiophores	Conidiophores	Conidiophores
	Mean length MEA		Mean length MEA
	Max. length MEA	Max. length MEA	
	Metulae	Metulae	Metulae
	% 5 days Cz		
	% 5 days MEA		% 5 days MEA
	Dominant form	Dominant form	Dominant form
	10 days Cz	10 days Cz	10 days Cz
	10 days MEA		
			Shape of heads MEA
			G aflatoxins

not among them. It also selected several others, such as the ranges of conidial and vesicle diameters, shape of the conidial heads, and presence of G aflatoxins. This sort of information could be utilized in further taxonomic work, or in developing identification systems.

The number of misidentified or misclassified isolates within each group was small. Group 2 had three misidentifications, and Groups 3 and 4 only one misidentification each. Group 1 had a smaller number of isolates identified, because the procedure eliminated any cultures with missing data, but, of those evaluated, only one was misidentified, excluding isolates of *A. flavus* reidentified as *A. flavus* var. *columnaris* and vice versa.

The "errors" made in reidentification of cultures can be useful to the taxonomist as flags for unusual isolates. One such isolate was SRRC 26, an *A. flavus*. It was the misidentified isolate in both groups 3 and 4 as *A. tamarii* and *A. parasiticus* respectively. It had slightly rougher conidia and longer conidiophores than the majority of the *A. flavus* isolates, and was the only A. flavus isolate that produced G aflatoxins, indicating affinities with both A. tamarii and A. parasiticus. Isolates we felt reasonably matched the description of A. flavus var. columnaris were identified in the data set. The most common error in reidentification in all four groups was misidentification of these isolates as the species and vice versa (Table 3). We feel that this adds weight to our argument that the variety is indistinguishable from the species.

Table 3. Discriminant analysis: problem isolates of A. flavus and A. flavus var. columnaris.

Initial Identification[a]	Group 1	Group 2	Group 3	Group 4
A. flavus	-----	*col*	*fla*	*fla*
A. flavus	*col*	*col*	*fla*	*fla*
A. flavus	*fla*	*col*	*fla*	*fla*
A. flavus	*fla*	*fla*	*tam*	*par*
A. flavus var. *columnaris*	*par*	*fla*	*fla*	*fla*
A. flavus var. *columnaris*	*fla*	*ory*	*col*	*col*
A. flavus var. *columnaris*	-----	*fla*	*fla*	*fla*

a SRRC culture numbers: 1000A, 1009, 42, 26, 1012, 252, 140.

fla = *A. flavus*
col = *A. flavus* var. *columnaris*
par = *A. parasiticum*
tam = *A. tamarii*
ory = *A. oryzae*

In our species list we left as unidentified the isolates which had been identified by others as A. toxicarius, because we had identified tham as either A. flavus or A. parasiticus (Table 4). Four of the six isolates had one "crossing over" to either A. flavus or A. parasiticus. These obviously have some affinities with the other species, but not necessarily in the sense of A. toxicarius. They had some unusual characters other than occasionally biseriate heads which could have led to the "cross over". All three of the A. parasiticus isolates that were misidentified in one group has unusually long conidiophores. One of these also had fairly large vesicles. The A. flavus isolate (SRRC 1063) with a Group 2 identification as A. parasiticus had a conidial roughening rating of one and around 10% metulae heads.

No A. sojae isolates were identified as A. parasiticus or vice versa in either the computer analysis or our own identifications. It does appear to be quite distinct. Two unusual A. sojae isolates were left out of the list of identified species in the computer analysis. They both possessed smooth conidia and one had a few biseriate heads. The isolate with smooth conidia and occasionally biseriate heads was classified as A. flavus in all but the stepwise analysis group which named it correctly. The other isolate was placed in either A. flavus or A. tamarii. We were also able to accurately distinguish

Table 4. Discriminant analysis: Aspergillus toxicarius.

Initial Identification[a]	Group 1	Group 2	Group 3	Group 4
A. par. (tox.)	-----	*A. par*	*A. par*	*A. fla*
A. par (tox.)	-----	*A. par*	*A. par*	*A. par*
A. par (tox.)	*A. par*	*A. fla*	*A. par*	*A. par*
A. par (tox.)	*A. fla*	*A. par*	*A. par*	*A. par*
A. fla (tox.)	*A. fla*	*A. col*	*A. fla*	*A. fla*
A. fla (tox.)	-----	*A. par*	*A. fla*	*A. fla*

a Culture numbers: SRRC 255, SRRC 77, SRRC 1039, FRR 2742, SRRC 1062, SRRC 1063.

A. par (tox.) = *A. parasiticus* (received as *A. toxicarius*)

A. par = *A. parasiticus*

A. fla (tox.) = *A. flavus* (received as *A. toxicarius*)

A. fla = *A. flavus*

A. col = *A. flavus* var. *columnaris*

between A. flavus and A. oryzae. In the computer analyses, only one A. flavus in one character group was misidentified as an A. oryzae. If A. oryzae and A. sojae were conspecific with A. flavus and A. parasiticus respectively, one would have expected many more "cross over" in the computer groupings, and at least some question about some of the isolates in our own identifications.

Functions such as the discriminant analysis can be useful in several ways to the researcher. First, using a stepwise analysis of the characters, potentially useful characters may be discerned. Second, it may be used to pinpoint unusual isolates that may warrant further study. Third, in a well developed system based on many isolates, the discriminant analysis may serve as an identification key. We hope we have also demonstrated that this procedure has limitations like any other research tool. Ultimately, taxonomic decisions must rely on the judgement of the taxonomist.

Acknowledgements

We would like to thank Dr. M.A. Chmielewski of USDA Southern Regional Research Center for her advice and assistance in performing the statistical analyses.

REFERENCES

CHRISTENSEN, M. 1981. A synoptic key and evaluation of species in the Aspergillus flavus group. Mycologia 72: 1056-1084.

FILTENBORG, O. and FRISVAD, J.C. 1980. A simple screening method for toxigenic moulds in pure cultures. Lebensm. Wiss. Technol. 13: 128-130.

KOZAKIEWICZ, M. 1982. The identity and typification of Aspergillus parasiticus. Mycotaxon 15: 293-305.

MURAKAMI, H. 1971. Classification of the koji mold. J. gen. appl. Microbiol., Tokyo 17: 281-309.

MURAKAMI, H., HAYASHI, K and USHIJIMA, S. 1982. Useful key characters separating three *Aspergillus* taxa: *A. sojae*, *A. parasiticus* and *A. toxicarius*. J. gen. appl. Microbiol., Tokyo 28: 55-60.
PITT, J.I. 1979. "The genus *Penicillium* and its Teleomorphic States *Eupenicillium* and *Talaromyces*" London: Academic press.
PITT, J.I., HOCKING, A.D. and GLENN, D.R.. 1983. An improved medium for the detection of *Aspergillus flavus* and *A. parasiticus*. J. appl. Bacteriol. 54: 109-114.
RAPER, K.B. and FENNELL, D.I. 1965. "The genus *Aspergillus*". Baltimore: Williams & Wilkins.

DIALOGUE FOLLOWING DR. KLICH'S PAPER

CHRISTENSEN: If you had assumed *Aspergillus toxicarius* was a legitimate taxon, would the isolates which you looked at have clustered together in the computer analysis? Did the technique you used force them into other recognized species?

KLICH: I don't think so. Some of the *A. toxicarius* isolates so obviously belonged in *A. flavus* and others so obviously in *A. parasiticus*, that I don't think it would have made any difference.

SIMILARITIES BETWEEN THE GENERA STILBOTHAMNIUM AND ASPERGILLUS

Marie-France ROQUEBERT and Jacqueline NICOT

Laboratoire de Cryptogamie, Museum National
d'Histoire Naturelle, 12 Rue Buffon
75005 Paris, France

SUMMARY

Species of the anamorph genus Stilbothamnium are seminicolous synnematous fungi which are similar to Aspergillus. A morphological and ultrastructural study of Stilbothamnium nudipes shows that this species differs from Aspergillus tamarii only by having septate phialides. The relationship between Stilbothamnium and Aspergillus is discussed.

INTRODUCTION

The species of Stilbothamnium belong to a group of Hyphomycetes which can be found in a specific ecological niche in equatorial rain forests of central Africa, South America and Indonesia. They grow on seeds of rotting fruit in litter and are characterized by abundant large simple, branched or thread-like synnemata. They are brightly coloured, greenish or golden yellow, and grow from the centres of the rotting fruits and emerge at the surface of the seed.

Conidiomata sometimes develop together with grey-brown sclerotia on short stipes. In some species, locules develop inside the sclerotia, in which globose, hyaline, bi-valved, crested ascospores develop. Such mature ascomata are related to the family Trichocomaceae and are classified in Penicilliopsis. However, most Stilbothamnium species do not form teleomorphs, and their taxonomy is essentially based on the morphology of the synnematous conidiomata.

Microscopically, conidiomata of Penicilliopsis species show many similarities. Conidiogenous cells are always phialidic and are either arranged in penicilli on short branches, or radiate from terminal vesicles like those of Aspergillus. However, assignment of these seminicolous anamorphs to either Penicillium or Aspergillus is difficult.

Hauman (1936) reviewed the seminicolous species described by previous authors, and supplemented this with studies of many collections from the Belgian Congo. Using the structure and arrangement of phialides he distinguished several morphological types on which he based four genera. Three of the four genera accepted by Hauman (1936) have *Penicillium*-like conidiophores: *Penicilliopsis* Solms-Laubach, *Stilbodendron* Syd. and *Pseudocordyceps* Hauman. The fourth, *Stilbothamnium* Henn., includes species with *Aspergillus*-like conidiophores.

Despite the precise information supplied by Hauman's monograph, the taxonomic position of such tropical seminicolous fungi has been confused and uncertain. R. Heim, J. Nicot and L. Faurel have collected numerous specimens of *Stilbothamnium* in the Central African Republic, and so we have been able to study this problem. Our attention has been particularly drawn by a noteworthy species, *Stilbothaminium nudipes*, which is frequent in Africa.

The history of the genus *Stilbothamnium*

Synnematous moulds growing on seed of exotic fruits were first included in the genus *Penicilliopsis*, the type species of which, *P. clavariaeformis* was observed by Solms-Laubach (1887) on seeds of *Diospyros macrophyla*. It produced a teleomorph and appeared to be similar to *Eurotium*. However, Patouillard (1892) collected *P. dybowskii* on seeds of *Ficus* from the French Congo, which he placed in *Penicilliopsis* although he did not find ascospores. In his description he noted that the anamorph was of the *Sterigmatocystis* type. The synnemata were composed of a unicellular axis surrounded by closely interwoven hyphae. The main stipe ended with a yellowish, globose vesiculate head. In old specimens the axial cell lost its thin hyphal cover. Evidently this new *Penicilliopsis* species did not possess the *Penicillium*-like morphology of the first species. The author did not draw attention to the organizational differences of these two *Penicilliopsis* anamorphs. Patouillard (1897) placed in *P. clavariaeformis* some specimens collected in Java on seeds of *Diospyros* and *Leucoxylon buxifolium* seeds. On one he observed, next to a synnematous form, a mononematous fungus with small clusters of raised stipes 1 to 2 mm high, terminated by a spherule. The stipes were unicellular, cylindrical, of granular texture, and inflated at the top. On each conidial heads identical to those of the synnemata were formed. The author explained this *Sterigmatocystis* form as a modification of the synnematous habit. He did not mention the sexual form or sclerotia.

Hennings (1897) segregated these anamorphs, creating the genus *Stilbothamnium* for a stilbaceous fungus from Togo with "cylindrical, ramified coremia bearing simple thread-like ramuli ending with conidiophore heads; cylindrical basidia monospored and clavulate; unicellular conidia, subglobose, inflated on top." The type species is *Stilbothamnium togoense* (Fig.1). The author did not mention the presence of a teleomorph, so *Stilbothamnium* was clearly applied only to an anamorphic fungus.

Penicilliopsis clavariaeformis sensu Patouillard (1897) is related to *Stilbothamnium penicilliopsis* described by Hen-

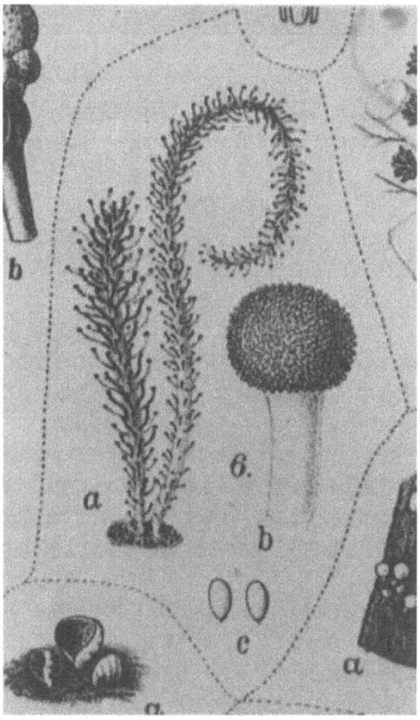

Fig. 1. Stilbothamnium togoense Henn. Type species.

nings (1899) on Diospyros seeds. In both of these fungi synnematous and mononematous forms grow together. Raciborski (1900) identified the mononematous form as an Aspergillus. This similarity was confirmed by Thom and Church (1926) who included A. penicillopsis (Henn.) Raciborski in the A. tamarii group.

The Stilbothamnium species described between 1897 and 1936 differed in the morphology of the Aspergillus-like heads. Hauman (1936) divided the genus into two sections. Section Eu-Stilbothamnium, with broad conidiophores surrounded by thin interwoven hyphae, included three biseriate species, S. dybowskii (Pat.) Henn., S. togoense Henn. and S. javanicum Henn. & Nym. Section Pseudo-Aspergillus, in which the conidiophore is bare and unicellular, included three biseriate species, S. usneoides Sydow, S. novoguineense Sydow and S. amazonense Henn., and a single uniseriate species, S. nudipes Hauman. This structural difference is not without interest, and why Hauman (1936) did not further subdivide section Pseudo-Aspergillus is unclear. On the other hand, the presence of a sterile cover around the stipe may not have been as important as Hauman (1936) believed. Hennings did not describe such a structure in his description of the type species, S. togoense. In P. dybowskii, Patouillard (1892) mentioned the inconstant character of this cover, which disappear with age.

Conidial heads of Stilbothamnium are characterized by their similarity to Aspergillus. We have made an ultrastructural and morphological analysis of S. nudipes to better understand its relationship to Aspergillus.

Morphological and structural analysis of S. nudipes

The fungus examined was collected on seeds of Landolphia (Apocynaceae)in the Republic of Central Africa by R. Heim in 1964. Isolated and identified by J. Nicot, it has been in cultivation since that date (strain LCP 159).

Colonies on Czapek or Malt Agar grow rapidly, attaining a diameter of 5 cm within one week and are golden yellow. The mononematous form is like an Aspergillus, mycelium is hyaline, with erect aseptate conidiophores, 1000-1500 x 15-20 µm, with verrucose walls; vesicles are spherical, 15-25 µm diam, with radiate phialides 10-13 x 6 µm. Synnemata are formed after a few days, are 6-8 cm long, simple or branched, phototropic, of hyaline, closely packed, erect, parallel hyphae. Before their functional differentiation, the conidiophores are distinct from the vegetative hyphae of the synnemata, being wider and verrucose. They diverge from the synnemata and terminate in vesicles, 50-75 µm diam, much larger than the mononematous type. These conidial heads are also Aspergillus-like. In the upper part of the synnema, the vesicles are covered with phialides 15-17 x 7 µm. In the heads located in the lower part of the synnema, a transverse septum is formed in the lower part of the phialide after it starts to produce conidia. After this septation occurs the conidial head looks biseriate, but no further phialides are formed. On either type of conidial head, phialides are cylindrical, slightly enlarged at the apex and with a narrow neck. Conidia are borne in chains, but soon separate; they are elongate, enlarged at the upper end, slightly constricted at the base and echinulate, 7-10 x 6-7 µm.

On natural substrates, S. nudipes produces subglobose grey sclerotia, 3-5 mm diam., smooth and more or less stipitate. The stipe may also be covered with conidial heads. Sclerotia have been obtained in culture on Czapek agar but they are not stipitate. Transverse sections of these sclerotia shows a homogenous plectenchymatous structure, denser near the periphery. Sclerotia in culture remain sterile. No fertile sclerotia have been found in herbarium specimens.

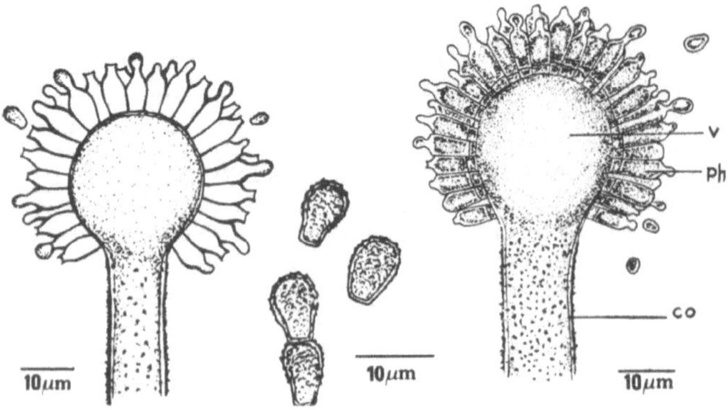

Fig. 2. Stilbothamnium nudipes Hauman. Conidial heads and conidia.

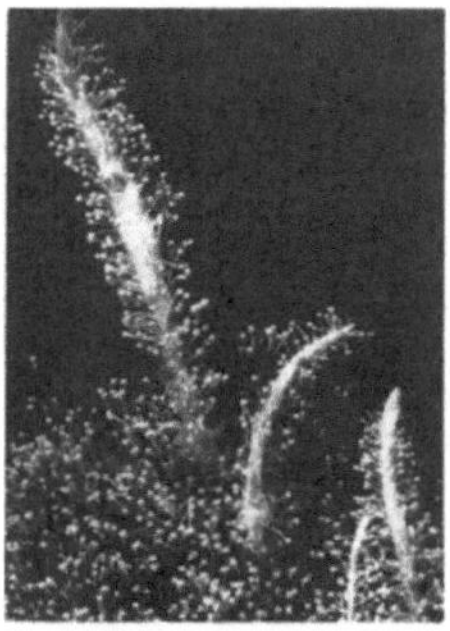

Fig. 3. Stilbothamnium nudipes. Synnemata developed in culture.

With its echinulate conidiophores and conidia as well as the uniseriate conidial heads, S. nudipes appears related to the A. flavus group, near A. tamarii Kita. Therefore, we have made a comparative analysis between S. nudipes and A. tamarii (strain LCP 1934). Differentiation of the conidiophores in the two species is similar (Roquebert 1973). However, the subsequent formation of a septum in the lower part of the phialide occurs only in S. nudipes, and makes it different not only from A. tamarii but from the other Aspergilli (Figs. 4 and 5).

There are differences in conidial wall differentiation between S. nudipes and A. tamarii. Conidia of A. tamarii are connected by a solid connective and the conidial walls mature slowly within the chain. In S. nudipes, however, conidial chains break apart soon after formation and their walls are less ornamented. This character suggests a relationship with the A. fumigatus group (Ghiorse and Edwards 1972).

From the morphological and structural point of view, S. nudipes is very close to Aspergillus, from which it differs only by secondary septation in the lower part of the phialides. The existence of large synnemata is the main distinguishing characteristic of Stilbothamnium. However, examples of synnematous Aspergilli are numerous, and we do not believe it is justified to exclude Stilbothamnium from Aspergillus, especially as both fasciculate and mononematous forms occur together frequently on the same host and in culture.

A. clavatoflavus Raper & Fennel is an example of a fasciculate Aspergillus already included in the genus (Raper and Fennel 1965). More recently, A. coremiiformis Bartoli & Maggi was included in the A. flavus group by Samson (1979), while Christensen (1982) suggested it had characteristics of both the A. flavus and A. ochraceus groups. This species resembles S. nudipes microscopically, but no longer forms synnemata in culture and was isolated from soil in the Ivory Coast, but it is possible that the fungus originated from decaying seeds present in the soil.

As long as teleomorphs in Hauman's section Pseudo-Aspergillus are unknown, nothing prevents Stilbothamnium usneoides, S. novoguineense and S. amazonense, like A. coremiiformis, from being considered synnematous Aspergilli. The

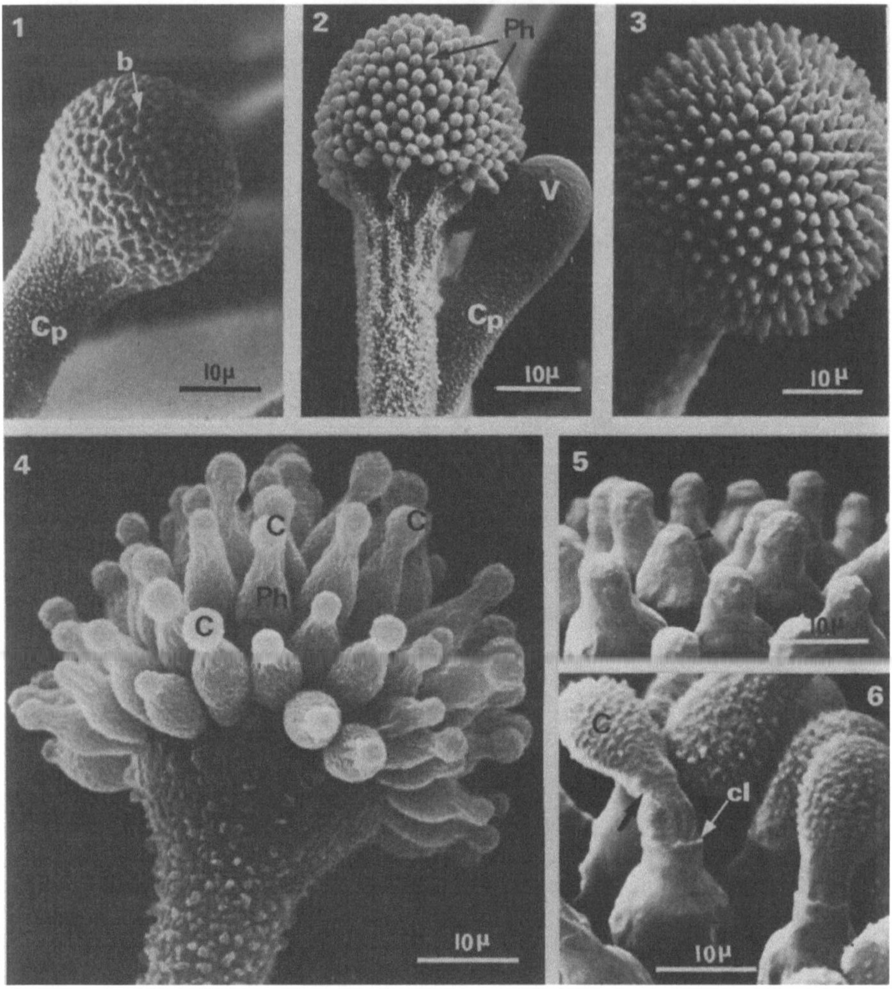

Fig. 4. Stilbothamnium nudipes. S.E.M. developmental morphology of the conidial head. Cp = conidiophore; Ph = phialide; C = conidia; Cl = collarette.

examination of fresh material will no doubt answer this question.

S. nudipes occupies a marginal position with respect to Aspergillus. The peculiar septation of the phialides is unique. However, the well defined ecology of this mould, the presence of sclerotia and the morphological stability in culture makes it a very well characterized species. The other species of Stilbothamnium which are typically biseriate are readily included in Aspergillus. Even though the observed sclerotia remain sterile we might speculate on their maturation into ascocarps of the Penicilliopsis or Petromyces type.

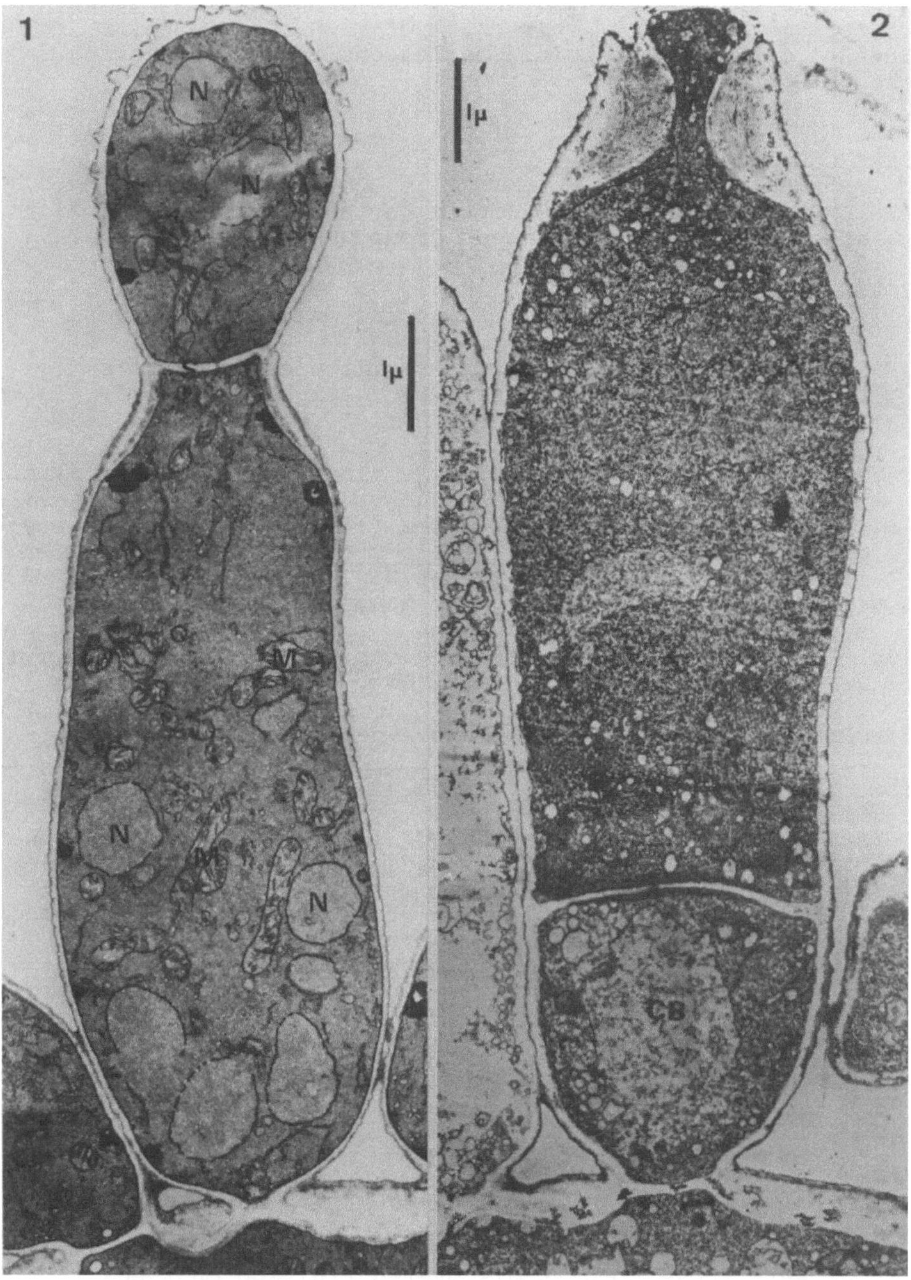

Fig. 5. Stilbothamnium nudipes. T.E.M. 1. One-celled young phialide with conidia. 2. Two-celled phialide. N = nucleus; S = septum; M = mitochondria; CB = Basal cell.

According to Malloch and Cain (1972), the evolution of conidial forms of Trichocomaceae would be towards the acquisition of more and more compact structures. To Penicilliopsis and Petromyces would correspond forms with complex branching while the more simple would show teleomorphs of the Dichlaena or Eurotium type. From this point of view, the peculiar structure of S. nudipes could correspond to an interesting phase in evolution.

REFERENCES

CHRISTENSEN, M. A. 1982. Synoptic key and evaluation of species in the A. ochraceus group. Mycologia 74: 210-225.

GHIORSE, W. C. and EDWARDS, M. R. 1973. Ultrastructure of A. fumigatus conidia: Development and maturation. Protoplasma 76: 49-59.

HAUMAN, L. 1936. Les champignons seminicoles des forêts tropicales. Bull. Soc. roy. Bot. Belg. 69: 96-129.

HENNINGS, P. 1897. Fungi camerunensis II. Bot. Jb. 23: 537-558.

KOBAYASI, Y. and INDOH, H. 1943. On the genus Penicilliopsis. Bull. Centr. Natl. Mus. Manchuokuo 5: 15-34 (in Japanese).

MALLOCH, D. and CAIN, R. 1972. The Trichocomataceae: Ascomycetes with Aspergillus, Paecilomyces and Penicillium imperfect states. Can. J. Bot. 50: 2613-2628.

PATOUILLARD, N. 1892. Champignons nouveaux extra-europeens. Bull. Soc. mycol. Fr. 8: 46-56.

PATOUILLARD, N. 1897. Enumeration des champignons recoltes a Java par Massart. Ann. Jard. Bot. Buitenzorg Suppl. 1: 124-125.

RACIBORSKI, M. 1900. Parasitische Algen und Pilzen Java's. II: 3-49. Buitenzorg.

RAPER, K. B. and FENNEL, D. I. 1965. "The genus Aspergillus". Baltimore: Williams & Wilkins.

ROQUEBERT, M. F. 1973. Modalites de la sporogenèse chez deux phialides: A. tamarii et S. nudipes. Rev. mycol. 38: 3-8.

SAMSON, R. A. 1979. A compilation of the Aspergilli described since 1965. Stud. mycol., Baarn 18:1-38.

SOLMS-LAUBACH, H. 1887. Penicilliopsis clavariaeformis, ein neuer javanischen Ascomycet. Ann. Jard. Bot. Buitenzorg 6: 53-71.

THOM, C. and CHURCH, M. B. 1926. "The Aspergilli". Baltimore: Williams and Wilkins.

DIALOGUE FOLLOWING DR.ROQUEBERT'S PAPER

WICKLOW: Were these species discovered fruiting on seeds in nature, or were they first isolated from seeds and then the fruiting structures found?

ROQUEBERT: They were originally found growing directly on seeds.

CHRISTENSEN: You discussed red ascospores mentioned in *Stilbothamnium*. Who found these?

ROQUEBERT: Roger Heim who collected this fungus and made notes of this observation. We haven't been able to find any ascospores in the specimens we have examined.

FRISVAD: Why do well-developed elaborate synnemata tend to occur only in the tropics?

SAMSON: These complex and beautiful structures are also common in tropical entomogenous fungi. When you compare similar species in domesticated substrata or in temperate zones, there is usually reduction in the size of the synnemata. Probably in the tropical rain forest, they have more time to develop and it may be necessary to make large synnemata to get above the litter layer.

CHRISTENSEN: The light bulb shape of the conidia in your electron micrographs (Fig. 4) would be unusual for *Aspergillus*.

SAMSON: How about *A. penicillioides* or the *A. restrictus* group?

KOZAKIEWICZ: I agree that the shape and ornamentation of these conidia seems quite different from the rest of *Aspergillus*. If you are going to combine *Stilbothamnium* with *Aspergillus*, go ahead, but put them in a separate group.

HAWKSWORTH: I think it might be a good idea to look at the chemistry of these species before making the new combinations.

Chapter 5:

NEW TECHNIQUES AND APPROACHES TO THE SYSTEMATICS OF PENICILLIUM AND ASPERGILLUS

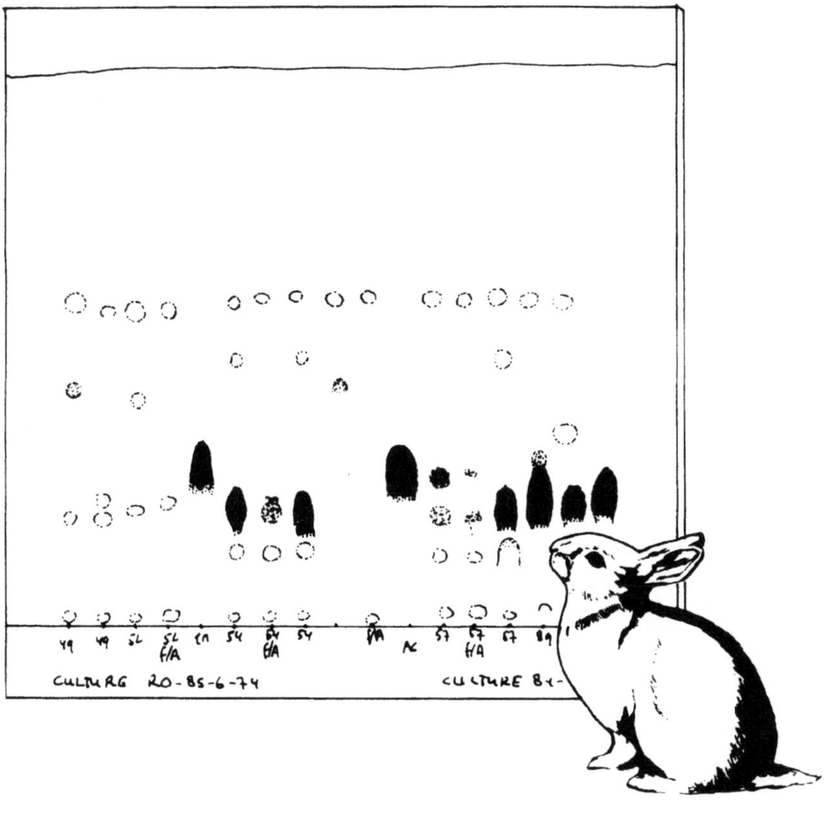

CLASSIFICATION OF FUNGI THROUGH NUCLEIC ACID RELATEDNESS

Clete P. KURTZMAN

Northern Regional Research Center, Agricultural Research Service, U.S. Department of Agriculture
Peoria, Illinois 61604, U.S.A.

INTRODUCTION

The separation of species in the genera Aspergillus and Penicillium is primarily based, as with many groups of fungi, on differences in morphology and ability to grow under various conditions. The genetic complexity of these phenotypic characteristics is generally unknown and consequently their phylogenetic significance can only be guessed. Because mutation as well as selection resulting from prolonged cultivation on common culture media sometimes provokes sufficient change in strains to allow reclassification as different species, we can begin to appreciate the ephemeral nature of the characteristics upon which we presently define taxa. Species should be viewed as populations which are reproductively isolated from other populations or species (Dobzhansky, 1976). Consequently, a species so defined becomes a product of nature and not of the taxonomist. Definition of imperfect taxa on the extent of their reproductive isolation would seem unrealistic but, as I hope to demonstrate, this is attainable with reasonable accuracy though comparisons of nucleic acid relatedness.

In this paper I wish to convey that mycologists who are not trained biochemists can learn to isolate and characterise nucleic acids. The fungal cell contains several types of nucleic acids which show differing amounts of sequence conservation, and because of this, we are able to resolve various levels of taxonomic hierarchy. I will point out what is presently known of the extent of resolution offered by the various nucleic acid comparisons. Data from nucleic acid studies are not always definitive, but they generally offer much more information than is available from conventional comparisons. Owing to a paucity of data from filamentous fungi, many of the examples discussed concern the yeasts. Nonetheless, I hope that these results are in general what we might expect if similar comparisons were made among other fungi such as Aspergilli and Penicillia.

The methodology for isolation and purification of fungal DNA has been dealt with in detail elsewhere, and the following account will serve only as a brief review to direct attention to some of the more pertinent references. For many fungi, DNA isolation is relatively straightforward. Cell wall breakage may be effected by mechanical or lytic means. Mechanical methods have the advantage of being relatively rapid, but unless cell breakage is carefully monitored, considerable shearing of the DNA may result. The Braun cell homogeniser, one of the more frequently used mechanical devices, causes breakage by its shaking of a buffered cell suspension in the presence of 0.45 mm glass beads (Bak and Stenderup 1969; Price et al., 1978; Kurtzman et al., 1980a). An inexpensive variation of this technique involves placing the buffered cell and bead mixture into a test tube and "vortexing" the mixture (Lipinski et al., 1976; Van Etten and Freer, 1978). Less frequently reported methods include grinding acetone-dried cells in a ball mill (Martini and Phaff 1973) and grinding cells in liquid nitrogen with a mortar and pestle (Storck and Alexopoulos 1970; Mendonca-Hagler et al., 1974). Sonication is seldom used for cell breakage because of the danger of extreme DNA shearing.

Lytic methods offer the possibility of isolating DNA with a higher molecular weight than might be obtained by mechanical cell disruption. Meyer and Phaff (1969) and Bicknell and Douglas (1970) used a modification of the Smith and Halvorson (1967) technique in which yeast cells undergo autolysis in the presence of saline EDTA, 2-mercaptoethanol, and sodium lauryl sulfate. The occasional failure of this method has prompted the widespread use of mechanical disruption. The enzyme preparations Glusulase and Zymolyase may be used to prepare osmotically fragile spheroplasts by cell wall digestion. Following this step, high-molecular-weight DNA is isolated from the lysed spheroplasts. Cryer et al. (1975) have given details of these procedures. Zymolyase generally promotes more rapid cell wall dissolution than Glusulase, but neither enzyme preparation is particularly effective with basidiomycetes. Brady (1981), however, reported the relatively rapid dissolution of *Rhodotorula* cell walls by an enzyme preparation from culture filtrates of *Paecilomyces (Penicillium) lilacinus*. Another success in this area was achieved by Rhodes and Kwon-Chung (1982), who produced protoplasts from blastospores of *Cryptococcus neoformans* with the multi-enzyme preparation mutanase.

Following rupture or lysis of the cells, DNA isolation and purification may be carried out by the methods of Marmur (1961) and Bernardi et al. (1970) as detailed by Price et al. (1978). Basically, this consists of making the cell suspension 1 *M* in sodium perchlorate and 1% in sarcosate, and emulsifying with an equal volume of chloroform-isoamyl alcohol (24:1, vol/vol). After incubation and phase separation by centrifugation, DNA is precipitated from the aqueous phase with cold ethanol. The DNA is treated with α-amylase, pancreatic RNase, and pronase. Following enzyme removal, the DNA is again ethanol precipitated, spooled, redissolved, and further treated with α-amylase, pancreatic RNase, and T1 RNase. Additional purification can include hydroxylapatite fractionation and banding in a cesium chloride gradient.

Some fungi have active DNAses that escape denaturation during the usual extraction procedures and cause sufficient degradation of the DNA to prevent spooling. Several methods have produced varying degrees of success in the deactivation of nucleases, and they merit testing if problems are encountered.

Diethyl pyrocarbonate used at 0.5-1.0% at the time of cell breakage gave encouraging results in studies of _Lipomyces_ spp. (Holzschu and Phaff, pers. commun.) and _Sporidiobolus_ (Holzschu et al., 1981). Ehrenberg et al. (1976) discussed the interaction of diethyl pyrocarbonate with nucleic acids and emphasised some of the precautions to be taken. Kurtzman and co-workers (unpublished data) found that the spermine-spermidine-sucrose buffer of Morris, as cited by Timberlake (1978), inactivated nucleases much more effectively in _Aspergillus_ spp. and _Neurospora_ spp. than the sucrose buffer frequently used in yeast studies (Price et al., 1978). Heating the cells to 60° C for 20 min was also reasonably effective in deactivating nucleases in _Aspergillus_ and _Neurospora_ (Kurtzman et al., unpublished data). Higher temperatures, if they do not result in denaturation of the DNA, may be even more effective.

Unless care is taken to isolate intact nuclei and mitochondria, the preceding methods will give a mixture of DNA from these two organelles. Because of its lower guanine + cytosine content, mitochondrial DNA may be separated from nuclear DNA through preparative CsCl density centrifugation. Methods for assessing DNA purity include chemical analyses as well as spectrophotometric measurements and have been summarised by Price et al. (1978) and Johnson (1981).

Techniques for isolation of RNA have been presented by Gillespie and Spiegelman (1965), Bicknell and Douglas (1970), and Johnson (1981). These references also include procedures for RNA-DNA reasociation experiments. Although the cellular RNA concentration is manyfold higher than that of DNA, the nearly ubiquitous occurrence of RNAses and their marked stability make isolation of undegraded RNA more difficult than the isolation of high-molecular-weight DNA.

DNA BASE COMPOSITION AND ITS TAXONOMIC USES

Nuclear DNA base composition, expressed as molar percentages of guanine + cytosine (G + C), is the simplest of the molecular techniques that allow resolution of taxa. Base composition may be determined from thermal denaturation profiles (melts), buoyant density in cesium salt gradients generated by ultracentrifugation, chemical analysis, absorbance ratios, or high-pressure liquid chromatography of nucleotides or free bases (Johnson 1981). Most investigators use thermal denaturation (Marmur and Doty 1962), but Bak (1973) showed that results obtained by this method can be greatly affected by sample impurities and/or minor DNA species, and must therefore be interpreted with caution. Cesium chloride buoyant density determinations (Schildkraut et al., 1962) generally show the greatest accuracy, since they are unbiased by the presence of contaminating RNA, mitochondrial DNA, and other impurities, such as carbohydrates and proteins.

The taxonomic uses of G + C values are mainly exclusionary because fungus species range in G + C content from approximately 30-70 mol% and overlap between unrelated species is inevitable. For example, *Debaryomyces hansenii* and *D. marama* have a G + C content of 39.0 and 39.1 mol%, respectively, yet they show only 8.4% base sequence complementarity (Price et al., 1978). Consequently, since similar G + C content does not necessarily indicate conspecificity, we must ask how dissimilar G + C values need to be in order to state with relative certainty that two strains represent different species.

The answer to this question depends to some extent on the method used for determination of base composition. When buoyant density is used, the span of G + C values among strains of a species is about 1% (Price et al., 1978; Kurtzman et al. 1980a, b), which is approximately coincident with the accuracy of the method, i.e., a standard deviation of 0.00-0.50 mol%. Thermal denaturation shows greater variation, and strains of a species may have a G + C range of as much as 2% (Meyer et al., 1978). Thus, strains showing a difference in G + C contents in excess of 1.0-1.5% by buoyant density or 2.0-2.5% by thermal denaturation may be expected to be different species.

An examination of values in Table 1 shows considerable overlap in G + C content among the subdivisions of fungi. Only among the yeasts is it possible to predict the subdivision from G + C content. Ascomycetous species range from 28-52%, whereas basidiomycetous species range from 49-68%. Thus, only a small region overlaps the yeasts, whereas the overlap among the higher fungi is much greater. Species within a genus frequently have G + C values that lie within a range of 10%.

DNA RELATEDNESS

The methods for assessing DNA relatedness vary, but short of actual sequencing rely on measuring the extent and stability of renatured DNA strands from the test pair, i.e., the fidelity of complementary base pairing. Depending on the method, the DNA may or may not need to be labelled with radioisotopes. Because of technical difficulties, heavy isotope labels are seldom used.

In vivo labelling of DNA is frequently done using ^{14}C, ^{3}H, or ^{32}P. Each isotope has its particular advantages and disadvantages. The DNA may also be labelled *in vitro* with ^{125}I (Commerford 1971; Mendonça-Hagler and Phaff 1975; Holzschu et al., 1979) or with ^{14}C, ^{3}H or ^{32}P by nick translation (Kelley et al., 1970; Maniatis et al., 1975). DNA labelled with ^{125}I *in vitro* must be highly purified, which requires the extra purification step of banding in CsCl in a preparative ultracentrifuge. Following labelling, DNA reassociation reactions may be carried out using the membrane method, which involves immobilising single strands of one of the DNA species onto nitrocellulose filters and allowing sheared single strands of

Table 1. Nuclear DNA base composition of selected fungus taxa.

Species	Mol% G+C
Mastigomycotina	
Phytophthora infestans	54[a]
Saprolegnia parasitica	59[b]
Allomyces arbuscula	62.2[c]
Zygomycotina	
Cunninghamella echinata	34[d]
Mucor rouxii	37[d]
Syncephalastrum racemosum	48[e]
Achlya racemosa	52[b]
Ascomycotina	
Pichia kluyveri	30[f]
Debaryomyces vanriji	33.2[g]
Debaryomyces hansenii	38.6[g]
Metschnikowia reukaufii	42.2[h]
Metschnikowia pulcherrima	48.3[h]
Pichia rhodanensis	51.9[i]
Neurospora crassa	54[d]
Gelasinospora calospora	55[a]
Ceratocystis ulmi	56[d]
Chaetomium globosum	58[d]
Basidiomycotina	
Filobasidiella neoformans	51.5[j]
Coprinus lagopus	53.0[a]
Lenzites saepiaria	54.0[k]
Daedalea confragosa	57[d]
Polyporus brumalis	59[k]
Rhodosporidium toruloides	60[l]
Schizophyllum commune	61[d]
Sporidiobolus salmonicolor	63[e]

[a]Storck and Alexopoulos (1970).
[b]Green and Dick (1972).
[c]Ojha *et al*. (1977).
[d]Villa and Storck (1968).
[e]Storck (1966).
[f]Nakase and Komagata (1968).
[g]Price *et al*. (1978).
[h]Meyer and Phaff (1969).
[i]Kurtzman *et al*. (1980a).
[j]Storck *et al*. (1969).
[k]Storck *et al*. (1971).
[l]Nakase and Komagata (1971).

the second DNA to react with the immobilised DNA (Denhardt, 1966). Alternatively, both DNA species may be allowed to react in free solution, and the extent of reassociation may be assessed by percent binding of the resulting duplexes to hydroxylapatite (Brenner, 1973) or resistance to hydrolysis by S1 nuclease (Crosa et al., 1973). Price et al. (1978) and Johnson (1981) provide references and detailed procedures for these methods.

In addition to following DNA reassociation with isotopically labelled preparations, the reactions may also be monitored spectrophotometrically by measuring the kinetics of duplex formation. This technique has been refined to the point where results are comparable to data obtained using isotopes (Subirana and Doty 1966; De Ley et al., 1970; Seidler and Mandel 1971; Kurtzman et al., 1980a). Relatedness may also be estimated from the thermal stability of the renatured heterologous DNA as compared with renatured homologous DNA. This may be done spectrophotometrically by monitoring thermal melts or with labelled DNA by thermal elution from hydrooxylapatite columns (Ullman and McCarthy 1973). Brenner et al. (1972) equated one degree of difference in thermal stability with 1% mismatch in base pairs. Lower thermal stability usually correlates well with lower DNA homology.

A number of factors affect DNA renaturation reactions and so may have a marked effect on the results obtained. Highly purified DNA preparations are essential. Contaminating protein or RNA can result in falsely high or low relatedness values, depending on whether the contaminants act as single-stranded or double-stranded DNA in the assay system (McCarthy and Church 1970; Brenner 1970).

Several classes of DNA occur in eukaryotes and their presence affects estimates of relatedness between taxa, since it is the intent of most studies to measure genomic similarity from complementarity of single-copy nuclear DNA sequences. Christiansen et al. (1971) reported up to 15% rapidly renaturing sequences in the nuclear DNA (nDNA) of yeasts, which may be interpreted as analogous to the repeated sequences found in higher eukaryotes (Britten and Kohne 1968). However, the presence of rapidly renaturing, plasmidlike 2-um circular DNA in Saccharomyces cerevisiae (Bak et al., 1972), with a buoyant density similar to nDNA, suggests the proportion of repeated sequences in nDNA to be smaller than first thought. Similar circular DNAs have been reported from Hansenula wingei and Candida (Torulopsis) glabrata (O'Connor et al., 1975, 1976). Linear DNA plasmids have been discovered in Kluyveromyces lactis (Gynge et al., 1981). These plasmids, which are lighter than the densities of both nuclear and mitochondrial DNA, appear to be associated with the killer character of this yeast. Among the filamentous fungi, kinetic studies also indicate the nuclear genome to be comprised primarily of unique sequences. Dusenbery (1975) reported 64% of the total cellular DNA of Phycomyces blakesleeanus to represent a slowly reassociating fraction that is probably made up of unique sequences. A similar figure, 60%, was reported for Allomyces arbuscula (Ojha et al., 1977). Hudspeth et al. (1977) determined 82% of the genome of Achlya bisexualis to be single copy DNA, whereas this class represented 97-98% for Aspergillus nidulans (Timberlake 1978) and 90% for Neurospora crassa (Krumlauf and Marzluf 1980).

Nuclear DNA represents the preponderance of cellular DNA in yeasts and has a kinetic complexity of 6.5-14 x 10^9 daltons or about two to five times that found in Escherichia coli (Christiansen et al., 1971; Petes et al., 1973). Genome sizes of the filamentous fungi cited above are estimated to range from 1.7-2.7 x 10^{10} daltons (Dusenbery 1975; Ojha et al., 1977; Hudspeth et al., 1977; Timberlake 1978; Krumlauf and Marzluf 1980). Between 5 and 20% of the total DNA in Saccharomyces cerevisiae is mitochrondrial in origin.

If comparisons of taxa are to be made from single-copy nDNA, mtDNA, plasmidlike circular DNA, and repeat sequence nDNA must be removed. Circular plasmidlike DNA as well as circular mtDNA may be relatively easily separated from nDNA by centrifugation in cesium chloride in the presence of ethidium bromide. Williamson and Fennelle (1975) emphasised that resolution is poor if mtDNA should break into linear fragments during isolation, a not uncommon happening. However, they point out that linear mtDNA may be effectively resolved in cesium chloride gradients when reacted with 4',6-diamidino-2-phenylindole (DAPI), a fluorescent dye that preferentially binds with AT-rich DNA. Repeated sequences may be removed from nDNA by partial reassociation of sheared DNA fragments followed by frationation with hydroxylapatite. Since the rate of reassociation is affected by concentration, multiple copies duplex first and can be removed because of their ability to bind to hydroxylapatite (Britten and Kohne 1968). Price et al. (1978) used this method to remove repeated sequences from yeast DNA for taxonomic studies. The fractionation undoubtedly also removes mtDNA and plasmidlike DNA because both occur as multiple copies, and their small molecular weight would allow rapid reassociation when in significant concentration. The same effect may be obtained in spectrophotometric determinations of relatedness if the time between melting and cooling to the reassociation temperature (Tm-25) is not too rapid. Under their conditions, Kurtzman et al. (1980a) found a cooling period of 12 min to allow duplexing of most repititive sequences as well as of mitochrondrial DNA. Relatedness values of strains common to the work of Price et al. (1978) and of Kurtzman et al. (1980a) were essentially identical showing that similar results can be obtained with rather different methodologies.

Optimal DNA reassociation occurs at 25-30°C below the midpoint of the melting curve (Tm), and this is dependent on base composition, since GC base pairs have greater thermal stability than do AT pairs (Marmur and Doty, 1962; Marmur et al., 1963). If reassociation occurs at a temperature higher than optimum, any mismatched sequences are less likely to pair, whereas temperatures under the optimum may allow the mismatch to appear complementary. Consequently, we must view DNA relatedness values obtained from whole genome comparisons as an estimate of overall base sequence similarity under defined conditions, rather than as a certain percentage of the DNA that has precisely the same base sequence.

EXTENT OF RESOLUTION CORRELATED WITH GENETIC HYBRIDIZATION

Guidelines establishing the extent of nuclear DNA relatedness exhibited between strains of microbial species have resulted primarily from studies of bacteria. For example, on the basis of extensive comparisons between species of enteric bacteria, Brenner (1973) proposed that strains showing 70% or greater DNA relatedness should be considered members of the same species. Johnson (1973) compared numerous anaerobic bacteria and arrived at essentially the same guidelines. Among the fungi, Price et al. (1978) examined species in four yeast genera and proposed that strains having 80% or greater DNA relatedness were conspecific, an opinion expressed earlier by Martini and Phaff (1973) from their studies in Kluyveromyces. However, these guidelines were based on strains that had been characterised by traditional morphological and physiological tests. Another approach to establishing guidelines would be to compare the fertility of sexually reproducing species with extent of DNA relatedness in order to correlate DNA complementarity with actual biological phenomena.

Van der Walt and Johannsen (1979) and Johannsen (1980) approached this problem through hybridization studies in the essentially homothallic genus *Kluyveromyces*, where they were able to compare their results with DNA relatedness data obtained earlier by Martini (1973) and Martini and Phaff (1973). Van der Walt and Johanssen mated auxotrophic yeast strains, attained by N-methyl-N'-nitro-N-nitrosoguanidine mutagenesis, on media that supported growth of prototrophic recombinant clones. Utilising this method, they observed various combinations of prototrophic colonies at a frequency of 10^{-2}-10^{-8}. Taxa with high DNA relatedness showed a higher frequency of recombination, but in a few cases taxa having low DNA complementarity also gave a considerable number of recombinants. Lack of crosses through successive generations of progeny has made interpretation of these data difficult.

Kurtzman et al. (1980 a, b) were able to correlate genetic relatedness and DNA complementarity by working with heterothallic strains, thereby circumventing problems inherent in the *Kluyveromyces* study where auxotrophic mutants of homothallic strains were employed. In addition, Kurtzman et al. (1980 a, b) isolated progeny from crosses by micromanipulation. In the first of these studies, *Pichia amylophila* and *P. mississippiensis* were shown to mate well, but ascospores were not viable. The two species showed only about 25% DNA relatedness, thus correlating lack of fertility with low DNA complementarity. However, the relationship between *Issatchenkia scutulata* var. *scutulata* and *I. scutulata* var. *exigua* was somewhat different. These two taxa also showed only about 25% DNA relatedness, but genetic crosses gave 3-6% viable ascospores. Furthermore, crosses between these F1 progeny gave 17% viability among F2 progeny. Reciprocal crosses among the F1 progeny were fertile, as were backcrosses to the parents. From this, it appears that all stocks had essentially homologous chromosomes and that progeny were neither amphidiploids nor aneuploids. Clearly then, the lower limit of DNA-DNA homology values suggesting species delimitation is not yet well defined, but base sequence divergence, as estimated from whole genome comparisons, may be as great as 75% before genetic exchange can no longer occur. Data in Table 2 summarise the preceding findings and provide further comparisons of other taxa. In these studies, progressively less DNA relatedness parallels decreasing mating competence and fertility. These data also indicate that resolution goes no further than the detection of sibling species. Beyond this, all species, regardless of extent of kinship, show essentially less than 10% complementarity.

However, despite the strong correlation between DNA relatedness and fertility exhibited in Table 2, exceptions may occur and whole genome DNA complementarity should therefore be regarded as a strong but not infallible indicator of biological relatedness. Exceptions to this trend would include chromosomal changes that affect fertility between strains such as inversions, translocations, autopolyploidy and amphidploidy (allopolyploidy). Most of these changes would not be detectable in whole genome DNA comparisons. However, amphidiploidy is recognisable if parental species have been included in the study or should be suspected if genome sizes vary between strains showing substantial relatedness. For example, *Saccha-*

Table 2. Correlation of mating reaction and DNA complementarity among closely related heterothallic ascomycetous and basidiomycetous yeasts.

Species	Mating reaction	% DNA relatedness
Filobasidiella neoformans X Filobasidiella bacillispora	Fair conjugation 0-30% basidiospore viability (F_1 progeny; F_2 not determined)	55-63[a]
Issatchenkia scutulata var. scutulata X Issatchenkia scutulata var. exigua	Good conjugation Ascospores viable: F_1 = 5%; F_2 = 17%	21-26[b]
Pichia amylophila X Pichia mississippiensis	Good conjugation Ascospores not viable	20-27[c]
Hansenula bimundalis X Hansenula americana	Poor conjugation Ascospores not produced	21[d]
Hansenula alni X Hansenula wingei	Poor conjugation Ascospores not produced	6[e]
Issatchenkia orientalis X Issatchenkia occidentalis	Infrequent conjugation Ascospores not produced	3-8[b]

[a]Aulakh et al. (1981).
[b]Kurtzman et al. (1980b).
[c]Kurtzman et al. (1980a).
[d]Kurtzman (1984a).
[e]Phaff et al. (1979); Fuson et al. (1979).

romyces carlsbergensis shows relatively high DNA homology with both S. cerevisiae and S. bayanus even though these latter two species demonstrate little relatedness (Fig.1). A comparison of genome sizes suggests S. carlsbergensis to be a partial amphidiploid which arose as a natural hybrid of S. cerevisiae and S. bayanus (Martini and Kurtzman, unpublished). In this case, we would predict S. carlsbergensis to be infertile with the proposed parents despite relatively high DNA relatedness.

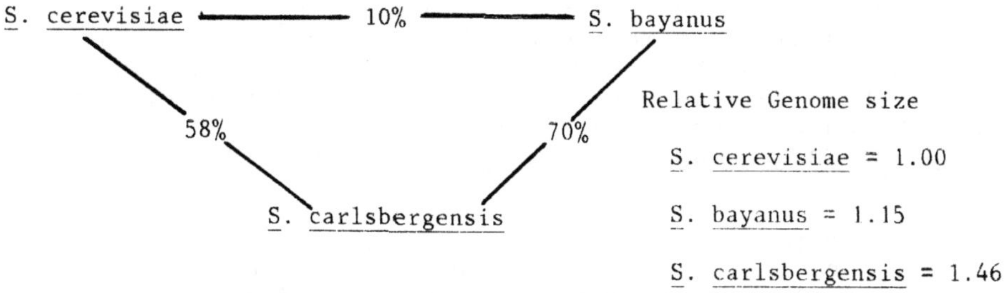

Fig. 1 DNA relatedness between type strains of Saccharomyces cerevisiae, S. bayanus and S. carlsbergensis (after Martin and Kurtzman, unpubl. results).

Comparisons of species

Bak and Stenerup (1969) appear to have been the first to make extensive use of DNA reassociation as a means for determining relatedness among fungi. Their study included several phenotypically similar Candida species of medical interest. Specifically, their data showed sufficiently high base sequence complementarity (ca. 65-81%) to regard C. albicans, C. stellatoidea, and C. claussenii as conspecific.

Additional advances in the clarification of imperfect yeasts taxa through DNA studies were made by Meyer and Phaff (1972), who found high degrees of homology between C. lusitaniae and C. obtusa and between C. salmonicola and C. sake; by Meyer et al. (1975), who demonstrated C. cloacae and C. subtropicalis to be synonyms of C. maltosa; and by Kurtzman et al. (1980 b), who showed C. krusei and C. sorbosa to represent, respectively, the imperfect states of Issatchenkia orientalis and I. occidentalis. Kurtzman et al. (1979) and Manachini (1979) demonstrated C. utilis to be the imperfect form of Hansenula jadinii. The work of Mendonca-Hagler and Phaff (1975) provided clear evidence that Saccharomyces telluris, Torulopsis bovina, T. pintolopesii, and C. sloofii are conspecific. This study also convincingly demonstrated that pseudomycelium formation provides no distinction between Torulopsis and Candida.

DNA relatedness studies have had an enormous impact on our assessment of the criteria used to define species and genera among the yeasts (Table 3). Yarrow and Meyer (1978) combined the genera Candida and Torulopsis when it was found that presence or absence of pseudohyphae might occur in a single species. Similarly, the 75% relatedness detected between Pichia lindneri and Hansenula minuta prompted Kurtzman (1984b) to propose that the two genera, which are separated on ability to assimilate nitrate, be combined.

Rather few DNA comparisons have been made among fungi other than yeasts. Of these, Mendonca-Hagler et al. (1974) examined species of Ceratocystis as possible teleomorphs of Sporothrix schenckii. The pairing of C. minor and S. schenckii gave 75% complementarity, thus indicating these two species to represent the perfect-imperfect forms of the same taxon.

Dutta and colleagues (Dutta 1976; Dutta et al., 1976; Williams et al., 1981) examined speciation in the genus Neurospora and reported the least related pair, N. crassa and N. dodgei, to show 61% complementarity. However, this figure is really quite high when viewed against our experience with the yeasts, where relatedness this great is indicative of conspecificity. Surprisingly, an apparent control pairing between N. crassa and Gelasinospora austosteira resulted in 56% complementarity.

In an effort to understand relationships among the agronomically and industrially important yellow-green Aspergilli, Kurtzman et al. (1984) compared Aspergillus flavus, A. parasiticus, A. oryzae, and A. sojae. The first two species commonly infest cereal grains and peanuts where they may produce the potent carcinogen aflatoxin. The latter two species, which are

Table 3. DNA relatedness between yeasts species differing in traditional taxonomic characteristics.

Species	Characteristic/+ or -	% DNA relatedness
Candida slooffii	Pseudohyphae/+	80[a]
Torulopsis pintolopesii	/-	
Hansenula wingei	True hyphae/+	78[b]
Hansenula canadensis	/-	
Debaryomyces formicarius	Glucose Ferm/+	96[c]
Debaryomyces vanriji	/-	
Schwanniomyces castellii	Lactose Assim/+	97[c]
Schwanniomyces occidentalis	/-	
Hansenula minuta	Nitrate Assim/+	75[d]
Pichia lindneri	/-	
Sterigmatomyces halophilus	Nitrate Assim/+	100[e]
Sterigmatomyces indicus	/-	

[a]Mendonça-Hagler and Phaff (1975).
[b]Fuson et al. (1979).
[c]Price et al. (1978).
[d]Kurtzman (1984b).
[e]Kurtzman et al. (1980c).

not known to produce aflatoxin (Wang and Hesseltine, 1982), are widely used throughout the Orient as koji moulds for fermentation of sake, miso, and soy sauce. Although mycologists have suggested that the four species are closely related, nonetheless, they usually have been retained as separate taxa (Blochwitz, 1929; Saito, 1943; Thom and Raper, 1945; Raper and Fennell, 1965).

The DNA comparisons showed the extent of base sequence complementarity between A. flavus and A. oryzae to be 100% (Fig.2). Similarly, A. parasiticus and A. sojae showed 91% relatedness. Complementarity between these two groups, demonstrated in the A. flavus-A. parasiticus pairing, was 70%. By contrast, A. tamarii showed 40-55% relatedness with the foregoing species, while A. leporis exhibited less than 10% similarity with them. Of further interest was the demonstration that certain strains initially identified as A. flavus represent a new species (Kurtzman et al., in prep.). This new species, which also produces aflatoxin (Hesseltine et al., 1970), was isolated predominantly from insects. The interpretation of these data is that A. flavus, A. oryzae, A. parasiticus and A. sojae represent variants of the same species. Owing to a few morphological differences, as well as to the need to keep toxic and non-toxic strains separate, the "species" will be maintained as varieties of A. flavus.

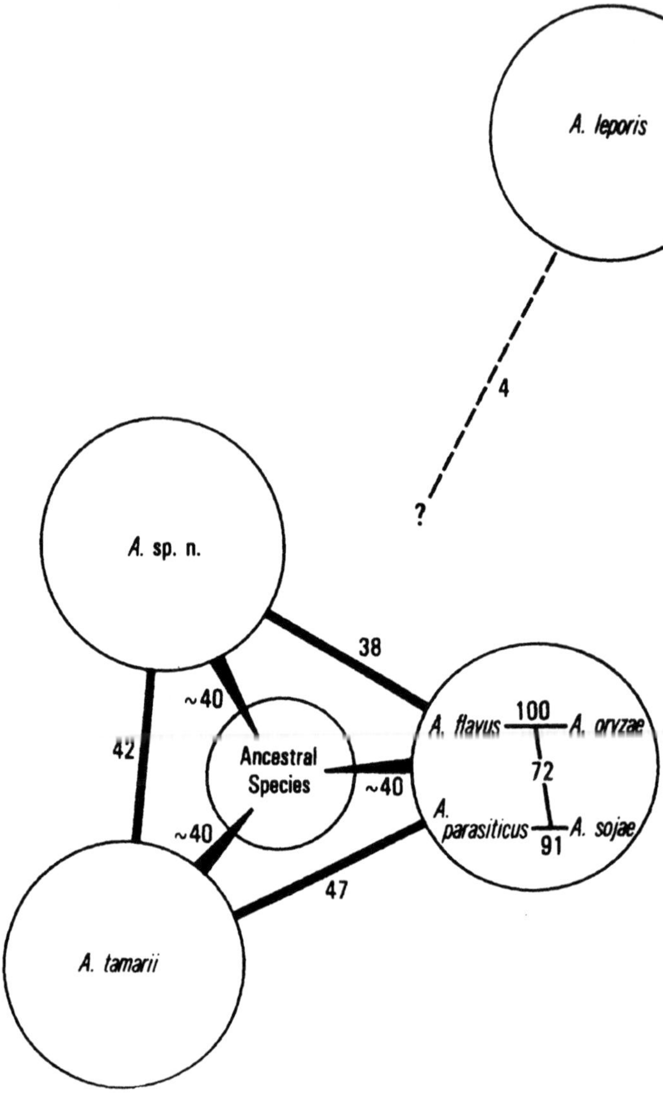

Fig.2. Diagrammatic representation of the extent of nuclear DNA complementarity between Aspergillus flavus, A. oryzae, A. parasiticus, A. sojae, and other phenotypically similar Aspergilli. Numbers along the connecting lines represent percent relatedness (standard deviation < 5%). Relatedness between these taxa and Neurospora crassa, included for comparative purposes, averaged 3% (Kurtzman et al., 1984)

The data are also consistent with the proposal (Blochwitz 1929; Saito 1943; Wicklow 1984) that A. oryzae and A. sojae developed from naturally occurring A. flavus and A. parasiticus following initial colonisation of koji starters. The domesticated strains show reduced sporulation (e.g. few and smaller, uniseriate conidial heads), produce conidiophores and conidia that are often less conspicuously roughened and more variable in size than those of the naturally occurring species, and do not produce sclerotia or aflatoxin. This is in keeping with the proposed transition from survival in nature to survival in koji fermentations, where improved fitness in the new environment would depend more on rapid conidium germination and ability to efficiently hydrolyse starches or proteins.

In general, we can conclude from the foregoing discussion that comparisons of nuclear DNA complementarity would offer resolution of the following relationships among the Aspergilli and Penicillia as it has with other fungi. We would expect high relatedness between:

1. Subcultures such as:
   a. Single-spore isolates
   b. Mycelial transfers
   c. Morphologically distinct sectors of a colony
   d. Mutants of a parent colony
   e. Mating types from heterothallic species
2. Toxigenic and non-toxigenic strains of the same species
3. Domesticated strains and their wild counterparts from nature
4. Anamorphs with their teleomorphs
5. Synonymous species

Using comparisons of nuclear DNA relatedness we would not expect to resolve:

1. Whether species belong in a particular genus
2. The hierarchy of species within a genus

Resolution of these questions awaits comparisons of ribosomal RNA and other highly conserved sequences and this will be discussed in later sections.

Mitochondrial DNA relatedness

In a few studies, relatedness among the fungi has been examined through comparisons of fragment patterns of mitochondrial DNA (mtDNA) generated with restriction endonucleases. The small size of the mitochondrial genome, as discussed earlier, makes such comparisons practical, whereas the larger nuclear genome would present more fragments than could be reasonably handled. McArthur and Clark-Walker (1983) used mtDNA restriction patterns to correlate perfect-imperfect relationships between the yeast genera *Dekkera* and *Brettanomyces*. They found identical restriction patterns for the pair *D. bruxellensis*/*B. lambicus*, as well as for the imperfects *B. abstinens*/*B. custersii* and *B. anomalus*/*B. clausenii*. This strongly suggests the pairs to be conspecific. Size differences in mtDNA among the other species assigned to these genera prevented an unambiguous assessment of their relationships.

An analysis of mtDNA from seven species of the genus *Aspergillus* was undertaken by Kozlowski and Stepien (1982). Of particular interest was the inclusion of *A. oryzae* and *A. tamarii*, species also compared by Kurtzman et al. (1984) through reassociation of nuclear DNA. Restriction patterns of mtDNA from these two species showed considerable similarity, as would be expected from the nuclear DNA study. One surprise in this study was the great similarity of mtDNAs from *A. tamarii* and *A. wentii*.

In both of the preceding studies, it was pointed out that one potential difficulty with restriction pattern analysis is that mtDNA polymorphisms, which arise from insertions or deletions, will give the erroneous appearance of greater sequence divergence than really exists. Because the rate of change in mtDNA in some organisms may be up to 10-fold more rapid than that of nuclear DNA, the resolution afforded by mtDNA patterns may not be sufficient to recognise the more divergent strains of a species (Groot et al., 1975; Brown et al., 1979).

Ribosomal RNA relatedness

The rather narrow resolution afforded by whole genome DNA reassociation and mtDNA restriction analysis does not allow verification of species assignments within genera or an understanding of intergeneric relationships. The DNA coding for ribosomal RNA (rRNA) appears to be among the most highly conserved sequences known, and it offers a means for assessing affinities above the species level. One of the commonest methods for measuring rRNA complementarity consists of immobilising single stranded DNA into nitrocellulose filters and incubating the filters in a buffer solution with radio-labelled rRNA (Bicknell and Douglas 1970; Kennell 1971; Johnson 1981). A modification of this method was developed by Baharaeen et al. (1983) in which complementary DNA was synthesised on 25S rRNA fragments and then allowed to hybridize with rRNA in solution. As we will discuss, another method that holds considerable promise for phylogenetic studies is the comparison of sequences from 5S rRNA sub-units.

Woese and collaborators (Fox et al., 1980) have effectively assessed phylogenetic relationships among the prokaryotes through catalogueing 16S rRNA oligonucleotides generated with T1 ribonuclease. Johnson and Harich (1983) reported good correlation of values obtained by this method with those from experiments in which rRNAs are hybridized to DNA bound on nitrocellulose. Application of the oligonucleotide catalogueing method to fungi is unlikely in the near future because of the greater number of fragments generated from the 18S ribosomal subunit of eukaryotes (G.E. Fox, person. commun.).

Bicknell and Douglas (1970) were among the first to employ rRNA comparisons to assess phylogenetic relatedness among fungi. Their study focussed on the genus *Saccharomyces*, some species of which now have been assigned to the genera *Zygosaccharomyces*, *Kluyveromyces*, and *Torulaspora*. In general, the more highly related species clustered into groups corresponding to present generic assignments. Similar results were obtained for this same group of species by Adoutte-Panvier et al. (1980) through electrophoretic and immunochemical comparisons of ribosomal proteins. Bicknell and Douglas (1970) also showed that 25S rRNA comparisons cannot resolve the phylogeny of closely related species because of the highly conserved nature of the sequences. In studies of other yeast species, Segal and Eylan (1974a, b, 1975) reached similar conclusions.

The preceding studies have estimated relatedness from the extent of complementarity of 25S rRNA. Another approach has been to compare the actual sequence of 5S rRNA nucleotides. Hori (1976) sequenced the 5S rRNA from a number of prokaryotes and eukaryotes including several yeasts. Using nucleotide substitution as an evolutionary clock, Hori suggested that prokaryotes and eukaryotes diverged approximately 2.5×10^9 years ago. Andersen et al. (1982) compared the 5S rRNA sequence of Phycomyces blakesleeanus with that of other eukaryotes and observed this fungus to show greater similarity with Tetrahymena thermophila than with certain Ascomycetes.

Walker and Doolittle (1982, 1983) also used 5S rRNA sequences to investigate relatedness among the Basidiomycetes. Their work identified two distinct clusters that correlated with the presence or absence of septal dolipores rather than with the traditional separation of these species into the classes Heterobasidiomycetae and Homobasidiomycetae. Further, the data suggest that capped dolipores evolved from capless dolipores which may have evolved from single septal pores.

CONCLUSIONS

As we have seen from numerous examples, application of molecular techniques has provided insight into evolutionary relationships among the fungi that were, in many cases, previously unsuspected. These new findings have changed many of our older concepts, and this has resulted in a considerable number of taxonomic revisions. While these name changes are inconvenient and may even seem capricious, they are necessary if our taxonomy is to reflect phylogeny, a supposition implicit in the assignment of species to genera and of genera to higher orders of classification.

REFERENCES

ADOUTTE-PANVIER, J., DAVIES, J.E., GRITZ, L.R. and LITTLEWOOD, B.S., 1980. Studies of ribosomal proteins of yeast species and their hybrids gel electrophoresis and immunochemical cross-reactions. Molec. Gen. Genet.179:273-282.

ANDERSON, J., ANDRESINI, W. and DELIHAS, N., 1982. On the phylogeny of Phycomyces blakesleeanus nucleotide sequence of 5S ribosomal RNA. J. Biol.Chem. 257:9114-9118.

AULAKH, H.S., STRAUS, S.E. and KWON-CHUNG, K.J., 1981. Genetic relatedness of Filobasidiella neoformans (Cryptococcus neoformans) and Filobasidiella bacillispora (Cryptococcus bacillisporus) as determined by deoxyribonucleic acid base composition and sequence homology studies. Int.J. Syst. Bacteriol. 31:97-103.

BAHARAEEN, S., MELCHER, U. and VISHNIAC, H.S., 1983. Complementary DNA-25S ribosomal RNA hybridization: an improved method for phylogenetic studies. Can.J. Microbiol. 29:546-551.

BAK, A.L., 1973. DNA base composition in mycoplasma, bacteria, and yeast. Curr.Top.Microbiol. Immunol. 61:89-149.

BAK, A.L. and STENDERUP, A., 1969. Deoxyribonucleic acid homology in yeasts. Genetic relatedness within the genus Candida. J.Gen. Microbiol. 59:21-30.

BAK, A.L., CHRISTIANSEN, C. and CHRISTIANSEN, G., 1972. Circular repetitive DNA in yeast. Biochim.Biophys. Acta. 269:527-530.

BERNARDI, G., FAURES, M., PIPERNO, G. and SLONIMSKI, P.P., 1970.Mitochondrial DNAs from respiratory-sufficient and cytoplasmic respiratory-deficient mutants of yeast. J.Mol.Biol. 48:23-43.

BICKNELL, J.N. and DOUGLAS, C., 1970. Nucleic acid homologies among species of *Saccharomyces*. J. Bacteriol. 101:505-512.

BLOCHWITZ, A., 1929. Die Gattung *Aspergillus*. Neue Spezies. Ann.Mycol. 2:205-240.

BRADY, R.J., 1981. A technique for preparation of spheroplasts of *Rhodotorula.* Abstr. Annu. Mtg., K65, p.148. Am.Soc. Microbiol.

BRENNER, D.J., 1973. Deoxyribonucleic acid reassociation in the taxonomy of enteric bacteria. Int.J. Syst. Bacteriol. 23:298-307.

BRENNER, D.J., STEIGERWALT, A.G. and FANNING, G.R., 1972. Differentiation of *Enterobacter aerogenes* from Klebsiellae by deoxyribonucleic acid reassociation. Int.J. Syst. Bacteriol. 22:193-200.

BRITTEN, R.J. and KOHNE, D.E., 1968. Repeated sequences in DNA. Science 161:529-540.

BROWN, W.M., GEORGE, M.Jr. and WILSON, A.C., 1979. Rapid evolution of animal mitochondrial DNA. Proc. Natl. Acad. Sci. USA: 76:1967-1971.

CHRISTIANSEN, C., BAK, A.L., STENDERUP, A. and CHRISTIANSEN, G., 1971. Repetitive DNA in yeast. Nature(London) New Biol. 231:176-177.

COMMERFORD, S.L., 1971. Iodination of nucleic acids *in vitro.* Biochemistry 10:1993-1999.

CROSA, J.H., BRENNER, D.J. and FALKOW, S., 1973. use of a single-strand-specific nuclease for analysis of bacterial and plasmid deoxyribonucleic acid homo- and heteroduplexes. J. Bacteriol. 115:904-911.

CRYER, D.R., ECCLESHALL, R. and MARMUR, J., 1975. Isolation of yeast DNA. *In* Methods in Cell Biology, Vol.12, Yeast Cells (D.M. Prescott, ed.), pp.39-44. New York: Academic Press.

DE LEY, J., CATTOIR, H. and REYNAERTS, 1970. The quantitative measurement of DNA hybridization from renaturation rates. Eur. J. Biochem. 12:133-142.

DENHARDT, D.T., 1966. A membrane filter technique for the detection of complementary DNA. Biochem. Biophys. Res.Commun. 23:641-646.

DOBZHANSKY, T., 1976. Organismic and molecular aspects of species formation. In Molecular Evolution (F.J. Ayala, ed.),pp.95-105. Sunderland Mass: Sinauer Associates.

DUSENBERY, R.L., 1975. Characterization of the genome of *Phycomyces blakesleeanus*. Biochim. Biophys. Acta 378:363-377.

DUTTA, S.K., 1976. DNA homologies among heterothallic species of *Neurospora*. Mycologia 68:388-401.

DUTTA, S.K., SHEIKH, I., CHOPPALA, J., AULAKH, G.S. and NELSON, W.H., 1976. DNA homologies among homothallic, pseudo-homothallic and heterothallic species of *Neurospora*. Molec. Gen. Genet. 147:325-330.

EHRENBERG, L., FEDORCSAK, I. and SOLYMOSY, F., 1976. Diethyl pyrocarbonate in nucleic acid research. Prog. Nucleic Acid Res. 16:189-262.

FOX, G.E., STACKEBRANDT, E., HESPELL, R.B., GIBSON, J., MANILOF, J., DYER, T.A., WOLFE, R., BALCH, W.E., TANNER; R.S., MAGRUM, L.J., ZABLEN, L.B., BALKEMORE, R., GUPTA, R., BONEN, L., LEWIS, B.J., STAHL, D.A., LEUHRSEN, K.R., CHEN, K.N. and WOESE, C.R., 1980. The phylogeny of prokaryotes. Science 209:457-463.

FUSON, G.B., PRICE, C.W. and PHAFF, H.J., 1979. Deoxyribonucleic acid sequence relateeness among some members of the yeast genera Hansenula. Int. J. Syst. Bacteriol.29:64-69.

GILLESPIE, D. and SPIEGELMAN, S., 1965. A quantitative assay for DNA-RNA hybrids with DNA immobilised on a membrane. J. Mol. Biol.12:829-842.

GREEN, B.R. and DICK, M.W., 1972. DNA base composition and the taxonomy of the Oomycetes. Can. J. Microbiol. 18:963-968.

GROOT; G.S.P., FLAVELL, R.A. and SANDERS, J.P.M., 1975. Sequence homology of nuclear and mitochondrial DNAs of different yeasts. Biochim. Biophys. Acta 378:186-194.

GUNGE, N., TAMURU, A., OZAWA, F. and SAKAGUCHI, K., 1981. Isolation and characterization of linear deoxyribonucleic acid plasmids from Kluyveromyces lactis and the plasmid-associated killer character. J. Bacteriol. 145:382-390.

HESSELTINE, C.W., SHOTWELL, O.L., SMITH, M., ELLIS, J.J., VANDEGRAFT, E. and SHANNON, G., 1970. Production of various aflatoxins by strains of the Aspergillus flavus series. In: "Toxic Micro-organisms" (M. Herzberg, ed.) pp. 202-210. Washington,D. C.: U.S. Govt.Printing Office.

HOLZSCHU, D.L., PRESLEY, H.L., MIRANDA, M. and PHAFF, H.J., 1979. Identification of Candida lusitaniae as an opportunistic yeast in humans. J. Clin. Microbiol. 10:202-205.

HOLZSCHU, D.L., TREDICK, J. and PHAFF, H.J., 1981. Validation of the yeast Sporidiobolus ruinenii based on its deoxyribonucleic acid relatedness to other species of the genus Sporidiobolus. Curr. Microbiol. 5:73-76.

HORI, H., 1976. Molecular evolution of 5S RNA. Molec. Gen. Genet. 145:119-123.

HUDSPETH, M.E.S., TIMBERLAKE, W.E. and GOLDBERG, R.B., 1977. DNA sequence organization in the water mold Achlya. Proc.Natl.Acad.Sci. USA 74:4332-4336.

JOHANNSEN, E., 1980. Hybridization studies within the genus Kluyveromyces van der Walt emend. van der Walt. Antonie van Leeuwenhoek 46:177-189.

JOHNSON, J.L., 1973. Use of nucleic acid homologies in the taxonomy of anaerobic bacteria. Int. J. Syst. Bacteriol. 23:308-315.

JOHNSON, J.L., 1981. Genetic characterization. In: "Manual of Methods for General Bacteriology" (P. Gerhardt, ed.) pp.450-472. Washington, D.C.: American Society for Microbiology.

JOHNSON, J.L. and HARICH, B., 1983. Comparisons of procedures for determining ribosomal ribonucleic acid similarities. Curr. Micriobiol. 9:111-120.

KELLY, R.B., COZZRELLI, N.R., DEUTSCHER, M.P., LEHMAN, I.R. and KORNBERG, A., 1970. Enzymatic synthesis of deoxyribonucleic acids. J. Biol. Chem. 245:39-45.

KENNELL, D.E., 1971. Principles and practices of nucleic acid hybridization. Prog. Nucleic Acid Res. Mol. Biol. 11:259-301.

KOZLOWSKI, M. and STEPIEN, P.P., 1982. Restriction enzyme analysis of mitochondrial DNA of members of the genus *Aspergillus* as an aid in taxonomy. J. Gen. Micriobiol. 128:471-476.
KRUMLAUF, R. and MARZLUF, G.A., 1980. Genome organization and characterization of the repetitive and inverted repeat DNA sequences in *Neurospora crassa*. J. Biol. Chem. 255:1138-1145.
KURTZMAN, C.P., 1984a. Resolution of varietal relationships within the species *Hansenula anomala, Hansenula bimundalis,* and *Pichia nakazawae* through comparisons of DNA relatedness. Mycotaxon 19:271-279.
KURTZMAN, C.P., 1984b. Synonomy of the yeast genera *Hansenula* and *Pichia* demonstrated through comparisons of deoxyribonucleic acid relatedness. Antonie van Leeuwenhoek 50:209
KURTZMAN, C.P., JOHNSON, C.J. and SMILEY, M.J., 1979. Determination of conspecificity of *Candida utilis* and *Hansenula jadinii* through DNA reassociation. Mycologia 71:844-847.
KURTZMAN, C.P., SMILEY, M.J., JOHNSON, C.J., WICKERHAM, L.J. and FUSON, G.B., 1980a. Two new and closely related heterothallic species. *Pichia amylophilia* and *Pichia mississippiensis:* Characterization by hybridization and deoxyribonucleic acid reassociation. Int. J. Syst. Bacteriol. 30:208-216.
KURTZMAN, C.P., SMILEY, M.J. and JOHNSON, C.J., 1980b. Emendation of the genus *Issatchenkia* Kudriavzev and comparison of species by deoxyribonucleic acid reassociation, mating reaction, and ascospore ultrastructure. Int. J. Syst. Bacteriol. 30:503-513.
KURTZMAN, C.P., SMILEY, M.J., JOHNSON, C.J. and HOFFMAN, M.J., 1980c. Deoxyribonucleic acid relatedness among species of *Sterigmatomyces.* Abstracts, 5th International Symposium on Yeasts, Y-5.2.5(L), p.246.
KURTZMAN, C.P., SMILEY, M.J., ROBNETT, C.J., AXT, A. and WICKLOW, D.T., 1984. DNA relatedness among the agronomically and industrially important fungi *Aspergillus flavus, A. oryzae, A. parasiticus,* and *A. sojae*. Abstracts, Annual Meeting,American Society for Microbiology O 11, p.190.
LIPINSKI, C., FERRO, A.J. and MILLS, D., 1976. Macromolecule synthesis in a mutant of *Saccharomyces cerevisiae* inhibited by *S*-adenosylmethionine. Mol. Gen. Genet. 144:301-306.
MANACHINI, P.L., 1979. DNA sequence similarity, cell wall mannans, and physiological characteristics in some strains of *Candida utilis, Hansenula jadinii* and *Hansenula petersonii*. Antonie van Leeuwenhoek 45:451-463.
MANIATIS, T., JEFFREY, A. and KLEID, D.G., 1975. Nucleotide sequence of the rightward operator of phage. Proc. Natl. Acad. Sci. USA 72: 1184-1188.
MARMUR, J., 1961. A procedure for the isolation of DNA from microorganisms. J. Mol. Biol. 3:208-218.
MARMUR, J. and DOTY, P., 1962. Determination of the base composition of DNA from its thermal denaturation temperature. J. Mol. Biol. 5:109-118
MARMUR, J., ROWND, R. and SCHILDKRAUT, C.L., 1963. Denaturation and renaturation of DNA. Progr. Nucl. Acid. Res. Mol. Biol. 1:231-300.
MARTINI, A., 1973. Ibridazioni DNA/DNA tra specie di lieviti

del genere Kluyveromyces. Ann. Fac. Agrar. Univ. Studi Perugia 28:1-15.
MARTINI, A. and PHAFF, H.J., 1973. The optical determination of DNA-DNA homologies in yeasts. Ann. Micriobiol. 23:59-68.
McARTHUR, C.R. and CLARK-WALKER, G.D., 1983. Mitochondrial DNA size diversity in the Dekkera/Brettanomyces yeasts. Curr. Genet. 7:29-35.
McCARTHY, B.J. and CHURCH, R.B., 1970. The specificity of molecular hybridization reactions. Annu. Rev. Biochem. 39:131-150.
MENDONCA-HAGLER, L.C. and PHAFF, H.J., 1975. Deoxyribonucleic acid base composition and DNA/DNA hybrid formation in psychrophobic and related yeasts. Int. J. Syst. Bacteriol. 25:222-229.
MENDONCA-HAGLER, L.C., TRAVASSOS, L.R., LLOYD, K.O. and PHAFF, H.J., 1974. Deoxyribonucleic acid base composition and hybridization studies on the human pathogen Sporothrix schenckii and Ceratocystis species. Infect. Immunity 8:674-680.
MEYER, S.A. and PHAFF, H.J., 1969. Deoxyribonucleic acid base composition in yeasts. J. Bacteriol. 97:52-56.
MEYER, S.A. and PHAFF, H.J., 1972. DNA base composition and DNA-DNA homology studies as tools in yeast systematics. In: "Yeasts, Models in Science and Technics" (A. Kochova-Kratochvilova and E. Minarik, eds.) pp.375-386. Prog. 1st Specialized Int.Symp. on Yeasts, Publ. House, Slovak Acad. Science, Bratislava, Czechoslovakia.
MEYER, S.A., ANDERSON, K., BROWN, R.E., SMITH, M.Th., YARROW, D. MITCHELL, G. and AHEARN, D.G., 1975. Physiological and DNA characterization of Candida maltosa, a hydrocarbon utilizing yeast. Arch. Micriobiol. 104:225-231.
MEYER, S.A., SMITH, M.T. and SIMIONE, F.P.Jr., 1978. Systematics of Hanseniaspora Zikes and Kloeckera Janke. Antonie van Leeuwenhoek 44:79-96.
NAKASE, T. and KOMAGATA, K., 1971. Significance of DNA base composition in the classification of yeast genera Cryptococcus and Rhodotorula. J. Gen. Appl. Micriobiol. 17:121-130.
OJHA, M., TURLER, H. and TURIAN, G., 1977. Characterization of Allomyces genome. Biochim. Biophys. Acta 478:377-391.
PHAFF, H.J., MILLER, M.W. and MIRANDA, M., 1979. Hansenula alni, a new.heterothallic species of yeast from exudates of alzder trees. Int. J. Syst. Bacteriol. 29:60-63.
PRICE, C.W., FUSON, G.B. and PHAFF, H.J., 1978. Genome comparison in yeast systematics: delimitation of species within the genera Schwanniomyces, Saccharomyces, Debaryomyces, and Pichia. Micriobiol. Rev. 42:161-193.
RAPER, K.B. and FENNELLE, D.I., 1965. The Genus Aspergillus. Baltimore: Williams & Wilkins.
RHODES, J.C. and KWON-CHUNG, K.J., 1982. A new efficient method for protoplast formation in Cryptococcus neoformans. Abstr. Annu. Mtg, F61, p.336. Am.Soc.Micriobiol.
SAITO, K., 1943. On the scientific name of Aspergilli isolated in Japan. Nippon Jozo Zasshi 38:412-414.
SCHILDKRAUT, C.L., MARMUR, J. and DOTY, P., 1962. Determination of the base composition of deoxyribonucleic acid from its buoyant density in CsCl. J. Mol. Biol. 4:430-433.
SEGAL, E. and EYLAN, E., 1974a. Genetic relatedness of Candida albicans to asporogenous and ascosporogenous yeasts as

reflected by nucleic acid homologies. Microbios 9:25-33.
SEGAL, E. and EYLAND, E., 1974b. Nucleic acid homologies between Candida albicans and Hansenula species. Microbios 10:133-138.
SEGAL, E. and EYLAN, E., 1975. Nucleic acid homology studies among Candida albicans, Syringospora albicans, and Leucosporidium species. Microbios 12:111-117.
SEIDLER, R.M. and MANDEL, M., 1971. Quantitative aspects of DNA renaturation: DNA based competition, state of chromosome replication, and polynucleotide homologies. J.Bacteriol. 106:608-614.
SMITH, D. and HALVORSON, H. O. 1967. The isolation of DNA yeasts. In Methods in Enzymology (L. Grossman and Moldave, K. eds.) vol 12 part A. pp. 538-541. New York: Academic Press.
STORCK, R. 1966. Nucleotide composition of nucleic acids of fungi. II. Deoxyribonucleic acids. J. Bacteriol. 91: 277-320.
STORCK, R. and C. J. ALEXOPOLOUS. 1970. Deboxyribonucleic acid of fungi. J. Bacteriol. Rev. 34: 126-154.
STORCK, R., ALEXOPOLOUS, C. J. and H. J. PFAFF. 1969. Nucleotide composition of deboxyribonucleotide acid of some species of Cryptococcus, Rhodotorula and Sporobolomyces. J. Bacteriol. 98: 1069-1072.
STORCK, R., NOBLES, M. K. and C. J. ALEXOPOLOUS. 1971. The nucleotide composition of deboxyribonucleotide acid of some species of Hymenochaetaceae and Polyporaceae. Mycologia 653: 38-49
SUBIRANA, J. A. and T. DOTY. 1966. Kinetics of renaturation of renatured DNA. I. Spectrophotometric results. Biopolymers 4: 171-187.
THOM, C. and K. B. RAPER. 1945. A manual of Aspergilli. Baltimore: Williams & Wilkins
TIMBERLAKE, W. E. 1978. Low repetitive DNA content in Aspergillus nidulans. Science 202: 973-975.
ULLMAN, J. S. and B. J. McCARTHY. 1973. The relationships between mismatched base pairs and the thermal stability of DNA duplexes. I. Effects of depurination and chain scission. Biochem. Biophys. Acta 294: 405-415.
VAN DER WALT, J. P. and E. JOHANNSON. 1979. A comparison of interfertility and in vitro DNA-DNA reassociation as criteria for speciation in the genus Kluyveromyces. Antonie van Leeuwenhoek 45: 281-291.
VAN ETTEN, J. L. and S. N. FREER. 1978. Simple procedure for disruption of fungal spores. Appl. Environ. Microbiol. 35: 622-623.
VILLA, V. D. and R. STORCK. 1968. Nucleotide composition of nuclear and mitochondrial deoxyribonucleotide acid of fungi. J. Bacteriol. 96: 184-190.
WALKER, W. F. and W. F. DOOLITTLE. 1982. Redividing the basidiomycetes on the basis of 5S rRNA sequences. Nature 299: 723-724.
WANG, H. L. and C. W. HESSELTINE. 1982. Oriental fermented foods. In Prescott & Dunn industrial Microbiology, 4th ed. (G. Reed ed.), pp.492-538. Westport: AVI Publishing Co.
WICKLOW, D. T. 1984. Adaptation in wild and domesticated yellow-green aspergilli. In Toxigenic Fungi-- Their toxins and health hazards (H. Kurata and Y. Ueno, ed.)pp. 78-86. Amsterdam: Elsevier.

WILLIAMS, N. P., MUKHOPADHYAY, and S. K. DUTTA. 1981. Homologies of Neurospora homothallic species using repeated and nonrepeated DNA sequences. Experientia 37: 1157-1158.
WILLIAMSON, D. H. and D. J . FENNELL. 1975. The use of fluorescent DNA-binding agent for detecting and separating yeasts mitochondria DNA. In Methods in cell biology, vol. 12. Yeast cells (D. M. Presscott, ed.) pp. 335-351. New York: Academic Press.
YARROW, D. and S. A. MEYER. 1978. Proposals for amendment of the diagnosis of the genus Candida Berkhout, nom. conv. Int. J. Syst. Bacteriol. 28: 611-615.

DIALOGUE FOLLOWING DR. WICKLOW'S PRESENTATION OF DR.KURTZMAN'S PAPER

KLICH: I have discussed some of Dr. Kurtzman's work with several molecular biologists and their opinion was negative. They wondered why he wasn't using more modern techniques specifically such as genetic mapping rather than just DNA relatedness.

WICKLOW: Kurtzman's techniques have enabled him to make anamorph-telemorph links in yeasts where this was unsuspected, and do away with some irrelevant physiological tests. It seems to be a very powerful tool. In the fast moving field of molecular biology what is modern or what is not is a naive question. Whole genome comparisons have been "calibrated" against genetic fertility in order to relate results to biological phenomena. Presumably, you mean gene sequencing when you say genetic mapping. While gene sequencing is certainly a much more sensitive technique, one must ask how many genes of the thousands present must be sequenced to reflect whether or not the strain are interfertile. In the human population we know that the genes for eye colour, hair texture and colour etc. have somewhat different sequences, yet this does not affect fertility.

KLICH: What does 100% relatedness actually mean?

WICKLOW: Just that; but you can have 100% relatedness and still have things that won't cross because one isolate may have a chromsomal inversion, for example: This is uncommon. Also differences in a few base sequences would not be detected.

KOZAKIEWICZ: Has Dr. Kurtzman looked at Aspergillus thomii? This appears to be a brown variant of A. flavus, and it would be interesting to check their DNA homology.

WICKLOW: He hasn't done that.

BRIDGE: Does 4% relatedness mean 4% hybridization?

WICKLOW: Yes.

SAMSON: Does this kind of work support the idea that industrially used strains like P. camemberti and Rhizopus oligosporus are domesticated variants of other known species?

WICKLOW: Yes, for example J.J. Ellis (Mycologia 77:243-247.1985) has used DNA homologies to show that Amylomyces rouxii is a variant of Rhizopus arrhizus.

PITT: I've seen wild type P. camemberti strains, so I do not agree that it is a domesticated variant of P. aurantiogriseum.

CHRISTENSEN: How can two species with 100% homology have visually different characteristics? As A.flavus and A. oryzae do?

WICKLOW: That's a good question. The answer is that many small mutations can cause morphological variation, because of small variations in base sequences, but the great majority of DNA base sequences is still conserved. When we measure DNA homology, we are studying a huge amount of DNA which tends to mask these small gene differences which we see as morphological differences. This question also finds an analogy in the human eye colour comment.

UDAGAWA: Is there a difference between A. parasiticus and A. toxicarius?

WICKLOW: We used strains that would have been classified by Murakami as A. toxicarius and the came out as A. parasiticus.

UDAGAWA: In Dr. Samson's opinion A. toxicarius is synonymous with A.flavus.

WICKLOW: This model would suggest that A. flavus is one large species including A. toxicarius and A. parasiticus. But there are sufficcient differences in ther ecology and economic importance which make it useful to distinguish them.

PITT: If, as you say, A. oryzae is actually a domesticated form of A. flavus, it should not be found in nature. However, I have recently seen some isolates from New Zealand from grain dust which I would place in A. oryzae. There is effectively no fermentation industry using A. oryzae in New Zealand, so these are clearly isolates from nature. Domesticated A. oryzae strains undoubtedly exist: I'm not certain that these isolates from New Zealand are taxonomically identical with the domesticated strains.

WICKLOW: Some of Raper and Fennell's isolates of A. oryzae came from soil cores.

ECOLOGICAL ADAPTATION AND CLASSIFICATION IN ASPERGILLUS AND PENICILLIUM

Donald T. WICKLOW

Northern Regional Research Center
Agricultural Research Service, U. S. Department of Agriculture, Peoria, Illinois, USA 61604

SUMMARY

Various adaptations of Aspergilli and Penicillia to their natural world have provided taxonomists the characters used in the classification of these moulds. From an ecologist's perspective, the potential adaptive value of such characters in nature is considered and their importance in placing these fungi within existing classification schemes is examined. The ecological data important in defining the niche of distinct populations of Aspergilli and Penicillia can be a strong predictor of genetic relatedness and taxonomic designation. This paper provides a sampling of the kinds of questions taxonomists should ask in attempting to identify clusters of characters that suggest a unique fungal niche and genetic divergence at the species level.

INTRODUCTION

Moulds that we classify as Aspergilli or Penicillia are frequently better known for their biochemical attributes than their ecology. Individual taxa are recognized as components of the soil mycoflora, as colonists of agricultural produce and products, as pathogens of insects and vertebrates, as domesticated forms used in food fermentations, and as commercial sources of enzymes, organic acids, antibiotics, and assorted other natural products. While considerable information has accumulated about the adaptive behavior of certain of these molds in environments where food or feed is stored, there has been little or no effort made to link different fungal adaptations to the distribution of these organisms in the wild (Raper and Fennell, 1965; Pitt, 1979). This paper attempts to assess the value of selected morphological, physiological, genetic, and biochemical adaptations in the life history and ecology of the Aspergilli and Penicillia. As taxonomic characters these fungal adaptations also impact differently in defining existing classification schemes. Kendrick (1978) has emphasized that, to define many of our taxa properly, we must not only consider the development, life history and morphology of the holomorph, but also place the information in an environmental context that specifies the ecological demands made by each species. For those who are considering the application of biochemical techniques for obtaining estimates of genetic variation between populations, it should be noted that most criteria traditionally used to separate yeasts are not very accurate predictors of DNA relatedness (Kurtzman et al., 1983). However, ecological data on host specificity and niche width of cactus yeasts is a strong predictor of genetic relatedness and taxonomic designation (Starmer et al., 1982). It is expected that through an understanding of the ecological requirements and niche determinants of Aspergilli and Penicillia one can organize a classification system based on adaptive characteristics relevant to the ecology and evolution of these moulds.

The ultimate goal of ecology is to understand how naturally occurring communities are organized at the population level and how these communities function as a system (McNaugthon, 1981). To say that a fungal isolate is from soil or colonizes agricultural produce does little to explain the organism's ecological requirements or functional role in nature. Unfortunately, little attention has been given to studies of the autecology of saprotrophic fungi (Frankland, 1984). We are able to characterize only a few obvious parameters defining the niche of Aspergilli and Penicillia from cereals (Wallace and Sinha, 1981). When considering patterns of variation among a collection of mould isolates, taxonomic mycologists seldom ask whether the characters in question are of any adaptive value to the organism in nature. If so, how might this help explain the natural distribution of isolates with or without such characters? I attempted this approach with the yellow-green Aspergilli (Wicklow, 1983; 1984 b, c). It was possible to logically explain why characters associated with 'wild' species (e.g, Aspergillus flavus Link; A. parasiticus Speare) isolated from nature might have been lost or altered through domestication in a koji environment. In this example, the domesticated koji moulds A. oryzae (Ahlburg) Cohn and A. sojae Sakaguchi & Yamada ex Murakami show reduced sporulation (e.g., fewer and smaller, uniseriate conidial heads), produce conidiophores and conidia that are less conspicuously roughened than those of the species, and no longer produce sclerotia or aflatoxin. At the same time, the lifting of natural environmental constraints (natural selection) was linked to greater variability in dimensions of conidia and conidiophores, as well as the production of a flocculent aerial mycelium. I also suggested that the elimination of costly chemical defenses, (e.g. mycotoxins) and survival structures such as sclerotia, improves the fitness of koji molds in a koji environment (Wicklow, 1984 b). Koji is a preparation of steam-cooked and mould-fermented cereals. Selection in a koji environment has favoured those individuals capable of rapid conidium germination and germ tube extension (Wicklow, 1984 c), and having a complement of enzymes suitably tailored for attack of either cooked rice or soybeans (Murakami, 1971).

Comparisons of nuclear DNA complementarity provide data supporting the view that domesticated yellow-green koji moulds are derived from the wild species, A. flavus and A. parasiticus (Kurtzman, this publication). The results of DNA comparisons involving other yellow-green Aspergilli provide an even more compelling evidence for emphasizing the ecological niche in approaches to classification. The discovery of a new species in the A. flavus group, represented primarily by strains isolated from diseased alkali bees in Wyoming (Hesseltine et al., 1970), prompted a closer examination of these strains.

In my laboratory B. W. Horn (unpublished data) has determined that the new species grows very slowly at 42 C producing a colony diameter of less than 1.5 cm on Czapek agar, while among strains of A.flavus growth was 2.3-3.7 cm. In addition, the new species produces sclerotia that elongate from an immature, white tip, which eventually hardens and forms a black rind. Otherwise, the new species is indistinguishable from A. flavus, which forms rounded sclerotia. The DNA comparisons also show A. leporis States & Christensen and A.flavus to be distinct species. Samson (1979) once considered placing A.leporis in synonymy with A.flavus. I have recently reported that A. leporis produces sclerotia on rabbit dung, an ability not shared by A. flavus or A. parasiticus (Wicklow, 1985). With the benefit of hindsight, might we not have anticipated that organisms from such unique habitats would be uniquely adapted to these habitats and thus represent potentially different species?

The following discussion considers the significance of different physiological, morphological, genetic and biochemical characteristics of Aspergilli and Penicillia as adaptations shaping the niche of individual species populations. There is evidence to suggest that some of these adaptive characteristics may be more useful than others in predicting species-level differences.

GROWTH HABIT AND REPRODUCTIVE STRATEGIES

The Penicillium growth pattern, as described by Sewell (1959) is one in which the fungus densely colonizes small individual substrates with no extension of the mycelium into the surrounding soil and then produces conidia heavily over the surface. In my opinion, this presents too narrow a view of the different mycelial growth patterns exhibited by many Aspergilli and Penicillia. Thom (1930) observed that among Penicillium species a submerged mycelium is usually massed within the first few millimeters below the agar surface, but in some species the mycelium "penetrates deeply into the mass infected". This pattern may be a speciality of Aspergilli and Penicillia, which invade seeds or other discrete, nutrient-rich substrates, enabling these fungi to exploit an entire resource. This could represent more than a casual observation by Thom (1930), if we consider the potential importance of such a characteristic in defining the niche of a species population.

Patterns of mycelial growth in culture also provides characters used in describing colony morphology. Are these cultural artefacts or evidence of fungal adaptation to growth in the wild? Thom (1930) believed that habit sketches made by direct observation of the colony were essential to any proper placement of the organism "among its fellows in the group". Some Aspergilli and Penicillia produce colonies that are dominated by conidiogenous structures arising directly from submerged hyphae. Other species produce their conidiophores from aerial mycelium, in the form of individual trailing hyphae, ropes of hyphae, or coremia. Coremia represent an obvious adaptation that promotes spore dispersal, particilarly when a fungus-infected insect pupa or seed lies beneath a layer of forest litter. Does funiculose growth have adaptive value or is it merely a condition from cultural deterioration of fasciculate or coremiform strains? While Thom (1930) states that "guesses as to the genetic significance of ropes of hyphae would be unprofitable", his habit sketches reveal that the rate of extension of these aerial hyphal strands keeps pace with the expanding colony margin. In the three-dimensional milieu of soil and litter, sparsely branched mycelial strands and rhizomorphs represent special multihyphal structures, which are capable of extending in one direction, enabling a fungus to invade discontinuously distributed nutrient sources (Smith and Berry, 1974; Cooke and Rayner, 1984). It is worth comparing habit sketches by Thom (1930) showing diffuse mycelial growth within an agar substrate and ropy growth above the colony surface, with W. Thompson's photographs (see Fig. 2.5 in Cooke and Rayner, 1984) contrasting the growth of wood-decaying Basidiomycotina in a non-sterile soil, where it forms cords, and a radiation-sterilized soil, where it grows as a diffuse mycelium. Cooke and Rayner point out that in sterile soil, nutrients are more freely available and established mycelium may "dwell", but where resources are more limited and localized (as in non-sterile soil), channeling of the mycelium occurs, so facilitating increased exploration of the environment through rapid mycelial extension and high 'inoculum potential' (sensu Garrett, 1970).

Deposits of yellow, orange or red pigment granules encrusted on solitary aerial hyphae or hyphal ropes contribute to colony color and thus are recognized in species descriptions of Penicillium (Raper and Thom, 1949; Pitt, 1979). Likewise, in the Basidiomycotina, hyphae on the outside of a mycelial cord are frequently heavily encrusted with calcium oxalate crystals, which are believed to have an excretory or antibiotic function (Sollins et al., 1981; Cooke and Rayner, 1984). Could deposits of toxic fungal secondary metabolites (i.e. oxalic acid) on the surface of ropes of hyphae serve to protect the hyphal aggregate from litter-inhabiting, fungus-feeding arthropods?

It often has been observed that Aspergilli and Penicillia like other fungi will produce larger and more complex conidial heads when originating from the substrate (culture medium) and smaller, simplified conidial heads from the aerial hyphae (Raper and Thom, 1949; Raper and Fennell, 1965). In this example a fungus has the advantage of being able to produce large numbers of conidia from a conidiophore 'rooted' in a nutrient-rich substrate. Malloch and Cain (1972) and Pitt (1979) have suggested that terverticillate Penicillia represent an advanced group, adapted to growth in cereals, food products, and other substrates of reduced water activity, and to producing maximal numbers of conidia per conidiophore.

There are rather obvious physical limits to the numbers of conidial heads a fungus can produce on a seed or other discrete nutrient-rich substrate. In this connection, I have suggested elsewhere (Wicklow, 1984 b) that the biseriate condition in A. flavus enables the fungus to support larger numbers of phialides from a single conidial head and thus convert substrate nutrients acquired by vegetative hyphae into greater numbers of dispersal propagules (conidia). Kozakiewicz (1978) estimated that biseriate heads of A. flavus produce at least four times as many phialides as related uniseriate forms. This biseriate condition is lost (or not expressed?) when wild strains of A. flavus become domesticated in a koji environment.

If we learn more about the ecology of these moulds it may be possible to explain, how complex (e.g. biseriate, terverticillate) versus simple (e.g. uniseriate, monoverticillate) conidial heads meet the organism's requirements for production of asexual dispersal propagules. Fungal ecologists should find it interesting that A. thomii G. Smith can be either biseriate or uniseriate, depending on the culture medium chosen (Raper and Fennell, 1965). In the classification of Aspergillus species, the importance of uniseriate or biseriate heads is treated with certain inconsistencies. For example, Raper and Fennell (1965) included both uniseriate and biseriate species in the A. niger-, A. flavus- and A. flavipes groups but excluded uniseriate forms such as A. flaschentraegeri Stolk from the A. wentii group, A. brunneo-uniseriatus Singh & Bakshi from the A. ustus group, and A. raperi Stolk from the A. versicolor group.

When we examine isolates of Penicillium or Eupenicillium from forest soils, the penicillus is commonly monoverticillate or divaricate with the conidiophores being produced as branches from aerial mycelium. This suggests to me that in forest litter these fungi may exhibit a diffuse pattern of mycelial growth with conidial heads produced from hyphae that form bridges between litter fragments. Thom (1930) observed that one series of related forms of Penicillium was constantly seen among cultures isolated from soil. He referred to these organisms as the "Soil Penicillia" and grouped them in the Lanata-divaricata and Asymmetrica-funiculosa sections of his monograph. Clearly, Thom recognized a relationship between morphology of the penicillus and habitat distribution of the fungus.

Conidiophore wall roughenings may protect the conidiophore from arthropod predation. In proposing this for A.flavus, I suggested that a critical stage in the fungal life cycle, where the loss of conidial heads would result in the greatest fitness loss to the fungus, would be at the point where conidiophores are produced from sclerotia that germinate at the soil surface (Wicklow, 1984 b; Wicklow and Donahue, 1984). Sclerotium-forming species of Penicillium are described with roughened walls, while some anamorphs of Eupenicillium species are illustrated as having smooth walls (Stolk and Samson, 1983). In Eupenicillium the sclerotioid ascostromata eventually ripen to form ascospores. We know nothing about the role of these ascospores in the fungal life cycle, although at some point they give rise to a mycelium, upon which the conidiophore is formed. Is it less critical to the survival of species producing an ascosporic stage (e.g. Emericella, Eurotium, Eupenicillium or Talaromyces) that thier anamorphs be protected from potential mycophagists?

In experiments to test the suitability of Aspergillus spp. as food for the foreign grain beetle Ahasverus advena (Waltl.), Hill (1978) made the interesting observations that beetle larvae consumed conidial heads of Eurotium but did not eat the cleistothecia. Eurotium cleistothecia are covered with a felt of sterile aerial hyphae encrusted with yellow, orange or red granules which, if toxic, could make such fungal structures less attractive to arthropods.

Fasciculate Penicillia which have strongly roughened conidiophores but smooth conidia (e.g. P. viridicatum Westling, P. aurantiogriseum Dierckx, etc.) are among the most commonly encountered toxigenic moulds in foods or feeds stored below 25°C. While the coarsely roughened conidiophore stipe could discourage arthropod predators, the smooth conidia should not. Fungus-feeding mites are very important in the dispersal of storage moulds within grain bulks (Sinha et al., 1969) and it seems reasonable to expect that the more successful storage moulds have adapted morphological features, which promote their dispersal.

Having an ecological basis from which to interpret reproductive strategies in Aspergillus, Penicillium and their related teleomorphs, will some day enable students of these fungi to more rationally explain correlations among certain taxonomic characters. In recent decades we have greatly expanded our knowledge of teleomorph-anamorph relationships in Eupenicillium (Scott and Stolk, 1967; Pitt, 1979; Stolk and Samson, 1983), because methods were developed to selectively isolate ascosporic fungi from soils (Warcup, 1951; Warcup and Baker, 1963; Samson and Mahoney, 1977). However, for only one of these taxa do we have any information about the role of the teleomorph in the fungal life cycle. E. ochrosalmoneum Scott & Stolk, a producer of the mycotoxin citreoviridin (Ueno and Ueno, 1972), forms immature ascostromata in preharvest corn in southern Georgia, and these propagules are dispersed onto the soil surface by combine harvesters (Wicklow et al., 1984). In our laboratory, we have recently shown (B. W. Horn and D.T. Wicklow, unpublished) that these detached ascostromata ripen to form ascospores when incubated from 4-6 weeks at 20-30° C on moist, non-sterile soil. Is the 1-6 week ripening time for E. ochrosalmoneum ascostromata important to this fungal disease cycle? Overwintering of ascostromata in their immature sclerotial form may be essential to fungal survival in soil and to the eventual dispersal of ascospores, as primary infective inoculum. When immature ascostromata were buried in moist, non-sterile soil for up to 16 weeks, they remained viable but did not develop ascospores. This unique example of 'soil mycostasis' may be an adaption preventing ascospores from forming below the soil surface, where opportunity for dispersal is limited.

An interesting correlation can be observed between rate of ascostromatal ripening in Eupenicillium and the complexity of the penicillus. Rapid rates of ripening are commonly found among species having a fragmentary penicillus, while ascostromata that are slow to ripen typically produce a larger penicillus with greater numbers of phialides (Pitt, 1979). This suggests to me that as the interval from ascospore to ascospore is reduced, conidia become less important in the fungal life cycle. It is noteworthy that there are apparently no naturally occurring anamorph-producing forms of Eurotium, which fail to produce cleistothecia. Apparently the formation of ascospores is so critical in enabling this group of moulds to occupy a specific niche, that loss of the ability to form cleistothecia is lethal.

Only a very few Aspergilli and no Penicillia have been found to be heterothallic. I have theorized elsewhere that an ephemeral existance in a limited resource, such as dung, has promoted self-fertility in coprophilous ascomycetes, because the chance of outcrossing is reduced (Wicklow, 1981). This must certainly also apply to life history strategies of Aspergilli and Penicillia.

TEMPERATURE AND WATER RELATIONS

Pitt (1979) has greatly advanced our understanding of the taxonomic relationships in the genus Penicillium because, in an effort to broaden the range of characters on which Penicillium taxonomy could be based, he chose to emphasize the techniques of incubating cultures at high and low temperatures (37°C and 5°C) and on a medium of reduced water activity. The data obtained showed high infraspecific correlation, but often with useful differences between species. I believe that most fungal ecologists would agree, moisture and temperature relationships rank among the most important parameters defining the niche of saprotrophic moulds such as Aspergillus and Penicillium. In choosing to emphasize moisture and temperature relations as taxonomic characters, Pitt (1979) has not only found stable characters distinguishing species, but fungal niches as well. I have difficulty with Stolk and Samson (1983) placing in synonomy teleomorphs and anamorphs with remarkably different growth patterns at the same temperature or substrate water activity. For example, growth data at 25°C and 37°C (Pitt, 1979) for taxa included in E. euglaucum (van Beyma) Stolk & Samson shows great variability (Fig. 1). To explain these differences as polymorphic variation within a species, would require that we reappraise our concept of the fungal niche in Aspergillus and Penicillium. Pitt (1979) has argued that in the absence of genetic controls the stability of 'natural species' must be under strict environmental control.

PREDATOR DEFENSE CHEMISTRY

All fungi produce natural products of low molecular weight referred to as secondary metabolites. Secondary metabolism is often distinguished from primary metabolism, because secondary metabolites are considered not essential for cellular growth and life of the producing organism. Most students of secondary metabolism now believe (1) that secondary metabolism should be viewed as chemical differentiation a process analogous to the acquisition of morphological differences in an organism; and (2) that the diversity of chemical structures and biological acticities demonstrated by fungal

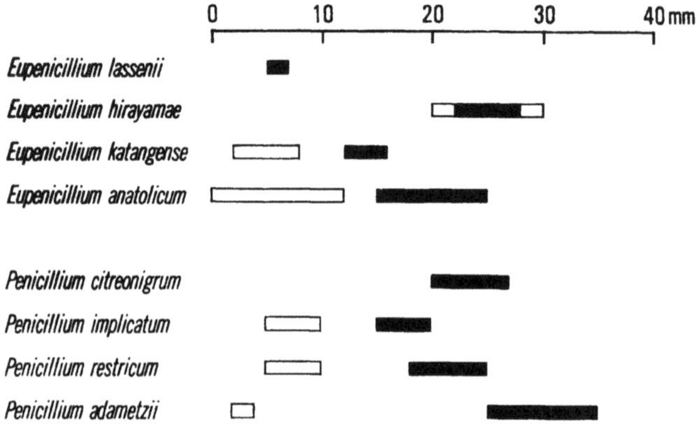

Fig. 1. Growth data (mm growth at 7 days) for species placed in synonymy with Eupenicilllium euglaucum (van Beyma) Stolk & Samson (1983).

metabolites suggest a diversity of ecological functions (Calam, 1979; Bu'Lock, 1980; Woodruff,1980; Bennett, 1983). It has been argued that fungi produce metabolites toxic to insects and vertebrates (mycotoxins) as a means of protecting themselves and the substrate they colonize from predation (Whittaker and Feeney, 1971; Janzen, 1977; Wicklow, 1984a). The ability of Aspergillus and Penicillium species to occupy a specific fungal niche may depend on their production of specific kinds of toxic secondary metabolites. Therefore, the relevance of secondary metabolite profiles to a classification system that emphasizes the fungal niche is considerable. Ciegler et al. (1973) and Frisvad and Filtenborg (1983) have emphasized profiles of secondary metabolites in their classification of terverticillate Penicillia. They made an important observation that conidial colour differences, which traditionally have had important taxonomic significance in this group, are based merely the ability to produce the yellow pigment brevianamide A. Frisvad and Filtenborg (1983) discovered that strains of P. viridicatum I that produce brevianamide A have bright yellow-green conidia, while non-producing strains have green conidia. Might the latter also resemble strains of P. aurantiogriseum Dierckx? Does loss in ability to produce brevianamide A explain why both Raper and Thom (1949) and later Pitt (1979) recognized the following strains as bright to dark yellow-green, but Samson et al. (1976) considered them blue-green and placed all in synonomy with P. verrucosum Dierckx var. cyclopium (Westling) Samson et al.: P. olivino-viride Biourge NRRL 959, NRRL 2028; P. casei Staub NRRL 844; P. palitans Westling NRRL 2033; P. crustosum Thom NRRL 968?

Brevianamide A is localized in the penicillus and may serve as a chemical deterrent to fungus-feeding arthropods. Fungal metabolites such as brevianamide A, which are not effective antibiotics or mammalian toxins have seldom been studied in insects.

REFERENCES

BENNETT, J.W. 1983. Differentiation and secondary metabolism in mycelial fungi, pp. 1-32. In: "Secondary Metabolism and Differentiation in Fungi." eds. J.W. Bennett and A. Ciegler. New York and Basel: Marcel Dekkker.

BU'LOCK, J.D. 1980. Mycotoxins as secondary metabolites, pp. 1-16. In: "The Biosynthesis of Mycotoxins, a Study in Secondary Metabolism", Edited by P.S. Steyn. New York: Academic Press

CALAM, C.T. 1979. Secondary metabolism as an expression of microbial growth and development. Folia Microbiol. 24: 276-285.

CIEGLER, A., D.I. FENNELL, G.A. SANSING, R.W. DETROY and G.A. BENNETT. 1973. Mycotoxin-producing strains of Penicillium viridicatum: classification into subgroups. Appl. Microbiol. 26:271-278.

COOKE, R.C. and A.D.M. RAYNER. 1984. "Ecology of Saprotrophic Fungi". London and New York: Longman.

FRANKLAND, J.C. 1984. Autecology and the mycelium of a woodland litter decomposer, pp. 241-26. In: The Ecology and Physiology of the Fungal Mycelium". Eds. D.H. Jennings and A.D.M. Rayner. London: Cambridge University Press.

FRISVAD, J.C.. and O. FILTENBORG 1983. Classification of terverticillate penicillia based on profiles of mycotoxins and other secondary metabolites. Appl. Environ. Microbiol. 46: 1301-1310.

GARRETT, S.D. 1970. "Pathogenic Root-Infecting Fungi". Cambridge: Cambridge Univ. Press.

HESSELTINE, C.W., O.L. SHOTWELL, M. SMITH, J.J. ELLIS, E. VANDEGRAFT, and G. SHANNON. 1970. Production of various aflatoxins by strains of the Aspergillus flavus series, pp. 202-210. In: "Toxic Micro-organisms". Ed. M. Herzberg. U.S. Gov.'t. Printing Office, Washington, D.C.

HILL, S.T. 1978. Development of Ahasverus advena (Coleoptera Silvanidae) on seven species of Aspergillus and on food molded by two of these. J. Stored Prod. Res. 14: 227-231.

JANZEN, D.H. 1977. Why fruits rot, seeds molds and meat spoils. Am. Nat. 111: 691-713.

KENDRICK, B. 1978. Recent developments in the systematics of pathogenic fungi, pp. 119-137. In: "Biosystematics in Agriculture". Ed. J.A. Romberger. Allanheld, Osmun, & Co., Montclair.

KOZAKIEWICZ, Z. 1978. Phialide and conidium development in the Aspergilli. Trans. Br. mycol. Soc. 70: 175-186.

KURTZMAN, C.P., H.J. PHAFF and S.A. MEYER. 1983. Nuclei acid relatedness among yeasts, pp. 139-166. In: "Yeast Genetics, Fundamental and Applied Aspects". Eds J.F.T.Spencer, D.M. Spencer and A.R..W. Smith. New York: Springer-Verlag.

MALLOCH, D. and R.F. CAIN. 1972. The Trichocomataceae: Ascomycetes with Aspergillus, Paecilomyces and Penicillium imperfect states. Can. J. Bot. 50: 2613-2628.

McNAUGHTON, S.J. 1981. Niche: definition and generalizations, pp.79-88. In: "The Fungal Community, Its Organization and Role in the Ecosystem". Eds D.T. Wicklow and G.C. Carroll. New York and Basel: Marcel Dekker.

MURAKAMI, H. 1971. Classification of the koji mold. J. Gen. Appl.Microbiol., Tokyo., 17: 281-309.

PITT, J.I. 1979. "The genus Penicillium and its teleomorphic states Eupenicillium and Talaromyces". London: Academic Press. London.

RAPER, K.B. and D.I. FENNELL. 1965. "The Genus Aspergillus". Baltimore: Williams & Wilkins.
RAPER, K.B. and C. THOM. 1949. "A Manual of the Penicillia". Baltimore: Williams and Wilkins.
SAMSON, R.A. 1979. A compilation of the Aspergilli described since 1965. Stud. Mycol., Baarn 18: 1-38.
SAMSON, R.A. and D.P. MAHONEY. 1977. Talaromyces galapagensis sp.nov. Trans. Br. Mycol. Soc. 69: 158-160.
SAMSON, R.A., A.C. STOLK and R. HADLOK. 1976. Revision of the subsection Fasciculata of Penicillium and some allied species. Stud. Mycol. Baarn 11: 1-47.
SCOTT, D.B. and A.C. STOLK. 1967. Studies on the genus Eupenicillium Ludwig. II. Perfect states of some Penicillia. Antonie van Leeuwenhoek 33: 297-314.
SEWELL, G.W.F. 1959. Studies of fungi in a Calluna-heathland soil II. By the complementary use of several isolation methods. Trans. Br. Mycol. Soc. 42: 354-369.
SINHA, R.N., H.A.H. WALLACE and F.S. CHEBIB. 1969. Principal component analysis of interrelations among fungi, mites and insects in grain bulk ecosystems. Ecology 50: 536-547.
SMITH, J.E. and D.R. BERRY. 1974. "An Introduction to biochemistry fungal development". London: Academic Press.
SOLLINS, P., K. CROMACK, R. FOGEL and C.Y. LI. 1981. Role of low-molecular-weight organic acids in the inorganic nutrition of fungi and higher plants, pp. 607-619. In: "The Fungal Commnity Its Organization and Role in the Ecosystem". Eds D.T. Wicklow and G.C. Carroll. New York and Basel: Marcel Dekker.
STARMER, W.T., H.J. PHAFF, MARY MIRANDA, M.W. MILLER, and W.B. HEED. 1982. The yeast flora associated with the decaying stems of columnar cacti and Drosophila in North America, pp. 269-295. In: "Evolutionary Biology" Vol. 14. Eds. M.K. Hecht, B. Wallace and C.T. Prance. New York: Plenum Publ.
STOLK, A.C. and R.A. SAMSON. 1983. The ascomycete genus Eupenicillium and related Penicillium anamorphs. Stud. Mycol., Baarn 23: 1-149.
THOM, C. 1930. "The Penicillia". Baltimore: Williams and Wilkins.
UENO, Y. and I. UENO. 1972. Isolation and acute toxicity of citreoviridin, a neurotoxic mycotoxin of Penicillium citreo-viride Biourge. Jap. J. Exp. Med. 42: 91-105.
WALLACE, H.A.H. and R.N. SINHA. 1981. Causal factors operative distributional patterns and abundance of fungi: A multivariate study, pp. 233-247. In: "The Fungal Community its Organization and Role in The Ecosystem". Eds D.T. Wicklow and G.C. Carroll. New York and Basel: Marcel Dekker.
WARCUP, J.H. 1951. Soil Steaming: A selective method for the isolation of Ascomycetes from soil. Trans. Br. Mycol. Soc. 34: 515-532.
WARCUP, J.H. and K.F. BAKER. 1963. Occurrence of dormant ascospores in soil. Nature Lond. 197: 1317-1318.
WHITTAKER, R.H. and P.F. FEENEY. 1971. Allelochemics: Chemical interactions between species. Science, N.Y. 171: 757-790.
WICKLOW, D.T. 1981. The coprophilous fungal community: a mycological system for examining ecological ideas, pp. 47-76. In: "The Fungal Community Its Organization and Role in the Ecosystem". Eds D.T. Wicklow and G.C. Carroll. New York and Basel: Marcel Dekker.
WICKLOW, D.T. 1983. Taxonomic features and ecological signifi-

cance of sclerotia, pp. 6-12. In: "Aflatoxin and Aspergillus in corn". Eds U.L. Diener, J.W. Dickens, and R.L. Asquith. Southern Cooperative Series Bulletin 279. Alabama Agricultureal Experiment Station, Auburn.
WICKLOW, D.T. 1984 a. Ecological approaches to the study of mycotoxigenic fungi, pp. 33-43. In: "Toxigenic Fungi--their Toxins and Health Hazard". Eds H. Kurata and Y. Ueno. Amsterdam: Elsevier.
WICKLOW, D.T. 1984 b. Adaptation in wild and domesticated yellow-green Aspergilli, pp. 78-86. In: "Toxigenic fungi--their toxins and health hazard". Eds H. Kurata and Y. Ueno. Amsterdam: Elsevier.
WICKLOW, D.T. 1984 c. Conidium germination rate in wild and domesticated yellow-green Aspergilli. Appl. Environ. Microbiol. 47: 299-300.
WICKLOW, D.T. 1985. _Aspergillus leporis_ forms sclerotia on rabbit dung. Mycologia (in press).
WICKLOW, D.T. and R.J. COLE. 1982. Tremorgenic indole metabolites and aflatoxins in sclerotia of _Aspergillus flavus_ Link: an evolutionary perspective. Can. J. Bot. 60: 525-528.
WICKLOW, D.T. and J.E. DONAHUE. 1984. Sporogenic germination of sclerotia in _Aspergillus flavus_ and _Aspergillus parasiticus_. Trans. Br. Mycol. Soc. 82: 621-624.
WICKLOW, D.T., B.W. HORN, W.R. BURG and R.J. COLE. 1984. Sclerotium dispersal of _Aspergillus flavus_ and _Eupenicillium ochrosalmoneum_ from maize during harvest. Trans. Br. Mycol. Soc. 83: 299-303.
WOODRUFF, H.B. 1980. Natural products from microorganisms. Science, N.Y. 208: 1225-1230.

DIALOGUE FOLLOWING DR. WICKLOW'S PAPER

GOCHENAUR: In looking at the development of _Eupenicillium ochrosalmoneum_ sclerotia you had some buried and some sitting on the surface. Did their development differ?

WICKLOW: Yes. Those which were buried in nonsterile soil stopped developing. When moved to the surface of the same soil, development continued.

KLICH: Do you have any idea why some _Aspergillus_ species produce metulae rather than just bigger vesicles?

WICKLOW: _Aspergillus clavatus_ does just that, produces a large vesicle with only phialides. It's probably just a way of producing more phialides from a restricted surface area, effectively increasing the surface area for sporulation.

SEROLOGICAL APPROACHES FOR IDENTIFICATION OF ASPERGILLUS AND PENICILLIUM SPECIES

L. POLONELLI *, M. CASTAGNOLA**, C. D'URSO**, and G. MORACE*

Istituto di Microbiologia* and Istituto di Chimica** Facoltà di Medicina e Chirurgia "A. Gemelli", Università Cattolica del Sacro Cuore 00168 Roma, Italy

SUMMARY

Eight species and one variety of Aspergillus i.e. A. flavus A. flavus var. columnaris, A. oryzae and A. parasiticus, A. fumigatus, A. niger, A. phoenicis, A. pulverulentus and A. tubigensis, and two species and three varieties of Penicillium, i.e P. expansum, P. verrucosum var. cyclopium, P. verrucosum var. melanochlorum and P. verrucosum var. verrucosum were antigenically studied in an attempt to identify them by different immunological procedures, i.e. immunodiffusion, two-dimensional immunoelectrophoresis, reversed phase high performance liquid chromatography and hybridoma techniques. The effectiveness and the potential of the different procedures is discussed.

INTRODUCTION

Traditional identification procedures for species in the genera Penicillium and Aspergillus are based primarily on the recognition of the characteristics of the colony and micromorphology. The wide number of species within the genera, and the varieties within the species, make identifications difficult even for an experienced mycologist. Consequently mycologists are attempting to evaluate other criteria for identifying Penicillia and Aspergilli rapidly and reliably. Serology might constitute an alternate method of identification and the potential of different procedures to identify various groups of Hyphomycetes has been investigated for many years.

In our Institute we developed the exoantigen procedure (Manych and Sourek, 1966; Standard and Kaufman, 1976, 1977) for the production of more specific reference antigens and adsorded antisera for the resolution of various mycological problems (Morace and Polonelli, 1981; Morace et al. 1984; Polonelli and Morace, 1982, 1984) 1985a; Polonelli et al., 1984). An immunodiffusion test has been developed to diagnose infections caused by Penicillium marneffei (Sekon, et al. 1982). In studies on exoantigens with 34 isolates of 10 species of Penicillium, only extracts from 6 isolates of P. marneffei and one isolate of an unknown Penicillium species reacted positively with the anti-P. marneffei rabbit serum, giving at least 2 lines of identity with the reference reagents. Based on this analysis the Penicillium species was identified as P. marneffei and the exoantigen test was considered to be a specific and rapid method for the identification and confirmation of P. marneffei isolates.

Analogously, the exoantigen technique has proved to be of value also for grouping Aspergilli of clinical interest (Sekon et al., 1985). In our experience, reference antigens prepared from reference cultures of Aspergillus flavus, A. fumigatus and A. niger used in the serodiagnosis of human aspergillosis proved to be specific when tested against the heterologous reference antisera.

The impressive specificity of exoantigens, enabling rapid and specifical identification of typical or atypical, sporulating or nonsporulating cultures of different fungi pathogenic to man (Kaufman and Lopez, 1980; Kaufman and Standard, 1978, b; Kaufman et al., 1980) prompted us to determine whether species specific exoantigens could be detected among related species within difficult series in Aspergillus and Penicillium. The eventual aim was to determine whether they could be used on a widespread and practical scale for diagnostic purposes.

The presupposition that certain Penicillium and Aspergillus series might be too difficult to be investigated overall by serology led us to select series with closely related but confused species or which were considered to be varieties or synonyms. Those chosen were A. flavus, A. flavus var. columnaris, A. oryzae and A. parasiticus, A. fumigatus, A. niger, A. phoenicis, A. pulverulentus, and A, tubigensis, P. expansum, P. verrucosum var. cyclopium, P. verrucosum var. melanochlorum and P. verrucosum var. verrucosum. These sets of species may effectively serve to illustrate the range and types of variation to be faced within series in Aspergillus and Penicillium.

We decided to evaluate the potential of new serological approaches, other than the exoantigen technique, to the identification of Aspergillus and Penicillium. The techniques used were two-dimensional immunoelectrophoresis (TDIE) with intermediate gel; reversed phase high performance liquid chromatography (RPHPLC) (Heftmann, 1984) and monoclonal antibodies (MAb).

Experience had shown that although the exoantigen technique repeatedly proved to be specific when applied to the identification of various cultures, it was most effective in monotypic genera. Distinguishing among species within the same genus is normally difficult because of extensive cross reactions. TDIE with intermediate gel (Mishra, 1984) provides the possibility of serologically comparing antigenically related species, by improving recognition of the occurrence of common and specific antigenic determinants between two species. Antiserum adsorption techniques make it possible to obtain specific antisera which react exclusively with their homologous reference antigens and with the exoantigens obtained from other isolates of the same species. The major problem in this system is the loss in sensivity of antisera due to the multiple adsorptions necessary for avoiding cross reactions. Conventional antisera are, moreover, heterogeneous mixtures of antibodies with different specificities and biological activities and they are difficult to duplicate. Hybridoma technology may solve these and other problems by providing monoclonal antibodies to species specific fungal antigens.

Recently, monoclonal antibodies produced against *Microsporum canis* and *Candida albicans* have been produced in our laboratory (Polonelli and Morace, 1985b.) for the identification of dermatophytes and yeasts. Optimally, monoclonal antibodies should be produced against the species to be investigated in this study. The specificity of monoclonal antibodies to single antigenic determinants, might make it possible to use them not only for such conventional procedures as immunodiffusion and enzyme linked immunosorbent assay, but also other techniques for studying the fine antigenic structure of heterologous species. Such sophisticated procedures as the Western blot analysis might be appropriately applied to either specific or nonspecific reactivities. Glycoproteins which have been denaturated, electrophoretically separated and transferred to nitrocellulose strips may be distinguished as single peptides at different molecular weights by monoclonal antibodies in immunoenzymatic reactions.

Hypothetically, different immunological responses to antigenically complex mixtures (glycoproteins) should reflect different chemical structures and thus it should be possible to characterize them by direct chemical analysis. Today, RPHPLC is one of the most sensitive methods for the separation of protein mixtures, based on the different polarity of the components.

MATERIALS AND METHODS

Fungal isolates. Thirteen *Aspergillus* and six *Penicillium* isolates, typical of the species investigated, were kindly supplied by Dr. R.A. Samson, Centraalbureau voor Schimmelcultures, Baarn, Netherlands (Tables 1 and 2).

Standard isolates. One culture from each of the 13 species investigated was selected as a standard, based on having typical morphology and used in the production of immunizing antigens. They were also used, along with the other isolates, to produce antigens for the ID, TDIE, RPHPLC and MAb studies where they were used as reference antigens.

Production of antigens. Reference antigens for the *A. flavus* group, *A. fumigatus* and *A. niger* cultures were obtained from acetone treated broth cultures of authentic isolates of the species studied (Morace and Polonelli, 1981). Reference antigens of the standard isolates of the species of *Penicillium* and the *A. niger* group were obtained directly from the supernatant of methiolated strain type cultures (Polonelli and Morace, 1982). The same procedure was adopted in preparing

Table 1. Isolates of *Aspergillus* species investigated.

ISOLATES	COLLECTION	NUMBER
Aspergillus flavus [a]	CBS [b]	564.65
A. flavus	"	569.65
A. flavus v. *columnaris* [a]	CBS	485.65
A. fumigatus [a]	UCSC [c]	--
A. niger [a]	CBS	102.12
A. niger	UCSC	--
A. niger	CBS	554.65
A. oryzae [a]	CBS	816.72
A. parasiticus [a]	CBS	103.13
A. phoenicis [a]	UCSC	--
A. pulverulentus [a]	CBS	115.48
A. pulverulentus	CBS	425.65
A. pulverulentus	CBS	558.65
A. tubigensis [a]	CBS	134.48

a. Standard isolate; b. Centraalbureau voor Schimmelcultures, Baarn, Netherlands; c. Universita Cattolica del Sacro Cuore, Rome, Italy

Table 2. Isolates of the *Penicillium* species investigated.

ISOLATES	COLLECTION	NUMBER
Penicillium expansum [a]	CBS	325.48
P. verrucosum v. *cyclopium* [a]	CBS	144.45
" " "	CBS	313.48
P. verrucosum v. *melanochlorum* [a]	CBS	487.75
P. verrucosum v. *verrucosum* [a]	CBS	390.48
" " "	CBS	603.74

a. Standard isolate; b. Centraalbureau voor Schimmelcultures, Baarn, Netherlands.

exoantigens from all the cultures. Chemical and serological data had previously shown that the two types of antigens were comparable.

Production of antisera. Reference antisera were produced from each reference antigen injected intramuscularly and intravenously into rabbits (Polonelli et al., 1984).

Immunodiffusion procedure. Both micro and macro immunodiffusion techniques were used either for evaluating whether the rabbits had produced antibodies to their homologous antigens or to visualize common and specific precipitin bands.

Adsorption of antisera. When necessary, adsorption procedures were performed by mixing equivalent amounts of heterologous antigens and antiserum. Mixtures were allowed to stand overnight at 4°C and then centrifuged. Supernatants were collected and then tested for evidence of adsorption by immunodiffusion procedures. Adsorptions were repeated until precipitin bands were observed between adsorbed sera and absorbing antigens.

Two dimensional immunoelectrophoresis with intermediate gel. The comparison of the antigenic pattern of each isolate was carried out by using the TDIE technique. In TDIE with intermediate gel, the antigen is subjected to an electric field in an agarose gel, then after the different antigens have migrated, the gel is removed and applied to another slide beneath two other gels. In the middle one, NaCl or heterologous serum can be diffused, whilst the homologous serum is diffused in the upper gel. When an orthogonal electric field is applied, the system with NaCl in the middle gel develops the antigenic pattern of the antigen-antiserum reference system investigated. The other system with heterologous serum in the middle gel reveals the common antigenic determinants that react with the heterologous antiserum and the specific antisera that react with the homologous serum (Mishra, 1984).

Reversed phase high performance liquid chromatography. For RPHPLC procedure, the antigens were lyophilized and dissolved in a lyophilized soluent; 0.010 ml of this solution was used for each HPLC analysis (Heftmann, 1984).

Hybridoma technique. Procedures for producing monoclonal antibodies have been described elsewhere (Polonelli and Morace, 1985b). Western blot analysis was performed as reported by Towbin et al. (1979).

RESULTS

All of the *A. flavus* group species investigated in this study proved to be closely related although distinguishable serologically (Fig. 1). The antigenic similarity was particularly demonstrated in the TDIE with intermediate gel test (Figs. 2 and 3). It showed common precipitinogens and a few specific ones between the two different species tested together. Adsorption of each reference antiserum with the heterologous antigens permitted the production of specific adsorbed antisera that reacted only with their homologous antigens (Fig. 4). Adsorbed antisera were used for serological identification of the isolates studied by producing their exoantigens directly from the mycelial cultures (Fig. 5).

RPHPLC studies analogously showed a very close antigenic relationship between *A. flavus* and *A. flavus* var. *columnaris*. The former produced a single chromatographic peak which was not observed in the latter.

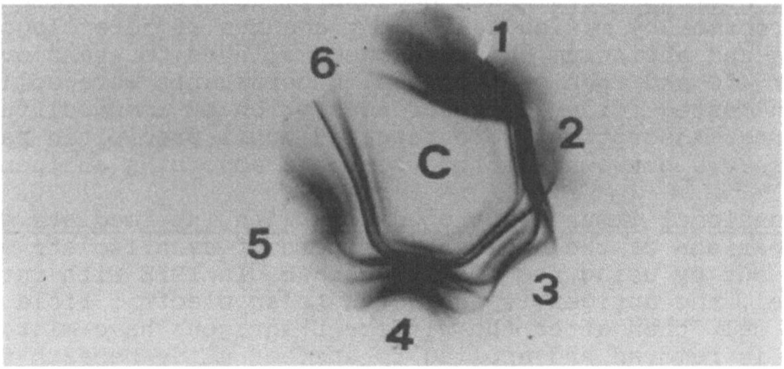

Fig. 1. Immunodiffusion technique for A. flavus. Well contents: c, reference antigen; 1 and 4, reference antiserum; 2, A. flavus var. columnaris reference antiserum;3, A. oryzae reference antiserum; 4, A. parasiticus reference antiserum.

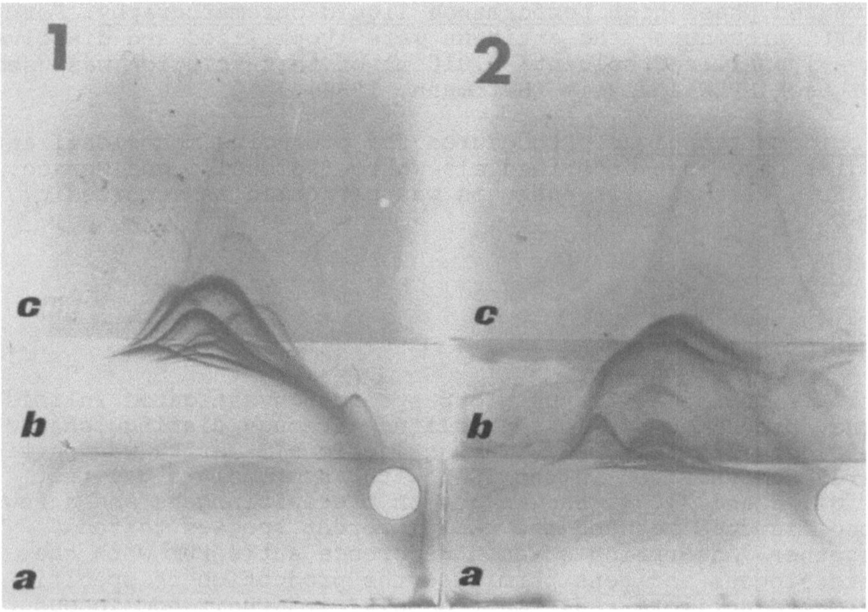

Fig. 2. Two dimensional immunoelectrophoresis with intermediate gel technique for A. flavus and A. parasiticus. Gel contents: a1 and a2, A. flavus reference antigen; b1, saline; b2, A. parasiticus reference antiserum; c1 and c2, A. flavus reference antiserum.

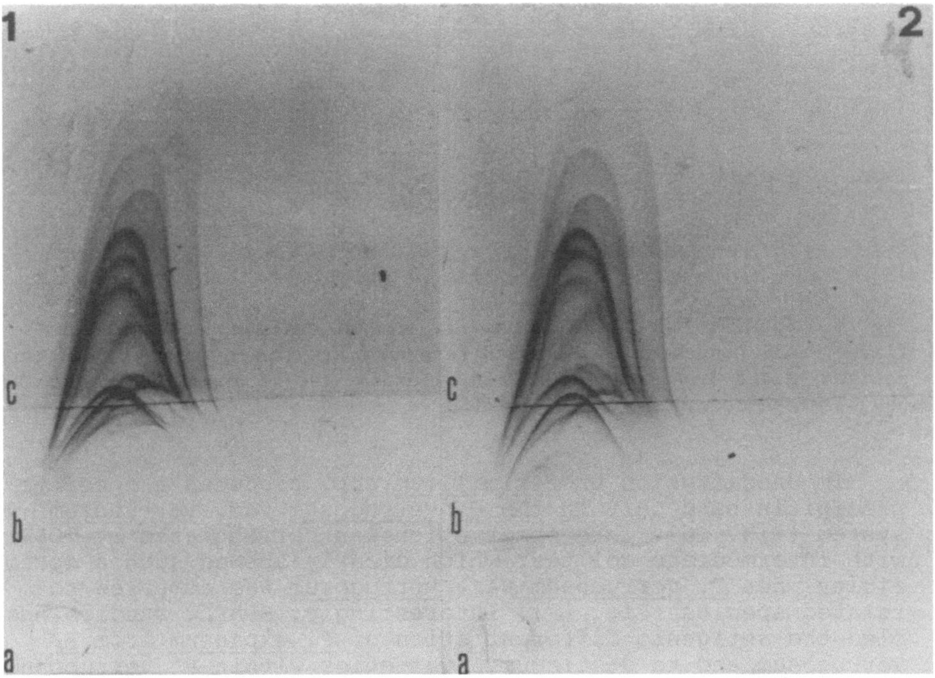

Fig. 3. Two dimensional immunoelectrophoresis with intermediate gel technique for A. flavus and A. oryzae. Gel contents: a1 and a2, A. flavus reference antigen; b1, saline; b2, A. oryzae reference antiserum; c1 and c2, A. flavus reference antiserum.

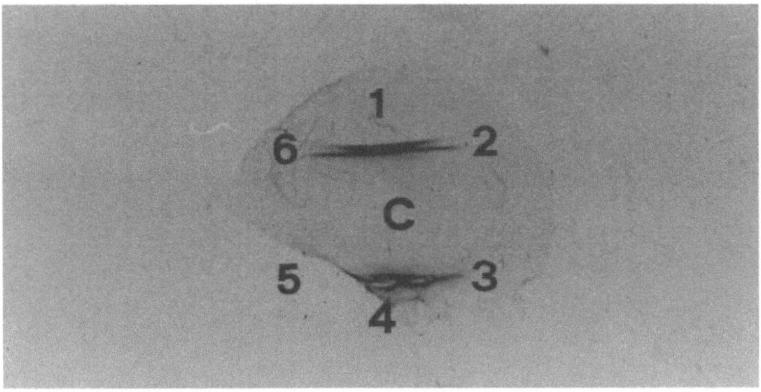

Fig. 4. Specificity of the immunodiffusion technique for A. flavus. Well contents: c, adsorbed reference antiserum; 1 and 4, reference antigen; 2, A. flavus var. columnaris reference antigen; 3, A. oryzae reference antigen; 5, A. parasiticus reference antigen; 6, empty.

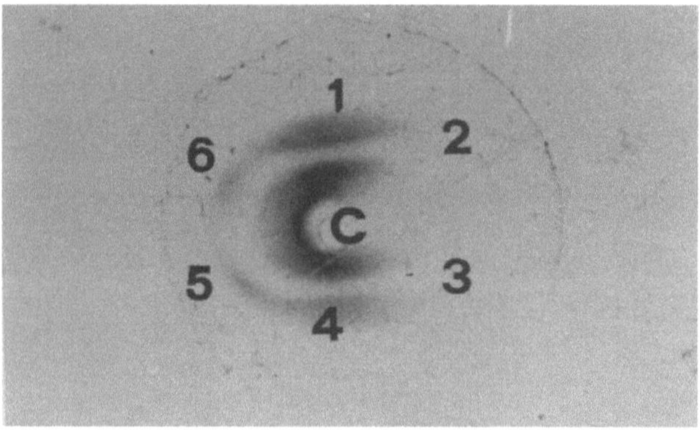

Fig. 5. Specificity of the immunodiffusion exoantigen technique for A. flavus. Well contents: c, adsorbed reference antiserum; 1 and 4, reference antigen; 2 and 3, A. oryzae CBS 816.62 exoantigen; 5 and 6, A. flavus CBS 569.65 exoantigen.

Immunodiffusion on Penicillium spp. produced a specific precipitin band only in the P. verrucosum var. verrucosum system (Fig. 6). This observation was corroborated by TDIE with intermediate gel test which clearly showed such a specificity when P. verrucosum var. verrucosum was compared to related species (Fig. 7). Interestingly, RPHPLC studies enabled the antigenic differentiation of P. expansum from P. verrucosum and to distinguish varieties within P. verrucosum.

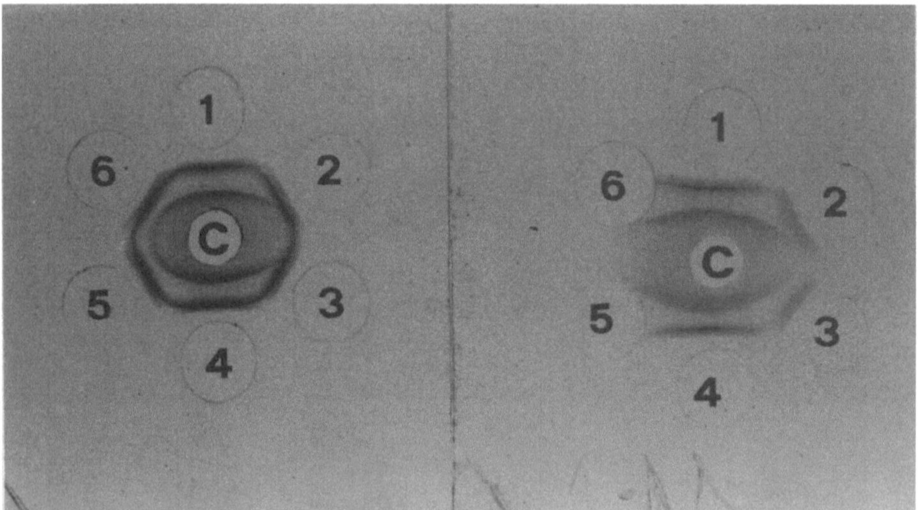

Fig. 6. Immunodiffusion technique for P. verrucosum var. verrucosum. Well contents: ac and bc, reference antiserum; a1, a4, b1 and b4, reference antigen; a2 and a3, P. expansum reference antigen; a5 and a6, P. verrucosum var. cyclopium reference antigen; b2 and b3, P. verrucosum var. melanochlorum reference antigen; b5 and b6, empty.

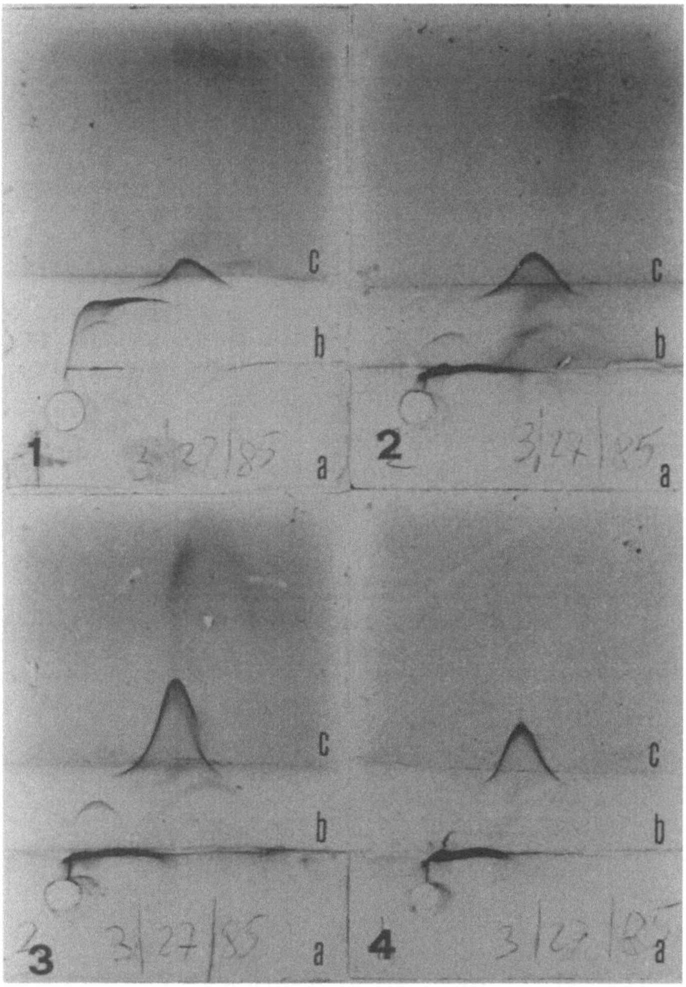

Fig. 7. Two dimensional immunoelectrophoresis with intermediate gel technique for P. verrucosum var. verrucosum. Gel contents: a1, a2, a3 and a4, reference antigen; b1, saline; b2, P. expansum reference antiserum; b3, P. verrucosum var. cyclopium reference antiserum; b4, P. verrucosum var. melanochlorum reference antiserum; c1, c2, c3 and c4, reference antiserum.

The situation in the A. niger group was more complex. Neither immunodiffusion nor TDIE with intermediate gel test permitted the easy differentiation of any of the species studied. However, RPHPLC antigenically differentiated single isolates of A. niger, A. phoenicis, A. pulverulentus and A. tubingensis. A. fumigatus was easily differentiated either by immunodiffusion or RPHPLC from the other Aspergillus species investigated (Figs. 8, 9, 10).

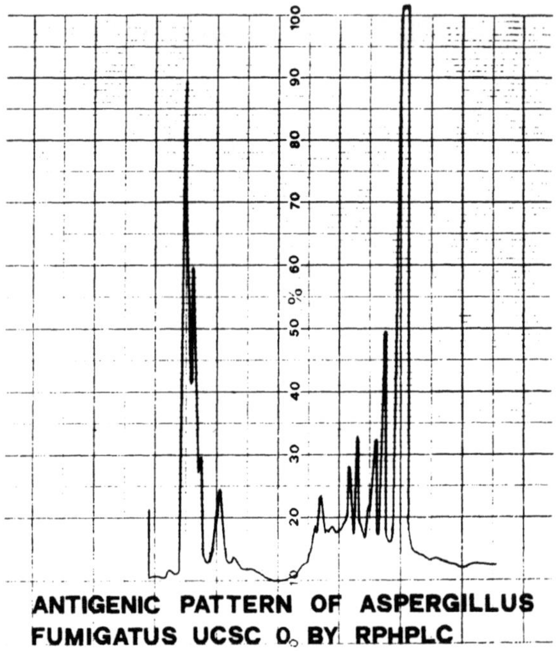

Fig. 8. Antigenic profile of A. fumigatus by reversed phase high performance liquid chromatography.

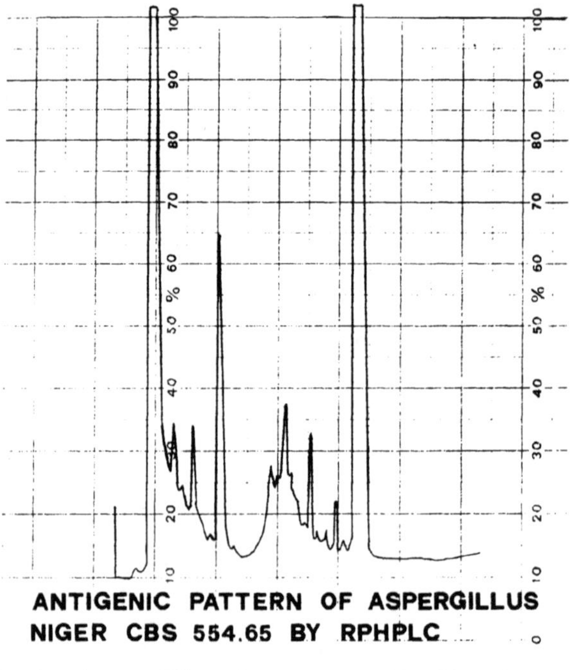

Fig. 9. Antigenic profile of A. niger, CBS 554.65, by reversed phase high performance liquid chromatography.

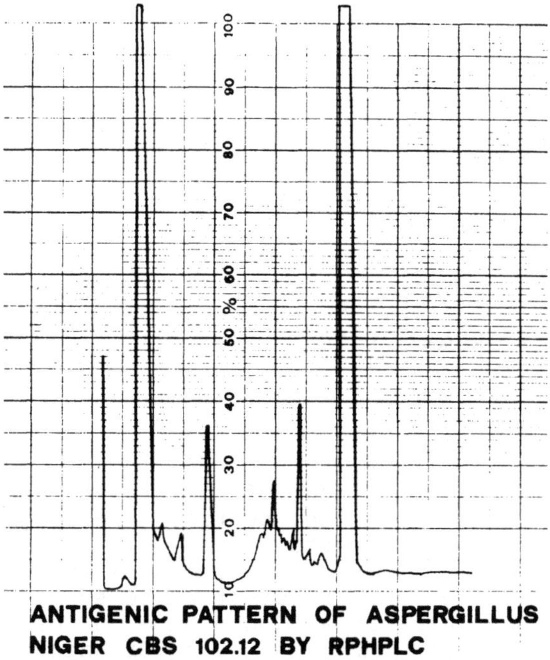

Fig. 10. Antigenic profile of A. niger, CBS 102.12, by reversed phase high performance liquid chromatography.

Immunodiffusion carried out with precipitating MAb produced against M. canis showed that different serotypes could be detected within the Penicillium and Aspergillus species studied, according to the ability of their exoantigens to react with the MAbs. Western blot analysis of all the exoantigens of the isolates studied confirmed that different serotypes exist within each species investigated.

DISCUSSION

The data obtained in this study clearly show that each of the three studied groups of Penicillium and Aspergillus contained antigenically homologous or similar species. Because of this close similarity only a few species within these series could be antigenically differentiated by using immunodiffusion TDIE procedures, i.e. A. flavus, A. flavus var. columnaris, A. oryzae, A. parasiticus, A. fumigatus, P. verrucosum var. verrucosum.

The more analytical RPHPLC procedure permitted the rapid differentiation of all the species investigated in this study. The experimental MAb procedure appears to be very promising for the differentiation of species within a genus as well as for the characterization of isolates within a species. Unfortunately the difficulty of obtaining numerous and reliably identified isolates of the species studied prevented us from evaluating whether specific antigenic determinants detected either by RPHPLC or MAb were consistently produced by all isolates of a single species. Intraspecific antigenic variations suggested the possible unstable nature of antigens of Penicillium and Aspergillus. Unless it can be shown that the antigenic profiles of many heterotypic isolates of a single species do not match the homotype, the diagnostic application of immunological procedures has potential.

Extensive studies to evaluate and standardize new immunological approaches for identification of fungi are now in progress in our Institute.

Acknowledgements

This study was supported by Grants of Ministero Pubblica Instruzione and Consiglio Nazionale delle Richerche (PF. MI. Contratto n. 84.02040.52)

REFERENCES

HEFTMANN, E. 1984. Proceedings of the 3rd International Symposium on HPLC of proteins, peptides and polynucleotides. J. Chromatogr. 296: parts I and II.

KAUFMAN, L. and LOPEZ, R.B. 1980. Immunodiffusion studies of morphologically similar dermatophyte species. Proceedings, 5th International Conference on the Mycoses, Washington D.C.: Pan American Health Organization. pp. 139-173.

KAUFMAN, L. and STANDARD, P.G. 1978a. Improved version of the exoantigen test for identification of _Coccidioides immitis_ and _Histoplasma capsulatum_ cultures. J. Clin. Microbiol. 7: 97-101.

KAUFMAN, L. and STANDARD, P.G. 1978b. Immunoidentification of cultures of fungi pathogenic to man. Curr. Microbiol. 1: 125-140.

KAUFMAN, L. and STANDARD P.G., and PADHYE,.A 1980. Serologic relationships among isolates of _Exophiala jeanselmei_ (_Phialophora jeanselmei_, _P. gougerotii_) and _Wangiella dermatitidis_. Proceedings, 5th International Conference on the Mycoses. Washington, D.C.: Pan American Health Organization, pp. 252-258.

MANYCH, J. and SOUREK, J. 1966. Diagnostic possibilities of utilizing precipitation in agar for the identification of _Histoplasma capsulatum_, _Coccidioides immitis_, _Blastomyces dermatitidis_, _Paraccocidioides brasiliensis_. J. Hyg. Epidemiol. Microbiol. Immunol. 10: 74-84.

MISHRA, S.K. 1984. Antigenic profile of some typical and septate phialide strains of _Aspergillus fumigatus_. Sabouraudia 22: 91-100.

MORACE, G. and POLONELLI, L. 1981. Exoantigen test for identification of _Petriellidium boydii_ cultures. J. Clin. Microbiol. 14: 237-240.

MORACE, G., ORSINI, D., CASTAGNOLA M. and POLONELLI, L. 1984. Exoantigen studies of _Phanaerochaeta chrysosporium_ and _Sporotrichum pruinosum_ cultures. Ig. Mod. 81: 314-321.

POLONELLI, L. and MORACE, G. 1982. Exoantigen studies of _Sporothrix schenckii_, _Ceratocystis minor_ and _Graphium penicillioides_ cultures. J. Clin. Microbiol. 15: 362-365.

POLONELLI, L. and MORACE, G. 1984. Antigenic relationships of _Basidiobolus_ species. Ig. Mod. 81: 1332-1344.

POLONELLI, L. and MORACE, G. 1985a. Antigenic characterization of Microsporum canis, M. distortum, M. equinum, M. ferrugineum and Trichophyton soudanense cultures. Mycopathologia, (in press).
POLONELLI, L. and MORACE, G. 1985b. Serological analysis of dermatophyte isolates with monoclonal antibodies produced against Microsporum canis. J. Clin. Microbiol. 21: 138-139.
POLONELLI, L., ORSINI, D., CASTAGNOLA M. and MORACE, G. 1984. Serological identification of aflatoxin potential producing Aspergillus species. Ig. Mod. 81: 1128-1136.
SEKON, A.S., LI, J.S.K. and GARG, A.K. 1982. Penicillium marneffei: serological and exoantigen studies. Mycopathologica 77: 51-54.
SEKON, A.S., STANDARD, P.G., KAUFMAN, L. and GARG, A.K. 1985. Value of exoantigens for grouping Aspergilli. Atlanta, Georgia: IX International Congress of the International Society for Human and Animal Mycology.
STANDARD, P.G. and KAUFMAN, L. 1976. Specific immunological test for the rapid identification of members of the genus Histoplasma. J. Clin. Microbiol. 3: 191-199.
STANDARD, P.G. and KAUFMAN, L. 1977. Immunological procedure for the rapid and specific identification of Coccidioides immitis cultures. J. Clin. Microbiol. 5: 149-153.
TOWBIN, H., STAHELIN, T. and GORDON, J. 1979. Electrophoretic transfer of proteins from polyacrylamide gels to nitrocellulose sheets: procedure and some applications. Proc. Acad. Sci. USA, 76: 4350-4354.

DIALOGUE FOLLOWING DR. POLONELLI'S PAPER

SAMSON: I asked Dr. Polonelli to come and speak to this workshop because he has shown to me that not only can one show relationships between taxa with these techniques, but one can make extremely rapid identifications, sometimes within 24 hours.

POLONELLI: In medical mycology rapid identification is very important.

SAMSON: Can you also use these techniques with dead specimens?

POLONELLI: I think so. Work at the Communicable Diseases Center in Atlanta has shown that the technique can be used for killed or contaminated cultures.

WICKLOW: How long would it take to develop antisera to 20 isolates in the Aspergillus flavus group?

POLONELLI: A conventional method such as producing antisera from rabbits doesn't take too long. If you have to compare 4-7 species you can hope to get a serological differentiation within one month. If you want to study a larger group, you have to make monoclonal antibodies. This takes much longer and requires a team of more experienced people. You have to allow one month to immunize the mice, and two weeks to segregate the specific monoclonal antibodies. Then you must check them for specificity before they can be used in tests.

SAMSON: Would it be worth while to set up some sort of monoclonal antibody bank?

POLONELLI: This requires liquid nitrogen for storage of the hybridomas.

PATERSON: What is the cost?

POLONELLI: Conventional antisera are not so expensive, because the only thing you need are rabbits. Monoclonal antibodies are expensive to produce, but once you have produced a good one it will last for ever. They can be stored and bought and sold.

KOZAKIEWICZ: Did you test Aspergillus parasiticus with the Western blott technique?

POLONELLI: No, we did not.

WICKLOW: Dr. Kurtzman will swear that 100% DNA relatedness means that two isolates are the same species. Yet, you have been able to show that Aspergillus flavus and A. oryzae can be distinguished using your techniques.

POLONELLI: These techniques have been used mostly in medical mycology, where they have been shown to be useful for identification of hundreds of isolates of Histoplasma and so on. The work I have reported on here was done using only a single isolate of each species, so it is possible that the difference we are seeing is a difference between two strains, not two species. In order to be sure that these are two different species, we would have to study many more isolates.

WICKLOW: That could be a problem, because we sometimes have only one isolate of a particular species.

POLONELLI: This is true. We've had lots of isolates to work with in medical mycology. We have never had a false positive identification.

PATERSON: How long does it take to determine if any, of say twenty isolates, were the same species using the Western block technique?

POLONELLI: If you already have your monoclonal antibodies, you can test about 40 samples at once in a vertical gel system, and it would take 18 hours. If you change the voltage you can do it faster, say in 8 hours. After you have made the gel you have to make the transfer to the nitrocellulose paper which takes three hours. The staining takes three more hours. So if you are in a hurry, you can do it in about 14 hours.

AN INTEGRATED APPROACH TO PENICILLIUM SYSTEMATICS

Paul D. BRIDGE, David L. HAWKSWORTH,
Zofia KOZAKIEWICZ, Agnes H. S. ONIONS,
Russel R.M. PATERSON and Michael J. SACKIN*

Commonwealth Mycological Institute, Kew
Surrey TW9 3AF and *Department of Microbiology
Leicester University, Leicester LE1 7RH, United Kingdom

SUMMARY

A multidisciplinary approach to critical groups of species in Penicillium has been initiated which considers a larger number of characters than has previously been employed in an integrated study. Some 300 selected isolates of fasciculate Penicillia are being examined and scored for 195 characters including ones derived from morphological, ultra-structural and biochemical attributes; many of these have not previously been employed in the taxonomy of the filamentous fungi. This contribution describes the techniques and standard procedures being adopted and outlines the numerical techniques utilised in the assessment of the results. Procedures for testing the validity of the groupings that will be produced are also described. The work is still in the data collection phase, but some preliminary results based on full data sets for 52 isolates are presented. These show that this integrated approach can be expected to make a major contribution to the clarification of species concepts in Penicillium.

INTRODUCTION

A satisfactory taxonomy for some sections within the genus Penicillium has not been evolved despite the application of a variety of morphological, physiological and biochemical approaches extending over several decades (Abe, 1956; Samson et al., 1976; Fassatiová, 1977; Pitt, 1979; Ramírez, 1982). This is particularly true of subsect. Fasciculata, for which several different schemes have been proposed since it was described by Raper and Thom (1949) (Samson et al., 1976; Pitt, 1979). Different authors have adopted different species concepts and criteria for delimitation and have employed different standard methods. The result has been the publication of significantly different taxonomic schemes (Onions et al., 1984; Samson and Gams, 1984). In some cases, a single isolate may be classifiable in different named species, depending on which system is followed. In other cases, two isolates assigned to different species in one scheme may be treated as a single species in another.

As the various systems have arisen through emphasis on particular attributes, it is not possible to recommend one with complete confidence on the basis of current data. However, a wide variety of characters not previously employed by mycologists but which can broaden the database in *Penicillium* have been used in the taxonomy of other groups of microorganisms. Physiological and biochemical tests used in bacteriology (e.g. Cowan, 1974) have been considered in some mycological schemes but, in general, the filamentous fungi show little specificity for different carbon soures.

The presence of particular enzymes has been used with some success for a few plant pathogens (e.g. Wong and Willetts, 1973). Enzymic activities have recently been used successfully to support species separations in *Monascus* (Bridge and Hawksworth, 1985) and enzymic studies with Penicillia have been encouraging, giving separations at above the species level in some cases (Bridge and Hawksworth, 1984). The production of particular secondary metabolites may provide useful taxonomic information (Hawksworth, 1976; Frisvad and Filtenborg, 1983), as in some cases single metabolites may be restricted to particular groups of organisms at or near the species level. In other cases particular secondary metabolites can occur in a diverse range of species (Frisvad and Filterborg, 1983; Harborne, 1984; Kubitzki, 1984). The use of electron microscopy techniques has proved a major asset in understanding fungal development and in providing specific taxonomic characters, especially of spore ornamentation (e.g. Hawksworth, 1979; Shipton and Lunn, 1980; Cannon and Hawksworth, 1982, 1984; Kozakiewicz, 1986).

While many of the above techniques have been used in studies on *Penicillium* species, such studies have usually been independent of each other, with resulting poor correlation either with each other or existing taxonomic schemes (Onions et al., 1984). To develop a more widely acceptable taxonomy, the evaluation of as large a number of phenotypic expressions of the genome as possible is necessary. The effectiveness of this approach depends on both a large number of characters and isolates, considered objectively. The characters selected must be reproducible and reliable. To be practical for a large number of isolates, the tests should be relatively simple and rapidly give clear cut results. This type of approach has been used for a wide range of organisms with the results being analyzed by numerical techniques (e.g. Bridge and Sneath, 1983, Williams et al., 1983 a)

In the fasciculate *Penicillia*, data collection assumes major importance. Relatively few taxonomic studies of microfungi have used a wide range of characters, although a large number of biochemical and ecological activities are reported in particular species in some genera (Domsch et al., 1980). New test methods and ways of recording attributes were necessary, and personnel with complementary expertise and experience were a prerequisite for such an integrated approach.

This contribution outlines the multi-disciplinary approach being adopted to studies of fasciculate Penicillia at the Commonwealth Mycological Institute. Emphasis is given to the standardisation of procedures, novel methods and applications, and techniques of analysing and testing the validity of results.

STANDARDISATION OF MEDIA AND CONDITIONS

Some characters of taxonomic value may vary with media used and incubation conditions (Raper and Thom, 1949; Samson and Gams, 1984). In a large scale systematic study it is important to be able to estimate and wherever possible eliminate the source of variation. Several pilot studies have been undertaken to ascertain the most appropriate conditions for use in this investigation.

Two isolates of each of five species of fasciculate Penicillia were grown on each of five different media (Czapek, Malt Czapek, Czapek Yeast Autolysate, Yeast Extract Sucrose and Glycerol Nitrate agars) at three temperatures (15°, 25°and 32°C). The resulting colony diameters are given in Table 1.

Three isolates received as _P. viridicatum_ Westling, _P. cyclopium_ Westling and _P. corymbiferum_ Westling were grown on each of the above media at 25°C in the dark for 10 days. The resulting colony morphologies showed variations in colony size, depth and profile, as well as varying conidial colours, diffusible pigments and droplet production.

Fourteen _Penicillium_ isolates were grown on a modified Czapek agar where sodium nitrate was replaced by ammonium oxalate as the sole nitrogen source. After 5 days incubation at 25°C, three of these cultures showed abnormal morphologies, producing bright yellow mycelia and a yellow diffusible pigment. Production of green conidia was delayed until up to 14 days.

These studies showed that large variations in media composition and incubation conditions could cause correspondingly large changes in colony size and appearance, mycelium, reverse colours, sporulation, diffusible pigments, and exudate production. All these characters have been used in taxonomic schemes, so variation of this nature must be minimised.

Media used in taxonomy must be reproducible both within and between laboratories. A further consideration is the ease of recording results, for example, an opaque coloured medium may not be suitable for noting colour changes. Medium constituents such as peptones, vegetable or yeast extracts, and digests may vary from batch to batch and certainly vary between manufacturers; they should therefore be avoided wherever possible, or standardised. Medium composition has also been shown to affect the production of specific enzymes, the degradation of substrates and secondary metabolite production (see Coutts and Smith, 1976; Odds et al., 1978). To estimate variability of this type, a pilot study on nitrogen assimilation was carried out.

Table 1. Colony diameter of 10 isolates of 5 species of fasciculate Penicillia grown on 5 different media at 4 different temperatures.

CZ

Temperature	15		RT		25		32	
Time (d)	7	10	7	10	7	10	7	10
P. expansum IMI 39761 (1)	28	43	34	50	37	52	-	10
P. expansum IMI 149551 (2)	28	42	34	52	37	48	10	10
P. cyclopium IMI 195050 (3)	23	33	27	34	26	34	10	11
P. cyclopium IMI 89372 (4)	26	37	32	40	32	28	4	7
P. viridicatum IMI 192905 (5)	25	28	26	34	27	38	7	8
P. viridicatum IMI 39758 (6)	25	35	25	33	29	40	11	14
P. crustosum IMI 91917 (7)	23	32	25	32	25	36	14	16
P. crustosum IMI 89384 (8)	25	40	29	45	25	52	18	20
P. corymbiferum IMI 40213 (9)	33	50	40	57	42	51	6	9
P. corymbiferum IMI 68414 (10)	34	50	33	45	21	25	-	-

MCZ

Temperature	15		RT		25		32	
Time (d)	7	10	7	10	7	10	7	10
P. expansum IMI 39761 (1)	37	55	45	62	48	66	-	-
P. expansum IMI 149551 (2)	40	65	43	62	56	76	11	17
P. cyclopium IMI 195050 (3)	31	50	33	49*	37	48	11	11
P. cyclopium IMI 89372 (4)	36	57	34	48	25	40	-	-
P. viridicatum IMI 192905 (5)	31	47	38	44	36	52	-	4
P. viridicatum IMI 39758 (6)	35	51	35	55	36	48	8	8
P. crustosum IMI 91917 (7)	28	44	35	49	33	52*	16	20
P. crustosum IMI 89384 (8)	28	42	33	49	33	48	-	11
P. corymbiferum IMI 40213 (9)	47	76	55	80*	49	72	-	-
P. corymbiferum IMI 68414 (10)	40	63	43	64	1	1	-	-

CYA

Temperature	15		RT		25		32	
Time (d)	7	10	7	10	7	10	7	10
P. expansum IMI 39761 (1)	33	50	38	55	40	55	-	-
P. expansum IMI 149551 (2)	41	56	46	65	45	60	11	14
P. cyclopium IMI 195050 (3)	28	41	31	41	32	44	16	20
P. cyclopium IMI 89372 (4)	32	47	34	48	32	44	2	2
P. viridicatum IMI 192905 (5)	29	40	34	46	29	40	9	13
P. viridicatum IMI 39758 (6)	30	41	32	45	30	40	7	10
P. crustosum IMI 91917 (7)	30	46	31	43	30	36	18	20
P. crustosum IMI 89384 (8)	25	36	33	50	37	51	14	15
P. corymbiferum IMI 40213 (9)	39	57	44	57	37	54	-	-
P. corymbiferum IMI 68414 (10)	37	51	42	56	31	39	-	-

YES

Temperature	15		RT		25		32	
Time (d)	7	10	7	10	7	10	7	10
P. expansum IMI 39761 (1)	39	58	42	78*	53	70	-	-
P. expansum IMI 149551 (2)	44	70	48	80*	62	73	18	23
P. cyclopium IMI 195050 (3)	31	40	32	45	30	37	15	17
P. cyclopium IMI 89372 (4)	36	56	41	59*	37	48	-	-
P. viridicatum IMI 192905 (5)	34	48	40	60*	35	50	10	12
P. viridicatum IMI 39758 (6)	36	54	40	60*	38	56	17	19
P. crustosum IMI 91917 (7)	30	43	32	44	35	50	22	25
P. crustosum IMI 89384 (8)	31	45	32	47	43	52*	12	12
P. corymbiferum IMI 40213 (9)	42	67	48	68	52	68	5	5
P. corymbiferum IMI 68414 (10)	43	66	50	75*	41	60	-	-

G25N

Temperature	15		RT		25		32	
Time (d)	7	10	7	10	7	10	7	10
P. expansum IMI 39761 (1)	34	52	39	56	41	56	-	-
P. expansum IMI 149551 (2)	37	58	42	64	46	62	12	12
P. cyclopium IMI 195050 (3)	8	13	9	12	16	22	7	7
P. cyclopium IMI 89372 (4)	9	15	5	11	10	14	3	4
P. viridicatum IMI 192905 (5)	5	9	11	17	16	21	-	2
P. viridicatum IMI 39758 (6)	9	13	7	15	10	19	7	7
P. crustosum IMI 91917 (7)	9	13	14	15	17	22	12	14
P. crustosum IMI 89384 (8)	7	11	8	10	9	14	2	4
P. corymbiferum IMI 40213 (9)	-	5	-	12	13	18	-	-
P. corymbiferum IMI 68414 (10)	8	13	2	10	-	4	-	-

*Medium and temperature giving greatest growth for each isolate.

1 YES RT; 2 YES RT; 3 MCZ RT; 4 YES RT; 5 YES 25°C; 6 YES RT; 7 MCE 25°C, 8 YES 25°C; 9 MCZ RT; 10 YES RT

A liquid Czapek medium was used, and utilisation of the sodium nitrate estimated after 7 days incubation by testing for the presence of nitrite and ammonia. It was found that the production of both nitrite and ammonia was not reproducible for the majority of isolates tested. Chemical analysis of uninoculated sterile media showed that some bottles contained the low levels of dissolved organic nitrogen (16 mg 100 ml $^{-1}$). This nitrogen may have been introduced into the medium from the laboratory atmosphere, the water in the medium or the autoclaving procedure, or a combination of these (see next section). This nitrogen was apparently sufficient to affect the reduction of the nitrate in the medium, probably as a reduced organic nitrogen source may be energetically more favourable.

Production of particular secondary metabolites may also be affected by the presence of precursors or catalysts in the medium. For example, Smith (1949) showed that copper and zinc ions were required for citrinin production.

As a result of these tests, specific media have been selected for the different sections of this study. Czapek agar (Cz) was chosen for morphological studies as it is chemically defined, colourless, and allows a full development of the organism within 14 days. For physiological tests, Cz was also used, but for the primary biochemistry specific media were used for each test. Yeast Extract Sucrose (YES) agar and Czapek Yeast Extract agar (CYA) were used for secondary metabolite production because they have been reported to encourage production of a wide range of secondary metabolites from the genus Penicillium (Scott et al., 1979; Frisvad and Filtenborg, 1983). Media were always made up as large batches and a series of particular tests carried out on media from one batch. Where more than one batch of medium was used, duplicate tests were run using both batches. Batch variation within the study is being estimated by repeating approximately 10% of all tests carried out.

PHYSIOLOGY AND PRIMARY METABOLISM

Forty-one tests involving physiology and primary metabolism have been either adapted or specifically designed for use in this study (Bridge, 1985). These tests provide 52 characters, 40 of which have so far given differential results between isolates. The 12 apparently ineffective characters will continue to be examined until a later stage in the study as it is conceivable that deviating isolates will be encountered. A further eight tests will be used from the 19 specific enzyme tests on the API ZYM strip (Bridge and Hawksworth, 1984). The full list of tests used physiological and primary metabolism tests are numbers 1-60 in Table 2. The major experimental methods for these tests have been reported elsewhere (Bridge, 1985). In our view, it is important that inocula are prepared from the same media and conditions, and that inocula are of the same size, and can be reproducibly repeated. In this study, we are using suspensions of spores in 0.2 % Tween 80 were made up at a turbidity equivalent to tube no. 6 on the MacFarland scale (Lennette et al., 1980), approx. 2 x 10^6 spores m^{-1}. It is important that the stability of the standard tube is checked and that the spore suspensions are

Table 2. Provisional list of characters to be included in numerical study.

Physiology

1. Growth at 4°C
2. Growth at 37°C
3. Growth at 40°C
4. Growth at pH 2
5. Growth at pH 10
6. Growth at pH 12
7. Growth on G25N agar
8. Survival of spores to heating at 75°C for 5 mins
9. Growth in the presence of formalin
10. Growth in the presence of pentachlorophenol
11. Growth in the presence of sodium hydrogen selenite
12. Growth in the presence of sodium azide
13. Size of inhibition zone with copper
14. Size of inhibition zone with prop ionic acid
15. Size of inhibition zone with acetic acid
16. Size of inhibition zone with benzoic acid
17. Size of inhibition zone with sorbic acid
18. Size of inhibition zone with crystal violet
19. Size of inhibition zone with malachite green

'Primary' biochemistry

20. Gelatin liquefaction
21. Polypectate liquefaction pH 6
22. Polypectate liquefaction pH 8
23. RNA hydrolysis
24. DNA hydrolysis
25. Aesculin hydrolysis
26. Casein hydrolysis
27. Cellulose hydrolysis
28. Tween 80 hydrolysis
29. Starch hydrolysis
30. Phosphatase
31. Selenite reduction
32. Tetrazolium reduction
33. Tellurite reduction
34. Growth on ammonium oxalate as sole carbon source
35. Growth on ammonium tartrate as sole carbon source
36. Growth on tannic acid as sole carbon source

37. Growth on citric acid as sole carbon source
38. Growth on lactic acid as sole carbon source
39. Growth on sodium nitrite as sole nitrogen source
40. Growth on creatin as sole nitrogen source (pH 4.5)
41. Growth on creatin as sole nitrogen source (pH 8)
42. Growth on urea as sole nitrogen source
43. Growth on glycine as sole nitrogen source
44. Growth on ammonium oxalate as sole nitrogen source
45. Production of pH rise during growth on ammonium tartrate
46. Production of pH rise during growth on citric acid
47. Production of pH rise during growth on lactic acid
48. Production of pH rise during growth on creatin (pH 4.5)
49. Production of subsequent pH rise during growth on creatin (pH 8)
50. Production of pH rise during growth on urea
51. Production of pH rise during growth on glycine
52. Production of yellow mycelia and pigment during growth on ammonium oxalate

API ZYM tests

53. Leucine arylamidase
54. Valine arylamidase
55. Cystine arylamidase
56. Chymotrypsinase
57. α -galactosidase
58. β -galactosidase
59. α -glucosidase
60. α -mannosidase

Secondary metabolites

61. Patulin
62. Mycophenolic acid
63. Viomellein
64. Penicillic acid
65. Roquefortine B
66. Chaetoglobosin C
67. PR toxin
68. Terrestric acid
69. Brefeldin A
70. G viridicatic acid
71. Brevianamid A
72. Kojic acid
73. Griseofulvin
74. Xanthomegnin
75. Citrinin
76. Gliotoxin
77. Cyclopiazonic acid
78. Roquefortine C
79. Ochratoxin A
80. Terrein
81. Viridicatum toxin
82. Palitantin

(Continued)

83. Cyclopenin
84. Xanthocillin
85. Unidentified Y
86. Unidentified Z
87. Unidentified AA
88. Unidentified AB
89. Unidentified AC
90. Unidentified AD
91. Unidentified AE
92. Unidentified AF
93. Unidentified AG
94. Unidentified AH
95. Unidentified AI
96. Unidentified AJ
97. Unidentified AK
98. Unidentified AL
99. Unidentified AM
100. Unidentified AN
101. Unidentified AO
102. Unidentified AP
103. Unidentified AQ
104. Unidentified AR
105. Unidentified AS
106. Unidentified AT
107. Unidentified AU
108. Unidentified AV
109. Unidentified AW
110. Unidentified AX
111. Unidentified AY
112. Unidentified AZ
113. Unidentified BA
114. Unidentified BB
115. Unidentified BC
116. Unidentified BD
117. Unidentified BE
118. Unidentified BF
119. Unidentified BG
120. Unidentified BH
121. Unidentified BI
122. Unidentified BJ
123. Unidentified BK
124. Unidentified BL
125. Unidentified BM
126. Unidentified BN
127. Unidentified BO
128. Unidentified BP
129. Unidentified BQ
130. Unidentified BR
131. Unidentified BS
132. Unidentified BT
133. Unidentified BU

Morphology

134. Colony diameter
135. Colony depth
136. Colony density
137. Colony flat/raised
138. Colony converse/domed
139. Colony gradual/umbonate
140. Colony wrinkled
141. Mycelium white
142. Mycelium yellow
143. Texture velvet
144. Texture floccose
145. Texture funiculose
146. Texture fasciculate
147. Margin submerged
148. Conidia blue/green
149. Conidia green
150. Conidia yellow/green
151. Conidia grey/green
152. Conidia dark/green
153. Conidia fawn
154. Conidia white
155. Exudate colourless
156. Exudate straw
157. Exudate yellow

158. Exudate pink

159. Exudate red/brown

160. Pigment yellow

161. Pigment brown

162. Pigment pink/red

163. Reverse white

164. Reverse brown

165. Reverse yellow

166. Reverse buff

167. Reverse pink/red

168. Reverse green

169. Black/grey areas in reverse

170. Head arrangement

171. Head size

172. Supporting hypha length

173. Supporting hypha apex swollen

174. Supporting hypha rough

175. Highest branch system: no branches

176. Highest branch system: level branched

177. Highest branch system: two or more levels

178. First level branch numbers

179. First level branch divergent/appressed

180. First level branch base cylindrical

181. First level branch rough

182. First level branch length

183. Maximum number of conidiogenous cells

184. Conidiogenous cells ampulliform

185. Conidiogenous cell length

186. Conidiogenous cell rough

SEM Characters

187. Spore mean length

188. Spore mean breadth

189. Ornamentation lobate

190. Ornamentation microtuberculate

191. Ornamentation reticulate

192. Ornamentation echinulate

193. Spore shape elliptical

194. Spore shape sub-globose

195. Spore shape elongate

prepared in the same size vessels as the standards. To prevent carry over of nutrients from the maintenance medium to the test, spore suspensions are washed by centrifugation.

In carbon and nitrogen assimilation tests (tests 34-44), two major problems may be encountered. First, most fungi utilise the more common carbon and nitrogen sources used by microbiologists (Bridge, 1985). A solution to this is to extend the range of substrates by considering the ecology of the fungi for example, growth on leather products may lead to the selectivity for tannic acid assimilation. Second, trace growth on negative controls occurs in carbon and nitrogen assimilation tests. Negative controls may be used, but where positive reactions are weak, differentiation between positive and negative results becomes difficult, and even subjective.

Measurement of pH changes associated with carbon and nitrogen assimilation tests may also be useful. We found that growth on a number of carbon and nitrogen sources can result in a large, 3-4 pH units, rise in the pH of the medium (tests numbers 45-51). Similar findings have been obtained by other workers (e.g. Thom, 1910; Frisvad, 1981), but the explanation remains obscure. Possibly it results from excretion of ammonia, from autolysis of protein compounds at the decline phase of growth. This has been detected in other fungi (Paterson, unpublished; Berry, 1975).

Growth in the presence of inhibitory compounds has been used extensively in bacterial and yeast taxonomy (Cowan, 1974; Kreger-van Rij, 1984). Tests of this kind have been evaluated in this study (tests 9-19), in most cases (tests 13-19) using disc diffusion methods. Filter paper discs impregnated with the compound to be tested are placed on dry agar plates seeded with 0.5 ml spore suspension in the agar. This results in confluent growth across the plate except in the area around the discs where a clear zone of inhibition may be seen. This method has several major advantages: (1) several discs containing different inhibitors may be used on the same plate, (2) the size of the zone of inhibition gives a measure of the inhibitory concentration (provided the diffusion co-efficient is known) (3) a large number of discs may be produced in a single batch and then stored easily for future use, (4) growth away from the inhibition zone provides a negative control, and (5) the large size of the inoculum enables results to be read after as short an incubation as three days.

The production of specific enzymes has been considered in a large number of taxonomic studies, with techniques ranging from isoelectric focusing and electrophoresis of isoenzymes to the degradation of particular substrates (e.g. Stout and Shaw, 1973; Jones and Noble, 1982). In this study, enzymic activities were investigated by two techniques, the degradation of particular substrates (tests 20-33) (Bridge, 1985) and the presence of specific enzymes (tests 33-60) (Bridge and Hawksworth, 1984). Media composition and pH are of prime importance, as, for example, polypectate gels may be liquefied by two different enzymes, one each acting at acid pH and alkaline pH (Gochenaur, 1984). Enzymic activities show promise as diagnostic taxonomic characters in the microfungi (e.g. Bridge and Hawksworth, 1985). Positive results may be easily read as a liquefaction of a gel (as for gelatin and polypectate), a colour change (as for Tween 80, tetrazolium, tellurite, selenite, phosphatase and API ZYM) or as the production of a clear zone (as for casein, RNAase, DNAase and cellulase).

The most important conclusion to date of our studies on physiology and primary metabolism in <u>Penicillium</u> is that a wide range of tests generally used only in bacterial and yeast taxonomy can be adapted for use with filamentous fungi. However, given the complex control mechanisms of fungal metabolism and their ability to use very low concentrations of nutrients, the care needed to control and standardise media and conditions assumes prime importance.

SECONDARY METABOLITES

Secondary metabolites are routinely used as important taxonomic characters in lichenized fungi (Hawksworth, 1976) and may therefore be expected to be of value in other fungi related to the Ascomycotina, including certain <u>Penicillium</u> species (Ciegler et al., 1973, 1981; Frisvad, 1981; Filtenborg et al., 1983; Frisvad and Filtenborg, 1983). In this study it was therefore essential to consider these compounds. The methods of Frisvad and Filtenborg (1983) have been used here because they permit large numbers of isolates to be studied relatively quickly. All plates were developed with griseofulvin as a standard and in some cases additional standard metabolites were employed as identification aids.

Statistical analysis of the secondary metabolite data from the studies of 52 isolates were also carried out. The Rf values of similar spots on the TLC plates from individual isolates were combined and the means and standard deviations (SD) calculated. In general a SD of equal to or less than 0.02 was common. The means of the Rf values were combined in groups of 0.06 Rf units on the basis that this range is equivalent to the SD of virtually all of the spots. Similar spots in each range from all 52 isolates were combined and the means and SD calculated. An example of the information being obtained in this way is shown in Table 3.

Table 3. Metabolites from isolate P18.

P18

	$\bar{x}$Rf.	SD	n	
System 1 (extracellular metabolites)				
A terrein	0.12	-	2	5 - brown 6 - dark
B brefeldin A or 'g'	0.25	0.01	5	5 - brown, 6 - dark, 7 - dark
C griseofulvin	0.32	0.01	4	2 - violet, 3 - blue, 6 - violet
D mycophenolic acid	0.44	0.01	4	2 - violet
System 2 (intracellular metabolites)				
a griseofulvin	0.62	0.01	4	2 - violet, 4 - violet
b ochratoxin A	0.68	0.01	14	2 - green
c unknown	0.80	0.02	3	2 - violet

2 - UV light (long μ)
3 - UV light (short μ)
4 - system 2 stain + UV light
5 - system 1 stain, visible
6 - system 1 stain + UV light

Table 4. Metabolite production from a Penicillium isolate P8. Numbers in column headings refer to time of incubation and associated letters to different production runs. The two separate runs (system 1) and CYA (system 2) are designated A and B. C refers to growth of C3 (system 1)

System 1

metabolite letter	6Ai	6Aii	8A	11A	13A	15A	7B	8C	z
A	-	-	-	-	-	-	-	+	roquefortine B ?
AI	-	-	-	-	-	+	-	-	terrestric acid ?
AZ?	-	-	-	-	+	+	-	-	griseofulvum & patulin ?

System 2

	6Ai	6Aii	8A	11A	13A	15A	7B
j	-	-	-	-	-	+f	-
aw	-	-	-	-	+	+	-
bi	-	-	-	+	+	+	-
bm	-	-	-	-	+	+	-

Table 5. Metabolite production from isolate P11 (Refer to table 4 for explanation of column heading).

metabolite letter	6Ai	6Aii	8A	11A	13A	15Ai	15Aii*	7B	8C	
A	-	-	-	+	+	-	-	-	+	roquefortine B?
U	-	-	-	-	-	+	-	-	+	I?
AC	-	-	-	-	-	-	-	-	+	
AV	-	-	-	-	-	-	-	-	+	
BF	-	-	-	-	-	-	-	-	+	
BO	-	-	-	-	-	-	-	-	+	
BX	-	-	-	-	-	+	-	-	+	
CC	-	-	-	-	-	-	-	-	+	
CG	-	-	-	-	-	-	-	-	+	

*15Aii colony was smaller and had little or no diffusion of red pigment when compared with 15Ai.

A standard system for recording the results of the analyses was also desirable as has become the practice in the lichen-forming fungi (Culberson and Kristinsson, 1970; Walker and James, 1980). The TLC data of all metabolites encountered at CMI from a wide range of microfungi is being retained for comparison with unknowns encountered in this study. New information on fungal metabolites is continually being added to our records.

Time of incubation of cultures also needs to be standardised. Metabolites from isolate P8 are detectable only after 13 days growth on YES and CYA (Table 4). Conversely we have data which indicates that metabolites which were detected initially could not be detected with increased time of incubation. Growth medium also effects metabolite production. Isolate P11 shows low productivity on YES medium, yet on Cz nine metabolites were detected after 8 days growth (Table 5). Cz often gives both increased production of metabolites and, in some cases, metabolites not detected from other media. The reasons for these variations are not fully understood.

Considerable effort is being devoted to the identification of unknown compounds. Requests have been made to many organisations for pure standards of *Penicillium* metabolites which can assist in the identification of unknown products. Approximately eighty metabolite standards are now held at CMI.

An important limitation of single dimensional TLC is that metabolites can be masked by spots with similar Rf values (e.g. citrinin and ochratoxin A in toluene/ethylacetate/formic acid). Two dimensional TLC can overcome this, but analysis time increases. High pressure liquid chromatography (HPLC) is an additional method which can be used to analyse metabolites. Our laboratory has successfully identified griseofulvin from *Penicillium* isolates using the method of Scott (1981). A data bank is being set up containing information on the HPLC characteristics of standards which will also be used in the identification of metabolites.

The retention of secondary metabolites in dried cultures preserved in herbaria has also been examined (Paterson and Hawksworth, 1985). This work is of significance because in some cases only dried material of a species exists. The secondary metabolites of nine living isolates, including four nomenclatural types, were compared with those present in cultures of the same isolate that had been dried and kept in the herbarium for 22-35 years. Most of the metabolites found in the living cultures were confirmed as present in the dried material.

MORPHOLOGICAL STUDIES

In an integrated study, a critical examination must be made of the value of all characters to be included. A preliminary list of morphological characters was prepared from current literature, and experience gained in the course of our work on *Penicillium*. A recording sheet was constructed and the final lists processed for inclusion in the data matrix for numerical analysis. For this the records must be adjusted to either "binary" (present or absent, + or -) or "numeric" (expressed as a degree of size on a graduated scale). This imposes limitations on the extent to which data can be included.

The characters to be examined in the morphological study (nos. 136-186) were divided into three groups, (1) macroscopic appearance of the cultures, including colony growth, texture, colours, exudate, reverse, etc; (2) appearance under low magnification using a stereomicroscope; and (3) microscopic appearance of mounted conidiophores examined in lactophenol cotton blue.

Numeric characters were measured using the light microscope with a calibrated eyepiece micrometer. The limit of reading for microscopes available to most observers is ca 1 µm, and as the majority of the Penicillium species under study form conidia in the range 2-4 µm averages lack significance and show little variation from one isolate to another. Similarly, the diameters of "stipes", "branches", "metulae" and condiogenous cells showed insufficient difference in the first study to warrant inclusion in the final matrix.

Structural characters are difficult to assess due to variation and the lack of clear lines of demarcation from one character to another. Depth of colony (test 135) is difficult to measure and has been recorded as thin (0), moderate (1) or deep (2), but all stages can occur even within the same culture. To obtain differential results, limits have been set to facilitate clear recording, but can tend to give a misleading impression of accuracy and are subject to error.

Care must be taken not to weight the significance of a character by including it more than once as part of several features of a single attribute, i.e. linked characters. For example, the presence or absence of branches is linked with branch shape (test 180) and length (test 182).

A problem in describing Penicillium isolates is the apparent lack of clear lines of demarcation; i.e., intergradations occur. For example, the description of colony textures (tests 143 to 146) as velvety, floccose, funiculose and fasciculate. These terms are well defined by Samson et al. (1976), but in many isolates vary from place to place over the colony. The extent of fasciculation is also difficult to see and so to standardise. In this project fasciculation (test 146) has been designated on a numeric scale none (0), slight (1), definite (2), and obvious synnemata (3; e.g. P. claviforme). However, P.expansum sometimes shows both fasciculation (2) and synnemata (3). Raper and Thom (1949) used texture for the separation of groups in Penicillium, and this is seen as one of the main criticisms of their work. Pitt (1979) excluded this character as far as possible, but we have attempted a compromise by including the feature but not weighting it heavily (i.e. as a single numeric feature).

Colours are also difficult to assess (tests 148-169). They can be defined using a colour code (e.g. Ridgway, 1912; Rayner, 1970; Kornerup and Warscher, 1978), but this would result in numerous entries because colours vary. Therefore, we decided to treat colour in a general sense.

As noted above, measurement of size becomes inaccurate for small structures, while the length of larger branches can vary widely in one culture, so a mean size is heavily dependent on the recorder's impartiality. A record of range seems desirable but would be difficult to assimilate into the data matrix. General descriptions prepared tend to include such statements as "usually" and "predominantly" which are not objective. Consequently, a maximum is recorded when recording numbers of branches (test 178) and conidiogenous cells (test 185) where all stages from one pair per level are likely to be seen as they are produced sequentially. "Metulae" are recorded in size ranges (test 182).

Appressed and spreading branched systems (test 179), although typical of some species, often occur in the same cultures and certainly vary with the stage of development of the penicillus. When recording branching systems (tests 175-177), the condition of the predominant highest branching state is recorded because all less-branched forms are to be found in most sporing cultures due to the sequential development.

In developing a system to take note of changing characters it must be remembered that a fungal culture is a growing and changing system with constant development and characters also change, such as colony size, colour, exudate, reverse pigments, texture, developing conidiophores and microscopic structures.

In order to reduce the variability of characters to a minimum and produce consistent results it is necessary to standardise all factors as far as is practicable to a level at which characters are stable and reproducible. Standard reproducible growth conditions are essential with the use of a standard, reproducible medium, pH, temperature, illumination, aeration, in standard dishes using a standard inoculum, and examination, after a fixed period. In this project, for morphological studies, growth on Czapek agar with added trace elements is being used. As this is prepared from Analar grade chemicals, it should be easily reproducible in other laboratories. Growth is at 25° in the dark, and the colonies are examined after 7 days and again at 14 days to allow for slow growing isolates and to record variation. Inoculation is by mass transfer of a loop of a conidial suspension standardised by turbidity. Despite this, further studies using single spores show that strain variation and possibly genetic factors may influence the system.

SCANNING ELECTRON MICROSCOPY

The scanning electron microscope (SEM) is being used to resolve minute differences in the ultrastructure of the conidium topography. Recent SEM studies in Aspergillus have shown that differences in spore topography, which are not resolvable with the light microscope, can be a valuable marker character in species identification (Kozakiewicz, 1986).

So far, four basic spore ornamentation types have been identified: echinulate, lobate, reticulate and microtuberculate (Fig.1). Where possible, several isolates referred to the same species have been used in these studies.

Using spore topography as a key character, several isolates donated to the project as *P. viridicatum* have been examined, including the ex-type and four different metabolite producing isolates. These yielded three distinct spore ornamentation types (echinulate, reticulate, and micro tuberculate) and represent three out of the four morpho-forms illustrating the ornamentation classes in Fig.1. The ex-type isolate of *P. viridicatum* possesses an echinulate spore, as do the majority of the isolates received as *P. viridicatum*. However, for the remainder of the isolates, including the biochemical strains, the situation is more complex. Together they include two distinct ornamentation morpho-forms within the reticulate, and two within the micro tuberculate classes. One of the micro tuberculate members represents *P. cyclopium* according to the numerical taxonomic studies and so this strain is probably misidentified. However, at present the status of the remaining three forms is uncertain and it is hoped that this will be resolved by considering the other characters investigated in this study.

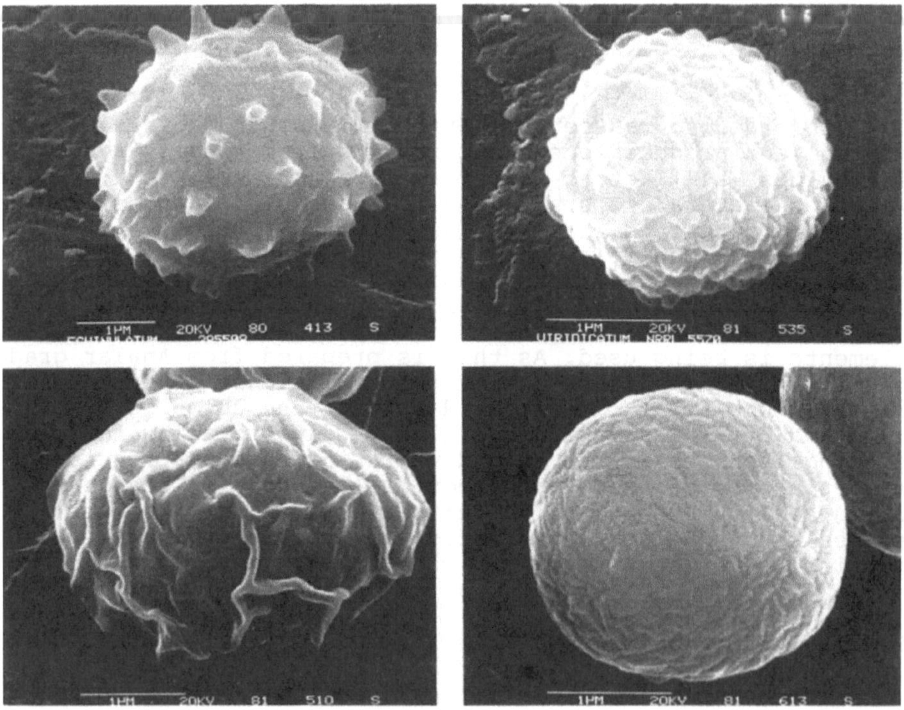

Fig. 1. SEM micrographs illustrating conidial ornamentation types in Penicillia. Top series: left - echinulate, right - lobate; Bottom series: left - reticulate, right - micro-tuberculate.

A character traditionally used in species determination and delimitation is the shape and size of the conidia. Interspecific and intraspecific variation exhibited by this character is also being evaluated using the SEM. The specimen stub is coated with a thin layer of spores by passing it across the surface of a mature colony. A series of replicate micrographs are taken at random across the stub, so that the spore images in each are equally magnified.

Spore diameters of a representative sample are being recorded automatically with a Summagraphics digitiser pen and statistically analysed with a computer program. To date, spore sizes of the majority of isolates examined have fallen within a range of 2.0-3.0 µm for both length and breadth, with isolates of the same species showing little variation.

The results were submitted to two statistical tests. First, the frequency of occurrence of each size measurement was plotted as a graph of frequency against size. This resulted in a normal curve for each isolate and for the mean results for all isolates. Secondly, the standard deviations of each set of results were calculated and compared to the range of results. In each case the range of results fell within 2 x S D of the mean, again indicating that the variation in individual results was normally distributed. The same was found for the mean results of all isolates tested (Fig.2). Consequently there does not appear to be any significant difference in conidium sizes. However, measurements will continue in order to arrive at an unequivocal view once the full data set is obtained, when more detailed statistical treatments will be applied as the preliminary set of strains studied may not be fully representative.

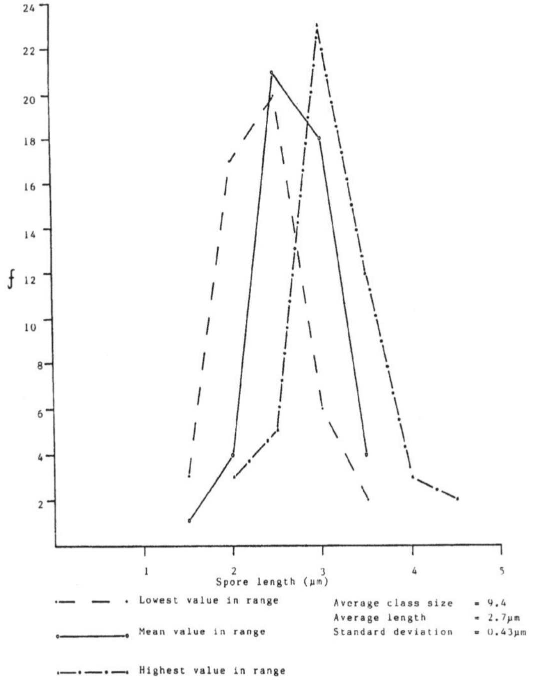

Fig. 2. Graph of range of spore lengths plotted against their frequency of occurrence (f)

A key feature in current classification schemes for the Penicillia is the pattern and number of the branches forming the penicillus. The variability of this character in respect of its interspecific or intraspecific variability has never been challenged. This is being investigated using the recently developed SEM cryosputter technique to study ontogenetic development of branching. The unique advantage of the cryotechnique is that it permits presentation to the electron beam of fresh material which has been plunged into liquid nitrogen, gold coated and introduced into the SEM whilst under vacuum at all times (Emscope); the preparation time lasts about five minutes. A large sample of heads is made available for study, and the life-like micrographs are not marred by shrinkage or other artefacts attributable to freeze-drying or critical point-drying techniques (Fig.3). The study is still in its early stages so that conclusive results cannot be given, but it is pertinent to note that branching does not occur synchronously (Fig. 3) unlike in <u>Aspergillus</u> (Kozakiewicz, 1978).

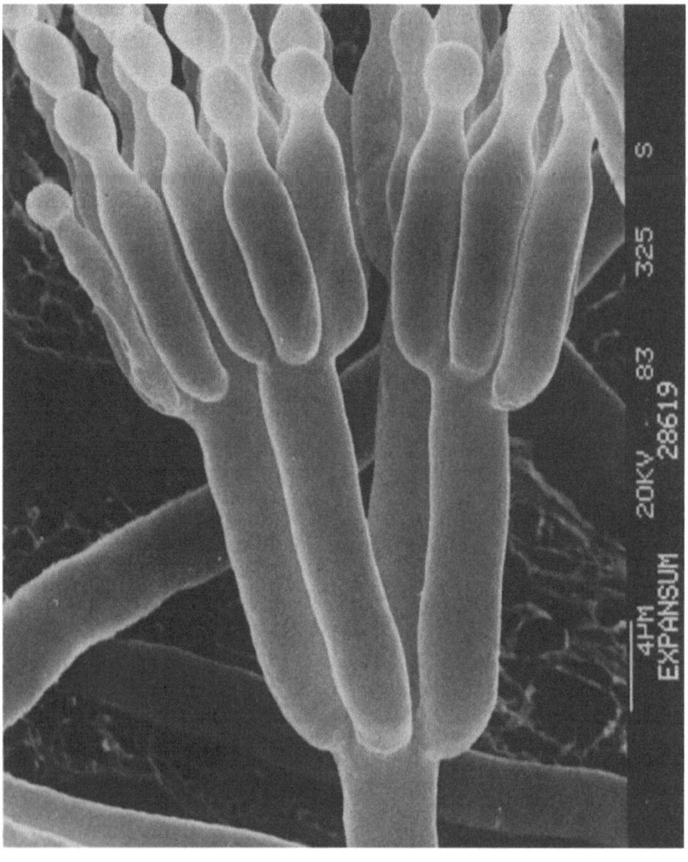

Fig. 3. Cryo-SEM micrograph illustrating the penicillus of <u>P. expansum</u> IMI 28619

STRAIN VARIATION

To determine the extent of variation in individual isolates, colonies derived from single conidia taken from single colonies are being investigated in a parallel study. Thirty single conidia are taken from each isolate selected for this study and the colonies derived from them are being studied with reference to both their morphological and biochemical characters.

For example, in the case of conidia derived from ex-type cultures of *Penicillium viridicatum*, two different stable morphological colony types are produced in more or less equal numbers from a fresh original culture, while from older stock cultures, lower numbers of a third colony type occurred which was unstable. Hyphal fusions did not occur between the types but there was no apparent inhibition in paired plates. In all cases the secondary metabolite profiles were identical at 7 days, but quantitative differences appear to give rise to visually different colours. The conidia are normally uninucleate, but in the unstable type abnormally large conidia with massive nuclei, which we suspect may be diploid or polyploid, can be encountered.

The explanation of this may be that the original culture is a heterokaryon so that different types of nuclei occur in more or less equal numbers in conidial heads. This hypothesis is being tested further and a full account of this study of variation and its implications for *Penicillium* systematics is prepared for separate publication.

Strain variation in the full study is being estimated by the deliberate inclusion of duplicate cultures of 10% of all isolates. These will be maintained and treated separately and any variation between them and their parent strains can be assessed.

NUMERICAL TAXONOMY

The construction of a taxonomy on the basis of a large number of independent characters can be expected to provide a more reliable system because of the amount of information considered (Sneath and Sokal, 1973; Sneath, 1978). In practice, the quality of the taxonomy obtained rises steadily with the number of characters considered up to a point, after which the addition of more characters has only a small effect on the overall outcome. In most cases for practical purposes this point is reached after about 150-200 independent characters. In numerical methods, similarity values between different isolates are calculated. In this study the similarity between each pair of isolates is being considered in turn and a representation made of the average similarities between all isolates using the unweighted pair group method of analysis (Sneath and Sokal, 1973). In order to consider each possible pair of isolates, a large number of comparisons are necessary. For the 300 isolates being considered, 44,850 comparisons must be made, excluding comparisons of each organism with itself; such large numbers of calculations can only be carried out quickly using a computer.

The results are most appropriately displayed as a dendrogram, and that from our preliminary study of 52 isolates is shown in Fig. 4. The horizontal scale represents similarity and each vertical link the average similarity between isolates and groups of isolates clustered above it. A number of similarity or difference coefficients may be used in this type of study and we have so far employed two: Gower's coefficient S_G (see Sneath and Sokal, 1973), which gives a measure of similarity from ranked data and Pattern Difference (D_p) (Sackin, 1981), a difference coefficient, which eliminates vigour differences between isolates. Dendrogram construction depends on converting true similarity values to average values; the final representation is consequently an approximation to the true values. In some cases this may result in the dendrogram distorting the results. Distortion that has occurred may be estimated by correlating the final average similarities between isolates implied by the dendrogram with the original similarities. This gives a cophenetic correlation measure, which should be greater than 0.7 and preferably above 0.8 (Jones and Sackin, 1980).

The clusters shown by the dendrogram are two dimensional and so may overlap. Overlap can be estimated by calculating distances between centroids (most typical isolates) and the spread of the clusters along this axis. The significance of any overlap can be determined statistically using a t-test. By using such techniques an estimate of the reliability of clusters may be obtained (Sneath, 1977).

The first stage in a numerical taxonomy is to construct a data matrix showing all test results for all isolates. Test results must be coded in a form that is acceptable for the coefficient used (Sneath, 1978). The most common methods of coding are either as binary (+/-) or as ranked (i.e. from 0 = negative, to 5 = very positive) data. Each character is given the same 'weight' or importance as each other; however, it is important that characters considered in this way are independent of each other, otherwise "weighting" of a particular feature is introduced (Sneath and Sokal, 1973).

In this study, characters from a number of separate areas are being considered. It is not known at this stage how well characters from one area correlate with characters from another area (e.g. physiology and morphology). In order to ascertain this the data from each section of the study is being considered both independently and in conjunction with the other data sets. In this way it will be possible to see if characters from one area give a significantly different clustering of organisms than characters from other areas.

The final numerical taxonomy results may be used in whole or in part in the production of simpler classification and identification schemes. For example, the most differential characters from the whole study may be selected to provide characteristics for diagnostic keys. These may be tested by using a further set of isolates and assigning them to the numerical taxonomy groups. As the initial data is held on the computer, the construction of identification schemes may also be computer-assisted and can also lead to probabilistic identifications from a percent positive table or to computer

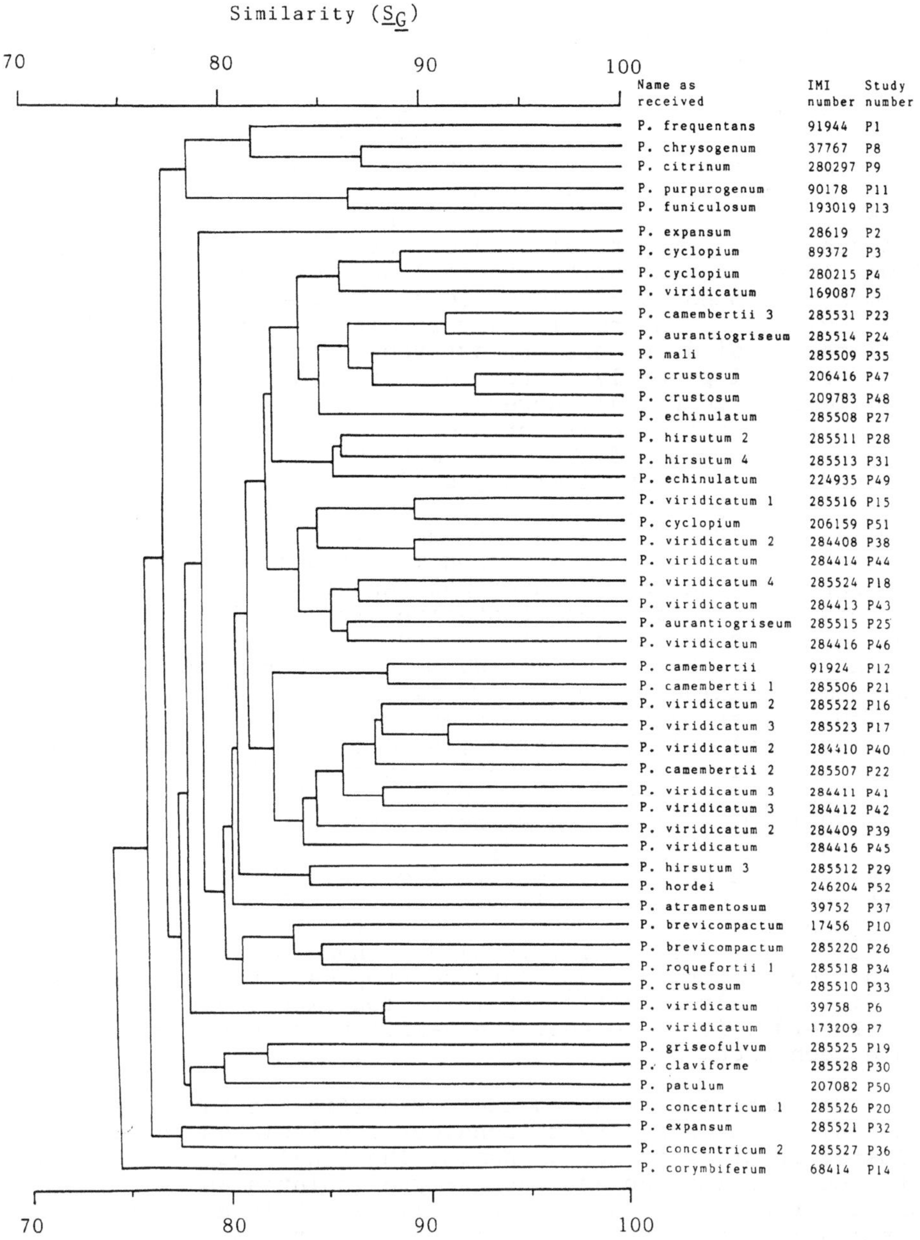

Fig. 4. Dendrogram from pilot study using 52 isolates and 198 characters. Clustered using the UPGMA alogrithm (Sneath and Sokal, 1973)

generated dichotomous keys (e.g. Sneath, 1979; Williams et al., 1983 b; Payne et al., 1982).

SYSTEM CONSTRUCTION

The results from the numerical study can be used to provide a taxonomic system. In order to do this the taxonomic (rather than statistical) significance of the clusters must be considered. The clusters from a numerical taxonomy dendrogram merely show groups of similar organisms; the taxonomic situation is a matter for interpretation. In general, numerical studies with bacteria have shown that members of the same species usually cluster at around the 80% similarity level (Sneath, 1978). Numerical studies with yeasts have shown a lower level of similarity within a species, with clusters formed at around 75-80% level (e.g. Campbell, 1973; Bridge and May, 1984). The preliminary results from the first set of isolates studied in this project showed that species level clusters will probably be formed at around the 78-82% similarity level (see Fig.4). However, the first 52 isolates contained representative marker species and the full range of species to be studied. The test methods have been revised and refined since this preliminary study, and these two factors may ultimately result in species level clusters at the 80-85% similarity level. Clusters formed by numerical methods can form the basis for further taxonomic studies that will either reinforce or weaken the numerical taxonomy. The usual way to do this is to select representative isolates from each cluster and to use these isolates for tests that are not so suitable for a large number of isolates. Methods that have been used in taxonomic studies are mol% G+C, DNA/RNA and DNA/RNA hybridization, electrophoresis of proteins and specific enzymes, ribosomal RNA 'fingerprinting', serology and cell wall/membrane analysis (see Hill, 1968; Mendonca-Hagler and Phaff, 1975; Fox et al., 1977; Azad and Kado, 1980; van Vuuren et al., 1981; Goodfellow and Pirouz, 1982; Hubalek, 1982). In this case a large scale study of mol% G+C values has been undertaken recently and results from this do not look to be particularly helpful (Galimova et al., 1982). Analysis of mitochondrial DNA has proved very useful in the genus *Aspergillus* (Kozlowski et al., 1982) and there may be an analogous situation in *Penicillium*. Outline proposals to undertake such studies and also somatic hybridization experiments (e.g. Anné and Peberdy, 1981) to test the validity of units recognised have been made.

Once the clusters and their significance have been ascertained, it is necessary to consider the isolates in each cluster in order to assign names. The final set of cultures studied will include as many ex-type isolates as are available and their presence in clusters will assist in determining the application of names in accordance with procedures required by the International Code of Botanical Nomenclature (see Hawksworth, 1974).

At this stage it will also be necessary to consider whether nomenclatural types for which ex-type cultures do not exist are pertinent. Studies on the metabolites of dried cultures may be crucial here (Paterson and Hawksworth, 1985).

This phase of our investigations will be greatly facilitated by the critical studies on the typification of names in Penicillium included in Pitt (1979). Until our full data set has been compiled, analysed numerically, and the application of pertinent names investigated, it is premature to consider these questions further here.

CONCLUSIONS

This study has so far concentrated on the development of standard procedures and recording systems using a sample of 52 isolates in order to evaluate this multi-disciplinary integrated approach. The results have been submitted to numerical analysis from which a final set of characters to be included in our study has been drawn up (Table 2). The phase now in hand consists of building up the number of strains to 300, consisting of ex-type cultures, key metabolite producers, strains regarded as important by previous monographers, and "atypical" isolates selected both from the CMI holdings of this genus and provided by colleagues.

As far as we are aware, this is the first time an integrated team approach has been utilised in the taxonomy of filamentous fungi. The results obtained so far are most encouraging and clearly merit widespread adoption in other groups where species circumscription is uncertain.

Acknowledgements

This study has been undertaken under Science and Engineering Research Council (SERC) contract no. SO/17/84 whose generous support is gratefully acknowledged. We are also grateful to the Ministry of Agriculture, Fisheries and Food (MAFF) for seconding Dr. Z. Kozakiewicz to the project, to Professor P.H.A. Sneath for making the numerical taxonomic programs at Leicester University available, and to Miss S. Wade and Miss R. Dunn for technical assistance.

REFERENCES

ABE, S. 1956. Studies on the classification of the Penicillia. J. gen.appl. Microbiol. 2:1-344.

ANNE, J. and PEBERDY, J.F. 1981. Characterisation of interspecies hybrids of Penicillium chrysogenum & P. roqueforti by isoenzyme analysis. Trans. Br.mycol. Soc. 77:401-408.

AZAD, H.R. and KADO, C.I. 1980. Numerical and DNA: DNA reassociation analyses of Erwinia rubrivaciens and other members of the Enterobacteriaceae. J. gen. Microbiol. 120:117-129.

BERRY, D.R. 1975. The environmental control of the physiology of filamentous fungi. In: "The Filamentous Fungi". Vol.1. Industrial Mycology, J.E. Smith and D.R. Berry, eds. London: Edward Arnold., pp 16-32.

BRIDGE, P.D. 1985. An evaluation of some physiological and biochemical methods as an aid to the characterisation of species of Penicillium subsection Fasciculata. - J. gen. Microbiol.: in press.

BRIDGE, P.D. and HAWKSWORTH, D.L. 1984. The API ZYM enzyme testing system as an aid to the rapid identification of Penicillium isolates. - Microbiol. Sci. 1:232-234.
BRIDGE, P.D. and HAWKSWORTH, D.L. 1985. Biochemical tests as an aid to the identification of Monascus species. - Letters Appl. Microbiol. 1: in press.
BRIDGE, P.D. and MAY, J.W. 1984. A numerical classification of fission yeasts of the genus Schizosaccharomyces Lindner. J. gen. Microbiol. 130:1921-1932.
BRIDGE, P.D. and SNEATH, P.H.A. 1983. Numerical taxonomy of Streptocossus. J. Gen. micobiol. 129: 565-597.
CAMPBELL, I. 1973. Numerical analysis of Hansenula, Pichia and related yeast genera. J. gen. Microbiol. 77:427-441.
CANNON, P.F. and HAWKSWORTH, D.L. 1982. A re-evaluation of Melanospora Corda and similar pyrenomycetes, with a revi sion of the British species. Bot. J. Linn. Soc. 84:115-160.
CANNON, P.F. and HAWKSWORTH, D.L. 1984. A revision of the genus Neocosmospora (Hypocreales). Trans. Br. mycol. Soc. 82:673-68.
CIEGLER, A., FENNELL, D.I., SANSING, G.A., DETROY, R.W. and BENNETT, G.A. 1973. Mycotoxin producing strains of Penicillium viridicatum: classification into subgroups. Appl. Microbiol.26:271-278.
CIEGLER, A., LEE, L.S. and DUNN, J.J. 1981. Production of napthoquinone mycotoxins and taxonomy of Penicillium viridicatum. Appl. Envir. Microbiol. 42:466-449.
COUTTS, A.D. and SMITH, R.E. 1976. Factors influencing the production of cellulases by Sporotrichum thermophile. Appl. Envir. Microbiol. 31:819-825.
COWAN, S.T., 1974 - Cowan and Steel's Manual for the Identification of Medical Bacteria. Second edition. Cambridge: Cambridge University Press.
CULBERSON, C.F. and KRISTINSSON, H. 1970. A standardised method for the identification of lichen products. J. Chromatography 46:85-93.
DOMSCH, K.H., GAMS, W. and ANDERSON; T.H. 1980. Compendium of Soil Fungi Vol.1. London: Academic Press.
FASSATIOVA, O. 1977. A taxonomic study of Penicillium series Expansa Thom emend. Fassatiova. Acta univ. Carol., Biol. 12:283-335.
FILTENBORG, O., FRISVAD; J.C. and SVENDSEN; J.A. 1983. Simple screening method for moulds producing intracellular mycotoxins in pure culture. Appl. Envir. Microbiol. 45:581-585.
FOX, G.E., MAGRUM, L.J., BALCH, W.E., WOLFE, R.S. and WOESE, C.R. 1977. Classification of methanogenic bacteria by 16S ribosomal RNA characterisation. Proc.natn. Acad. Sci. 74:4537-4541.
FRISVAD, J.C. 1981. Physiological criteria and mycotoxin production as aids in identification of common asymmetric penicillia. Appl. Envir.Micriobiol.41:568-579.
FRISVAD, J.C. and FILTENBORG, O. 1983. Classification of terverticillate penicillia based on profiles of mycotoxins and other secondary metabolites. Appl. Envir. Microbiol. 46:1301-1310.
GALIMOVA, A.L., SIZOVA, T.P. and ZLOCHEVSKAYA, I.V. 1982. Use of biochemical criterion in the taxonomy of the genus Penicillium. Vest.mosk. gos. Univ., Biol., 16:47-52.
GOCHENAUR, S.E. 1984. Fungi of a Long Island oak - birch forest.II. Population dynamics and hydrolase patterns for the soil Penicillia. Mycologia 76:218-231.

GOODFELLOW, M. and PIROUZ, T. 1982. Numerical classification of sporoactinomycetes containing meso-diaminopimelic acid in the cell wall. J.gen.Microbiol. 128:503-527.
HARBORNE, J.B. 1984. Chemical data in practical taxonomy. In: "Current concepts in plant taxonomy" (V.H. Heywood and D.M.Moore eds.): 237-262.- London: Academic Press.
HAWKSWORTH, D.L. 1974. Mycologist's Handbook. Kew: Commonwealth Mycological Institute.
HAWKSWORTH, D.L. 1976. Lichen chemotaxonomy. In:" Lichenology: progress & problems (D.H. Brown, D.L. Hawksworth & R.H. Bailey, eds.):139-184.
HAWKSWORTH, D.L. 1979. Ascospore sculpturing and generic concepts in the Testudinaceae (syn. Zopfiaceae). Can. J. Bot. 57:91-99.
HILL, L.R. 1968. The determination of deoxyribonucleic acid base compositions and its application to bacterial taxonomy. In: "Identification methods for microbiologists, part B (B.M. Gibbs and D.A. Shapton, eds.): 177-186. London: Academic Press.
HUBALEK, Z. 1982. A review: Numerical comparitive serology - the methods. J. appl. Bact.52: 307-318.
JONES, D. and SACKIN, M.J. 1980. Numerical methods in the classification and identification of bacteria with especial reference to the Enterobacteriaceae. In: "Micriobiological classification and identification (M. Goodfellow and R.G.Board, eds.): 73-106 - London: Academic Press.
JONES, M.G. and NOBLE, W.C. 1982. An electrophoretic study of enzymes as a tool in the taxonomy of the dermatophytes. J. gen. Microbiol. 128: 1101-1107.
KORNERUP, A. and WANSCHER, J.H. 1978. "Methuen Handbook of Colour". 3rd edn. London: Eyre Methuen.
KOZAKIEWICZ, Z. 1978. Phialide and conidium development in the Aspergilli. Trans. Br. myol. Soc. 70: 175-186.
KOZAKIEWICZ, Z. 1986. Identification of Aspergillus species of stored food products, using the scanning electron microscope, in press
KOZLOWSKI, M., BARTNIK, E. and STEPHEN, P.P. 1982. Restriction enzyme analysis of mitochondrial DNA of members of the genus Aspergillus as an aid in taxonomy. J.gen. Microbiol. 128: 471-476.
KREGER-VAN RIJ, N.J.W., ed. 1984. "The yeasts. A taxonomic study. 3rd edn." Amsterdam: Elsevier.
KUBITZKI, K. 1984. Phytochemistry in plant systematics and evolution. In: "Current Concepts On Plant Taxonomy" (V.H. Heywood and D.M. Moore, eds.):263-280. London: Academic Press.
LENNETTE, E.H., BALOW, A. and HAUSLER, W.J. 1980. Manual of Clinical Microbiology. 3rd edn. American ,Society for Micriobiology, Washington.
MENDONCA-HAGLER, L.C. and PHAFF, H.J. 1975. Deoxyribonucleic acid base composition and deoxyribonucleic acid/ deoxyribonucleic acid hybrid formation in psychrophobic and related yeasts. Internat. J. Syst.Bact.25:222-229.
ODDS, F.C., HALL, C.A. and ABBOTT, A.B. 1978. Peptones and mycological reproducibility. Sabouraudia 16:237-246.
ONIONS, A.H.S., BRIDGE, P.D. and PATERSON, R.R. 1984. Problems and prospects for the taxonomy of Penicillium. Microbiol. Sci.1:185-189.
PATERSON, R.R.M. and HAWKSWORTH, D.L. 1985. The detection

of secondary metabolites in dried cultures of Penicillium preserved in herbaria. Trans.Br.mycol. Soc.84: in press.

PAYNE, R.W., YARROW, D and BARNETT, J.A. 1982. The construction by computer of a diagnostic key to the genera of yeasts and other such groups of taxa. J. gen. Microbiol. 128:1265-1277.

PITT, J.I. 1979. The Genus Penicillium and its teleomorphic states Eupenicillium and Talaromyces. London: Academic Press.

RAMIREZ, C. 1982. Manual and Atlas of the Penicillia. Amsterdam: Elsevier Biomedical Press.

RAPER, K.B. and THOM, C. 1949. A manual of the Penicillia. Baltimore: Williams and Wilkins.

RAYNER, R.W. 1970. A Mycological Colour Chart. Kew: Commonwealth Mycological Institute.

RIDGWAY, R. 1912. Color standards and color nomenclature. Washington: Ridgway

SACKIN, M.J. 1981. Vigour and pattern as applied to multistate quantitative characters in taxonomy. J. gen. Microbiol. 122:247-254.

SAMSON, R.A. and GAMS, W. 1984. The taxonomic situation in the hyphomycete genera Penicillium, Aspergillus and Fusarium.Antonie van Leeuwenhoek, 50: 815-824.

SAMSON, R.A., STOLK, A.C. and HADLOK, R. 1976. Revision of the subsection Fasciculata of Penicillium and some allied species. Stud.Mycol., Baarn 11:1-47.

SCOTT, P.1981. Liquid chromatography in analysis of mycotoxins. In: "Trace Analysis" (J.F. Lawrence, ed.) 1:194-266. London: Academic Press.

SCOTT, P.M., LAWRENCE, J.W. and VAN WALBECK, W. 1979. Detection of mycotoxins by thin layer chromatography: applications to screening of fungal extracts. Appl.Microbiol. 20:839-842.

SHIPTON, W.A. and LUNN, J.A. 1980. Sporangiole morphology and species separation in Cunninghamella. Trans. Br. mycol. Soc.74:483-491.

SMITH, G. 1949. The effect of adding trace elements to Czapek-Dox medium. Trans.Br.mycol. Soc. 32:280-283.

SNEATH, P.H.A. 1977. A method for testing the distinctness of clusters: a test of the disjunction of two clusters in Euclidean space as measured by their overlap. J. Internat. Ass.Math.Geol. 9:123-143.

SNEATH, P.H.A. 1978. Classification of micro-organisms. In: "Essays in Microbiology" (J.R. Norris and M.H. Richmond, eds.): 9/1-9/31. Chichester: John Wiley.

SNEATH, P.H.A. 1979. BASIC program for identification of an unknown with presence-absence data against an identification matrix of percent positive characters. Computers & Geosciences 5: 195-213.

SNEATH, P.H.A. and SOKAL, R.R. 1973. Numerical Taxonomy. 2nd edn. San Francisco: W.H. Freeman.

STOUT, D.L. and SHAW, C.R. 1973. Comparative enzyme patterns in Thamnidium elegans and T. anomalum. Mycologia 65:803-808.

THOM, C. 1910. Cultural studies of species of Penicillium. Bull.Anim.Ind., U.S.D.A. 118:1-109.

VAN VUUREN, H.J.J., KERSTERS, K., DELEY, J. and TOERIEN, D.F. 1981. The identification of Enterobacteriaceae from breweries: combined use of and comparison of API ZOE system, gel electrophoresis of proteins and gas chromato-

graphy of volatile metabolites. J.appl. Bact.51:51-66.
WALKER, F.J. and JAMES, P.W. 1980. A revised guide to microchemical techniques for the identification of lichen products. Bull.Br. Lichen Soc.46:13-29.
WILLIAMS, S.T., GOODFELLOW, M., ALDERSON, G., WELLINGTON, E.M.H., SNEATH, P.H.A., and SACKIN, M.J.1983a. Numerical classification of Streptomyces and related genera. J.gen.Microbiol. 129:1743-1813.
WILLIAMS, S.T., GOODFELLOW, M., WELLINGTON, E.M.H., VICKERS, J.C., ALDERSON, G., SNEATH, P.H.A., SACKIN, M.J. and MORTIMER, A.M. 1983 b. A probability matrix for identification of some Streptomycetes. J. gen.Micriobiol. 129:1815-1830.
WONG, A.L. and WILLETTS, H.J. 1973. Electropheretic studies of soluble proteins and enzymes of Sclerotinia species. Trans. Br. mycol. Soc. 61:167-178.

DIALOGUE FOLLOWING CMI PRESENTATION

WICKLOW: Dr Bridge, you used the temperatures 4°, 25°, 37° and 40°C. What was the rationale behind chosing these temperatures?

BRIDGE: The refrigerator used runs at 4° rather than 5°C. Thirty-seven degrees has been used on a number of occasions before and we also wanted a temperature in excess of 37°C.

WICKLOW: You only used one character for water relations, which was growth on G25N agar. Given the importance of this character in the taxonomy of these fungi, don't you think this deserves more tests in your matrix? You could trade off some of your temperature data from those of water relations.

BRIDGE: We are experimenting with some other water relations media and we are looking for tests that can be easily performed in other laboratories, especially biochemical and physiological tests.

KLICH: Now that you have completed the pilot study, can you tell us which tests were of most value? Which tests provided correlations with the morphological data? Can you give us some results? Have you done a cluster analysis to show which characters are splitting the groups apart?

ONIONS: We put a limit on the number of tests because of the limit of computer space. We limited the number of tests to 200 and the number of isolates to 300. After our pilot study, we agreed on the characters which we thought were differential.

GAMS: You inoculate your API ZYM strips with ungerminated conidia, and incubate them only for four hours. What will happen in these four hours? Are you testing only a diffusion of the enzymes that are present in the conidia. If so, are you not missing a great proportion of the enzymes which may be expected?

BRIDGE: It is possible that we are missing a great deal. Certainly the results seem to be quite reproducible. The spore suspension results generally correlate with the results obtained with culture filtrates. We haven't checked to see if the conidia germinate in 4 hours. One criticism of enzyme test in bacteria is that they were developed for inducible enzymes, but we would prefer them to be for constitutive enzymes. We believe, that most of the tests we are using are for constitutive enzymes. If any are inducible enzymes, we are at least carrying out the tests under standardised conditions.

GAMS: Dr Kozakiewicz, both Dr Samson and I are worried about the possible differences in spore measurements between what you see under the SEM and what you see under the light microscope. Can you standardize what you measure with the SEM against measurements with the light microscope?

KOZAKIEWICZ: We are planning a separate study at CMI to check this point. We are also going to compare the effect of different stains and mounting media on measurements under the light microscope.

SAMSON: Although I think the SEM gives very beautiful photographs, its use for taxonomic studies is limited. During preparation, and observation under vacuum, there are so many factors which you cannot control and standardize. The problem of shrinkage is very well documented, so you can't use SEM for measurements. As for surface ornamentation, you can sometimes be certain what you are really seeing, such as spiny conidia, but in other cases, such as the reticulate pattern you showed us, to me this is a sign that your conidia have shrunk. Are you detecting shrinkage, or are you really detecting ornamentation?

KOZAKIEWICZ: This is only an outline of what we've done so far. Obviously, at the end of the three years it may be a different story.

SAMSON: But you are putting all these data in your data matrix before evaluating them?

PITT: Have you attempted to correlate the spininess you see under the SEM with what we see under the light microscope? To me, the useful distinction in Penicillium lies between genuinely rough conidia and those which are smooth to finely roughened under the light microscope.

KOZAKIEWICZ: Yes, I can correlate what I see under the SEM with light microscopic observations.

HAWKSWORTH: I think it is important to point out that these SEM characters are not necessarily going to be used in a key, but they are being used to help establish what the groups are as satisfactorily as possible.

ONIONS: We are attempting to find correlations among a whole range of characters. If we see consistent effects under the SEM this is useful information, no matter what you actually label the characters.

RAMIREZ: I think electron microscopy is the greatest improvement for allowing us to interpret the morphological characters of conidia.

SAMSON: But, Dr. Ramírez, are the data from the SEM for the taxonomist or for somebody trying to identify a specimen or culture?

RAMÍREZ: It is for the specialist.

POLONELLI: How many isolates do you check for each species?

BRIDGE: At the moment we are just checking isolates as isolates, we are not attempting to correlate them with classical species.

POLONELLI: In medical mycology where time is important for antimycotic therapy, we find that resistance varies from strain to strain. We need to know more than just the species, we need to know the strain. Do you see any strain specific biochemical characters which may be of use in a situation like this?

BRIDGE: It may be in this case that we will find strain specific characters which are not necessaily constant characters for a whole species.

PITT: One more thing, as shown in Table 9 of Pitt (1979) only one species in the subgenus you are studying ever grows at 37° C. There is no point in using 37° or 40° C. If you were to use 30° C you might very well find useful results.

BRIDGE: We are using 37° C as a check for the purity of our isolates, and also because some of our marker strains grow at this temperature.

PROFILES OF PRIMARY AND SECONDARY METABOLITES OF VALUE IN CLASSIFICATION OF PENICILLIUM VIRIDICATUM AND RELATED SPECIES

Jens C. FRISVAD

Food Technology Laboratory
Technical University of Denmark
2800 Lyngby, Denmark

SUMMARY

Species of Penicillium can often be identified by morphological methods, but additional criteria such as profiles of secondary metabolites can be useful for confirmation, especially in species occurring on domesticated plants used for foods and feedstuffs. Fungi may also be classified by polyphenic methods based on quantitative profiles of primary metabolites, e.g. gas chromatographic determination of pyrolysis products or carbohydrates. The use of these methods is exemplified in this paper on species and isolates which are especially difficult to identify by traditional taxonomical methods, i.e. the P. viridicatum, P. cyclopium, P. expansum, and P. ochraceum series of Raper and Thom. It is shown that Penicillium isolates with rough-walled conidia, isolated from seed caches and cheek pouches of the North American desert kangaroo rat, are P. aurantiogriseum II and not P. echinulatum.

INTRODUCTION

Many species in Penicillium maintain a definite taxonomy, despite different species concepts used by authors of monographs. The connection between Penicillium species and profiles of secondary metabolites (Bu'Lock, 1980) has long been obscured, often because of misidentification. The main reason for these misidentifications is probably the lack of sufficiently distinctive diagnostic criteria in Penicillium. Frisvad and Filtenborg (1983) and Frisvad (1981) have shown that the terverticillate Penicillia can be classified according to a combination of chemical and physiological criteria, and the resulting experimental taxa are closely related to species based on classical morphological characters (Frisvad, 1985; Samson and Gams, 1984). A number of problems have to be solved, however, before species based on morphology and profiles of secondary metabolites can be routinely identified in

Penicillium taxonomy. After taxometric studies on a great number of isolates of Penicillium, simplification and standardization of the identification methods has to be carried out. Some species of Penicillium are especially difficult to classify and among these are the members of the P. cyclopium, P. viridicatum, P. expansum and P. ochraceum series of Raper and Thom (1949). The classification of these species has been treated differently by various taxonomists (Fassatiová, 1977; Samson et al., 1976; Pitt, 1979; Onions et al., 1981; Raper and Thom, 1949; Frisvad and Filtenborg, 1983) and the appearance of new subgroups and morphological variants has caused further complications. The aim of this study was to classify isolates belonging to the series mentioned above into homogeneous clusters, based on different sets of characters (morphology, some physiological characters, the profile of secondary metabolites and the quantitative profile of carbohydrates and pyrolysis products).

MATERIAL AND METHODS

The cultures listed by Frisvad and Filtenborg (1983), Frisvad (1981), Söderström and Frisvad (1984), Hermansen et al. (1984) and Frisvad (1983), cultures ex type from each species, and new cultures from different substrates were analyzed for mycotoxins and other secondary metabolites using the methods of Filtenborg et al. (1983) and Filtenborg and Frisvad (1980). The cultures used for more detailed biochemical analysis are listed in Table 1.

Some of the more controversial species, subgroups and isolates were also analyzed for carbohydrates and polyols after growth on YES agar for 16 days at 25°C (Frisvad and Olsen, in prep.). The contents of one Petri dish was homogenized using methenol-water as solvent. A neutral water fraction was obtained using the methods of Bjerg et al. (1984) and the carbohydrate fraction was analyzed using gas liquid chromatography (GLC) after silylation.

The logarithms of the peak areas of ten carbohydrates and polyols common to all chromatograms were used for statistical analysis. The data were normalized to give a constant total sum area of all peaks. At least three replicates of each analysis were performed. The data were subjected to SIMCA principal component analysis (Wold and Sjöstrom, 1977; Derde et al., 1984; Wold et al., 1984; Söderström and Frisvad, 1984). The inital plot of the principal components for all isolates and replicates of P. viridicatum I, P. aurantiogriseum II, P. aurantiogriseum I, P. echinulatum and P. viridicatum IV showed three obvious classes in the data. Cross validation of these disjoint classes showed that they were well described by few component models. A PLS classification was therefore appropriate, because the classes are better separated using this method than an ordinary principal component analysis (PLS = partial least square analysis, see Wold et al., 1984) PLS analysis the kind of data for here uses "dummy variables" for each class (1 for class one, 10 for class two, 100 for class 3 etc.), and the object - variable matrix is reduced in dimensionabality in such a way that the reduced data matrix approximates the original data matrix and predicts the

Table 1. Experimental groups, based on secondary metabolites, for the classification of species in the *Penicillium viridicatum*, *P. cyclopium*, *P. ochraceum* and *P. expansum* series of Raper and Thom (1949).

---

1. *Penicillium aurantiogriseum* I

   P. aurantiogriseum - NRRL 971 (NT)[a]; IMI 285514
   P. aurantiovirens - CBS 573.68
   P. lanosocoeruleum - CCF 308
   P. martensii - CCF 198; CBS 744.74 (=NRRL 3608)
   P. ochraceum - Fujimoto 1
   P. puberulum - ATCC 15683

2. *Penicillium aurantiogriseum* II

   P. aurantiogriseum - NRRL 1888 (T, *P. cyclopium*); NRRL 1899; IMI 92199 (T, *P. brunneoviolaceum*); IMI 285515; IMI 293186
   P. cordubense - CBS 162.81 (T)
   P. crustosum - ATCC 48414
   P. echinulatum - NRRL A-27147; NRRL A-27160; NRRL A-27003[b]
   P. olivicolor - IMI 92264 (T)
   P. martensii - NRRL 3612
   P. puberulum - NRRL 1889 (NT); LC 283bis; CCF 1275

3. *Penicillium viridicatum* I

   P. expansum - FRR 1347
   P. olivicolor - NRRL 870; NRRL 882
   P. viridicatum - NRRL 963 (NT); NRRL 5569; NRRL A-26909; NRRL A-27011; IMI 285516; IMI 293203; NRRL 2028 (T, *P. olivinoviride*), NRRL 961; NRRL 962; NRRL A-14307; NRRL A-19118; NRRL A-15505; NRRL A-14502

4. *Penicillium viridicatum* IV

   P. verrucosum var. *corymbiferum* - Leistner Sp 1448
   P. crustosum - CCM F-398
   P. verrucosum var. *cyclopium* - IJFM 7791
   P. viridicatum - CSIR 1375 (first known isolate, also identified as *P. expansum*); IMI 285525 (also identified as *P. crustosum*/*P. chrysogenum*); IMI 293194

5. *Penicillium viridicatum* II

   P. cyclopium - NRRL 6065
   P. nordicum - IJFM 7813 (T)
   P. palitans - CBS 223.71
   P. viridicatum - NRRL 1160; NRRL 5571; NRRL 5583; NRRL 3712; LC 3044; IMI 285522; IMI 293203

5a. *Penicillium viridicatum* III

   P. casei - CBS 302.48
   P. mediolanensis - IJFM 7812

(Continued)

P. verrucosum - NRRL 965 (NT); NRRL 5573; NRRL 5574; NRRL A-19179b; IMI 285523; Leistner Sp 134 & 991

6. Penicillium camembertii II

P. biforme - NRRL 6497
P. camembertii - FRR 2160
P. commune - NRRL 890 (T); CBS 261.69, IMI 285507; IMI 293179; CBS 169.44
P. cyclopium - LC 3239; Leistner Sp 603 & 608; Marth 25 & 40; IJFM 7793
P. lanosoviride - NRRL 879 (T), Marth 44
P. ochraceum var. macrosporum - NRRL 873 (T)
P. puberulum - NRRL 845; LC 218; Marth 33; Marth 37
P. terrestre - ATCC 32028
P. viridicatum - Leistner Sp 119

7. Penicillium camembertii III

P. biforme - NRRL 885 (T)
P. casei var. compactum - CBS 427.65
P. crustosum - LC 1628; Leistner Sp 607
P. cyclopium - LC 189; Leistner Sp 605 & 613; Marth 8
P. cyclopium var. album - CBS 343.51 (T)
P. lanosogriseum - IMI 39822 (T); LC 2183
P. lanosoviride - NRRL 930
P. palitans - NRRL 2033 (T), IMI 89378; IMI 285531; IMI 293201
P. psittacinum - NRRL 932 (T)
P. puberulum - FRR 168; FRR 1703; FRR 2014
P. roqueforti var. punctatum - IMI 68234 (T)
P. solitum - NRRL 837 (T); IMI 92225
P. verrucosum - NRRL 948 (T, P. flavoglaucum); FRR 1830

8. Penicillium crustosum

P. casei - CBS 483.75
P. commune - AUA 827
P. crustosum - IMI 91917 (T); NRRL 968; NRRL 1983; NRRL 5186; Engel 6842; LC 2069; LC 3045; IMI 285510; IMI 293182
P. cyclopium - NRRL 3476; NRRL 3477; NRRL 6093; ATCC 32014; Leistner Sp 1191 & 458
P. farinosum Novobranova - IMI 174717
P. griseum Bonorden - ATCC 20068
P. lanosocoeruleum - ATCC 32017; ATCC 32020
P. meleagrinum - ATCC 32021
P. palitans - NRRL 3468
P. roqueforti - ATCC 32027
P. terrestre - IMI 89384; IMI 89385
P. verrucosum var. melanochlorum - Leistner Sp 2304
P. viridicatum - LC 212

9. Penicillium mali

- P. cyclopium - IMI 89374
- P. mali - CBS 500.73 ("T")
- P. melanochlorum - CBS 487.75 (T); IMI 285509; IMI 293181
- P. palitans - NRRL 914; NRRL 966; NRRL 1164
- P. viridicatum - IMI 49096; IMI 154731

10. Penicillium echinulatum

- P. crustosum var. spinulosporum - FRR 1621 (T)
- P. echinulatum - NRRL 1151 (T); CBS 255.55; Leistner Sp 1894; IMI 285508; IMI 293180
- P. palitans var. echinoconidium - CBS 337.59 (T)
- P. verrucosum - FRR 1963
- P. verrucosum var. album - CBS 203.57

11. Penicillium hirsutum IV

- P. aurantiovirens - ATCC 32000; ATCC 32002
- P. hirsutum IV - IMI 285513

12. Penicillium hirsutum I

- P. corymbiferum - NRRL 2032 (T), NRRL 999, CCM F-240; CCF 1445

13. Penicillium hirsutum II

- P. hirsutum II - IMI 285511; IMI 293183; IMI 293206

14. Penicillium hirsutum III

- P. hordei - CBS 701.68 (T); IMI 285512; 293184

15. Penicillium expansum

- P. expansum - NRRL 976 (NT), CP 554; CCF 1443; IFO 5854, IMI 285521; IMI 293192; Leistner Sp 248; 454;529; 558; 559; 560; 693; Frank 597; 599; 603
- P. hirsutum - FRR 48
- P. resticulosum - NRRL 2021 (T)

---

[a] NT: cultures ex neotype, T: cultures ex type

[b] These isolates are from seed caches and seed pouches of USA kangaroo rats (Wicklow, Filtenborg & Frisvad, in prep.). Two isolates had smooth walled conidia (NRRL A-26843 & NRRL A-26846), but the other 35 isolates tested had rough walled conidia.

dummy variable matrix well at the same time.

The SIMCA-MACUP 3B package is written in Basic by S. Wold, Umea University and can be obtained from SEPANOVA AB, Östrandsvagen 14, S-122 43 Enskede, Sweden. The program was run on a 16 bit computer (IBM PC/XT).

RESULTS

Micromorphology

The species and subgroups listed in Table 1 could be divided into four anatomical species: (1) P.expansum, with smooth stipes and conidia on MEA; (2) P. verrucosum, with finely to conspicously roughened stipes, or occasionally a few smooth stipes, and smooth conidia on MEA; (3) P. echinulatum, with stipes rough-walled or occasionally smooth-walled and conidia conspicously rough walled; and (4) P. hordei, with rough-walled stipes and finely roughened conidia on MEA (see Samson et al., 1976).

The conidia of P. echinulatum FRR 1963 (= P. verrucosum sensu Pitt, 1979) and FRR 1621 (= P. crustosum fide Pitt, 1979) were less conspicously roughened that the other isolates in this species. P. hirsutum IV produced divaricate penicilli.

Macromorphology

P. expansum, P. echinulatum and P. hirsutum I to IV produced distinct coremia on MEA or PDA containing the fungicide botran (dichloran; 2,6-dichloro-4-nitroanaline). P. resticulosum NRRL 2021 produced coremia on MEA after prolonged incubation at 0°C. P. crustosum, P. viridicatum I to IV and P. aurantiogriseum produced granular to fasciculate colonies on the substrates mentioned above but the colonies of P. mali and P. camemberti II and III were velutinous, mealy or lanose. Old culture collection strains were often lanose.

Conidial colour

Conidial colour in Penicillium is primarily caused by melanin production (Wheeler, 1983; Ha-Huy-Ke and Luckner, 1979; Frisvad, unpublished results). Cu-ion deficiency or tricyclazol inhibition clearly caused a olive brown conidial colour, i.e. the colour seen in the naturally occurring mutants given the names P. ocharaceum, (= P. olivicolor) and P. carneolutescens, in all of the species listed in Table 1. The conidial colour of P. echinulatum, P. mali, P. hirsutum IV. P. camamberti III and a strain of P. crustosum (NRRL 3468) was dark green with a greyish tint in P. camemberti III and P. crustosum. The conidia of P. expansum, P. camemberti and P. crustosum were dull green with a greyish tint in the two latter species. The conidia of P. viridicatum I to IV, P. hirsutum III and P. hirsutum 1 were green though in different shades: P. viridicatum I and IV and P. hirsutum III had conidia with a yellow greenish tint, P. viridicatum II and III were more grey and P. hirsutum I had a bluish tint. P. hirsutum II, and P. aurantiogriseum I and II had bluish green conidia. Some strains of P. viridicatum (those not producing the yellow metabolite brevianamide A) had green conidia with a bluish tint, but they were not as blue as those of P. aurantiogriseum.

Secondary metabolites

Nearly all isolates produced colourless or pale yellow to orange brown exudate droplets. Droplets were often copiously produced by P. hirsutum I, P. viridicatum II and III, and P. aurantiogriseum II. The exudate of P. hirsutum I was coloured deep orange brown to violet brown. Soluble pigments were often produced by isolates of P. expansum, P. hirsutum I and III, P. aurantiogriseum I and II, and P, viridicatum I, and the colours of the diffusible pigments ranged from yellow to orange or red brown on CYA and yellow to yellow brown on MEA. Colony reverse colours on CYA was cream, though with a brown center, in P. echinulatum, P. hirsutum IV, P. echinulatum, P. mali, P. camemberti II and III and P. viridicatum II and III. It was yellow in P. viridicatum IV, and pale yellow to yellow or orange brown in the other species. The reverse of P. viridicatum I and P. aurantiogriseum was occasionally pink. Reverse colours on YES agar were violet brown in P. viridicatum II, orange red in P. hirsutum IV, orange yellow in P. mali and P. expansum, and yellow in the other species. The reverse of P. viridicatum III and P. camemberti II and III was often cream coloured, but the reverse colour of P. camemberti II became blackish brown in age. P. camemberti II, P. hirsutum II and III, P. aurantiogriseum II and P. viridicatum I to III produced few conidia on YES agar, and the colony obverse of P. aurantiogriseum II and P. viridicatum I was typically yellow, while the obverse colour of P. expansum and P. mali on YES agar was orange yellow.

The production of specific known secondary metabolites by the species and subgroups treated are listed in Table 2. Further unknown secondary metabolites, determined by TLC and HPLC, indicated that the species or subgroups listed in Table 1 are all distinct taxa except P. viridicatum III which can be incorporated in P. viridicatum II, P. aurantiogriseum I and II and P. viridicatum I are very closely related and P. hirsutum I, II and III are rather closely related, though distinct.

Isolates from the North American desert kangaroo rats obtained from D. Wicklow (NRRL, Peoria, USA)(P. aurantiogriseum II, with rough walled conidia) did not produced nephrotoxic naphtoquinones (xanthomegnin and viomelein), but they produced cyclopenin, viridicatin, penicillic acid and six unknown metabolites in common with other isolates of P. aurantiogriseum II. Two isolates of P. aurantiogriseum II from the kangaroo rats had smooth walled conidia and produced the same profile of secondary metabolites as those with rough walled conidia. According to the taxonomic systems of Samson et al. (1976) or Pitt (1979) these rough walled isolates would have been classified as P. echinulatum, but apart from the production of cyclopenin and viridicatin, the latter species produces a completely different profile of secondary metabolites.

Table 2. Known secondary metabolites produced by the *P. cyclopium* and *P. viridicatum* complex and related species[a].

	viridicatin	roquefortine	terrestric acid	penicillic acid	citrinin	xanthomegnin	patulin	cyclopiazonic acid	griseofulvin	viridicatum-toxin	penitrem A	ochratoxin	palitantin	rugulovasin	chaetoglobosin
P. aurantiogriseum I	+	-	-	+	-	-	-	-	-	-	-	-	-	-	-
P. viridicatum I	+	-	-	+	-	+	-	-	-	-	-	-	-	-	-
P. aurantiogriseum II	+	-	(+)	+	-	+	-	-	-	-	(+)	-	-	-	-
P. hirsutum I	+	+	+	-	-	-	-	-	-	-	-	-	-	-	-
P. hirsutum II	+	+	+	(+)	+	-	-	-	-	-	-	-	-	-	-
P. hirsutum III	+	+	+	-	-	-	-	-	-	-	-	-	-	-	-
P. crustosum	+	+	+	-	-	-	-	-	-	-	+	-	-	-	-
P. mali	+	-	-	-	-	-	-	-	-	-	-	-	-	-	-
P. echinulatum	+	-	-	-	-	-	-	-	-	-	-	-	-	-	-
P. hirsutum IV	+	-	-	-	-	-	-	-	-	-	-	-	-	-	+
P. camembertii III	(+)	-	-	-	-	-	-	+	-	-	-	-	+	-	-
P. camembertii II	[illegible]	[illegible]	[illegible]	[illegible]	[illegible]	[illegible]	[illegible]	+	[illegible]	[illegible]	[illegible]	[illegible]	[illegible]	+	[illegible]
P. expansum	-	+	-	-	+	-	+	-	-	-	-	-	-	-	-
P. viridicatum II	-	-	-	-	+	-	-	-	-	-	-	+	-	-	-
P. viridicatum III	-	-	-	-	-	-	-	-	-	-	-	+	-	-	-
P. viridicatum IV	-	-	-	-	-	-	-	-	+	+	-	-	-	-	-

[a] Additional data: *P. viridicatum* I - brevianamide A and B; *P. aurantiogriseum* I - S-toxin (Leistner, 1984); *P. aurantiogriseum* II - puberulic and puberulonic acids (Birkenshaw and Raistrick, 1932), known producers, NRRL 1889 and IMI 34846; *P. camemberti* III - cyclopiazonic and cyclopaldic acid (Birkenshaw et al., 1952), known producers, IMI 89312 and IMI 89375.

Overproduced primary metabolites

Most isolates in the species treated, the exceptions being isolates of *P. viridicatum* II and III and a few isolates of *P. hirsutum* I and *P. aurantiogriseum* II, produced acids on creatine sucrose agar. The strongly creatine positive isolates, however, produced basic products after the production of acids. Among the acidic and basic products produced the most important seems to be citric acid in the former group and ammonia in the latter group (Frisvad, unpubl. results).

A number of isolates in each of the species treated were also analysed for presence of carbohydrates and polyols in YES agar after 16 days of incubation at 25 C(Frisvad and Olsen, in prep.) The profile of carbohydrates and polyols contained glycerol, erythritol, arabitol, mannitol, fructose, glucose, sucrose and two unidentified compounds. Principal component analysis of the quantities of the ten compounds in P. aurantiogriseum I and II, P. viridicatum I and IV and P. echinulatum (SIMCA) showed that P. aurantiogriseum I, II and P. viridicatum I formed one class, P. viridicatum IV a second class and P. echinulatum a third class. Cross validation of the classes showed that the classes were well described with no outliers and no alien objects. P. aurantiogriseum I (except one isolate: IMI 285514) seemed to form a subclass somewhat separated from P. aurantiogriseum II and P. viridicatum I. The P. aurantiogriseum I subgroup produced many conidia on YES agar, while IMI 285514 produced few conidia on YES agar, similar to P. aurantiogriseum II and P. viridicatum I. A PLS projection of these data is shown in Fig. 1. The classes are clearly separated, and isolates from North American kangaroo rats with rough walled conidia clearly belong to P. aurantiogriseum II and not P. echinulatum. P. viridicatum I and IV are also clearly separated.

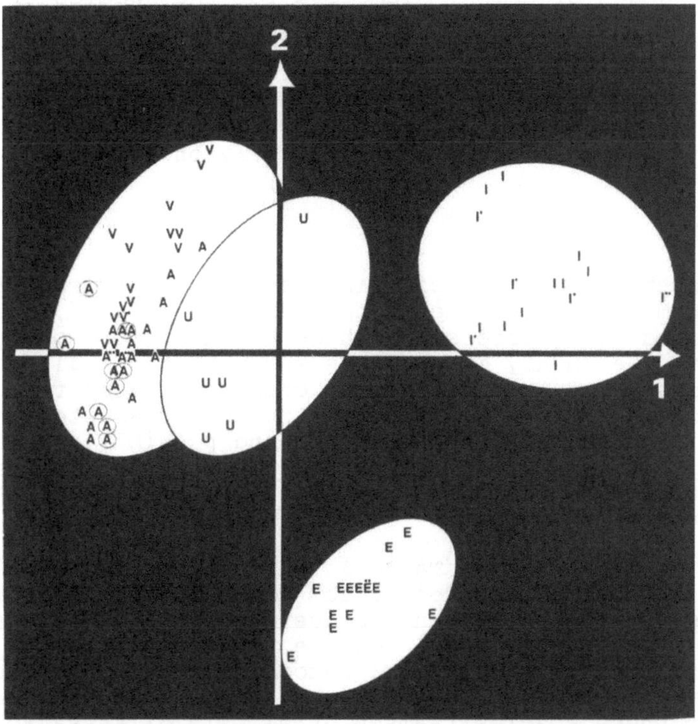

Fig. 1. A PLS projection of the first two PLS dimensions for the carbohydrate and polyol data for P. viridicatum I (V), P. aurantiogriseum II, P. aurantiogriseum I (U), P. echinulatum (E) and P. viridicatum IV (I). The stars indicate the additional number of superimposed data projections points. The points in small ellipses are P. aurantiogriseum II isolates with rough walled conidia from the North American kangeroo rat caches.

Principal component analysis of pyrolysis products (as determined by quantitative GLC) of P. crustosum and P. mali showed that these two species are distinct (Söderström and Frisvad, 1984).

Physiological characters

All isolates of P. expansum produced an extensive rot in apples and all isolates of P. crustosum a limited rot. Isolates of P. hirsutum I caused a rot in one or more types of onions and/or bulbs.

Isolates of P. echinulatum, P. mali, P. hirsutum IV , P. camemberti II and III, P. crustosum and P. expansum grew very well on creatine sucrose agar, while the other species were clearly inhibited. P. viridicatum II and III grew well on nitrite sucrose agar, but P. aurantiogriseum I and II and P. viridicatum I does not grow at all on this medium. The other species cannot grow on nitrite sucrose agar, except some isolates of P. camemberti II and III, P. mali, P. hirsutum I and II. Isolates of P. viridicatum II and III do not tolerate moderate concentrations of propionic acid or combinations of benzoic and sorbic acid (Frisvad, 1981), while most of benzoic and sorbic isolates of the other species grow on such substrates.

The ranges of colony diameters for the species treated on CYA, MEA and YES agar are listed in Table 3. Most ranges of colony diameters overlap, but some species are consistently slow growing (P. hirsutum IV and P. viridicatum II and III). P. mali and P. crustosum may be separated based on colony diameter on CYA. Colony diameters of P. echinulatum and P. aurantiogriseum II only overlap slightly on YES agar.

DISCUSSION

Even though many species in the P. cyclopium and P. viridicatum series may seem rather closely related, they can be separated using dfferent methods based on morphology, primary and secondary metabolites and physiological characters. The selection of diagnostic characters used for species separation should preferably await further characterisation, a taxometric treatment of the results and possibly character weighting based on how fundamental the characters are and whether the characters can be used in practical daily taxonomical work.

It is apparent from Table 1 that many culture collection isolates have been "misidentified" according to a taxometrically based differentiation species concept (see Frisvad, 1984), but most of the isolates are probably correctly identified according to the original taxonomic authority. Even though several new approaches have been used in the classification of these species (Samson et al., 1976; Fassatiová, 1977; Pitt, 1979), identification of species in the P. viridicatum and P. cyclopium series is still difficult (Onions et al., 1984). The situation is further complicated by the appearance of a new subgroup of P. viridicatum (IV) and P. aurantiogriseum II isolates with rough-walled conidia. These series includes species which are rather similar micromorphologically but very different in physiological and biochemical characters (Table 2, Fig. 1, Table 3). Fig. 2 is a suggested guide to the relationships among these fungi based on morphology, profiles of secondary metabolites, conidium colour, growth on

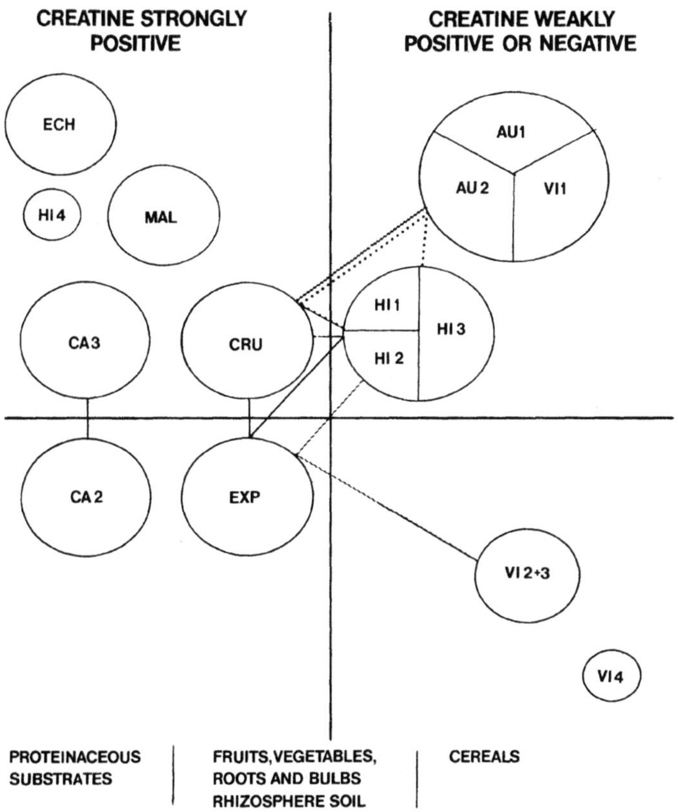

Fig. 2. Suggested relationships in the P. viridicatum and P. cyclopium series based on profiles of secondary metabolites, morphology, growth rates, growth on creatine sucrose agar and habitats. Species over the horizontal line are producers of cyclopenin and/or viridicatin. The lines between the species indicate production of the same secondary metabolite (see Table 2): the upper wavy line represents penitrem A (only produced by a few isolates of P. aurantiogriseum II), the lower wavy line citrinin, the solid line between P. camemberti II (CA 2) and P. camemberti III (CA 3) is cyclopiazonic acid, the other solid lines represents roquefortine C and the dotted line represents terrestric acid (only produced by a few isolates of P. aurantiogriseum II (AU 2). ECH = P. echinulatum, MAL = P. mali, HI 1 to 4 is P. hirsutum I to IV, CRU = P. crustosum, EXP = P. expansum, VI 1 to 4 is P. viridicatum I to IV and AU 1 is P. aurantiogriseum I.

creatine sucrose agar and preferred habitats. The species in the center of Fig. 2 are plant pathogens (P. expansum, P. crustosum and P. hirsutum I). P. mali, P. hirsutum II, P. italicum and P. digitatum are often found in the rhizosphere soil of domesticated plants (Frisvad, unpubl.data). P. echinulatum, P. mali and P. camemberti II and III are very common on lipid containing proteinaceous substrates (meat, liver paste,

Table 3. Range of colony diameter of CYA, MEA and YES agar after one week of growth at 25°C for members of the _P. viridicatum_ and _P. cyclopium_ series and allied species.

Experimental name	Colony diameter on CYA	MEA	YES
P. viridicatum II, III	11-23	7-22	17-28
P. hirsutum IV	22-27	28-36	34-46
P. mali	18-33	22-34	30-44
P. echinulatum	28-38	27-35	33-44
P. viridicatum I	(17-)24-35[a]	(21-)28-39[a]	(21-)26-40[a]
P. aurantiogriseum II	22-27(-34)	(14-)26-36	23-35
P. aurantiogriseum I	(28-)32-38	31-38	32-46
P. camembertii II	24-35(-38)	(23-)25-36	34-48
P. camembertii III	22-34(-37)	(21-)25-37	(31-)35-48
P. hirsutum III	26-45	25-48	35-45
P. hirsutum II	32-48	38-50	38-56
P. hirsutum I	25-34	39-48	(27-)44-52
P. expansum	32-43(-50)	26-46(-60)	34-55(-62)
P. crustosum	34-48	33-48	39-59

[a] The slow growing isolates of _P. viridicatum_ I were from the North American kanagaroo rats (Wicklow et al., in prep.)

sausages, cheese, mouldy bakers yeast, together with _P. crustosum_ and _P. roqueforti_, and nuts and on substrates with high lipid content such as butter, margarine, salads etc.(Frisvad, unpubl. data). _P. aurantiogriseum_ I and II, _P. viridicatum_ I and II and _P. hirsutum_ III (= _P. hordei_) are very common and dominant on cereals. These species are probably the most important toxin producers among the Penicillia. They are also found on fruits and vegetables, but in lower frequency than _P. crustosum_ etc. _P. viridicatum_ III isolates are found on proteinaceous substrates, but they appear to be deteriorating cultures of _P. viridicatum_ II, having lost the ability to produce citrinin and violet brown reverse on YES agar, and they have probably contaminated substrates such as meat from raw materials (wheat flour or other cereals). These differences in preferred habitats, profiles of mycotoxins and other secondary metabolites and physiological characters in the _P. viridicatum_ and _P. cyclopium_ series indicates that 15 taxa are necessary to achieve homogenous species. Pitt (1979), Fassatiová (1977), Samson et al. (1976) used 9 taxa and Raper and Thom used 13 taxa for the species treated in this paper, but these authors have not evaluated the taxonomic status of

P. viridicatum IV, P. hirsutum IV and P. hirsutum II. P. cyclopium var. album (= P. verrucosum var. album, P. cyclopium f. album). P. ochraceum Bainier and P. carneolutescens (= P. verrucosum var. ochraceum, P. olivicolor) are here regarded as mutants, completely or partially blocked in their melanin synthesis ability, of other species (see Table 1).

Most of the experimental species listed in Table 1 were homogeneous morphologically. P. aurantiogriseum II, however, includes isolates with rough-walled conidia and no ablility to produce nephrotoxins. These isolates are common invaders of grains in the North American kangaroo rat seed caches and cheek pouches (Wicklow, Filtenborg and Frisvad, in prep.) and the special environmental conditions may have selected for this type. Whether these differences from the original cereal strains of P. aurantiogriseum II are sufficient to erect a new taxon is dubious, but further investigations on the physiology and chemistry of this important group of Penicillia may give the answer. Profiles of secondary metabolites and carbohydrates (Table 2 and Fig. 1) indicate that these variants of P. aurantiogriseum II and the original cereal strains are rather flexible organisms with a wide range of possible differentiation characters.

REFERENCES

BIRKINSHAW, J.H. and RAISTRICK, H. 1932. Studies in the biochemistry of microorganisms. XXII. Puberulic acid $C_8H_6O_6$ and puberulonic acid $C_8H_4O_6$, new products from glucose by Penicillium puberulum Bainier and Penicillium aurantiovirens Biourge. Biochem. J. 26: 441-453.

BIRKINSHAW, J.H. RAISTRICK, H., ROSS, D.J. and STICKINGS, C.E. 1952. Studies in the biochemistry of microorganisms. 85. Cyclopolic and cyclopaldic acids, metabolic products of Penicillium cyclopium Westling. Biochem. J. 50: 610-628.

BJERG, B., OLSEN, O., RASMUSSEN, K.W. and SORENSEN, H., 1984. New principes of ion-exchange techniques suitable to sample preparation and group separation of natural products prior to liquid chromatography. J. Liq. Chromatogr. 7: 691-707.

BU'LOCK, J.D. 1980. Mycotoxins as secondary metabolites. In "The biosynthesis of mycotoxins. A study in secondary metabolism", P. Steyn, ed. London: Academic Press. pp. 1-16.

DERDE, M.P., COMANS, D. and MASSART, D.L. 1984. SIMCA (Soft Independent Modelling of Class Analogy) demonstrated with characterisation and classification of Italian olive oil. J. Assoc. Off. Anal. Chem. 67: 721-727.

FASSATIOVÁ, O. 1977. A taxonomic study of Penicillium series Expansa Thom emend. Fassatiová. Acta Univ. Carol. Biol. 12: 283-335.

FILTENBORG, O. and FRISVAD, J.C. 1980. A simple screening method for toxigenic moulds in pure cultures. Lebensm. Wiss. Technol. 13: 128-130.

FILTENBORG, O., FRISVAD, j.C. and SVENDSEN, J.A. 1983. Simple screening method for molds producing intracellular mycotoxins in pure cultures. Appl. environ. Microbiol. 45: 581-585.

FRISVAD, J.C. 1981. Physiological criteria and mycotoxin pro-

duction as aids in identification of common asymmetric Penicillia. Appl. environ. Microbiol. 41: 568-579.
FRISVAD, J.C. 1983. A selective and indicative medium for groups of *Penicillium viridicatum* producing different mycotoxins in cereals. J. appl. Bacteriol. 54: 409-416.
FRISVAD, J.C. 1984. Expressions of secondary metabolism as fundamental characters in *Penicillium* taxonomy. In "Toxigenic fungi - their toxins and health hazard", H. Kurata and Y. Ueno, eds. Amsterdam: Elsevies. pp. 98-106.
FRISVAD, J.C. 1985. Aids in the identification of important food-borne species of filamentous fungi. In "Methods for mycological examination of foods", A.D. King, J.I. Pitt, L.R. Beuchat and J. Corry, eds. New York: Plenum Press. In press.
FRISVAD, J.C. and FILTENBORG, O. 1983. Classification of terverticillate penicillia based on profiles of mycotoxins and other secondary metabolites. Appl. environ. Microbiol. 46: 1301-1310.
HA-HUY-KE and LUCKNER, M. 1979. Structure and function of conidiospore pigments of *Penicillium cyclopium*. Z. Allg. Mikrobiol. 19: 117-122.
HERMANSEN, K., FRISVAD, J.C., EMBORG, C. and HANSEN, J. 1984. Cyclopiazonic acid production by submerged cultures of *Penicillium* and *Aspergillus* strains. FEMS Microbiol. Lett. 21: 253-261.
LEISTNER, L. 1984. Toxigenic Penicillia occuring in feeds and foods. In - "Toxigenic fungi - their toxins and health hazard", H. Kurata and Y. Ueno, eds. Amsterdam: Elsevier pp. 162-171.
ONIONS, A.H.S., ALLSOPP, D. and EGGINS, H.O.W. 1981. "Smith's Introduction to Industrial Mycology". London: Edward Arnold.
ONIONS, A.H.S., BRIDGE, P.D. and PATTERSON, R.R. 1984. Problems and prospects for the taxonomy of *Penicillium*. Microbiol. Sci. 1: 185-189.
PITT, J.I. 1979. "The Genus *Penicillium* and its Teleomorphic States *Eupenicillium* and *Talaromyces*". London: Academic Press.
RAPER, K.B. and THOM, C. 1949. "A Manual of the Penicillia". Baltimore: Willliams and Wilkins.
SAMSON, R.A. and GAMS, W. 1984. The taxonomic situation in the Hyphomycete genera *Penicillium, Aspergillus* and *Fusarium*. Antonie van Leeuwenhoek 50: 815-824.
SAMSON, R.A. STOLK, A.C. and HADLOK, R. 1976. Revision of the subsection *Fasciculata* of *Penicillium* and some allied species. Stud. Mycol., Baarn 11: 1-47.
SÖDERSTROM, B. and FRISVAD, J.C. 1984. Separation of closely related asymmetric penicillia by pyrolysis gas chromatography and mycotoxin production. Mycologia 76: 408-419.
WHEELER, M.H. 1983. Comparison of fungal melanin biosynthesis in ascomycetous, imperfect and basidiomycetous fungi. Trans. Br. mycol. Soc. 81: 29-36.
WOLD, S. and SJOSTROM, M. 1977. SIMCA: a method for analysing chemical data in terms of similarity and analogy. In "Chemometrics: Theory and Application", B.R. Kowalski, ed. Washinghton D.C.: American Chemical Society. pp. 243-282.
WOLD, S., ALBANO, C., DUNN, W.J., III, EDLUND, U., ESBENSEN, K., GELADI, P., HELLBERG, S., JOHANSSON, E., LINDBERG, W. and SJOSTROM, M. 1984. Multivariate data analysis in

chemistry. In "Proceedings of the NATO Advanced Study Institute on Chemometrics", B.R. Kowalski, ed. Dordrecht: Reideli Publ. Co.

For a discussion of Dr. Frisvad's paper see also page 334.

CLASSIFICATION OF ASYMMETRIC PENICILLIA USING EXPRESSIONS OF DIFFERENTIATION

Jens C. FRISVAD

Food Technology Laboratory
Technical University of Denmark
2800 Lyngby, Denmark

SUMMARY

In Subgenus Penicillium sect. Penicillium macromorphology, micromorphology, profiles of secondary metabolites and colony diameters are valuable criteria for the correct identification of isolates. Even though colours are used extensively in fungal taxonomy, secondary metabolites have not gained general acceptance as taxonomical characters. It is recommended that individual secondary metabolites (known or partially characterized) be used in species descriptions and keys in taxonomic works. For experimental names of Sect. Penicillium isolates, based on morphology and profiles of secondary metabolites, traditional epithets are suggested.

INTRODUCTION

Penicillium taxonomy has been based on micromorphology, macromorphology, colours and some physiological characters (Frisvad, 1984; Pitt, 1984). Fundamentally any character can be used in taxonomy, but they should be stable and reproducible and at least a proportion of the character states should persist in mutants and variants. Characters used in taxonomy should preferably, but not necessarily, be based on many genes and be adaptable to routine use. Based on a battery of diverse characters, a taxonomy can be built upon a personal judgement of the characters and their correlations i.e. an omnispective species concept (Blackwelder, 1964), and subsequent weighting or a statistical treatment of the unweighted characters, i.e. a taxometric species concept (Sneath and Sokal, 1973). In this paper, characters based on differentiation are discussed, with emphasis on isolates of the section Penicillium.

Micromorphology

A pure micromorphological species concept has seldomly been used in fungal taxonomy. However, Samson et al. (1976) and Stolk and Samson (1983) have used this approach in Penicillium taxonomy. Even though they reject characters such as pigmentation and growth data because of low stability and unreliability, they also state that most morphological characters, except the structure of ascogenous hyphae, are variable within one species (Stolk and Samson, 1983). The resulting classification in the terverticillate species of Penicillium (Samson et al., 1976, 1977a, 1977b) agrees rather closely with classifications, such as Pitt (1979), based on further characters (Frisvad and Filtenborg, 1983; Samson and Gams, 1984). However, their classification of Eupenicillium and related Penicillium anamorphs is significantly different from that of Pitt (1979), the latter being based on a wider range of criteria. To reach agreement on the classification in Eupenicillium and related Penicillium species, one can either treat the overlapping quantitative morphological characters statistically or use such a wide range of characters that diagnostic clear cut criteria can be chosen for deteminative taxonomy. A multivariate statistical treatment of overlapping characters is necessary, because such mathematical methods can extract information based on correlation between characters that can not be extracted looking at one variable at a time (Wold et al., 1983). Agreement on species concepts in Penicillium, based on micromorphology, profiles of secondary metabolites and colony growth seems to be within reach (Samson and Gams, 1984) and such characters are rather easily recorded (Frisvad and Filtenborg, 1983; Pitt, 1979).

Profiles of secondary metabolites

Colours (pigments in conidia, stipes of conidiophores, ascospores, exudate, excreted in the growth medium, in reverse and obverse mycelium etc.) have been extensively used in fungal taxonomy, especially in Aspergillus and its teleomorphs (Raper and Fennell, 1965; Samson, 1979), Penicillium (Raper and Thom, 1949; Pitt, 1979; Ramírez, 1982) and Penicillium teleomorphs (Pitt, 1979), Monascus (Hawksworth and Pitt, 1983), Arthroderma (Ghani et al., 1975), Dermocybe (Keller and Ammirati, 1983) and other ascomycetous and basidiomycetous genera. These colours are also integral parts of keys in these genera. Even though these mixtures of secondary metabolites have been used in the taxonomy of most fungi, plants and animals, taxonomists have been reluctant to use individual secondary metabolites in species descriptions. Ciegler et al. (1973, 1981) were able to separate P. viridicatum into three groups based on the production on different nephrotoxic pigments, but they recommended maintaining P. viridicatum as a single species, based on the advise of K.B. Raper. Later work (Samson et al., 1976; Pitt, 1979; Frisvad, 1981, 1983; Frisvad and Filtenborg, 1983) indicate that these groups may represent different species.

Bridge and Hawksworth (1984) noted that the technique of thin-layer chromatography is not within the traditional mycologist' sphere, but the methods described by Filtenborg and Frisvad (1980) and Filtenborg et al. (1983) are very simple, and inexpensive and can be performed rapidly. Lack of standards may cause some problems, but for example, griseofulvin is readily available as an external standard and other secondary metabolites may be characterized by their rates of development on plates and colours after different treatments. Furthermore, known culture collection strains may be used as standards (see below).

Most secondary metabolites in Penicillium seem to be consistently produced in all isolates in a species (Frisvad and Filtenborg, 1983). Older culture collection strains may loose the ability to produce some secondary metabolites, but this is often followed by general degeneration (Bennett et al., 1981) and the ability to produce secondary metabolites may sometimes be revived in degenerate cultures (Kole et al., 1982). Furthermore, as the production of all secondary metabolites in the fungi seems to be encoded within the nucleus (Mohammed et al., 1984), certain hormone like substances may indirectly enhance conidiation and production of secondary metabolites if they are added during conidial germination (Zendem et al., 1982). The ability to produce extracellular secondary metabolites is more readily lost (Frisvad, unpublished observations) than the ability to produce intracellular secondary metabolites. Cyclopiazonic acid, for example, is produced even in deteriorated culture collection strains and domesticated species (Hermansen et al., 1984).

Morphological structures and secondary metabolites may be regarded as very important survival agents for the fungi (Wicklow, 1984; Janzen, 1977) and as such can be considered to be primary structures and products (Swain, 1977), of fundamental taxonomic significance.

Section Penicillium and secondary metabolites

Frisvad and Filtenborg (1983) and Frisvad (1981, 1983, 1984) showed that secondary metabolites can be used in the classification of the terverticillate Penicillia. Most of the species can be identified directly using thin layer chromatography of secondary metabolites extracted with chloroform/methanol. All species can be identified using high pressure liquid chromatography of such metabolites (Frisvad, in prep.). Identification based solely on secondary metabolites is not recommended, however. Morphological, physiological and growth rate characters should always be used in Penicillium taxonomy as well. Further clear cut physiological criteria must be found before a key based solely on morphology and physiological criteria, wich excludes criteria based on secondary metabolites, can be made.

The experimental names used by Frisvad and Filtenborg (1983) are listed in Table 1 with proposed taxonomic epithets. Three groups of isolates can probably be described as new species (P. hirsutum II, P. hirsutum IV and P. viridicatum IV), but this should await further characterization of these strains. For one of the groups termed there P. mali, a new combination is suggested:

Table 1. Experimental groups of Frisvad and Filtenborg (1983) related to taxonomic designations in Subgen. *Penicillium* (Pitt, 1979).

Experimental group	Taxonomic name
P. camemberti I	*P. camemberti* Thom
P. camemberti II	*P. commune* Thom
P. camemberti III	*P. palitans* Westling
P. echinulatum	*P. echinulatum* Raper & Thom ex Fassatiova
P. mali	*P. melanochlorum* (Samson et al.) Frisvad
P. crustosum	*P. crustosum* Thom
P. hirsutum I	*P. hirsutum* Dierckx
P. hirsutum II	
P. hirsutum III	*P. hordei* Stolk
P. hirsutum IV	
P. viridicatum I	*P. aurantiogriseum* Dierckx
P. aurantiogriseum I	*P. aurantiogriseum* Dierckx
P. aurantiogriseum II	*P. aurantiogriseum* Dierckx
P. chrysogenum	*P. chrysogenum* Thom
P. roqueforti I	*P. roqueforti* Thom
P. roqueforti II	*P. roqueforti* Thom
P. brevicompactum	*P. brevicompactum* Dierckx
P. expansum	*P. expansum* Link
P. viridicatum II	*P. verrucosum* Dierckx
P. viridicatum III	*P. verrucosum* Dierckx
P. viridicatum IV	
P. griseofulvum	*P. griseofulvum* Dierckx
P. concentricum I	*P. concentricum* Samson
P. concentricum II	*P. granulatum* (atypical)
P. granulatum	*P. granulatum* Bain.
P. claviforme	*P. claviforme* Bain.
P. clavigerum	*P. clavigerum* Demelius
P. isariiforme	*P. isariiforme* Stolk et Meyer
P. italicum	*P. italicum* Wehmer
P. atramentosum	*P. atramentosum* Thom
(*P. digitatum*	*P. digitatum* (Pers. ex Fr.) Sacc.)
(*P. olsonii*	*P. olsonii* Bain et Sartory)
(*P. arenicola*	*P. arenicola* Chalabuda)

P. MELANOCHLORUM (Samson, Stolk et Hadlok) Frisvad, comb. nov.

=*P. verrucosum* Dierckx var. *melanochlorum* Samson, Stolk et Hadlok, Stud. Mycol., Baarn 11: 41 (1976) (basionym).
=*P. mali* Novobranova, Biol. Nauki. 10: 105 (1972), (*nom. inval* Art. 36, 37, 38)
=*P. mali* Gorlenko et Novobranova, Mikol. Fitopatol. 17: 464 (1983), (*nom. inval*. Art. 64)

Typus: Herb. CBS 487.75, from meat product, Giessen, Germany, *R. Hadlok*.

Cultures examined: CBS 487.75 (culture ex type); CBS 500.73 (=VKM F-1751 = ATCC 24727), ex type of *P. mali*, from apple, Alma-Ata, USSR, *T.I. Novobranova*; IMI, 89374, contaminant of culture, Croyden, Surrey, U.K. *R.L. Hodges*; IMI 49096; IMI

154731, from chocolate sponge biscuit, U.K.; NRRL 966 laboratory contaminant, Washington D.C.; NRRL 914, from pear, Pullman, Washington, Heald; NRRL 1164, from pulp-mill waste, Canada, G.A. Ledingham; IMI 285509, from Jerusalem artichoke, Lyngby, Denmark, J.C. Frisvad; IMI 293181, from apple, Denmark, J.C. Frisvad and numerous other isolates from pomaceous fruits, vegetables, mouldy bakers yeast, salami, horse and pig feed, the aerial flora of factories and many other substrates.

P. melanochlorum is proposed as a new combination, because the type culture of P. palitans, an earlier valid and typified name, is representative of another group of isolates (P. camemberti III, producing cyclopiazonic acid). P. melanochlorum is characterized by dark green conidia, low growth rates on CYA,, orange yellow obverse and reverse mycelium on YES agar, smooth conidia, production of cyclopenin and often viridicatin and good growth on creatine sucrose agar. P. melanochlorum is a very common species, especially in the temperate regions of the world (Denmark, Sweden, Norway, Faroe Islands, United Kingdom, Northern part of U.S.A., Canada and U.S.S.R.).

Table 2. Culture collections strains producing the full and typical profile of secondary metabolites in sect. Penicillium

Experimental name	
P. camemberti I	CBS 299.48, ATCC 42009, IMI 285506
P. camemberti II	IMI 285507, IMI 293179, ATCC 32028, NRRL 6497
P. camemberti III	NRRL 2033, IMI 285531, IMI 293201
P. echinulatum	NRRL 1151, IMI 285508, CBS 337.59
P. mali	CBS 487.75, NRRL 914, IMI 293181
P. crustosum	NRRL 968, CBS 483.75, IMI 293182
P. hirsutum I	NRRL 2032, NRRL 999, CCM F-240, CCF 1445
P. hirsutum II	IMI 285511, IMI 293183, IMI 293206
P. hirsutum III	CBS 701.68, IMI 293184, IMI 285512
P. hirsutum IV	ATCC 32000, ATCC 32002, IMI 285513
P. viridicatum I	NRRL 963, IMI 293187, NRRL 5569
P. aurantiogriseum I	CBS 744.74, CBS 573.68, IMI 293185
P. aurantiogriseum II	IMI 293186, IMI 285515, ATCC 48414
P. chrysogenum	IMI 293188, IMI 285517, NRRL 1984
P. roqueforti I	NRRL 849, IMI 293189, IMI 285518
P. roqueforti II	ATCC 46836, ATCC 46837, (IMI 293190)
P. brevicompactum	IMI 285520, IMI 293191, NRRL 862
P. expansum	IMI 293192, IMI 285521, CCF 1443
P. viridicatum II	CBS 223.71, NRRL 5571, IMI 285522
P. viridicatum III	NRRL 5574, IMI 285523, NRRL 5573
P. viridicatum IV	IMI 293194, IMI 285524, CCM F-389
P. griseofulvum	NRRL 2159-A, IMI 285525, IMI 293195
P. concentricum I	CBS 473.75, IMI 285526, IMI 293196
P. concentricum II	NRRL 2034, IMI 285527, IMI 293197
P. granulatum	CBS 494.75, NRRL 2036, NRRL 3480
(P. claviforme	CBS 305.63, NRRL 1002, IMI 285528)
(P. clavigerum	NRRL 1003, S.M. MacDonald 176/170 B/2A & C/3
(P. isariiforme	NRRL 2638, IMI 205074)
P. atramentosum	NRRL 795, FRR 1446, IMI 293200, IMI 285530

Standards of secondary metabolites may be difficult to obtain at present. Culture collection strains, producing the typical profiles of secondary metabolites, can be obtained from CBS, IMI and ATCC (see table 2). Such cultures should be available as standards in cases where identification of section Penicillium isolates is important.

In conclusion it is recommended that parts of profiles of secondary metabolites (including mycotoxins) be included in species descriptions, taxonomic keys and identifications of new isolates.

REFERENCES

BENNETT, J.W., SILVERSTEIN, R.B. and KRUGER, S.J. 1981. Isolation and characterization of two nonaflatoxigenic classes of morphological variants of Aspergillus parasiticus. J. Am. Oil Chem. Soc. 1981: 952A-955A.

BLACKWELDER, R.E. 1964. Phyletic and phenetic versus omnispective classification. In "Phenetic and phylogenetic classification", V.H. Heywood and J. McNeill, eds. London: Systematics Association. pp. 17-28.

BRIDGE, P.D. and HAWKSWORTH, D.L. 1984. The Api ZYM enzyme testing system as an aid to the rapid identification of Penicillium isolates. Microbiol. Sci. 1: 232-234.

CIEGLER, A., FENNELL, D.I., SANSING, G.A., DETROY, R.W. and BENNETT; G.A. 1973. Mycotoxin producing strains of Penicillium viridicatum: classification into subgroups. Appl. Microbiol. 26: 271-278.

CIEGLER, A., LEE, L.S. and DUNN, J.J. 1981. Production of naphtoquinone mycotoxins and taxonomy of Penicillium viridicatum. Appl. Environ. Microbiol. 42: 446-449.

FILTENBORG, O. and FRISVAD, J.C. 1980. A simple screening method for toxigenic moulds in pure cultures. Lebensm. Wiss.Technol. 13: 128-130.

FILTENBORG, O., FRISVAD, J.C. and SVENSON, J.A. 1983. Simple screening method for molds producing intracellular mycotoxins in pure cultures. Appl. environ. Microbiol. 45: 581-585.

FRISVAD, J.C. 1981. Physiological criteria and mycotoxin production as aids in classification of common asymmetric Penicillia. Appl. environ. Microbiol. 41: 568-579.

FRISVAD, J.C. 1983. A selective and indicative medium for groups of Penicillium viridicatum producing different mycotoxins in cereals. J. appl. Bacteriol. 54: 409-416.

FRISVAD, J.C. 1984. Expressions of secondary metabolism as fundamental characters in Penicillium taxonomy. In Toxigenic Fungi - their Toxins and Health Hazard", H. Kurata and Y. Ueno, eds. Amsterdam: Elsevier. pp. 98-106.

FRISVAD, J.C. and FILTENBORG, O. 1983. Classification of terverticilate Penicillia based on profiles of mycotoxins and other secondary metabolites. Appl. environ. Microbiol. 46: 1301-1310.

HAWKSWORTH, D. L. and Pitt, J. I. 1983. A new taxonomy for Monascus species based on cultural and microscopical data. Aust. J. Bot.. 31: 253-261.

GHANI, H.M., LANCASTER, J.H and LARSH, H.W. 1975. Chromatographic separation of pigments produced by Arthroderma ben- Sabouraudia 13: 51-61.

HERMANSEN; K., FRISVAD; J.C., EMBORG, C. and HANSEN, J. 1984.

Cyclopiazonic acid production by submerged cultures of Penicillium and Aspergillus strains. FEMS Microbiol. Lett. 21: 253-261.
JANZEN, D.H. 1977. Why fruits rot, seeds mold, and meat spoils. Am. Naturalist 111: 691-713.
KELLER, G. and AMMIRATI, J.F. 1983. Chemotaxonomic significance of anthraquinone derivates in North American species of Dermocybe, section Sanguineae. Mycotaxon 18: 357-377.
KOLE, H.K., SAMANTA, A.K., GOSWAMI, S.K. and BOSE, S.K. 1982. Attempt at gene reversion of a desired phenotype as a rescue against strain degeneration. J. appl. Bacteriol. 53: 163-167.
MOHAMMED, Z.S., TODOROVA-DRAGANOVA, T.W. and LUCKNER, M. 1984. Nuclear inheritance of the biosynthesis of cyclopenin and cyclopenol in Penicillium cyclopium. Z. Allg. Mikrobiol. 24: 615-618.
PITT, J.I. 1979. "The Genus Penicillium and its Teleomorphic States Eupenicillium and Talaromyces". London: Academic Press.
PITT, J.I. 1984. The value of physiological characters in the taxonomy of Penicillium - In "Toxigenic Fungi - their Toxins and Health Hazard" H. Kurata ad Y. Ueno, eds. Amsterdam: Elsevier. pp. 107-118.
RAMÍREZ, C., 1982. "Manual and Atlas of the Penicillia". Amsterdam: Elsevier.
RAPER, K.B. and FENNELL, D.I. 1965. "The genus Aspergillus". Baltimore: Williams and Wilkins.
RAPER, K.B. and THOM. C. 1949. "Manual of the Penicillia". Baltimore: Williams and Wilkins.
SAMSON, R.A., STOLK, A.C. and HADLOK, R. 1976. Revision of the subjection Fasciculata of Penicillium and some allied species. Stud. mycol., Baarn 11: 1-47.
SAMSON, R.A. ECKHARDT, C. and ORTH, R. 1977a. The taxonomy of Penicillium species from fermented cheeses. Antonie van Leeuwenhoek 43: 341-350.
SAMSON, R.A., HADLOK, R. and STOLK, A.C. 1977b. A taxonomic study of the Penicillium chrysogenum series. Antonie van Leeuwenhoek 43: 26-274.
SAMSON, R.A. and GAMS, W. 1984. The taxonomic situation in the Hyphomycete genera Penicillium, Aspergillus and Fusarium. Antonie van Leeuwenhoek 50: 815-824.
SNEATH, P.H.A. and SOKAL, R.R. 1973. "Numerical taxonomy". San Fransisco: W.H. Freeman.
STOLK, A.C. and SAMSON, R.A. 1983. The ascomycete genus Eupenicillium and related Penicillium anamorphs. Stud. Mycol., Baarn 23: 1-149.
SWAIN, T. 1977. Secondary compounds: primary products. Nova Acta Leopoldina Suppl. 7: 411-421.
WICKLOW, D.T. 1984. Ecological approaches to the study of mycotoxigenic fungi. In "Toxigenic Fungi - their Toxins and Health Hazard", H. Kurata and Y. Ueno, eds. Amsterdam: Elsevier. pp. 33-43.
WOLD, S., ALBANO, C., DUNN, W.J., III, ESBENSEN, K., HELLBERG, S., JOHANSSON, E. and SJOSTTROM, M. 1983. Pattern recognition: finding and using regularities in multivariate data. - In "Food Research and Data Analysis", H, Marttens and H. Russworm Jr., eds. London: Applied Science.
ZENDAM, Z., KHALIL, S. and LUCKNER, m. 1982. P-factor, a developmental hormone of Penicillium cyclopium. Phytochemistry 21: 839-842.

PATERSON: When should we examine a culture for secondary metabolites? Some are produced during logarithmic growth, others are biphasic, or are produced at the end of logarithmic growth. So to look only at the end of logarithmic growth is not valid.

FRISVAD: I agree, we should look at 7 days, 14 days and more.

SAMSON: As a taxonomist I am unclear about these subspecific groups you have designated. Is it not possible to give these nomenclaturally correct names centred around type isolates?

FRISVAD: It is possible and I've been waiting for concensus about what these things should be called before I name the formally.

SAMSON: You use P. camemberti I, II and III and maybe I is the real P. camemberti. What are the others?

FRISVAD: In my opinion, P. camemberti II is the wild type of P. camemberti I, but I think they are the same species. It is easily distinguished from other species such as P. aurantio-griseum.

SAMSON: Are you waiting for us to provide names?

FRISVAD: I could easily make new names for these, but I am waiting for some agreement before I do it.

GAMS: You are far enough in your work to select the differential characters. What we really need to decide is whether to recognize your groups as species, varieties or whatsoever.

SAMSON: Do you find any differences between P. italicum and P. concentricum?

FRISVAD: Yes, several. P. concentricum is quite a common species which we find on a number of different substrates.

SAMSON: In the Netherlands and other European countries P. roqueforti is becoming a major problem in rye bread and corn silos, therefore I am interested how you have subdivided P. roqueforti.

FRISVAD: There are no morphological differences, but the important point is that P. roqueforti I produces PR-toxin which is not produced by P. roqueforti II, but the latter always produces patulin and botryodiploidin. Both produce mycophenolic acid and roquefortine. So they have some common and some different metabolites. P. roqueforti II has darker green conidia than P. roqueforti I. P. roqueforti II does not occur on cheese, but particularly on rye bread.

SAMSON: Could you also comment on P. hirsutum and P. hordei? Are they separate species?

FRISVAD: Yes, I believe they are separate species because we find *P. hordei* almost exclusively on barley and it has rough conidia. *P. hirsutum* is found in soil in the rhizosphere of carrots, potatoes, bulbs, onions and so on. It has smooth conidia. It also has some different metabolites. They are closely related species but should be kept separate.

PITT: If you found a *P. hordei* that did not come from barley would you be able to recognize it?

FRISVAD: Yes, we have seen a few isolates from other sources.

PITT: You recently wrote a paper discussing *P. crustosum*. What do you think about that species now?

FRISVAD: I still think *P. crustosum* is a good species. I don't like the dissimilar forms that have been put into it, so I have removed some, particularly those which I call *P. camemberti* II, which is near Raper's concept of *P. palitans*.

NIRENBERG: How can you give separate names to a wild type and its domesticated variant?

SAMSON: How do you know that you have the wild type of *P. camemberti*?

FRISVAD: I don't know for certain. I do have Pitt's culture FRR 2160 which he called a *P. camemberti* wild type, and I agree with this. I call this *P. camemberti* II.

PITT: I think some of the other isolates you call *P. camemberti* II are something different. The *P. camemberti* wild type of mine has grey conidia. Your other isolates have green conidia, and the fact that they all produce cyclopiazonic acid is probably not significant.

I think you have confused your work in our minds by attaching your groups to know species names. If you had designated these as unknowns, "A, B, C, etc" it would have made it easier to accept what you are doing. I do appreciate, however, your desire not to give formal names to your groups until we can agree what these groups represent.

RAMIREZ: As I understand it, you are trying to make some kind of metabolic species.

FRISVAD: No. I would never do that. They must be correlated with morphology.

PITT: I think you must be very careful about using carbohydrate profiles for taxonomy. Ailsa Hocking in my laboratory is currently doing her Ph.D. on the internal solutes used by Penicillia to balance external water activity, i.e. glycerol, mannitol, erythritol and glycitol, etc. If the water activity of the medium changes, the proportions of these solutes change.

FRISVAD: However, I can take a fresh isolate and identify it correctly with its carbohydrate profile. Of course, my conditions are standardized; the results are reproducible.

TAXONOMIC IMPLICATION OF HYBRIDIZATION OF PENICILLIUM PROTOPLASTS

Jozef ANNÉ

Rega Institute, Katholieke Universiteit
B-3000 Leuven, Belgium

SUMMARY

Protoplasts of distinct Penicillium species can be fused and their fusion products eventually give rise to interspecific heterokaryons and hybrids. The morphology, stability and segregation pattern of the heterokaryons and their nuclear fusion progenies varies depending on the species involved in the cross.

Fusions between related species resulted in heterokaryotic and diploid-like progeny resembling that of normal parasexual cycle. Heterokaryons produced between less related species were slowly growing or lethal and recombination rarely occurred. Such variation in viability and recombination ability of different interspecific fusion progeny imply gradations in somatic compatibility and genome homology between species and they suggest the applicability of interspecific protoplast fusion experiments in Penicillium taxonomy.

INTRODUCTION

Comparative morphology, physiology and biochemistry are the main criteria used for classification of fungal species. Although phenotypic characters are the expression of the genotype and, hence, the degree of phenotypic difference between species is a reflection of the genetic relationship, genetic studies following conventional sexual or parasexual processes have been used only rarely for classification purposes (Lewis, 1969; Hastie, 1973; Typas and Heale, 1976; Cullen et al., 1983). In those instances the possibility of heterokaryon formation by natural means and the extent of subsequent diploidization and hybridization have been used as criteria to determine species affinity and genetic homology. However, the existence within species of systems of vegetative heterokaryon incompatibility (Caten, 1972) could restrict this approach.

Protoplast fusion applied as an alternative method to induce the parasexual cycle (Anné and Peberdy, 1975, 1976;

Ferenczy et al., 1975) circumvents heterokaryon incompatibility which originates from the prevention of hyphal anastomosis. Using protoplast fusion, heterokaryons can be produced at high frequency, also between otherwise vegetatively incompatible isolates (Dales and Croft, 1978) and between isolates of different species. As a result, protoplast fusion has been extensively applied to filamentous fungi in recent years as a means to induce intra- and interspecific heterokaryosis and thus to achieve diploid and recombinant or hybrid strains (Anné, 1983; Croft and Dales, 1983; Kevei and Peberdy, 1984).

In this review the results of attempted protoplast fusions amongst several closely related and taxonomically diverse Penicillium species will be discussed. From the morphology and segregation patterns of the heterokaryons and hybrids it can be noted that some species showed an extensive somatic affinity and genetic homology, whilst other species were somatically incompatible.

Genetic processes in the parasexual cycle

In fusion experiments carried out with Penicillium, fusion progenies were recovered as balanced heterokaryons on selective media. Estimates of the frequency of fusion were made by comparison of the number of colonies developing on minimal and complete media after fusion of complementary auxotrophic mutants. In all cases heterokaryons were recovered, the frequencies of fusion ranged between 0.03 and 3.0% of the surviving protoplasts. These values were not influenced by the taxonomic relationships of the species involved. Furthermore, they were in the same range as for intraspecific crosses (Anné, 1977).

In the standard (i.e. intraspecific) parasexual cycle (Pontecorvo, 1956), the majority of the fusion progeny dissociates on non-selective media to the constitutive parental strains. On prolonged incubation and under selective pressure nuclear fusion occurs at a low frequency. This nuclear fusion product is usually stable, but after exposure to certain chemical and physical agents (Lhoas, 1961; Hastie, 1970; Sermonti, 1969) a high frequency of segregants including recombinants is usually obtained. Diploid progeny can be selected as outgrowths bearing conidia with an increased size and prototrophic character when applying complementary auxotrophic strains. Using spore colour markers, such as yellow or white, diploid conidia show the wild type spore colour because of complementation of the genes coding for spore pigmentation. Segregation of the diploids can be observed by the formation of irregular sectors or patches on the diploid colony.

In interspecific hybridization the immediate product after protoplast fusion is supposed to be the heterokaryon as in the intraspecific parasexual cycle, although this could not always be proven. In some instances, primary fusion products did not follow the standard parasexual cycle. The heterokaryons selected for by balanced complementary auxotrophic markers grew extremely slowly on the selective plates and on sporulation only one of the parental spore types could be recovered. Interspecies nuclear fusion products - indicated by the term "hybrid" or "allodiploid" (Clausen et al., 1945) -

arose from the heterokaryons as more vigorously growing sectors with a compact morphology. Sporulation of this diploid-like progeny was in most instances very poor. The spores had a larger spore size and were prototrophic, but complementation for spore colour markers usually did not occur. The absence of pigmentation in interspecific hybrids could be caused by the existence of differences in regulatory mechanisms or due to pleiotropic effects resulting in the suppression of genes controlling spore pigmentation; it could also be a reflexion of differences in metabolites or metabolic pathways involved in pigmentation. Depending on the species used in the crosses, the hybrids showed a varying degree of stability and several types of hybrid segregants could be recovered. However, the genetic nature of the hybrids and the genetic processes involved to produce them is not understood

The slow growth of the heterokaryons probably reflected some kind of somatic incompatibility originating, for example, from different regulatory systems for the common biosynthetic pathways or from some kind of mitochondrial incompatibility as observed in mammalian cells (Ziegler and Davidson, 1981) and yeast (Whittaker and Leach, 1978) and as also suspected in *Aspergillus* (Croft and Dales, 1983). Anomalous segregation patterns of the hybrids or the absence of genetic recombination observed in several hybridization experiments probably indicated genetic differences between species on the chromosomal level. To what extent the viability of interspecific heterokaryons and the occurrence of genetic recombination correlates with taxonomic relationships will be discussed hereafter.

Interspecific somatic hybrids produced between closely related species

In several experiments protoplast fusions have been attempted between members of closely related species. The term "closely related" refers here to species classified in the same taxonomic series as described by Raper and Thom (1949). In this respect, fusions were executed between members of the *P. chrysogenum* series (Anné and Peberdy, 1976; Anné, 1977; Peberdy et al., 1977), the *P. camemberti* series (Raymond and Fèvre, 1985) and *P. cyclopium* series (Anné, unpublished results). These experiments are shown in Table 1.

Table 1. Crosses between closely related *Penicillium* species.

P. chrysogenum series (Subsection Velutina)

P. chrysogenum + P. notatum

P. chrysogenum P. cyaneofulvum

P. notatum + P. cyaneofulvum

P. camemberti series (Subsection Lanata)

P. caseicola + P. album

P. cyclopium series (Subsection Fasciculata)

P. cyclopium + P. puberulum

In the classification of Raper and Thom (1949), P. chrysogenum, P. notatum and P. cyaneofulvum are members of the P. chrysogenum series. Hybridization experiments between these species resulted in the formation of heterokaryons which grew well, though less vigorously than intraspecific heterokaryons of P. chrysogenum or P. cyaneofulvum. Colonies were irregular in shape and the heterokaryons produced spores of both the complementing parents.

Nuclear fusion in the heterokaryons resulted in the formation of prototrophic "diploid-like" progeny having a normal colony shape and size. Spores of these hybrids did not show the wild type spore colour, but were white or pale green which indicated that, contrary to intraspecific diploids, complementation for spore pigmentation did not occur. This was a more common phenomenon in interspecific hybrids, e.g. no complementation for spore colour markers occurred in sexually produced interspecific hybrids in the genus Sordaria (Lewis, 1969) or in hybrids of Aspergillus nidulans + A. rugulosus obtained after protoplast fusion (Kevei and Peberdy, 1977).

The interspecific hybrids produced between members of the P. chrysogenum series were rather stable, but could be broken by treatment with haploidizing agents. Genetic analysis of the segregants, combined with spore measurements, revealed that several of these segregants were non-parental in genotype and contained a haploid set of chomosomes. It also appeared that there was no preferential loss of chromosomes, because the markers of both species involved in the cross were recovered at almost equal frequency. Such data imply that between the chromosomes of these species considerable chromosomal homology may exist at least for those chromosomes on which the markers were located. Peculiar to these crosses was the high number of "aneuploid-like" segregants and also the isolation of an unusual multiple auxotroph, probably having a haploid set of chromosomes. These multiple auxotrophs did not grow on selective media supplemented with all requirements of both parents. The failure to grow without supplementation of additional amino acids may be a result of deregulation of biosynthetic pathways after the combination of chromosomes of both species.

Samson et al. (1977b) reconsidered the taxonomy of the P. chrysogenum series and placed P. notatum and P. cyaneofulvum in synonymy with P. chrysogenum. However, following the observations made on the morphology of the heterokaryons and on the basis of the segregation behaviour of the hybrids, it must be concluded that if these fungi are all of one species, they probably have at least some compatibility differences either on the cytoplasmic or on the genetic level, in the latter case, e.g. due to chromosomal translocation. The view of cytoplasmic differences is supported by the fact that heterokaryons between these species could not be obtained by hyphal fusions (Anne and Peberdy, 1976). It does not exclude, however, that these species might be considered as variants of the same species, but that they have a genetically controlled system for vegetative heterokaryon incompatibility, as observed in A. nidulans (Croft, 1985).

P. caseicola and P. album strains isolated from dairy industry sources readily formed heterokaryons after protoplast fusion and, subsequently, "diploid-like" progeny (Reymond and Fèvre, 1985). Recombination occurred between the genomes of these species which are both members of the P. camemberti-series (Raper and Thom, 1949). In addition, haploid nonparental-like recombinants could be isolated. Because heterokaryon formation, diploidization and recombination between these species occurred without apparent compatibility problems, these strains must be genetically closely related, thus confirming the view of Samson et al. (1977a), who placed P. caseicola and P. album in synonymy with P. camemberti on physiological and morphological grounds. The frequency of intraspecific protoplast fusions in P. caseicola was somewhat higher than that of the interspecific fusion with P. album. When these differences are not the consequence of experimental variations, it could indicate that some kind of cytoplasmic incompatibility between these species should exist. P. cyclopium and P. puberulum (CBS 123.14 = P. verrucosum var. cyclopium) (Samson et al., 1976) have been classified in the P. cyclopium series according to Raper and Thom (1949). Pitt (1979) placed P. cyclopium, but not P. puberulum, in synonymy with P. aurantiogriseum. As described earlier (Anné, 1982 b) and also discussed in this chapter (see next section) P. cyclopium and P. puberulum showed a different compatibility with regard to P. chrysogenum. On the other hand, fusions between P. cyclopium and P. puberulum resulted in well growing heterokaryons producing spores of both the constitutive strains, and diploid-like progeny was also recovered. Moreover, in these hybrids complementation for spore colour markers did occur. These results indicate that, although these species showed a different compatibility towards P. chrysogenum, they are mutually compatible. To what extent they are genetically homologous is still to be investigated.

Hybridization between species classified in different series of subsection Velutina

During an investigation of the influence of taxonomic relationship on the possibility of obtaining heterokaryons and hybrids between taxonomically less related species, protoplasts from members belonging to different taxonomic series of subsection Velutina were fused (Table 2).

Hybridization experiments of P. chrysogenum, P. notatum or P. cyaneofulvum with P. citrinum resulted in well developing heterokaryons producing - as in the intraspecific parasexual cycle - spores of both the complementing parents (Anné and Eyssen, 1978). From the heterokaryons, "diploid-like" progeny with a morphology and stability similar to that described for P. chrysogenum + P. cyaneofulvum (cfr supra) arose and recombinants of non-parental genotype and multiple auxotrophs were also isolated. This implies that a similar somatic and genetic homology exists between P. citrinum and each of the members of the P. chrysogenum series as between P. chrysogenum and P. cyaneofulvum or P. notatum.

The close relationship between members of the P. chrysogenum series and P. citrinum was further confirmed by their similar behaviour in hybridization attempts with P. roqueforti (Anné, unpublished results). Heterokaryotic progeny recovered from each of these crosses differed from normal heterokaryons:

Table 2. Crosses between Penicillium species belonging to subsect. Velutina.

Cross	Sporulation of heterokaryons	Allodiploid formation	Haploid recombinant or parental segrements
P. chrysogenum + P. cyaneofulvum	both parents	+	+
P. chrysogenum + P. citrinum	both parents	+	+
P. cyaneofulvum + P. citrinum	both parents	+	+
P. roqueforti + P. chrysogenum	*P. roqueforti*	+	*P. roqueforti*
P. roqueforti + P. cyaneofulvum	*P. roqueforti*	+	*P. roqueforti*
P. roqueforti + P. citrinum	*P. roqueforti*	+	*P. roqueforti*
P. roqueforti + P. stoloniferum	*P. roqueforti*	not yet obtained	-
P. stoloniferum + P. chrysogenum	*P. chrysogenum*	+	*P. chrysogenum*
P. stoloniferum + P. cyaneofulvum	*P. cyaneofulvum*	+	*P. cyaneofulvum*
P. stoloniferum + P. citrinum	*P. citrinum*	not yet obtained	–

they grew very slowly in atypically sparse colonies and they produced only P. roqueforti conidia. Conidia of the other parent could not be detected. Progeny containing fused nuclei occasionally arose from the heterokaryons. This hybrid progeny developed well and showed a regular morphology. On sporulation white prototrophic conidia were produced that were larger than the parental conidia and they showed different stabilities.

As studied in detail by Anné et al. (1976) the unstable protrophic spores of P. chrysogenum + P. roqueforti produced on germination prototrophic colonies with a dissimilar morphology and eventually segregated P. roqueforti-like sectors (Fig. 1). P. chrysogenum segregants were never recovered. Primary nuclear fusion progeny produced a red pigment on a glucose rich medium. Such pigment production was also observed for interspecific hybrids of Aspergillus: in these cases the pigment was supposed to be melanin (Croft and Dales, 1983) and its production could be the result of the deregulation of metabolic pathways. Genetically stable forms were unaffected when grown on non-selective media and in the presence of haploidizing agents and they did not produce a red pigment. The genetic constitution of these hybrids is unknown, but the fact that amongst these hybrids several phenotypes were isolated suggests at least differences in chromosome constitution.

Crosses of P. stoloniferum with P. roqueforti also resulted in slowly growing heterokaryons and, even so, on sporulation only P. roqueforti conidia were released. From the heterokaryons nuclear fusion products were never obtained. On the other hand, following protoplast fusion and subsequent heterokaryon development, brownish or white sporulating unstable hybrids could be isolated from P. stoloniferum + P. chrysogenum; P. stoloniferum + P. notatum; P. stoloniferum + P. cyaneofulvum; P. stoloniferum + P. citrinum (Anné, 1982a, b). Stable hybrids were not detected, however. On non-selective media, the brownish or white hybrids segregated to a variety of other unstable forms and finally to the parental strain

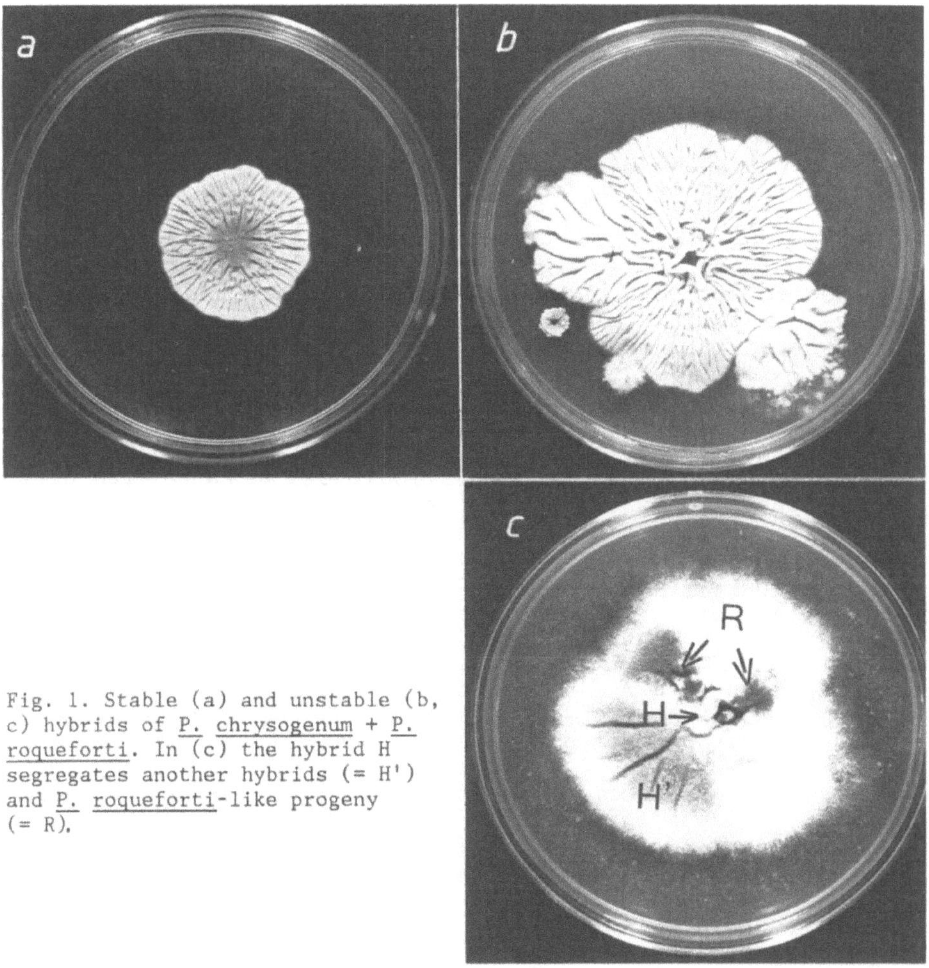

Fig. 1. Stable (a) and unstable (b, c) hybrids of P. chrysogenum + P. roqueforti. In (c) the hybrid H segregates another hybrids (= H') and P. roqueforti-like progeny (= R).

which complemented P. stoloniferum. P. stoloniferum-like conidia could not be detected, from neither the hybrids nor the heterokaryons. The segregation pattern of P. stoloniferum + P. chrysogenum implied that the instability of the hybrids appeared to be the consequence of the selective loss of P. stoloniferum chromosomes, probably during mitotic events. That no stable hybrids or recombinants were recovered from these crosses might be an indication of the extent of genetic nonhomology between these species. On the other hand, these results show again that P. chrysogenum, P. notatum, P. cyaneofulvum and P. citrinum gave a similar response in various hybridization experiments. It is clear, therefore, that on the somatic and genetic level these species have to be very closely related. It confirms the supposed affinity and relationship found between the members of the P. chrysogenum series and P. citrinum as based on morphological and physiological grounds (Raper and Thom, 1949; Pitt, 1979), and by numerical taxonomy (Bridge et al., this volume).

Hybridization experiments between species classified in different subsections

In the previous section it was stated that when less related species are involved in crosses the segregant progenies can be different from those in the normal parasexual cycle. In most instances one species seems to be dominant over the other in terms of release of parental conidia: only conidia of one species can be recovered, not only in the first order fusion progeny (heterokaryons) but also following segregation of the hybrids. In such crosses haploid recombinant hybrids were generally not detected. Similar anomalies were observed for progeny obtained from protoplast fusions between species classified in different taxonomic subsections, in cases where viable heterokaryons were produced (Table 3).

P. patulum (= P. griseofulvum), classified in subsection Fasciculata could be hybridized with P. chrysogenum of subsection Velutina (Anne, 1982b; Liang and Liu, 1982). Slowly growing heterokaryons produced allodiploid progeny with different morphology and stability. Stable hybrids were probably aneuploid in chromosome number as concluded from spore measurements. Unstable hybrids as well as the heterokaryons released only P. chrysogenum conidia on segregation.

Fusions between P. chrysogenum and P. puberulum also classified in subsect. Fasciculata, resulted in slowly growing heterokaryons (Fig 2 a),which on sporulation released P. chrysogenum conidia. On prolonged incubation the heterokaryons produced outgrowths of allodiploid mycelium with varying stability (Fig. 2 b, c). A similar array of hybrid progeny was isolated as described for other crosses between less related Penicillium species. Unstable hybrids normally segregated the P. chrysogenum parent, but a few segregants resembling P. puberulum were also obtained. As mentioned earlier, P. cyclopium can form robust heterokaryons with P. puberulum. Nevertheless, both species showed a different compatibility towards P. chrysogenum, because protoplast fusions between P. chrysogenum + P. cyclopium produced no viable heterokaryons, in contrast to P. puberulum. The differences in heterokaryon compatibility with respect to the same species indicate that there is at least a cytoplasmic difference between P. cyclopium and the tested isolate of P. puberulum. A similar phenomenon of species which are compatible with each other, but only one of them with respect to a third species, was observed for P. chrysogenum, P. roqueforti and P. puberulum. Whereas P. chrysogenum could be hybridized with P. roqueforti and with P. puberulum, P. roqueforti could not be hybridized with P. puberulum.

Species in subsect. Velutina, including P. chrysogenum and P. roqueforti, were somatically incompatible with species tested from subsect. Lanata. No viable heterokaryons were obtained from crosses between P. chrysogenum + P. lanosum and P. cyaneofulvum + P. lanosum (Anné, unpublished results) or from P. roqueforti + P. caseicola or P. roqueforti + P. album (Reymond and Fèvre, 1985). In sect. Divaricata, Raper and Thom (1949) have classified sexually as well as asexually reproducing species such as e.g. respectively, P. baarnense, P. nigricans. No viable fusion products could be isolated from fusion attempts between P. chrysogenum or P. cyaneofulvum and

Table 3. Crosses between <u>Penicillium</u> species belonging to different subsections.

Cross	Sporulation of heterokaryon	Allodiploid formation	Haploid recombinants or parental segregants
P. chrysogenum + *P. patulum*	*P. chrysogenum*	+	*P. chrysogenum*
P. chrysogenum + *P. puberulum*	*P. chrysogenum*	+	*P. chrysogenum* (*P. puberulum*)[a]
P. chrysogenum + *P. cyclopium*	no viable heterokaryons	-	-
P. roqueforti + *P. puberulum*	no viable heterokaryons	-	-
P. roqueforti + *P. caseicola*	no viable heterokaryons	-	-
P. chrysogenum + *P. lanosum*	no viable heterokaryons	-	-
P. chrysogenum + *P. nigricans*	no viable heterokaryons	-	-
P. chrysogenum + *P. baarnense*	*P. baarnense* (*P. chrysogenum*)[a]	+	+
P. chrysogenum +			*P. baarnense* (*P. chrysogenum*)[a]

[a] in a few instances.

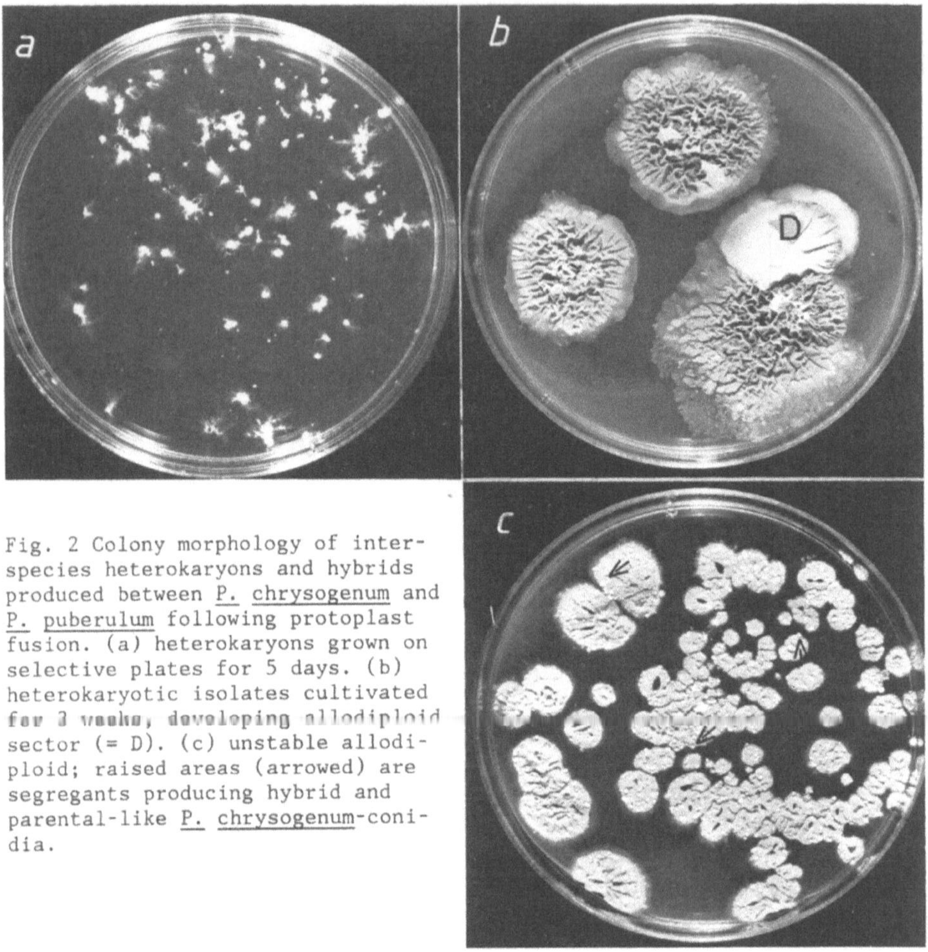

Fig. 2 Colony morphology of interspecies heterokaryons and hybrids produced between *P. chrysogenum* and *P. puberulum* following protoplast fusion. (a) heterokaryons grown on selective plates for 5 days. (b) heterokaryotic isolates cultivated for 3 weeks, developing allodiploid sector (= D). (c) unstable allodiploid; raised areas (arrowed) are segregants producing hybrid and parental-like *P. chrysogenum*-conidia.

P. nigricans (Anné, unpublished results). In contrast, from crosses between *P. chrysogenum* and *P. baarnense*, after heterokaryosis a range of stable and unstable fusion products were isolated (F.A. Mellon and J.F. Peberdy, to be published). Stable fusion products showed no indication of breakdown. Unstable types normally segregated one or other of the two parents, but also haploid recombinant auxotrophic phenotypes were isolated at low frequency. Mitochondrial DNA characterization using restriction digestion analysis on the hybrids *P. chrysogenum* + *P. baarnense* showed that the mitochondria of both species could coexist in a common cytoplasm, and also that mitochondrial DNA recombination could occur (F.A. Mellon and J.F. Peberdy, to be published). The *P. chrysogenum* + *P. baarnense* fusions are interesting in that they demonstrate that nuclear fusions can occur also between nuclei of asexually and sexually reproducing *Penicillium* species. In addition, the recombination events and the isolation of haploid recombinants imply that a reasonable chromosome homology may exist between these two species.

CONCLUSIONS

A survey of interspecific crosses between *Penicillium* species shows that hybridization can occur between several species. It is clear, however, that the possibility of genomic interaction depends on the somatic compatibility and genetic homology between the species. The morphological appearance and genetic behaviour of the heterokaryons and hybrids may show something about the morphological and genetic relatedness of the species involved. Fusions between taxonomically closely related species described in this chapter resulted in heterokaryotic and diploid-like progeny resembling that of the normal parasexual cycle. Heterokaryons released the spores of both parents and haploid interspecific recombinants were produced from the allodiploids. Genetic analysis of these segregants showed that the parents have a great deal of genetic homology. The high number of aneuploids observed in several crosses may indicate some genetic differences between the species involved. The nature of these differences can be investigated when fully marked strains are used. Comparing the behaviour of the interspecific hybrids of closely related species and that of some of the *P. citrinum* hybrids such as *P. chrysogenum* + *P. citrinum* or *P. cyaneofulvum* + *P. citrinum*, *P. citrinum* could be placed in the same somatic compatibility group and showed the same genetic homology as the members of the *P. chrysogenum* series.

Less related species showed a more extensive somatic incompatibility as concluded from heterokaryon morphology. Heterokaryons were slowly growing colonies and they produced spores of only one of the parents. In most instances, different types of aneuploid-like prototrophic hybrids were produced. Clearly, these different types reflected differences in genotype, as verified by the analysis of iso-enzyme patterns (Anné and Peberdy, 1981). The instability of interspecific progeny was probably the consequence of selective loss of chromosomes of the recessive species from the hybrid nuclei, whereas stable hybrids may represent new genome assemblies resulting from the concurrent action of mitotic crossover and selective loss of chromosomes. In most instances absence of genetic recombination suggests a lack of genetic homology between these species. However, it has to be taken into account that different results of segregation patterns for markers located on different chromosomes might be obtained, depending on the homology of the particular chromosomes.

The lack of heterokaryon development after protoplast fusion as observed for several crosses of non-related species is probably a reflection of profound physiological or biochemical differences between the species involved. In this respect, a striking observation made for *Aspergillus* interspecific hybridization was that all species of the *A. nidulans* group which could be hybridized with each other had mitochondrial genomes which share a similar overall organization, though they might differ widely in size due to the presence or absence of inserts in the mitochondrial DNA (Croft and Dales, 1983). To what extent this phenomenon occurs in *Penicillium* has still to be investigated.

Generally, the potential for hybridization in Penicillium species provides clear indications of taxonomic relationships, which can be related to somatic incompatibility and/or genetic non-homology. Somatic and genetic homology between different species can be investigated using protoplast fusion, and this will considerably aid our knowledge of the taxonomy of Penicillium.

Acknowledgment

The author wishes to thank Drs J.F. Peberdy and M. Fèvre for allowing him to report on their unpublished results.

REFERENCES

ANNÉ, J. 1977. Somatic hybridization between Penicillium species after induced fusion of their protoplasts. Agricultura 25:1-117.

ANNÉ, J. 1982a. Genetic evidence for selective chromosome loss in interspecies hybrids from Penicillium chrysogenum + P. stoloniferum. FEMS Microbiol. Lett. 14:191-196.

ANNÉ, J. 1982b. Comparison of penicillins produced by interspecies hybrids from Penicillium chrysogenum. Eur. J. Appl. Microbiol. Biotechnol. 15:41-46.

ANNÉ, J. 1983. Protoplasts of filamentous fungi in genetics and metabolite production. In "Protoplast 1983", I. Potrykus, C.T. Harms, A. Hinnen, R. Hutter, P.J. King and R.D. Shillito, eds. Basel: Birkhauser Verlag. pp. 167-178.

ANNÉ, J. EYSSEN, H. 1978. Isolation of interspecies hybrids of Penicillium citrinum and P. cyaneo-fulvum following protoplast fusion. FEMS Microbiol. Lett. 4:87-90.

ANNÉ, J., EYSSEN, H., DE SOMER, P. 1976. Somatic hybridization of Penicillium roqueforti with P. chrysogenum after protoplast fusion. Nature, Londen 262: 719-721.

ANNÉ, J. and PEBERDY, J.F. 1975. Conditions for induced fusion of fungal protoplasts in polyethylene glycol. Arch. Microbiol. 105:201-205.

ANNÉ, J. and PEBERDY, J.F. 1976. Induced fusion of fungal protoplasts following treatment with polyethylene glycol. J. Gen. Microbiol. 92:413-417.

ANNÉ, J. and PEBERDY, J.F. 1981. Characterisation of interspecies hybrids of Penicillium chrysogenum + P. roqueforti by iso-enzyme analysis. Trans. Br.Mycol. Soc. 77:401-408.

CATEN, C.E. 1972. Vegetative incompatibility and cytoplasmic infection in fungi. J. Gen. Microbiol. 72:221-229.

CLAUSEN, J., KECK, D.D. and HIESEY, W.M. 1945. Experimental studies on the nature of species. II. Plant evolution through amphiploidy and autoploidy with examples from the Madiinae". Publs. Carnegie Instn.No. 564.

CROFT, J.H. 1985. Protoplast fusion and incompatibility in Aspergillus. In "Fungal Protoplasts: Their Uses in Biochemistry, Genetics and Physiology" J.F. Peberdy and L. Ferenczy eds. New York: Marcel Dekker, pp.225-240.

CROFT, J.H. and DALES. R.B.G. 1983. Interspecies somatic hybridization in Aspergillus. In "Protoplasts 1983" I. Potrykus, C.T. Harms. A. Hinnen, R. Hutter, P.J. King and

R.D. Shillito eds. Basel: Birkhauser Verlag, pp. 179-186.
CULLEN, D., SMALLEY, E.B. and DIMOND, R.L. 1983. Heterokaryosis in *Fusarium tricinctum* and *F. sporotrichioides*. J. Gen.Microbiol. 129:3035-3041.
DALES, R.B.G. and CROFT, J.H. 1977. Protoplast fusion and the isolation of heterokaryons and diploids from vegetatively incompatible strains of *Aspergillus nidulans*. FEMS Microbiol. Lett. 1:201-204.
FERENCZY, L., KEVEI, F. and SZEGEDI, M. 1975. High-frequency fusion of fungal protoplasts. Experientia 31:1028-1029.
HASTIE, A.C. 1970. Benlate-induced instability of *Aspergillus* diploids. Nature, London 226:77.
HASTIE, A.C. 1973. Hybridization of *Verticillium albo-atrum* and *Verticillium dahliae*. Trans. Br. mycol. Soc.60:511-523.
KEVEI, F. and PEBERDY, J.F. 1977. Interspecific hybridization between *Aspergillus nidulans* and *Aspergillus rugulosus* by fusion of somatic protoplasts. J. Gen. Microbiol. 102:255
KEVEI, F. and PEBERDY, J.F. 1984. Further studies on protoplast fusion and interspecific hybridization within the *Aspergillus nidulans* group. J.Gen. Microbiol. 130:2229-2236.
LEWIS, L.A. 1969. Genetic evidence for hybridization in an interspecific cross in the genus *Sordaria*. J.Gen. Microbiol. 59:359-367.
LHOAS, P. 1961. Mitotic recombination by treatment of *Aspergillus niger* with p-fluorophenylalanine. Nature, London 190:744.
LIANG, P. and LIU, H. 1982. Interspecific somatic hybridization between *Penicillium patulum* and *Penicillium chrysogenum.* I. Protoplast fusion hybrid development and induced segregation. Acta Microbiol. Sin. 22:248-256.
PEBERDY, J.F., EYSSEN, H. and ANNE, J. 1977. Interspecific hybridization between *Penicillium chrysogenum* and *P. cyaneo-fulvum* following protoplast fusion. Mol. Gen.Genet. 157: 281-284.
PITT, J.I. 1979. "The Genus *Penicillium* and its teleomorphic states *Eupenicillium* and *Talaromyces*. London: Academic Press.
PONTECORVO, G. 1956. The parasexual cycle in fungi. Ann. Rev. Microbiol. 10:393-400.
RAPER, K.B. and THOM, C. 1949. "A Manual of the Penicillia" Baltimore: Williams and Wilkins.
REYMOND, P. and FÈVRE, M. 1985. Recombination following protoplast fusion of *Penicillium* strains used in the dairy industry. Enzyme Microbiol. Technol. (in press).
SAMSON, R.A., ECKARDT, C and ORTH, R. 1977a. The taxonomy of *Penicillium* species from fermented cheeses. Antonie van Leeuwenhoek 43:341-350.
SAMSON, R.A., HADLOK, R. and STOLK, A.C. 1977b. A taxonomic study of the *Penicillium chrysogenum* series. Antonie van Leeuwenhoek 43:169-175.
SAMSON, R.A., STOLK, A.C. and HADLOK, R. 1976. Revision of the subsection *Fasciculata* of *Penicillium* and some allied species. Stud.Mycol., Baarn 11:1-47.
SERMONTI, G. 1969. "Genetics of antiobiotic producing microorganisms". London: Wiley Interscience.
TYPAS, M.A. and HEALE, J.B. 1976. Heterokaryosis and the role of cytoplasmic inheritance in dark resting structure

formation in *Verticillium* spp. Molec. Gen. Genet. 146:17-26.

WHITTAKER, P.A. and LEACH, S.M. 1978. Interspecific hybrid production between the yeast *Kluyveromyces lactis* and *K. fragilis* by protoplast fusion. FEMS Microbiol. Lett. 4:31-34.

ZIEGLER, M.L. and DAVIDSON, R.L. 1981. Elimination of mitochondrial elements and improved viability in hybrid cells. Somat. Cell Genet. 7:73-88.

DIALOGUE FOLLOWING DR. ANNÉ'S PAPER

GOCHENAUR: How do you know that your mycelial segregates are diploid? Are you looking at nuclear volume?

ANNÉ: No. The conidial size of the diploids is larger by about 1.2-1.5 times.

GOCHENAUR: When you had your heterokaryon formed and you got back one parental type, could this mean that you started out with a fusion between a protoplast with a single nucleus and one with several nuclei, so that your heterokaryon had a much larger number of nuclei from one parent type?

ANNÉ: The average number of nuclei per protoplast in a fresh protoplast preparation is about 1. To what extent, however, there is an imbalance between the number of complimenting nuclei of the heterokaryons has not been investigated.

GOCHENAUR: Did you find that your diploids normally grew faster than the parental types?

ANNÉ: Yes, always.

BRIDGE: Do you have any idea what the diffusible red or brown pigments produced by your heterokaryons are?

ANNÉ: Probably melanin.

FRISVAD: Did you suggest that *P. chrysogenum* and *P. roqueforti* are closely related? Morphologically they are quite different, but they do have two secondary metabolites in common, roquefortine-C and PR-toxin.

ANNÉ: I can only say that they are somatically compatible. It is difficult to speculate on their degree of relatedness from this.

NIRENBERG: Are you only fusing protoplasts? Strains which may form compatible protoplasts may originally have hyphae which will not anastomise because of differences in their cell walls. The ability of two protoplasts to fuse may not necessarily indicate such a close relationship.

ANNÉ: It allows you to study the genetic relationship by seeing to what extent recombination occurs, or whether there are the same number of chromosomes or if the genes are located at the same place on the chromosome.

SOLUTIONS TO SOME PROBLEMS IN ASPERGILLUS TAXONOMY USING THE SCANNING ELECTRON MICROSCOPE

Zofia KOZAKIEWICZ

Ministry of Agriculture, Fisheries & Food, Slough Laboratory Slough, SL3 7HJ. Present address: Commonwealth Mycological Institute, Kew, Surrey TW9 3AF, United Kingdom

SUMMARY

Data obtained from a scanning electron microscope (SEM) study of intraspecific and interspecific variation in Aspergilli associated with post-harvest storage are used to resolve some difficulties in species identification and delimitation. Solutions are provided to problems of identifying isolates of Aspergillus flavus, A. parasiticus and A. oryzae, while the status of A. sojae is resolved. The morphological features of both the conidia and ascospores of A. glaucus are evaluated as marker characters and are shown to be equally useful either complimenting one another or acting indepedently. The long-standing, intricate puzzle surrounding A. melleus, A. quercinus and A. petrakii is explained satisfactorily. Finally, mycologists employing SEM techniques are warned of the wide variability that arises from the ontogenetic development of conidia, and cautioned to ensure that micrographs of mature conidia are obtained.

INTRODUCTION

Problems relating to species delimitation and determination in the Aspergilli arise from a variety of factors. Foremost are (1) the almost unwavering confidence with which mycologists rely upon subjective decisions to differentiate colony characteristics, and (2) the limited ability of the light microscope to adequately resolve micro-morphological features. These difficulties are compounded by the considerable inherent, as well as non-inheritable variability exhibited by isolates belonging to the same taxon.

In a recently completed revision (Kozakiewicz, in press) of Aspergillus species associated with post harvest storage, the use of the scanning electron microscope (SEM) has eliminated the problem of inadequate microscope' resolution. A feature of this work has been to use the genetic integrity exhibited by the morphology of the conidium to assess the importance and implications of the considerable variability associated with commonly emphasised characters such as colony appearance. As a result, new data have become available which enables certain mistakes and contentious points in Aspergillus taxonomy to be unequivocably resolved.

METHODS

Conidial samples were prepared by passing an aluminium specimen stub (Agar Aids) across the surface of a mature Petri dish plate colony. The stub was then gold coated (Emscope Sputter coater), examined in a Cambridge Scientific S250 scanning electron microscope at an accelerating voltage of 20kv and photographed using Ilford FP4 35 mm film.

RESULTS

Aspergillus flavus group

An important group within the Aspergilli is the A. flavus group, which includes A. parasiticus, A. sojae, A. flavus and A. oryzae. The identity and separation of these has concerned mycologists for many decades. It has been stated that there are no sharp line of demarcation between, on the one hand A. flavus and A. parasiticus and on the other A. flavus and A. oryzae (Thom and Raper, 1945, Raper and Fennell, 1965). Koza-kiewicz (1982) has considered the difficulties of differentiating A. flavus and A. parasiticus, but it is pertinent to reiterate some of those findings here. Differentiation of these two species depends on colour, stipe length, presence or absence of metulae and conidial ornamentation. SEM studies of conidia taken from ex-type cultures and various strains labelled as one of these two species revealed two distinct spore forms. Conidia taken from a particular isolate always consisted of one of these two forms, never a mixture of both. One form occurred in the ex-type isolate of A. parasiticus, and a neotype for A. flavus based on the second conidium form was selected (Fig.1). Use of this new diagnostic character revealed that the ex-type cultures of A. parasiticus at CBS, ATCC and some of those at NRRL were in fact A. flavus. A similar history existed for some dried cultures in herbaria labelled as A. flavus which actually belonged to A. parasiticus.

Problems assigning isolates to either A. flavus or A. oryzae are equally acute because it is considered that these species are 'bridged almost completely by a series of intermediate forms' (Raper and Fennell, 1965: 70).

I have examined isolates which have been crucial in the development of the taxonomy of A. oryzae, including one (= NRRL 447) believed to be derived from that used in the original description (Cohn, 1884) and ones originating from the soy sauce industry; all possess conidia ornamented as in Fig. 1. With the exception of one isolate (WB 453) originating from Brazil nuts (Thom and Church, 1926: 201), all others I have examined from sources outside these parameters proved to be misidentifications of A. parasiticus or A. flavus. In general, colonies of A. oryzae isolates can be recognized by

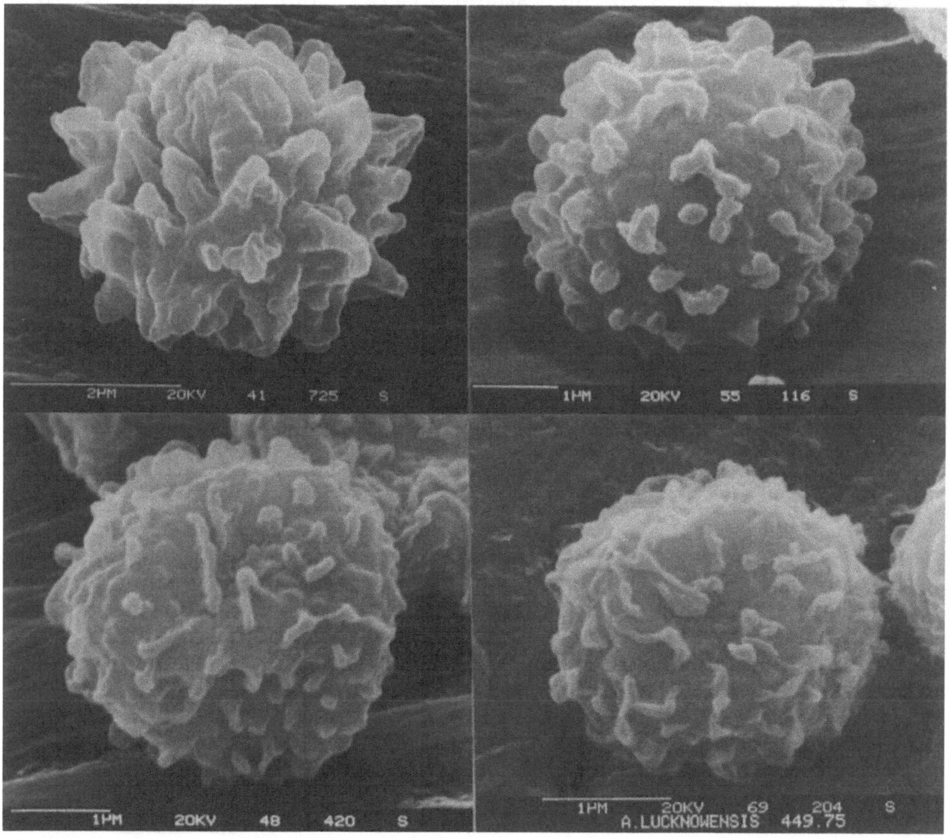

Fig. 1. Species identification using conidial topography in the A. flavus group. Top: left, A. parasiticus IMI 15957vi, right, A. flavus IMI 124930. Bottom: left, A. oryzae IMI 126842, right, A. lucknowensis CBS 449.75.

floccose growth with olive-brown conidial heads borne on very long stipes. However, some isolates with this same conidial form are less floccose, greener, and sporulate more heavily.

It has long been argued that A. oryzae represents a domesticated race derived from A. flavus (see also Wicklow; and Kurtzman, this publication). Equally, it could be assumed that these 'greener' colonies of A. oryzae may represent the wild type of the species. This would obviate the need to explain changes in spore ornamentation, albeit not excessive, which must have occurred if A. flavus is considered as the ancestor. Additional evidence which may support the case for the existence of a wild type of A. oryzae emanates from an SEM examination of A. lucknowensis (Rai et al., 1968). This species, like A. oryzae, is characterized by colonies which are deeply floccose and in which conidial heads are sparsely distributed. Further, the conidial ornamentation (Fig. 1) is identical to that of A. oryzae. Known only from one isolate taken from soil in India, it differs from A. oryzae in that the mycelium is bright yellow. It has been considered to be close to A. versicolor by Samson (1979), but on the basis of conidial ornamentation it has an affinity with A. oryzae.

Finally, there is the question of intraspecific variation within the A. flavus group. Examination of all key isolates of A. sojae, including a strain of the ex-type isolate, has revealed that without exception the morphology of the conidium is identical to that of A. parasiticus. This fact, together with consideration of the traditional characteristics exhibited by a range of isolates leads me to consider that the two taxa are conspecific. However, differences in their biochemical nature (in particular the anisaldehyde reaction), and to a lesser extent minor variations in cultural characteristics, are such that A. sojae should be considered as a variety of A. parasiticus.

Sakaguchi and Yamada (1944) described A. sojae as a new species without reference to A. parasiticus. Nehira (1955) reduced it to a form of A. parasiticus, without discussion. Ten years later, again without discussion, Raper and Fennell (1965) treated it as a synonym of A. parasiticus. Murakami (1971) reinstated A. sojae as a species mainly on the basis of biochemical data. Christensen (1981) concurred with this view, but based her conclusions solely on morphological criteria, namely the floccose colony habit, and conidium size. However, Sakaguchi and Yamada (1944) made no reference to floccose growth, and ex-type cultures grown by me were not floccose. Moreover, some strains representing both A. sojae and A. parasiticus can be floccose . This difference of opinion may be explained by the fact that Christensen (1981) examined only two isolates, neither of which was representative of the type. Further, whilst conidia of A. sojae tend to be larger than those of A. parasiticus, the degree of overlap is such that conidium size alone cannot serve as a conclusive character for species determination. I conclude that the similarities in the morphological features of A. parasiticus and A. sojae, suggest they are conspecific, but that biochemical differences and the use of A. sojae isolates as Koji moulds justifies placement of A. sojae in an infraspecific rank.

Aspergillus glaucus-group

The A. glaucus group is characterized by its Eurotium teleomorph. Because it arises after karyogamy, some mycologists believe that its morphology reflects genotypic differences which are more reliable as diagnostic features than are those associated with the anamorph. This belief is aided by the fact that certain of the ascospore features are more easily perceived by light microscopy. Data obtained using the SEM shows that with improved resolution the characteristics of both types of spores are equally valuable (Fig. 2, top and middle series).

Reliance on the topography of ascospores alone may lead to erroneous judgements. An example is taken from the work of Blaser (1976), who on the basis of phase contrast microscopy of the ascospores reduced E. rubrum and E. umbrosum to synonymy with E. herbariorum. SEM micrographs of both the

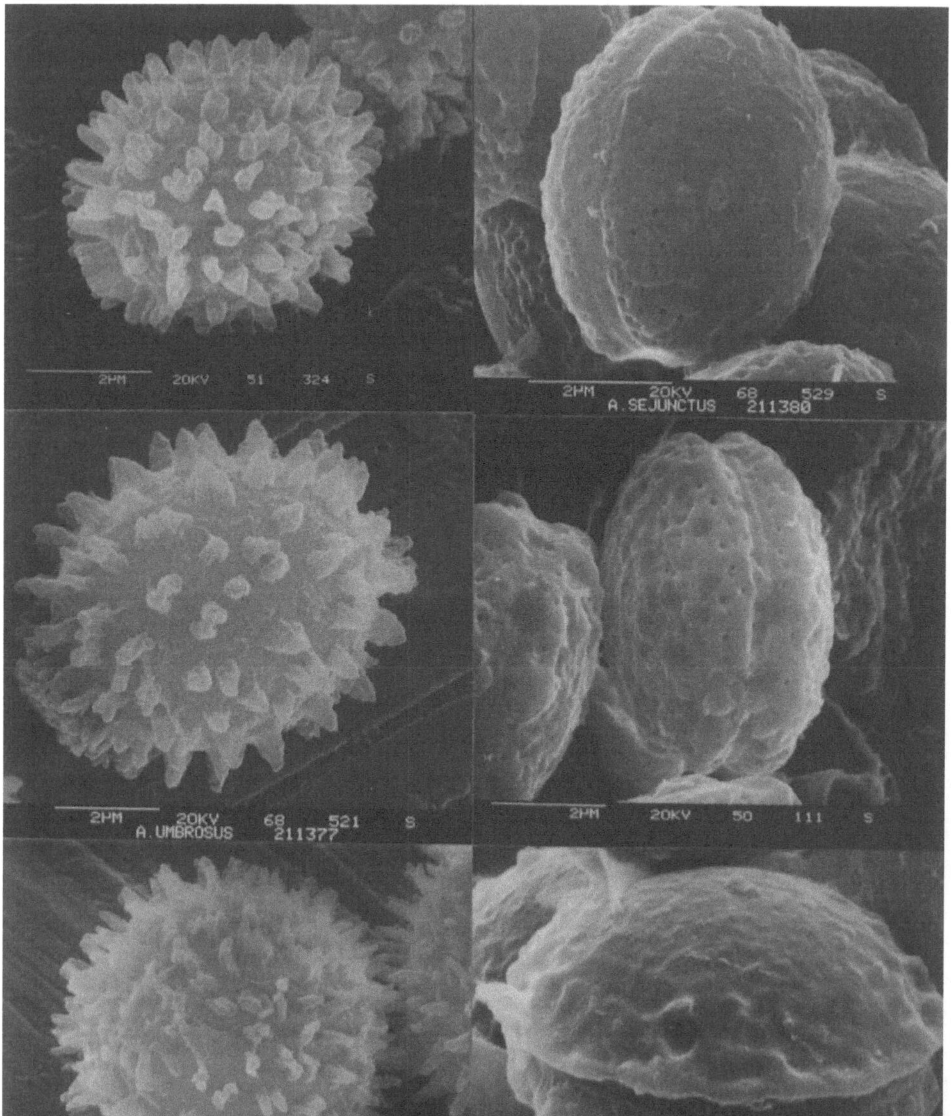

Fig. 2. High resolution micrographs used to validate the synonymy of E. umbrosum IMI 211377 (CBS 117.46), with E. rubrum IMI 211380 (CBS 530.65), but rejecting their synonymy with E. herbariorum IMI 211383 (CBS 516.65).

ascospores and conidia of these species (Fig. 2) shows that morphological similarities between the conidia of E. umbrosum and E. rubrum complement those of their asospores enabling one with certainty to agree to their synonymy. However, a similar comparison with the micrographs of E. herbariorum conidia (Fig. 2, bottom series) negates Blaser' recommendation. Whilst the ascospore morphology is only very slightly different and therefore unresolved in his micrographs, the ornamentation of the conidium is distinct. Thus the specific ranking of E. herbariorum is dependent upon features of its conidium.

A further example in which the ornamentation of the conidium provides a more valuable marker character is illustrated in Fig. 3. On the basis of similarities of their ascospores, E. heterocaryoticum and E. amstelodami are usually treated as conspecific. However, micro-morphological differences in their conidia suggest otherwise.

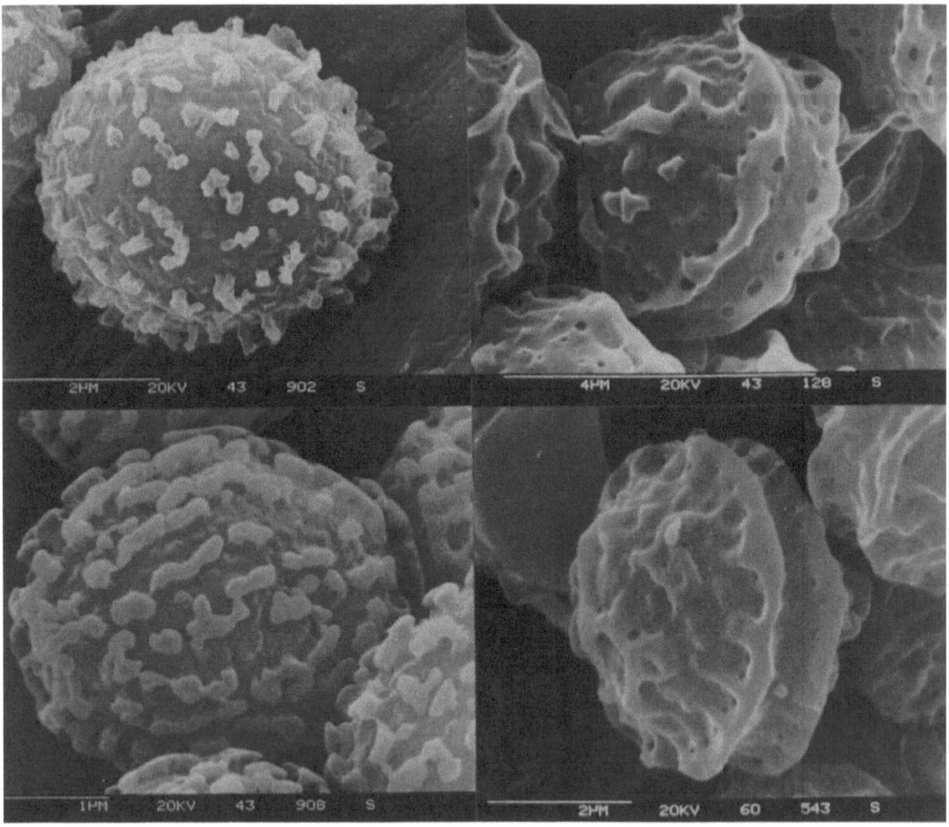

Fig. 3. Micrographs used to reject the synonymy of Top, E. heterocaryoticum IMI 115158 with bottom, E. amstelodami IMI 211806.

Aspergillus ochraceus-group

My final example is of a puzzle dormant in the literature for many years and revealed only when conidia of isolates said to represent *A. melleus* and *A. quercinus* were examined with the SEM. *A. quercinus* Bain. (1881) and *A. melleus* Yukawa (1911), no original material of which is extant, have been redescribed in subsequent revisions of the genus. In this process a series of mistakes has arisen, culminating in the synonymy of *A. quercinus* under *A. melleus*.

Thom and Church (1926) redescribed both species. Their description of *A. quercinus* was based on 3 isolates (4754.86, 4186.1 and 4078K-2), none of which came from Bainier. Since these isolates possessed an abundance of sclerotia, this feature was now included in the redescription. *A. melleus* was redescribed on the basis of Hanzawa'isolate(Thom 4291.6 = WB 416), believed to represent a subculture of Yukawa's orginal strain.

Thom and Raper (1945) included both species in their revision but now, because the three isolates ascribed to *A. quercinus* by Thom and Church (1926) were lost, they based their description on isolate WB 394, from date fruits. The description for *A. melleus* and the isolate on which it was based remained unchanged.

Raper and Fennell (1965) stated that some of the strains which they were including under the specific epithet *A. melleus* had been referred to previously under *A. quercinus* by Thom and Raper (1945). They mentioned WB 394 as one of these, so that the same plate of micrographs of WB 394 used in 1945 to represent *A. quercinus* were now re-labelled *A. melleus*. They stated: " A careful reconsideration of the older literature have led us to conclude that *Sterigmatocystis quercina* Bain. and *A. melleus* Yukawa are probably synonyms and that Yukawa's (1911) *A. melleus* should be selected as the specific designation because Bainier's (1881) description fails to mention sclerotia. Thom and Church's (1926) emendation of *St. quercina* Bain. is antedated by Yukawa's *A. melleus*".

Their description of *A. melleus* was based upon isolate WB 394 and one other new isolate WB 5103 from India. The isolate WB 416 (previously used with valid reasons, for *A. melleus*) was not mentioned in this part of the text. However, it reappeared as one of the isolates used to support the redescription of *A. petrakii* Voros with the brief annotation that " WB 416 fails to conform with Yukawa's description of *A. melleus*".

Using the SEM the conidia of the four pertinent isolates WB 416, 394, 5103 and the ex-type culture of *A. petrakii*, WB 4396 were examined. They represent three distinct forms - one for WB 416, a second for WB 5103 and a third shared by WB 394 and WB 4396, ex type of *A. petrakii* (Fig. 4). The most appropriate solution to this problem appears to be:

(1) Since WB 416 is unique and is said to be derived from a type strain of *A. melleus*, its selection as the neotype of *A. melleus* and the re-establishment of the species is recommended.

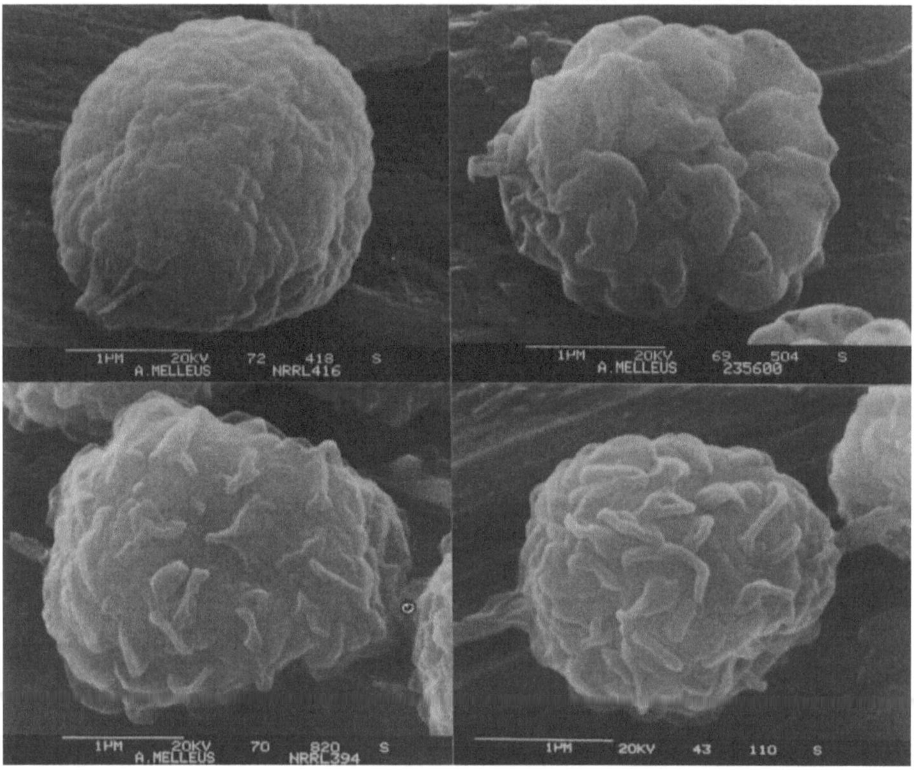

Fig. 4. The A. melleus/A. quercinus complex. Micrographs depicting the three distinct morphotypes. Top: left, A. melleus NRRL 416, right, A. melleus NRRL 5103. Bottom: left, A. quercinus NRRL 394, right, A. petrakii NRRL 4396.

(2) Since the species A. quercinus Bain. has no satisfactory type, nor for that matter any extant isolates, it should be regarded as of uncertain application and excluded from current usage.

(3) Isolate WB 394, previously used in description of A. quercinus, with conidia identical to those of A. petrakii, becomes merely an isolate identified as A. petrakii.

(4) Isolate WB 5103, on conidial topography, represents a taxon distinct from the above and may represent an undescribed species.

CONIDIUM ONTOGENY

The SEM has significantly improved our understanding of Aspergillus taxonomy, but in so doing it has introduced an additional problem, the extreme changes in external morphology exhibited by the conidium during ontogenetic development.

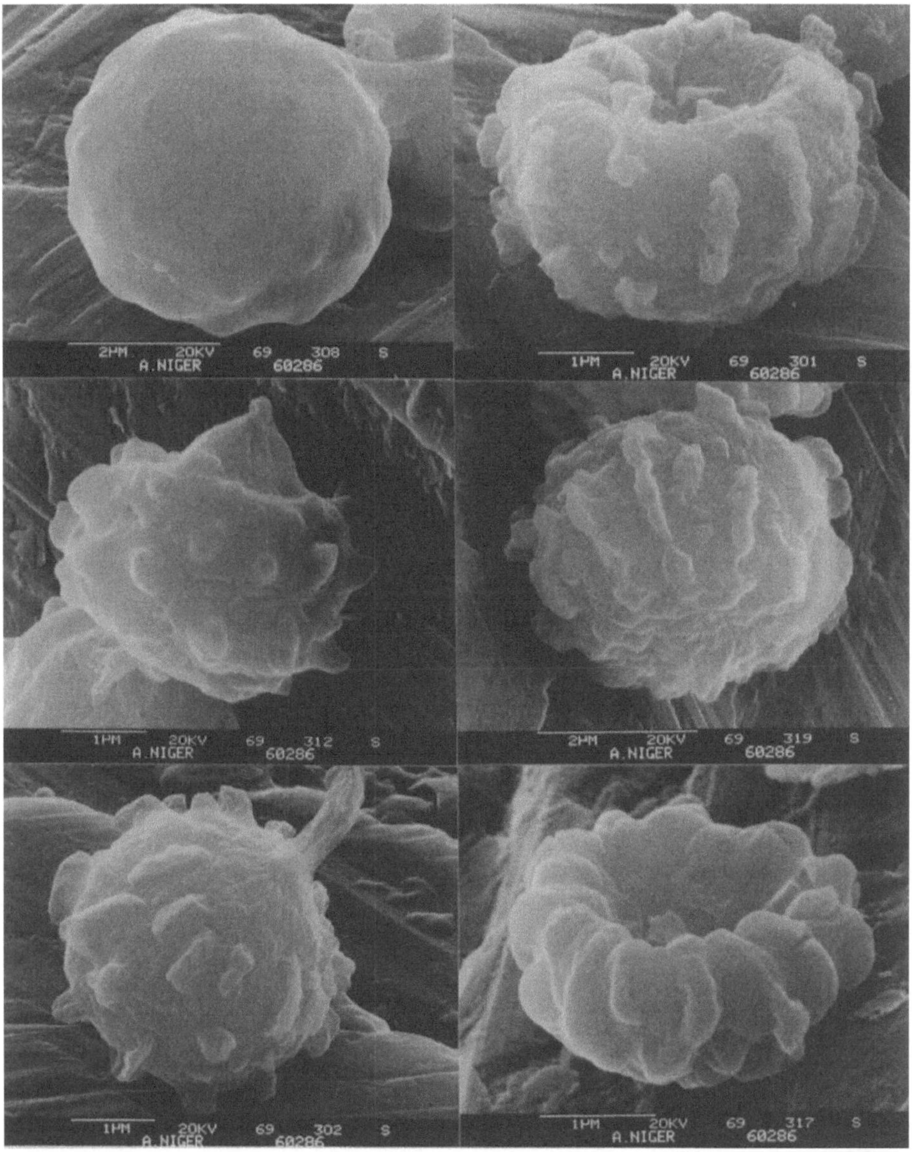

Fig. 5. A. niger IMI 60286, SEM micrographs illustrating how the topography of the conidium changes during development, starting, top left as a smooth conidium, descending with the mature verrucate conidium at bottom right.

In Fig. 5 the development of the conidium of A. niger is shown to proceed from an initially smooth conidium through lobate, striate to an ultimately verruculose condition. When using the proposed time-saving technique of in situ examination,selection of any one of these micrographs other than that illustrating the verruculose mature form may lead to misidentification of the species.

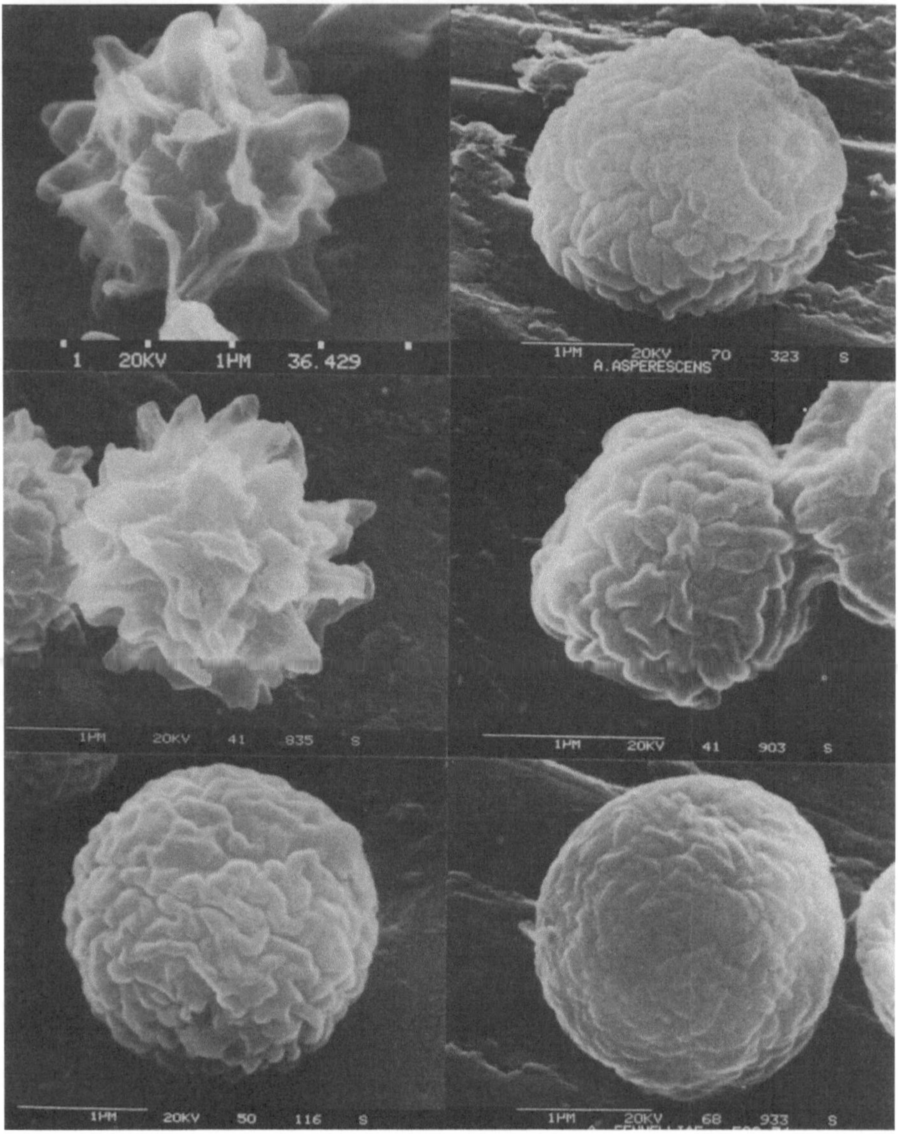

Fig. 6. Demonstrating the need to employ micrographs of mature conidia for identification purposes. Top and middle series from left to right, mature and immature conidia of A. asperescens IMI 46813 (T) and A. microcysticus IMI 139275 (T) respectively. Bottom series, mature conidia of, left, N. stramenia IMI 172293 (T), right, A. fennelliae CBS 598.74.

An example is given in Fig. 6 in which immature forms differ markedly from the mature state, in which condition they equate with mature forms found in other species groups.

Acknowledgments

I wish to thank Miss B. Higgins for excellent scanning microscopy and Mr. M. Brown for photography; and Dr A.H.S. Onions and Professor D. L. Hawksworth for reading and commenting on my original manuscript.

REFERENCES

BLASER, P. 1976. Taxonomische und physiologische untersuchungen uber die Gattung Eurotium Link ex Fries. Sydowia 28: 1-49.

CHRISTENSEN, M. 1981. A synoptic key and evaluation of species in the Aspergillus flavus group. Mycologia 73: 1056-1084.

COHN, F. 1884. Berichte uber die Tatigkeit der botanischen sektion der Schlesischen gesellschaft im Jahr 1883. Jber. schles. Ges. varterl. Kult. 61: 226-230.

KOZAKIEWICZ, Z. 1982. The identity and typification of Aspergillus parasiticus. Mycotaxon 15: 293-305.

MURAKAMI, H. 1971. Classification of the koji mould. J. gen. appl. Microbiol., Tokyo 17: 281-309.

NEHIRA, T. 1955. Sporulation of A. oryzae IV. Comparison of various media. J. Ferment. Technol., Osaka 33: 313-317.

RAPER, K. B. and FENNELL, D. I. 1965. "The genus Aspergillus". Baltimore: Williams and Wilkins.

SAKAGUCHI, K. and YAMADA, K. 1944. Morphology and classification of koji moulds. Parts I and II. J. agric. Chem. Soc. Japan 20: 65-73, 141-154.

SAMSON, R. A. 1979. A compilation of the Aspergilli described since 1965. Stud. Mycol., Baarn 18: 1-38.

THOM, C. and CHURCH, M. B. 1926. "The Aspergilli." Baltimore: Williams and Wilkins.

THOM, C. and RAPER, K. B. 1945. "Manual of the Aspergilli." Baltimore: Williams and Wilkins.

DIALOGUE FOLLOWING DR. KOZAKIEWICZ'S PAPER

SAMSON: I think your SEM photographs all show shrinkage, and that it is difficult to judge ornamentation of conidial walls. I would hesitate to use this information for taxonomic purposes.

KOZAKIEWICZ: I respect your opinion, but I disagree.

Chapter 6:

TELEOMORPHS AND THEIR RELATIONSHIPS

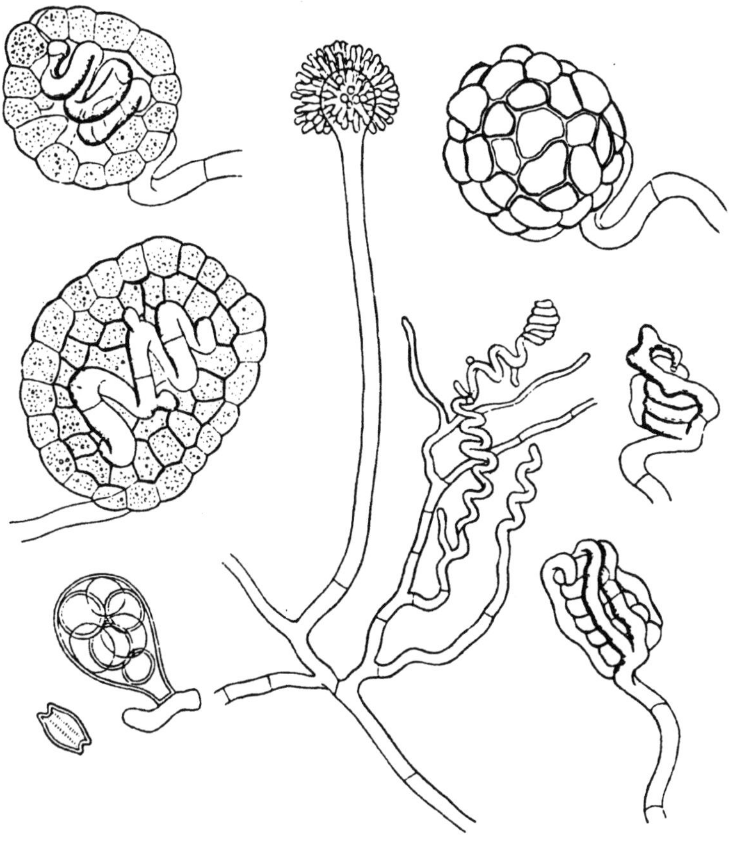

THE TRICHOCOMACEAE:
RELATIONSHIPS WITH OTHER ASCOMYCETES

David MALLOCH

Department of Botany, University of Toronto
Toronto, Ontario M5S 1A1, Canada

SUMMARY

The Trichocomaceae is a relatively large family of Ascomycetes with members frequently impinging upon human activities. Because of this, they have received considerable taxonomic attention, focussed mainly on their conspicuous anamorphs. When the holomorphs are examined closely two taxonomic groups or subfamilies become apparent: 1) the Dichlaenoideae, characterized by bivalved ascospores, frequent presence of true cleistothecial peridia, production of various anamorphs including Aspergillus, and a marked affinity for starchy or oily substrata, and 2) the Trichocomoideae, characterized by non-bivalved ascospores, lack of a true cleistothecial peridium, largely Paecilomyces and Penicillium anamorphs, and an affinity for cellulosic substrata. Most morphological features of the Trichocomaceae resemble those of Ascomycetes in other families. In spite of this, an independent terminology has developed in the Trichocomaceae that is often inconsistent with more general usage. Some recommendations are offered to resolve these differences.

INTRODUCTION

The Trichocomaceae are unusual among Ascomycetes in being most commonly observed in their asexual forms. This is true in part because they do indeed produce conidia more often and more abundatly than ascospores but it also reflects the way in which they intrude on human activities. Most members of the family are mycological opportunists, rapidly invading uncolonized substrata and reproducing in the fastest manner possible, ie. asexually. Many do, in fact, regularly reproduce sexually, but do so at the end of the proliferative stage when their activities no longer interest the observer, as when the substrate is so completely occupied that it is considered to be "spoiled" and is discarded. Their characteristic life cycles and ecology have thus had a sociological effect upon mycology. Of all fungi these are the most likely to be considered in terms of their anamorphs; either exclusively or in large part. This emphasis on anamorphs generally has been a practical one and represents an intuitive approach to a taxonomically difficult group. To the more general taxonomist, though, the family appears to have been treated as though it is quite apart from more "conventional" Ascomycetes. To be fair, it should also be pointed out that both Ascomycete and Hyphomycete taxonomists have tended to dismiss this group as a collection of annoying moulds that are best left to the industrial microbiologist.

A close look at the Trichocomaceae reveals that in most ways they are not especially unusual Ascomycetes. Although their anamorphs are indeed complex and conspicuous they are probably no more so than those of many species of Ceratocystis Ellis & Halst., Nectria Fr. or Puccinia Pers. The teleomorphs appear to be extremely simple and unlike any others, but in fact they take a great variety of forms and can be related to those of other ascomycetes structure for structure. In this paper I have attempted to examine in the Trichocomaceae all of the major features used by Ascomycete taxonomists and to use these to draw some conclusions as to the relationships between this family and others. Considerable stress is placed on relationships within the family as well, particularly in reference to the two subfamilies Trichocomoideae and Dichlaenoideae (Table 1).

The modern taxonomist makes use of a large number of characters in classifying Ascomycetes, but still lays particular stress on the following:

1. Cleistothecial or perithecial initials;
2. Structure and development of the cleistothecial centrum;
3. Structure of the cleistothecial peridium;
4. Ascus structure;
5. Stromatic structures;
6. Ascospore morphology;
7. Anamorphs.

To determine how each of these characters applies to the Trichocomaceae I shall examine each in its turn.

Cleistothecial initials

The term "cleistothecial initials" is used here for those structures that give rise to a cleistothecium and is not used for the early stages of stromatal or sclerotial development. It is presumed that one or possibly more cells of the cleistothecial initials serve as an ascogonium but I have purposely avoided the use of that term as it applies quite precisely to the cell or cells in which nuclear migration and dikaryotization occur. We know yet little about these early stages of development in the Trichocomaceae and it would be misleading to apply to them any anatomical terms with precise meanings that we cannot defend.

Table 1. Subfamilies of Trichocomaceae.

TRICHOCOMOIDEAE Malloch (1985)

Characteristics:
1) Asci borne within a tuft or layer of loosely interwoven to parallel hyphae that may or may not be borne on a sterile stipe or base;
2) Pseudoparenchymatous or Hulle cell stromata lacking;
3) Ascospores sphaerical to ellipsoidal, roughened, lacking slits or furrows;
4) Anamorphs *Penicillium* or *Paecilomyces*;
5) Usually associated with wood or cellulose or possibly parasitic on other fungi.

Genera: *Byssochlamys* Westling, *Dendrosphaera* Pat., *Sagenoma* Stolk and Orr, *Talaromyces* C. Benj., *Trichocoma* Junghhuhn.

DICHLAENOIDEAE Malloch (1985)

Characteristics:
1) Cleistothecia present or absent; when present having pseudoparenchymatous or very rarely hyphal peridia and arising from distinct cleistothecial initials;
2) Stromata with pseudoparenchymatous walls or represented by Hulle cells;
3) Ascospores bipolarly compressed (oblate) to ellipsoidal, nearly always with a furrow or slit along which they split upon germination;
4) Anamorphs *Aspergillus*, *Penicillium*, *Merimbla*, *Paecilomyces* or *Polypaecilum*;
5) Frequently associated with starchy or oily seeds and fruit, often proteolytic;

Genera: *Chaetosartorya* Subr., *Cristaspora* Fort and Guarro, *Dichlaena* Dur. and Mont., *Dichotomomyces* Scott, *Eupenicillium* Ludw., *Edyuillia* Subr., *Emericella* Berk. and Br., *Eurotium* Link, *Fennellia* Wiley and Simmons, *Hamigera* Stolk and Samson *Hemicarpenteles* Sarbhoy and Elphick, *Hemisartorya* Rai and Chowdheri, *Neosartorya* Malloch and Cain, *Penicilliopsis* Solms-Laubach, *Petromyces* Malloch and Cain, *Sclerocleista* Subr., *Thermoascus* Miehe, *Warcupiella* Subr.

The most commonly encountered form of cleistothecial initials is the helix. This can be very regular and barrel-shaped as in species of Eurotium* and Dichotomomyces or it may be rather irregular as in Neosartorya. Helical initials, whether regular or irregular, appear to be the rule throughout the subfamily Dichlaenoideae except in those genera (e.g. Eupenicillium) where there is no true cleistothecial peridium.

In the subfamily Trichocomoideae cleistothecial initials are much more variable in form and appear to be more irregularly coiled or not helical at all. Initials in this family, where known, remain quite constant for each species. They are best described in species of Talaromyces (Stolk and Samson, 1972) and have long served as major taxonomic features. Raper and Thom (1949) called attention to the differences in cleistothecial initials between T. vermiculatus Dangeard and T. wortmannii Klöcker. In T. vermiculatus the initials are rather long and slightly clavate. These are septate and become surrounded by a second coiled hyphae. Dangeard (1907) labelled the first an ascogonium and the second a trophogonium. In T. wortmannii the cleistothecial initials are branched and quite unlike those of T. vermiculatus. It should be pointed out that in some collections of Talaromyces species it is difficult to find cleistothecial initials. It appears that the so-called peridium in this genus is in fact stromatic as it is in Warcupiella and Hamigera (Subramanian and Rajendran, 1979b) and that initials appear later in development. According to Stolk and Samson (1972) the initials occur singly or in groups within the developing hyphal tufts.

The helical initials of the Dichlaenoideae are not substantially different from those of many species of Ascomycetes from a great variety of families. Helically coiled initials are known in the Pezizales (e.g. Durand, 1973) and in various Diaporthales (Chadefaud, 1982), indicating that these structures offer litte evidence of interfamilial relationships. The most we can say is that the Dichlaenoideae are rather "typical" Ascomycetes in this regard. The Trichocomoideae, on the other hand, seem to follow so little pattern in the morphology of their cleistothecial initials that is quite difficult to relate them to any other group.

Cleistothecial centrum

As pointed out by Malloch and Cain (1972), Subramanian (1979) and Subramanian and Rajendran (1979b, 1980, 1981a,b), many species of Trichocomaceae produce cleistothecia within stromatic structures or produce asci within stromatic structures without any cleistothecial wall at all. If we take into account both ascostromata and "true" perithecia (or cleistothecia) then we must recognize at least two centrum types in the Trichocomaceae. No species of the Trichocomaceae studied to date has sterile filaments within the centrum. Instead, the ascogenous hyphae appear to proliferate throughout the centrum and produce a uniform mass of spherical to broadly clavate

---

* Taxonomic authorities for names of members of the Trichocomaceae are given in Table 1.

asci (Benjamin, 1955; Brefeld, 1874; de Bary, 1854, Subramanian and Rajendran, 1977, 1979a, b, 1980, 1981a, b). When the ascoma is a true cleistothecium arising from the hyphae surrounding the cleistothecial initials then, according to Malloch (1981), it has a centrum development of the Ophiostoma type of Luttrell (1951). When the ascoma is an ascostroma, as in Eupenicillium, then the centrum type is close to the Phaeotrichum type of Barr (1956), although in that type the early development of the ascostroma should be entirely pseudoparenchymatous rather than hyphal. In an earlier paper (Malloch, 1981) I pointed out that many cleistothecial Ascomycetes from apparently diverse origins have a very simple kind of centrum development in which sterile tissues are lacking and in which the asci come to fill the centrum uniformly at maturity. This type of centrum probably represents a gradual simplification from many unrelated taxa and is of very little use to the taxonomist. Thus centrum type in the Trichocomaceae tells us little more than that these fungi use the very simplest means possible to fit a mass of asci into the centrum.

Cleistothecial peridium

The cleistothecial peridium in the Trichocomaceae varies considerably from genus to genus and is one of the most important characters in their delimitatation. In those species with a true peridium, this arises from the hyphae that come to surround the cleistothecial initials after they have been formed. As pointed out above, many species have asci produced wthin a stromatic structure, and this may or may not contain a peridiate cleistothecium. It is important to the taxonomist therefore to distinguish between those species that produce their asci in a small stroma which does not contain a peridiate cleistothecium, such as Eupenicillium species, and those that produce a peridiate cleistothecium but no stroma, as in Eurotium species. It is probably also important to adhere to a more precise terminology for these structures than has been used in the past, and at the end of paper I have made some suggestions as to how this might be done.

In the subfamily Trichocomoideae there is no well defined cleistothecial peridium. The massive basal structures in Trichocoma and Dendrosphaera are undoubtedly stromatic in origin, athough no developmental studies have confirmed this. The small and often spherical hyphal mass enclosing the asci in Talaromyces has been called a peridium but often appears to be independent of, and to preceed, cleistothecial initials. It is possible that in this whole subfamily there are no true cleistothecia and that the ascomata, whether atromatic or not, have entirely different origins from those of the Dichlaenoideae.

Ascomata in the subfamily Dichlaenoideae are usually easier to interpret. When present, cleistothecia arise from the hyphae surrounding the cleistothecial initials. The peridium can form either a textura angularis or a textura epidermoidea (see Korf, 1973, for an explanation of these terms). Development of cleistothecia within a stroma is characteristic of several families of Hypocreales, Diaporthales, Clavicipitales and others and is not clearly indicative of evolutionary origins. Thus the ascomata in this subfamily appear to be quite "conventional" and not useful taxonomically above the generic level. The species that lack a true cleistothecium have been interpreted to have lost this structure during their evolution (Malloch, 1981).

Ascus structure

Asci in the Trichocomaceae range from nearly spherical to broadly clavate. Walls appear to be uniformly thin and are evanescent at ascospore maturity. Asci of this type are known throughout most of the families of cleistothecial ascomycetes and are of little taxonomic use. Asci in the Trichocomaceae often can be demonstrated to arise from croziers; where they do not, this feature has been given taxonomic recognition (e.g. Talaromyces). It is unknown whether they are unitunicate or bitunicate because, as in most cleistothecial species, when active ascosores discharge was lost all of the features associated with discharge were lost. On the other hand, we are fairly certain that they are derived from unitunicate types; the developmental features of the ascomata and the highly developed phialidic anamorphs are characteristics of Hymenioascomycetes not Loculascomycetes.

Stromatic structures

The Trichocomaceae may have evolved from an ancestral form having numbers of perithecia embedded in a stroma (Malloch and Cain, 1972). If this is so, then the evolutionary history of the family has been one of continual simplification. The most complex and hence possibly most primitive members are in the genera Dichlaena, Petromyces and Saitoa. In all of these the stromata are fairly large and contain one to four cleistothecia with well developed peridia. The difference between these genera and members of several Pyrenomycete families is not great, resting mainly in the Ophiostoma type of centrum development in the former and the more complex types of the latter. A parallel situation occurs in the genus Pulveria Malloch and Rogerson (Xylariaceae)(Malloch and Rogerson, 1977) which produces spreading stromata containing numerous cleistothecia. The differences between Pulveria and Hypoxylon Bull. are few and, again, mainly involve details of centrum development and ascus morphology.

Reduction from the complex stromatic forms in the Trichocomaceae seems to have taken two paths: 1) reduction of the stroma and retention of the cleistothecium; and 2) reduction of the cleistothecium and retention of the stroma. Reduction of the stroma has probably occurred in more than one taxonomic line of the family and has resulted finally in taxa having no stroma at all (eg. Eurotium and Neosartorya). Earlier stage of stroma breakdown can be see in Dichlaena where, at maturity, the stroma is reduced to a thin papery covering around the cleistothecia (Malloch and Cain, 1972). In the genera Emericella and Fennellia the cells of the stroma have become disarticulated to form a loose pile of thick-walled spore-like bodies called Hülle cells. As in more solidly stromatic forms, species of Emericella produce the stromatic tissues well in advance of the cleistothecium formation and surround them at maturity. Within that genus occur species with massive coverings of Hülle cells and others in which the Hülle cells are fairly sparse. Cristaspora arxii Fort and Guarro (1984) has a stroma that is hyphal to plectenchymatous and envelopes several peridiate cleistothecia. Because it seems to lack an anamorph, its inclusion in the Trichocomaceae might be questioned (Fort and Guarro placed it in the Onygenaceae), but it otherwise fits in the Dichlaenoideae quite well. Throughout the family there is an abundance of forms demonstrating the transition from stromatic to non-stromatic ascomata.

Reduction of the cleistothecium and retention of the stroma has also occurred in more than one line but we lack the intermediate forms that would make clear the reduction of stromata. Penicilliopsis and Thermoascus are genera in which the stromata are still quite large and are composed of more than one locule or cluster of asci. Reductions from the large to the small stromata are seen in the genus Eupenicillium where the stromata are small but often contain more than one group of asci (Brefeld, 1874).

The subfamily Trichocomoideae poses a problem for this discussion. There is really no indication in this group that its members were ever derived from genera like Petromyces or Dichlaena; instead the most complex members of the subfamily are the massive Dendrosphaera eberhardtii and Trichocoma paradoxa, which produce large stipitate to substipitate fruiting bodies bearing little resemblence to any group of Pyrenomycetes. Because of their loose hyphal "stromata" or massive fruiting bodies, the Trichocomoidea stand quite apart from the rest of the family and may in fact be only rather distantly related.

Before going further it is necessary to deal with the term "sclerotium" which has been used in the Trichocomaceae to denote small, hard cellular structures produced by many species. It is abundantly clear in all cases that the "sclerotia" are entirely homologous with fertile stromata and should not be labelled in a way that implies that they are not. In some groups of fungi (e.g. Typhula Pers., Sclerotinia Fckl.) the term "sclerotium" is used for bodies giving rise to ascus- or basidium-bearing structures, but in these cases the ascogenous or basidiogenous structures are produced externally rather than internally as in the Trichcomaceae. As noted above, the stromata in the Trichocomaceae are quite comparable to those of many Pyrenomycetes and should bear a terminology consistent with those groups.

The structure of ascomata and stromata in the Trichocomaceae is the most important taxonomic character for separating genera. It is clear from Table 1 why this is so; the genera break down into several well-defined groups that are rather easily identified. On the other hand, those familiar with the Trichocomaceae mainly by its Aspergillus and Penicillium anamorphs will notice that the groups do not seem to reflect anamorph taxonomy at all. This question is dealt with below.

Ascospore morphology

Ascospores in the Trichocomaceae are all hyaline to brightly coloured and single celled. Many species in the family are characterized by ascospores having a bivalved structure. Such spores have an equatorial or lateral line of thin wall material that cracks open upon germination. Germinating spores of this type rarely swell but crack open like clam shells, allowing the germ tube to extend. While this type of ascospore is not entirely unknown outside the Trichocomaceae it certainly is highly characteristic of the family. Onygenaceae (including the Gymnoascaceae) may have ascospores resembling those of the Trichocomaceae but as far as I know, all of these swell considerably on germination and then break open irregularly. Most of the bivalvate spores in the Trichocomaceae are strongly flattened at the poles and have rotational symmetry around the short axis, a shape usually described as oblate. The flattened appearance is often further accentuated by a pair of thin ridges that extend the diameter of the spore one or two micrometres more. In addition to this, the faces of the spores may be sculptured in various ways. A few species, such as Petromyces alliaceus Malloch and Cain and Warcupiella spinulosa (Warcup) Subramanian, have bivalvate ascospores that are ellipsoidal rather than oblate, but this condition is unusual. Thermoascus aurantiacus Miehe has been illustrated as germinating through a pore-like opening (Subramanian and Rajendran, 1980), which is apparently unique in the family.

Although the bivalvate ascospore is very characteristic of the Trichocomaceae it is in fact present only in the Dichlaenoideae, where it is almost universal. Ascospores of the Trichcomoideae are never bivalvate and merely swell upon germination until they burst. They are spherical to ellipsoidal in shape and appear never to be oblate. Thus again the Trichocomaceae appear to be rather unlike the Dichlaenoideae.

Ascospores in the Trichocomaceae are little help in finding relationships with other families. On the one hand, the bivalvate ascospore is so characteristic of the Dichlaenoideae and so very rare elsewhere that we cannot find a good connection to any other group of fungi. On the other hand, the ellipsoidal, often scultured ascospores of the Trichocomoideae are so undistinctive that they offer no more clues to relationships than the highly distinctive ones of the Dichlaenoideae.

Anamorphs

As stated at the beginning, the Trichocomaceae are best known for their anamorphs. Anamorphs in this family are produced in abundance and great variety but can be classified in just a few genera. Three form genera dominate the family: Aspergillus Mich., Penicillium Link, and Paecilomyces Bain. It is significant that all of these genera are known to occur only in the Trichocomaceae. Even Paecilomyces, a rather simple genus of phialidic species, is not known elsewhere. There have been reports of Paecilomyces anamorphs in Cordyceps Fr. and Pyxidiophora Bref. & von Tavel (listed in Kendrick, 1979) but these have all been discredited, by Lundqvist (1980) for _Pyxidiophora_ and by Samson (1974) for _Cordyceps_. _Polypaecilium_ G. Smith, a rather curious phialidic form genus, is known only in _Dichotomyces_ and _Thermoascus_, while _Sagenomella_ W. Gams is associated only with _Sagenoma_, and _Raperia_ Subr. and Rajendran with _Warcupiella_.

Malloch and Cain (1972) and Subramanian (1979) have suggested that the nearest relative to the Trichocomaceae should be sought in the Hypocreales. Like the Trichocomaceae, the Hypocreales often produce conidia on complex phialide-bearing structures, are mostly brightly coloured and tend to be mycological weeds. As Subramanian (1979) pointed out, though, conidia in the Trichocomaceae always occur in dry massess while those in the Hypocreales always occur in wet masses. While the distinction may not seem great, this character is so uniform throughout the two groups that it demands the taxonomist's attention.

As Table 2 indicates, the structure of the ascomata does not appear to correlate well with anamorph structure. We know that the presence of metulae in _Aspergillus_ can be variable at times, even among closely related species, so that is not entirely surprising to find species such as _Dichlaena lentici_ Mont. and Dur., with ametulate vesicles, similar to metulate speciessuch as _Petromyces alliaceus_. More disturbing at first glance is the presence of _Penicillium_ and _Aspergillus_ anamorphs in taxa that appear very similar in the teleomorph morphology. It must be understood however that, in such simplified organisms as these are, it is very likely to find many examples of convergent evolution. Thus a complex stromatic form with a _Penicillium_ anamorph and one with an _Aspergillus_ anamorph might, on reduction, both end up with small stromata that lack cleistothecia as in _Eupenicillium_ and _Hemicarpenteles_. The other possibility is that there may be more than one evolutionary link between _Aspergillus_ and _Penicillium_ and that species in one genus may have given rise to species in the other. There does not seem to be much evidence for this latter hypothesis, but it should be borne in mind by anyone attemping to study relationships among the Trichocomaceae.

The Trichocomoideae again cause some difficulty; the anamorphs here are rather unlike those of the Dichlaenoideae in producing often rather loose penicilli, long tapering phialides, and narrow conidia. It is usually not difficult to separate the two subfamilies on the anamorphs alone and a good case could be made for placing the Penicillia of the Trichocomaceae in a separate genus. A second characteristic of the anamorphs in the Trichocomoideae is the complete lack of _Aspergillus_ and _Aspergillus_-like forms. This too suggests a long separation of the two subfamilies or even separate origins followed by some degree of parallel evolution.

DISCUSSION

This paper so far has stressed the difficulties of dealing with the subfamily Trichocomoideae. They seem to fit poorly into any system that is based on the much larger Dichlaenoideae yet are within themselves a rather distinctive

Table 2. Genera of Trichocomaceae group according to ascomatal type, with anamorphs in parentheses.

Stroma present; cleistothecia present

1. Cristaspora (Unknown)
2. Dichlaena (Aspergillus)
3. Emericella (Aspergillus)
4. Fennellia (Aspergillus)
5. Petromyces (Aspergillus)

Stroma present; cleistothecia absent;

1. Eupenicillium (Penicillium)
2. Hamigera (Merimbla)
3. Hemicarpenteles (Aspergillus)
4. Penicilliopsis (Aspergillus, Sarophorum, Stilbodendron)
5. Thermoascus (Paecilomyces or Polypaecilum)
6. Warcupiella (Raperia)

Stroma absent; cleistothecia present

1. Chaetosartorya (Aspergillus)
2. Dichotomomyces (Polypaecilum)
3. Eurotium (Aspergillus)
4. Hemisartorya (Aspergillus)
5. Neosartorya (Aspergillus)
6. Sclerocleista (Aspergillus)

Stroma absent; cleistothecia absent

1. Edyuillia (Aspergillus)

Asci borne in hyphal masses or tufts; cleistothecia absent

1. Byssochlamys (Paecilomyces)
2. Dendrosphaera (Penicillium)
3. Sagenoma (Sagenomella)
4. Talaromyces (Penicillium, Paecilomyces)
5. Trichocoma (Penicillium)

group. In the remaining sections I will concentrate on the Trichocomoideae and in particular on Trichocoma, which may hold the key to the taxonomy of this interesting group.

Taxonomic controversy in the Trichocomaceae involves even the family name itself. Most members of the family have been classified in the Eurotiaceae, a family proposed by Clements and Shear (1931). Before recognition of the Eurotiaceae most mycologists used the family name Aspergillaceae Link (Link, 1826). This latter name is based on Aspergillus, a form genus, and is thus not valid for Ascomycetes. The Eurotiaceae has appeared to most mycologists to be a good name, based on well-known and very "typical" members of the group and has become thoroughly established in mycological tradition. The one disadvantage to the use of this name is that it has been applied to nearly all cleistothecial Ascomycetes, which includes forms now known to be not even remotely related to the group under consideration here.

The more recently used family name Trichocomaceae is in fact much older than Eurotiaceae, dating back to Fischer (1897). Its use has been very restricted because it is based on the obscure species *Trichocoma paradoxa* Junghuhn, and might have remained obscure had Kominami et al. (1952) not discovered that *T. paradoxa*, in spite of its large and unusual ascomata, has much in common with the familiar Ascomycetes having *Aspergillus* and *Penicillium* anamorphs. These authors grew *T. paradoxa* in pure culture and pointed out that the anamorph was a *Penicillium* typical of the biverticillate type. In addition, they showed that the ascospores are nearly identical with those of *Talaromyces luteus* (Zukal) C. Benj.. So impressed were these authors with the similarities between them that they suggested the two might be ecological forms of the same species. Given this evidence it was difficult to justify a separate family for *Trichocoma* and Malloch and Cain (1972) argued that the Trichocomaceae should be the correct name for all of the Ascomycetes with *Aspergillus* and *Penicillium* anamorphs. Although the evidence for inclusion of *Trichocoma* in this group is very convincing and the family name Trichcomaceae is clearly the oldest, some authors are still reluctant to use it.

It might be convenient to dismiss *Trichocoma* as an obscure and very uncharacteristic member of the family but to the taxonomist it suggests some potential clues to the origins of the entire group. Of all Trichocomaceae it seems furthest from the activities of man or the kind of opportunism we associate with the Trichocomaceae. It has been found in a number of localities around the world, usually growing on highly decayed stumps of woody angiosperms in warm, humid forests. It is rather large and conspicuous and has invariably fallen into the hands of the herbarium mycologist rather than the microbiologist who works with *Penicillium*.

Trichocoma paradoxa, apparently the only species of *Trichocoma*, has been described and illustrated amany times (Boedijn, 1935; Ciferri, 1957; Fischer, 1980; Kobayasi and Indoh, 1943; Kominami et al., 1952, Malloch and Cain, 1972). The ascomata are characterized by a fairly massive goblet-shaped basal portion up to 12 mm high bearing a soft "brush" up to 30 mm high of densely parallel hyphae from its upper surface. In the field these resemble small yellow camel-hair brushes arising from rotting wood of the substrate. The basal portion is dark red-brown and is woody to leathery in appearance. The soft inner portion arises from the inside of the "goblet" and is entirely enclosed by it at first. During maturation the inner tissue extends upward to form the elongated brush. Ascogenous hyphae and asci are produced at the base of the brush (called a capillitium by Fischer, 1890). The production of asci is evidently prolonged so that the entire brush becomes filled with ascospores that are pushed up by those developing below.

In spite of a number of studies on it, *T. paradoxa* remains somewhat enigmatic. Cleistothecial initials have not been observed, so that it is not clear what the various tissues represent. It seems almost certain that the brown goblet-shaped portion of the ascomata is stromatic in origin and that cleistothecial initials, if they occur, are produced later within it.

Dendrosphaera eberhardtii Pat., a species similar to T. paradoxa, is known from soil or buried wood and roots in southeast Asia. It produces very similar brushes of soft hyphae bearing asci and ascospores, but the brushes are borne in small clusters at the tops of long stipe-like structures. Kobayashi and Yokoyama (1981) reported on the germination of ascospores of D. eberhardtii and described the anamorph as a biverticillate Penicillium. These observations have not been repeated; material of D. eberhardtii kindly given to me by Dr. Kobayashi yielded many ascospores but these failed to germinate. When placed side by side, ascomata of T. paradoxa and D. eberhardtii are remarkably similar in appearance. It should also be noted that the two species have an overlapping geographical range. Kobayashi and Yokoyama (1981) cited the presence of the tree Machilus thunbergii Sieb. and Zucc. (Lauraceae) in D. eberhardtii localities on Iriomote Island (Japan). It is perhaps significant that dead wood of this tree is known as a substrate for T. paradoxa.

Structurally complex and biologically far less opportunistic in nature that other Trichocomaceae, Trichocoma and Dendrosphaera seem likely to be relatively primitive members of the family. If we take this point of view, then we might imagine members of this family becoming more and more specialized as rapid colonizers of disturbed substrata and to be exchanging a type of complex long-term dispersal structure for a simple and profilic one. Such a shift from stablity (k-selection) to weediness (r-selection) might also result in greater dependence on the rapidity and low energy commitment of asexual reproduction.

Searching for a likely extrafamilial relative of Trichocoma or Dendrosphaera one might take the following characteristics into consideration:

a) Wood decay;
b) Bright colours or absence of melanin-like pigments;
c) Single celled ascospores;
d) Probable Hymenioascomycete type of development with accompanying unitunicate asci;
f) Ascomata borne in or on stromata.

Malloch and Cain (1972) suggested the Hypocreales as the most likely ancestor to have this combination of characters, although they considered Trichocoma to be rather strongly modified from any Hypocrealean ancestor. As was discussed previously, if the Hypocreales are considered ancestral to the Trichocomaceae it is through species rather different from Dendrosphaera and Trichocoma. A second possible ancestor might be found in the Helotiales, which also possess most of the characters given above and which show a degree of specificity to plant material similar to that of Trichocoma and Dendrosphaera. It is not completely unlikely that these genera might have arisen from an apothecial ancestor in which the hymenium first became angiocarpic and then the asci became spherical and evanescent. The basal hymenium in Trichocoma might be the remnant of such apothecial origins.Stromatic tissues in Helotiales are far less developed than those of the Hypocreales; but, in fact, neither Trichocoma nor Dendrosphaera exhibit the pseudoparenchymatous stroma found in the Hypocreales. Highly developed phialidic anamorphs similar to biverticillate Penicillia are produced by a number of Helotiales (Hennebert and Bellemer, 1979)

We must of course bear in mind that Trichocoma and Dendrophaera might represent rather advanced and specialized fungi that are further from a common ancestor than are the structurally simple members of the family. Many authors have noted the similarities between the certain Onygenaceae Schroet. and Trichcomaceae such as Talaromyces and view all of these as very primitive and related forms. Most authors today, though, recognize the connections between many cleistothecial Ascomycetes and more complex non-cleistothecial forms and generally view these as simplifications rather than as ancestors. While the Trichocomaceae may represent a primitive group of Ascomycetes quite unlike those cleistothecial Ascomycetes with obvious connections to more complex forms, I seriously doubt that such a ubiquitous and ecologically successful group represents a vestige of an earlier mycota.

When Malloch and Cain (1972) focussed attention on evolution in the Trichocomaceae, they turned to the members of the family having cleistothecia embedded in stromatic tissues. As they pointed out, genera such as Dichlaena and Petromyces are not unlike many Hypocreales having perithecia embedded in a stroma. In Dichlaena, a hard stroma develops on the substrate and later on one to three coiled cleistothecial initials appear inside. Each of these gives rise to a cleistothecium. At maturity most of the stromatic tissue disintegrates so that the thin-walled cleistothecia remain surrounded only by the papery stromatal peridium. Petromyces alliaceus has a very similar development except that the stromatic tissue remains thick at maturity.

Relating Trichocoma or Dendrosphaera to such "primitive" genera as Dichlaena, Petromyces or Saitoa is problematic; if we accept Malloch and Cain's thesis that the ancestral group of the family resembled these latter genera then the former are either rather specialized and derived as suggested above or are part of a quite independent line. The idea of two major lines of evolution in the Trichocomaceae is an attractive one that overcomes some of the problems of trying to relate Trichocoma to such genera as Dichlaena and Petromyces. It is necessary to bear in mind, though, that the presence of the distinctive Penicillium and Paecilomyces anamorphs in both groups suggests that the two subfamilies are related. As already discussed these anamorphs occur only in the Trichocomaceae and represent the one feature that unites them as a group.

CONCLUSIONS

The use of anamorphs as primary taxonomic features in the Trichocomaceae and the great economic importance of the group has tended to keep it outside the mainstream of taxonomic thought. Some terminology has thus arisen that is not consistent with conventional usage and which should be abandoned or modified. I suggest the following:

1. The term ascogonium should be replaced by the term cleistothecial initials except in those cases where it applies specifically to cell in which dikaryotization occurs.

2. The term cleistothecium should apply to those sructures arising from the hyphae that come to surround the cleistothecial initials after they are formed and should not be applied to stromatic structures. The term ascoma is more general in nature and can apply to either cleistothecia or stromata when the nature of the structure is not clear or where a general term is needed.

3. The term stroma should be applied to structures forming before the production of the cleistothecial initials and bearing cleistothecia, asci or ascospores. When the stromata do not contain cleistothecia as in *Eupenicillium*, they can be called ascostromata. Sterile structures clearly homologous to these should also be called stromata. The term sclerotium should be abondaned as inappropriate.

REFERENCES

BARR, M.E. 1956. The development of the ascocarp in *Phaeotrichum hystricinum*. Can. J. Bot. 34: 563-568.

BARY, A. de. 1854. Uber die Entwicklung und den Zusammenhang von *Aspergillus glaucus* und *Eurotium*. Bot. Zeit. 12: 425-471.

BENJAMIN, C.W. 1955. Ascocarps of *Aspergillus* and *Penicillium*. Mycologia 47: 669-687.

BOEDIJN, K.B. 1935. On the morphology and cytology of *Trichocoma paradoxa*. Ann. Jard. Buitenzorg 44: 243-256.

BREFELD, O. 1874. "Botanische Untersuchungen über Schimmelpilze. Heft. II. Die Entwicklungsgeschichte von *Penicillium*". Leipzig: Felix.

CHADEFAUD, M. 1982. Les principaux types d'ascocarps: leur organisation et leur evolution. Troisieme partie: les pyrenocarps. Cryptog., mycol. 3: 199-235.

CIFERRI, R. 1957. *Trichocoma paradoxa* in Santo Domingo and the order Onygenales. Atti Ist. Bot. R. univ. R. Lab. Crittog. Pavia 14: 1-4.

CLEMENTS, F.E. and SHEAR, C.L. 1931. "The genera of Fungi". New York: H.W. Wilson.

DANGEARD, P.A. 1907. L'orgine du perithece chez les ascomycetes. Botaniste 10: 1-385.

DURAND, F. 1973. Sur le developpement d'une Ascobolacee du genre *Ascophanus*. Bull. trimest Soc. mycol. Fr. 89: 113-125.

FISCHER, E. 1890. Beiträge zur Kenntnis exotischer Pilze.

Trichocoma paradoxa Jungh. Hedwigia 29: 161-171.
FISCHER, E. 1897. Plectacineae. In "Die naturlichlichen Pflanzenfamilien. Vol. I", A. Engler and K. Prantl, eds. Leipzig: Engelmann.
FORT, F. and QUARRO, J. 1984. Cristaspora a new genus of Eurotiales. Mycologia 76: 1115-1118.
HENNEBERT, G.L. and BELLEMERE, A. 1979. Les formes conidiennes des discomycetes. Essai taxonomique. Rev. mycol. 43: 259-315.
KENDRICK, B., ed. 1979. "The Whole Fungus". Ottawa: National Museums of Canada. 2 Vol.
KOBAYASI, Y. and YOKOYAMA, T. 1981. On the genus Dendrosphaera and its conidial state. Bull. Nat. Sci. Mus., Tokyo 7: 15-22.
KORF, R.P. 1973. Discomycetes and Tuberales. In "The Fungi. Vol. IV a". G.C. Ainsworth, F.K. Sparrow and A.S. Sussman, eds. New York: Academic Press.
LINK, H.F. 1826. Entwurf eines physiologischen Pflanzensystems, nebst einer Anordnung der Kryptophyten. Abh. deutsch. Akad. Wiss. Berlin 1824: 145-194.
LUNDQVIST, N. 1980. On the genus Pyxidiophora sensu lato (Pyrenomycetes). Bot. Not. 133: 121-144.
LUTTREL, E.S. 1951. Taxonomy of the Pyrenomycetes. Univ. Missouri Studies 24(3): 1-120.
MALLOCH, D. 1981. The Plectomycete Centrum. In "Ascomycete Systematics: The Luttrellian Concept". D.R. Reynolds, ed. New York: Springer-Verlag.
MALLOCH, D. 1985. Taxonomy of the Trichocomaceae. In "Filamentous microorganisms". T. Arai ed. Tokyo: Japan Scientific Society Press.
MALLOCH, D. and CAIN, R.F. 1972. The Trichocomaceae: Ascomycetes with Aspergillus, Paecilomyces and Penicillium imperfect states. Can. J. Bot. 50: 2613-2628.
MALLOCH, D. and ROGERSON, C.T. 1977. Pulveria, a new genus of Xylariaceae (Ascomycetes). Can. J. Bot. 55: 1505-1509.
RAPER, K.B. and THOM. C. 1949. "A Manual of the Penicillia". Baltimore: Williams and Wilkins..
SAMSON, R.A. 1974. Paecilomyces and some allied Hyphomycetes. Stud. Mycol. 6: 1-149.
STOLK, A.C. and SAMSON, R.A. 1972. The genus Talaromyces. Stud. Mycol., Baarn 2: 1-65.
SUBRAMANIAN, C.V. 1979. Phialidic hyphomycetes and their teleomorphs - an analysis. In "The Whole Fungus". B. Kendrick, ed. Ottawa: National Museums of Canada.
SUBRAMANIAN, C.V. and RAJENDRAN, C. 1977. Developmental morphology of Ascomycetes. I. Edyuillia athecia. Rev. Mycol. 41: 223-231.
SUBRAMANIAN, C.V. and RAJENDRAN, C. 1979a. Developmental morphology of Ascomycetes. IV. Chaetosartorya chrysella Rev. Mycol. 43: 193-204.
SUBRAMANIAN, C.V and RAJENDRAN, C. 1979b. Developmental morphology of Ascomycetes. V. Warcupiella spinulosa and Hamigera avellanea. Rev. Mycol. 43: 351-371.
SUBRAMANIAN; C.V. and RAJENDRAN, C. 1980. Developmental morphology of Ascomycetes. VI. Thermoascus aurantiacus. Cryptog. Mycol. 1: 175-185.
SUBRAMANIAN, C.V. and RAJENDRAN, C. 1981a. Developmental morphology of Ascomycetes. VII. Fennellia flavipes. Cryptog. Mycol. 2: 85-99.
SUBRAMANIAN, C.V. and RAJENDRAN, C. 1981b. Developmental

morphology of Ascomycetes. VIII. Petromyces alliaceus Cryptog Mycol. 2: 189-208.

DIALOGUE FOLLOWING DR. MALLOCH'S PAPER

MALLOCH: Can I conclude my paper with a question? Based on the two distinct teleomorph types associated with Penicillium, is it conceivable that what we currently see as the genus Penicillium could have evolved from two separate origins?

PITT: I think that you could well be right. If I look at a Penicillium under a good quality microscope, I can tell the difference between the phialides and penicillus of species from Subgenus Biverticillium and all the others. These differences are minute and not always easily described, but they are there. There are only two or three species where there might be some disagreement over disposition in this subgenus. One is P. claviforme, which I suspect may be unrelated to any others. Apart from this, I see no interface between Subgenus Biverticillium and the rest of the genus. From a practical point of view, however, I see no reason to separate these two entities into separate genera.

MALLOCH: Penicillium is a form-genus, and there is no reason to split up a form genus, provided you recognize it on purely anatomical grounds.

SAMSON: I agree with Dr. Pitt that Subgenus Biverticillium is different, but as shown in my paper with Miss Stolk, we see more intermediates between these two entities and therefore, don't think it is really possible to satisfactorily split them.

You haven't mentioned Penicilliopsis. In our paper we showed that Penicilliopsis has three anamorphs, one of which we consider to be Stilbothamnium, although we have not seen ascospores in the ascocarps associated with it.

MALLOCH: I think Penicilliopsis belongs fairly comfortably in the Dichlenoideae. It has the stromatic type development similar to Eupenicillium, it has the Aspergillus-like anamorph, it grows on oily seeds which seems to be a typical habitat for Dichlenoideae and not typical of genera in Trichocomoideae, which seems to be cellulolytic or mycopathogenic.

SAMSON: The Sarophorum anamorph of Penicilliopsis clavariaeformis is similar to Penicillium, and if you grow it in culture it looks very like P. digitatum. But it tends to produce polyphialides in cultures, so it may be closer to Polypaecilum than Penicillium.

I see that in your table you have placed Saitoa in the Dichlanoideae. I examined the type culture soon after it was described, and found it was a mixed culture. It has sclerotia of the Aspergillus niger type, but there was also what I believe was a Neosartorya. When I purified it, it was only an anamorphic fungus. The dried type specimen could not be traced.

GOCHENAUR: Why do so many different teleomorphs develop similar anamorphs?

MALLOCH: There are actually quite a variety of anamorphs, which could be separated into more genera than are presently recognized. In many cases it is possible in Aspergillus to recognize the teleomorph genus by looking at the anamorph.

SEIFERT: The yellow to red or purple KOH reaction which occurs in some species of Nectria also occurs in Emericella, Thermoascus and Penicilliopsis. Do you know if it occurs in Trichocoma or in the Trichocomoideae?

MALLOCH: I am familiar with it in Emericella. I've mounted Trichocoma in KOH, and haven't seen any colour change. I also haven't seen it in Talaromyces.

SAMSON: What is your opinion of two similar teleomorphs, which produce two very different anamorphs, one in Aspergillus, one in Penicillium? For example, Hamigera with a Merimbla anamorph, is very similar to Warcupiella with a Raperia anamorph. A similar situation occurs with Neosartorya and Talaromyces.

MALLOCH: If you accept the idea that there is a primitive type, a stroma containing cleistothecia, and that it may reduced down in various ways, either by loss of the cleistothecium, or by loss of the stroma, then why not accept two genera being reduced in parallel way, so that both finish similar cleistothecia, but one with a Penicillium anamorph and one with an Aspergillus?

I don't think Talaromyces has a true cleistothecium; there is no peridium. The mycelium forms first, and then the ascogonia form within. In Neosartorya, cleistothecia are initiated with a nice coil.

SAMSON: In Thermoascus, there are also three anamorphs: Polypaecilum, Paecilomyces and one chlamydosporic type.

MALLOCH: Thermoascus seems a very nondescript and possibly polyphyletic genus.

SAMSON: What do we do with the term sclerotium?

MALLOCH: I don't believe that what we call a sclerotium in Penicillium or Aspergillus is functionally the same as a sclerotium in the genera Sclerotium, Typhula, Sclerotinia, Cordyceps or Claviceps. In these fungi, the sclerotium germinates to form a stipe which grows and produces the teleomorph externally.

ONIONS: The only difference between these and what we call sclerotia in Eupenicillium is that in this case the asci are produced internally.

WICKLOW: In Aspergillus flavus the sclerotia germinate by producing conidiophores on the surface.

ONIONS: Isn't this the same thing? Functionally, as potentially producing teleomorphs they are certainly the same.

GAMS: I agree that sclerotium is heterogeneous and a poorly defined term. Some of us want to replace it with stroma, which is also a poorly defined term. I would like to draw the term bulbil into the discussion.

MALLOCH: Isn't that a term sometimes used for immature ascomata of *Melanospora*? And the genus *Burgoa*? They are not ascogenous, are they?

GAMS: No, a bulbil is just a globose structure composed of homogeneous cells.

MALLOCH: The sclerotium in *Penicillium* is functionally similar to immature stromata of *Pleospora*, which later develop asci inside with no ascomal peridium. Most people would call this an ascostroma.

GAMS: When we talk of sclerotia in *Penicillium* we mean those structures which do not develop into ascomata.

MALLOCH: But they are functionally exactly the same as the ones that do.

GAMS: Potentially!

MALLOCH: What you are suggesting is, that we call them sclerotia if they do not develop asci, and something else if they do.

WICKLOW: I have used the term sclerotioid cleistothecium for such structures, but Dr. Malloch suggested I should use the term immature sclerotioid ascostromata for sclerotia of *Eupenicillium*. When mature, these become sclerotioid ascomata. Dr. Malloch has pointed out that the term cleistothecium is inappropriate for these structures because the development is different.

MALLOCH: The ascoma of an *Eupenicillium* is developmentally not a cleistothecium, and I don't like this term used for it. The term I prefer is sclerotioid ascostroma.

PITT: The structure really under discussion is that produced by *Penicillium thomii* or *Aspergillus flavus*, which, as far as we know, never produce asci. With these we either have to stay with the term sclerotium, which is generally accepted, or coin a completely new word because neither stroma nor bulbil is appropriate here.

MALLOCH: It doesn't bother me if you call what *P. thomii* produces a sclerotium.

NOMENCLATORIAL AND TAXONOMIC PROBLEMS IN THE GENUS EUROTIUM

John I. PITT

CSIRO, Division of Food Research
North Ryde, N.S.W. 2113 Australia

SUMMARY

The nomenclature of the genus Eurotium is discussed in relation to changes brought about by the recent revision of the Botanical Code. Other nomenclatorial problems, relating to errors in typification, species with closely similar anamorphs or teleomorphs, and old valid names, are also discussed. The taxonomy of the genus is partially revised.

INTRODUCTION

Eurotium Link 1809 is a well circumscribed genus of Ascomycetes characterised by yellow or rarely white cleitothecia with glabrous, pseudoparenchymatous walls and Aspergillus anamorphs in which stipes bear phialides only. All species are xerophilic. The genus was typified by Malloch and Cain (1972), with E. herbariorum Link as neotype. Eurotium was monographed most recently by Blaser (1975), but the taxonomy of Raper and Fennell (1965), where it is described under the name "Aspergillus glaucus group", is still widely used. Samson (1979) has provided a very useful compilation and critique of species described since 1965.

This study was carried out to critically reappraise the nomenclature and taxonomy of the genus, especially in the light of changes to the Internatonal Code of Botaical Nomenclature (Voss et al., 1983) made at the International Botanical Congress in Sydney in 1981 (Korf, 1982; Hawksworth, 1984). A second aim which developed during the study was to highlight other nomenclatorial and taxonomic problems in Eurotium, and to seek concensus on the most acceptable approach to them.

Two changes in the Sydney Code particularly affect the nomenclature of Eurotium: first, the revision to the 1753 starting point date (Korf, 1982) and second, the clarification of nomenclature of teleomorphic fungi - on the one hand where the name originally given was in an anamorph genus but was applied to a teleomorph as well as to the anamorph, and on the other, the correct anamorph name for a species originally described as an Ascomycete. Changes brought about by the earlier starting point date are, I believe, uncompilated,. Names such as *Eurotium*, *E. herbariorum* and *Aspergillus glaucus* can now be ascribed to Link alone. However some explanation of the effects of the second change are necessary.

To circumvent the difficulties inherent in earlier Codes, where up to three separate conclusions could be drawn about names for a single telemorph-anamorph species (Hawksworth and Sutton, 1974), the Sydney Code revised the principles of Article 50 to produce unequivocal names for such species, at the expense of renaming a number of anamorphs, especially in *Penicillium* and *Aspergillus* (Hawksworth, 1984). If a species named as a *Penicillium* or *Aspergillus* includes the teleomorph, and is subsequently transferred to a teleomorph genus such as *Eupenicillium* or *Eurotium*, the original epithet cannot be retained for the anamorph, because it originally referred to the teleomorph as well. For example, the description of *P. javanicum* van Beyma included both anamorph and teleomorph, and was correctly transferred to *Eupenicillium* as *E. javanicum* (van Beyma) Stolk and Scott (1967). The name *P. javanicum* is thus no longer available for the *Penicillium* state alone, because the protologue includes the teleomorph; Pitt (1979) provided the new name *P. indonesiae* for this species.

On the other hand, where a species named as a *Eurotium* includes the *Aspergillus* anamorph, the *Eurotium* name can be adopted for the anamorph by publishing it as a new species, provided that the name has not already been combined, inclusive of the teleomorph, into *Aspergillus*. For example, Mangin (1909: 361) published *E. chevalieri*, including both anamorph and teleomorph. The name *A. chevalieri* sp. nov. can be described for the anamorph, provided that the name *A. chevalieri* (Mangin) does not exist. In this case, Thom and Church (1926: 111) made the combination *A. chevalieri* (Mangin) Thom & Church, clearly intending the *Aspergillus* name to include the *Eurotium* state. The name *Aspergillus chevalieri* is therefore not correct under the Sydney Code and must be replaced.

Situations of these two types occur throughout *Eurotium*, necessitating name changes for a number of *Aspergillus* anamorphs. Some of the existing synonyms provide an available name; others, however, include the teleomorph and are unavailable. New names will have to be introduced for some species.

Various other problems have been found as this study has progressed. First, the nomenclature of the teleomorph name may be confused; for example Table 1 shows the authorities for *Eurotium repens* in common usage. Second, names in common use do not always relate to the original description, e.g. *E. amstelodami*. Third, some *Eurotium* species posses virtually identical anamorphs, and some *Aspergillus* species produce identical teleomorphs. How should the taxonomy of such species be handled? Fourth, old, valid names exist for some accepted species. Thom and Church (1926: 101) listed some 35 names of species which they regarded as "doubtfully identifiable and inadequately described". What should be done about these? Some do not refer to *Eurotium* at all and many should undoubtedly

Table 1. Authorities for *Eurotium repens* in current and recent use

Thom and Church (1926)	*Eurotium repens* (Corda) de Bary & Woronin
Thom and Raper (1945)	*Aspergillus repens* (Corda) de Bary
Raper and Fennell (1965)	*A. repens* de Bary
Benjamin (1955) and Blaser (1975)	*E. repens* de Bary
CMI catalogue	*E. repens* Corda
CBS catalogue	*E. herbariorum* (Wiggers:Fr.) Link
ATCC catalogue	*E. repens* Biourge

remain forgotten. However a few may exist in herbaria and, if so, should be examined.

This study also aimed to reappraise currently recognised species, i.e. those accepted by Raper and Fennell (1965), Blaser (1975), Samson (1979) and Domsch et al. (1980), to provide a new taxonomy.

Some of these aims remain incomplete, particularly the appraisal of old names and the resolution of problems areas, the latter because the author sought the consensus of the participants at the workshop where this paper was presented. This paper must be viewed as a progress report, not a final document. In concentrating on nomenclature here, only a brief description of the species discussed has been given. For more complete descriptions, see Raper and Fennell (1965), Samson (1979) or, for commonly occurring species, Pitt and Hocking (1985).

The following species are discussed below, and accepted: *Eurotium amstelodami* Mangin (although the name itself is doubtful); *E. chevalieri* Mangin; *E. cristatum* (Raper and Fennell) Malloch & Cain; *E. echinulatum* Delacroix; *E. herbariorum* (Wiggers) Link; *E. repens* (Corda) de Bary; and *E. rubrum* Konig et al.

The following species are discussed and reduced to synonymy: *E. appendiculatum* Blaser; *E. intermedium* Blaser [(synonym *E. chevalieri* Mangin var. *intermedius* (Thom and Raper) Malloch & Cain)]; *E. montevidense* (Talice & Mackinnon) Malloch & Cain; *E. spiculosum* Blaser; and *E. umbrosum* (Bain. & Sart.) Malloch & Cain. The taxonomy of *E. glabrum* Blaser, *E. heterocaryoticum* Christensen et al., and *E. pseudoglaucum* (Blochwitz) Malloch & Cain is discussed. The following valid species are not considered here, as being both rare and noncontroversial: *E. carnoyi* Malloch & Cain; *E. halophilicum* Christensen et al.; *E. leucocarpum* Hadlok & Stolk; *E. medium* Meissner; *E. niveoglaucum* (Thom & Raper) Malloch & Cain; *E. tonophilum* Ohtsuki; and *E. xerophilum* Samson & Mouchacca.

METHODS

Type or neotype isolates of all currently accepted species were examined, with the exceptions of *E. acutum* Blaser and *E. niveoglaucum* (Thom & Raper) Malloch & Cain. Other authentic cultures, and isolates from the FRR collection, were also studied. Cultures were grown comparatively on a variety of media (Pitt and Hocking, 1985); descriptions are given here on Czapek Yeast Extract Agar with 20% sucrose (CY20S), which has the following formulation: yeast extract, 0.5%; $NaNO_3$, 0.3%; K_2PO_4, 0.1%; $MgSO_4.H_2O$, 0.05%; KCL, 0.05%; $FeSO_4.7H_2O$, 0.001%; sucrose, 20%; agar 1.5%, all (v/v) in distilled water. The medium is autoclaved for 15 min. at 121°C.

Cultures were incubated at 25°C and examined at 7 and 14 days. Where necessary, incubation was continued for longer periods until mature ascospores were observed.

CY20S is identical with the medium used by Raper and Fennell (1965) for Eurotium taxonomy, except for the addition of yeast extract, which promotes vigorous colony development with little or no effect on colony morphology. Colonies grown on this medium can be readily compared with descriptions given by Raper and Fennell (1965) if desired.

Capitilised colour names and colour terminology are taken from Kornerup and Wanscher (1978).

EUROTIUM AMSTELODAMI Mangin - Annls Sci. nat. botan., Ser. 9, 10: 360, 1909.

Anamorph: Aspergillus amstelodami (Mangin) Thom & Church - Aspergilli: 113, 1926 (invalid name, includes teleomorph)

= Eurotium montevidense (Talice & Mackinnon) Malloch & Cain, Can.J. Bot. 50: 64, 1972.
= Aspergillus montevidensis Talice & Mackinnon, Compt. rend. Soc. biol. Fr. 108: 1007, 1931 (nom. inval., includes teleomorph).

Colonies on CY20S at 7 days usually 45-55 mm diam, at 14 days 60-70 mm, at both times plane, low and velutinous, usually with a layer of yellow cleistothecia near the agar surface surmounted by a dense layer of radiate conidial heads, Dull Green (27-28E3-4) at 7 days, becoming darker green in age; reverse in similar colours to the observe, i.e. yellow beneath cleistothecia, and pale grey green under conidial areas.
Cleistothecia bright yellow, mostly spherical, 110-150 µm diam, not usually enveloped in sterile hyphae, maturing in 9-12 days; ascospores yellow, ellipsoidal, 4.5-5.0 µm long, with rough walls, and with two conspicuous, often sinuous, broad, longitudinal ridges of low but irregular heigh and spacing.
Conidiophores borne from aerial hyphae, stipes relatively short, 300-400 µm long, vesicles spherical to spathulate, 18-30 µm diam, fertile over the upper two-thirds to three quarters, bearing phialides only; phialides 5-7 µm long; conidia spherical to subspheroidal 4.0-5.0 µm diam, with densely spinulose walls.

Typification. Blaser (1975) designated CBS 518.65 as neotype of this species. This strain was derived from NRRL 90 (=WB 90), regarded by Raper and Fennell (1965) as representative of this species.

Distinguishing characteristics. The principal diagnostic of *E. amstelodami* is the conspicuous ascospore ornamentation: low, wide, irregular flanges and rough walls. Mycelium remains white or conspicuous, and colony colours are made up of yellow from the cleistothecia and dull green from the conidia.

Taxonomy. On examination of the protologue of Mangin (1909) for this species, it is immediately clear that the fungus he was describing is not that which has more recently come to be regarded as *E. amstelodami*. Mangin's species appears to be closely related to *E. repens*: he described the ascospores as "lenticulaires, hyalines, lisse", i.e. smooth walled. The description by Thom and Church (1926) states "ascospores with furrow only", and indicates that only the presence of the furrow separates the species from *E. repens*. Moreover, they stated that *A. scheelei* Bain. & Sart., a teleomorphic species placed by Raper and Fennell (1965) in synonymy with *E. repens*, is "close if not identical with *A. amstelodami*". However, when Thom and Raper (1941) described *A. amstelodami*, ascospores were stated to have a "prominent V-shaped equatorial furrow and broad irregular ridges". One of the cultures they cited as representative was the same number (109) as that examined by Thom and Church (1926). It can only be speculated that the original had been replaced since 1926 by the species they described in 1941. Although culture no. 109 was not cited by Raper and Fennell (1965), the error of 1941 has been perpetuated. Neotypification is in error also, as a result. Both name and neotype must be rejected, and a search made for alternative names.

A. montevidensis, a teleomorphic species, has been accepted by Thom and his coworkers over the years on the grounds that it was of medical significance. *A. montevidensis* was isolated from a human ear drum and was reputed to be pathogenic. Thom and Raper (1941) and subsequent authors have pointed out that it is only a minor variant of *E. amstelodami*, with particularly heavy production of the anamorph. The name has rarely been used for other than the original type culture, and should be placed in synonymy. *E. montevidense* is a possible replacement name for *E. amstelodami*, but has not been used here because of the possibility of earlier valid and identifiable names. *A. montevidensis* cannot be used for the anamorph, because it was described as a holomorph.

EUROTIUM CHEVALIERI Mangin - Annls Sci. nat. botan., Ser. 9, 10: 361, 1909.

Anamorph: *Aspergillus chevalieri* (Mangin) Thom & Church - Aspergilli: 111, 1926 (nom. inval., includes teleomorph).

Colonies on CY20S at 7 days 45-60 mm diam, at 14 days similar, plane, low and dense; colony character varying, in isolates with sparse conidial heads, mycelium white at the margins, then Bright to Deep Yellow (3A6; 4A8), often becoming Deep Orange (5A8) centrally, enveloping abundant cleistothecia, in isolates with heavy conidia, Dull Green near Cactus Green (28-29E3-4); reverse pale, yellow, orange brown or dull green.

Cleistothecia bright yellow, spherical, 100-140 µm diam, enveloped in yellow to orange vegetative hyphae, maturing at colony centres in 8-10 days; ascospores yellow, ellipsoidal, shaped like pulley wheels, 4.5-5.0 x 3.5-3.8 µm, with two prominent, parallel, sometimes sinuous, narrow, longitudinal crests, and smooth walls.

Conidiophores borne from aerial hyphae, stipes mostly 500-800 µm long, thin walled, already collapsing in 7 days, broadening to vesicles 25-35 µm diam, fertile over the whole area, bearing phialides only; phialides ampulliform, 5-8 µm long, conidia ellipsoidal or doliiform (barrel-shaped), 4.0-5.5 µm long, with spinose walls.

Typification. Blaser (1975) designated CBS 522.65 as neotype of *E. chevalieri*.

Distinguishing characteristics. As with most other *Eurotium* species, the ascospores is the prime distinguishing feature: in *E. chevalieri* these can characteristically shaped like minute pulley wheels. Colonies grow quite rapidly on CY20S, and have yellow to orange hyphae.

Taxonomy. This species was originally described as a *Eurotium*. The *Aspergillus* anamorph could also be correctly known as "*chevalieri*" except that Thom and Church (1926) combined *E. chevalieri* into *Aspergillus*, for both anamorph and teleomorph. A valid anamorph name is being sought.

EUROTIUM REPENS de Bary - Abhandl. Senckenberg. naturforsch. Ges. 7: 379, 1870.

Anamorph: *Aspergillus repens* (de Bary)Fischer - Abhandl. Senckenberg. naturforsch. Ges. 7: 379, 1870.

= *Aspergillus glaucus* var. *repens* Corda -- Icon. Fung. 5: 53.1839*

Colonies on CY20S at 7 days 45-65 mm diam, at 14 days 60 mm or more, at both times plane, low and dense or somewhat floccose; mycelium white to yellow or orange, overall colour varying from yellow with scattered dull green areas in predominantly cleistothecial isolates to dull green or bluish green in those with predominant conidial heads; reverse dull green or brigh yellow to orange or both.

Cleistothecia bright yellow, spherical, 75-100(-125) µm diam, maturing at colony centres in 10-14 days; ascospores yellow, ellipsoidal, 5.0-5.5 x 4.0-4.5 µm, smooth walled, with a trace of a longitudinal furrow, but without ridges or crests.

Conidiophores borne from aerial hyphae, stipes mostly 500-1000 µm long, broadening to a vesicle 15-30 µm diam, fertile over the whole area, bering phialides only; phialides ampulliform, 7-0 µm long; conidia spherical to subspheroidal, 5-7 µm diam, with densely spinulose to spinose walls.

Typification. CBS 529.65 was designated by Blaser (1975) as neotype of *Eurotium repens*.

---

* see also *Aspergillus reptans*, nom. nov. on page 48.

Distinguishing characteristics. The prime distinction of *Eurotium repens* from other species in the genus is the production of ascospores with only a trace of longitudinal furrow, no crests or ridges, and smooth walls. Colonies grow rapidly on CY20S; hyphal and reverse colours are yellow to orange, never red.

Taxonomy. The correct citation of *E. repens* out of those in current use (Table 1) appears to be *E. repens* de Bary. Corda's original description of *Aspergillus glaucus* var. *repens* appears to refer only to the anamorph. Under the Sydney Code, if an author introduces a new element into a description, and consequently introduces a new generic name, the name is regarded as new even if presented as a combination. The anamorph is correctly treated as a combination, *A. repens* (Corda) de Bary.

EUROTIUM RUBRUM König et al.- Z. Untersuch. Nahrung. Genussm. 4: 726, 1901.

Anamorph: *Aspergillus sejunctus* Bain. & Sart.- Bull. trimest. Soc. mycol. Fr. 27: 346, 1911 (invalid name, includes teleomorph).

= *Aspergillus ruber* (König *et al.*) Thom & Church, Aspergillus: 112, 1926 (invalid name; includes teleomorph)

Colonies on CY20S at 7 days 45-60 mm diam, plane or lightly sulcate, usually low, dense and velutinous; mycelium conspicuous, at the margins yellow (2A6), becoming Orange to Deep Orange (6A6-8) or more reddish elsewhere, enveloping abundant yellow cleistothecia, and surmounted by sparse to abundant dull green radiate heads: reverse in colours similar to the mycelium, though rarely so bright, or deep yellow brown; in age, from 10 days to many weeks, developing bright red brown hyphal colours, Brick Red or Venetian Red (7-8D-E7-8) over the whole plate or in patches, not developed under densely conidial areas or rarely not at all, remaining orange; reverse darkening to deep brown or almost black.

Cleistothecia yellow, borne from and enveloped in sterile orange to red hyphae, spherical, 100-140 µm diam, maturing in 9-12 days; ascospores ellipsoidal, yellow, 5.0-6.0 x 4.2-4.8 µm, with a shallow longitudinal furrow flanked by low minutely roughened ridges, otherwise with smooth walls.

Conidiophores borne from aerial hyphae, stipes mostly 500-700 µm long, terminating in spherical vesicles, fertile over the upper two-thirds, bearing phialides only; phialides 7-9 µm long; conidia subspheroidal to ellipsoidal. less commonly spherical or pyriform, comonly 6-7 µm long, larger in occasional isolates, with spinose walls.

Typification. This species lacks a type isolate, and has not been neotypified.

Distinguishing characteristics. Colonies *E. rubrum* on CY20S usually show areas of brilliant red or rusty colours after 10 days or more incubation. Ascospores are 5.5-6.0 µm long, rarely otherwise, and have a definite longitudinal furrow and low ridges, minutely roughened.

Taxonomy. The earliest name for the anamorph of this taxon appears to be *A. sejunctus*. However it is not available because it was described inclusive of the teleomorph. No other valid anamorph name has yet been found, but searches are incomplete.

EUROTIUM HERBARIORUM (Wiggers) Link - Mag. Ges. Naturf. Fr. Berlin 3. 31. 1809.

Anamorph: *Aspergillus glaucus* Link - Mag. Ges. Naturf. Fr. Berlin 3. 16. 1809.

= *Mucor herbariorum* Wiggers, Primitiae Florae Holsatiae: 111, 1780.
= *Eurotium umbrosum* (Bain. & Sart.) Malloch & Cain, Can. J. Bot.50: 64, 1972.
= *Eurotium herbariorum* ser. *minor* Mangin, Annls Sci. nat. bot., Ser. 9, 10: 365, 1909.
= *Aspergillus umbrosus* Bain. & Sart., Bull. trimest. Soc. mycol. Fr. 28: 267, 1912.
= *Aspergillus herbariorum* (Link) Fischer ser. minor (Mangin) Thom & Church, Aspergilli: 110, 1926 (*Nom. inval.*, includes teleomorph).
= *Aspergillus minor* (Mangin) Thom & Raper, US Dep. Agric. Misc. Publ. 426: 27, 1941 (*nom. inval.*, includes teleomorph).
= *Aspergillus mangini* Thom & Raper, man. Aspergilli: 127, 1945 (*nom, inval*., Arts 36, 59).

Colonies on CY20S 25-35 mm in diam in 7 days at 25°C, plane, rather sparse, very deep and floccose, with developing cleistothecia surrounded by Yellowish Brown (5D-E6) or Orange to Brownish Orange (6-7A-C6) hyphae and surmounted by scattered radiate heads, dull bluish green; reverse uncoloured or pale yellow; at 14 days 60 mm or more diam, plane, deep and floccose, mycelium yellow (2A6) to Orange (5A6-7), of intermingled cleistothecia and conidial heads at several levels, or with cleistothecia surmounted by conidial heads; reverse uncoloured, yellow brown or centrally deep brown.

Cleistothecia yellow, 140-180 μm diam, maturing in 12-16 days; ascospores ellipsoidal, (5-)6-8 x (4.2)5-6 μm, with walls smooth to very finely roughened, and with a well defined though shallow longitudinal furrow flanked on either side by a low ridge, sometimes irregularly roughened.

Conidiophores usually 700-800 μm long, colourless or brown, smooth-walled, enlarging to vesicles 25-40 μm diam, bearing phialides over the upper half; phialides 8-10 μm long; conidia usually ellipsoidal, 7-9 μm long, with spinose walls.

Typification. DAOM 137960 was designated as neotype of *E. herbariorum*, and hence of the genus *Eurotium*, by Malloch and Cain (1972).

Distinguishing characteristics. *E. herbariorum* grows relatively slowly on CY20S, less than 40 mm in 7 days, and produces yellowish brown colonies. Ascospore mature in about 2 weeks and are similar in appearance to those of E. rubrum, or have very small irregular crests on the longitudinal ridges. They are larger than those of E. rubrum, commonly being 6-8 µm long.

Taxonomy. Blaser (1975) reduced E. umbrosum to synonymy with E. herbariorum, a disposition with which I agree. Examination of several cultures representative of both species showed that isolates labelled E. herbariorum had smaller ascospores, while those of E. umbrosum generally were browner and less orange in colour. However, differences were slight and not consistent.

Blaser (1975) placed E. rubrum in synonymy with E. herbariorum. Domsch et al. (1980) disposed of both E. rubrum and E. repens in a similar fashion. However, there are consistent and usually clear cut differences among these three species. Colonies of E. herbariorum grow more slowly on CY20S, and take longer to produce mature ascospores, which are consistently larger than those of the other two species. E. herbariorum lacks the red hyphae of E. rubrum, and its hyphal colours are more brown than those of E. repens.

EUROTIUM CRISTATUM (Raper and Fennell) Malloch & Cain - Can. J. Bot. 50: 64, 1972.

Anamorph: Aspergillus spiculosus Blaser - Sydowia 28: 42, 1975.

= Eurotium spiculosum Blaser, Sydowia 28: 42, 1975.
= Eurotium intermedium Blaser, Sydowia 28: 41, 1975 = Eurotium chevalieri var. intermedium (Thom and Raper) Malloch & Cain, Can. J. Bot. 50: 64, 1972 (nom.inval.; basionym not validy published) = Aspergillus chevalieri var. intermedius Thom & Raper, US Dep. Agric.Misc. Publ. 426: 21, 1941 (nom.inval., includes teleomorph)
= Aspergillus cristatus Raper & Fennell, gen. Aspergillus: 169, 1965 (nom.inval., includes teleomorph).
= Aspergillus intermedius Blaser, Sydowia 28: 41, 1975.

Colonies on CY20S in 7 days at 25°C 40-60 mm diam, plane, low and dense; mycelium Vivid Yellow to Deep Yellow (2A8-4A-B8), varying from sparse wefts to dense layers surrounding or enveloping developing cleistothecia; conidial heads radiate or irregular, usually sparse, sometimes entirely absent, Dark Green (27E4); reverse occasionally bright yellow, usually dull yellow brown; in age colours deepening to yellow brown (4 1/2B4-5).

Cleistothecia yellow, spherical, 100-200 µm diam, maturing rapidly, usually within 7 days; ascospores ellipsoidal, yellow, bodies 4.0-5.5 x 3.5-5.0 µm, ornamented with two fairly broad, conspicuous crests, but lacking a furrow between, and with walls irregularly bearing small but conspicuous spikes.

Conidiophores varying with isolate, 200-800 µm long, with smooth walls; vesicles 20-35 µm diam, bearing phialides only; phialides 5-7 µm long; conidia ellipsoidal to short cylindroidal, 3.5-4.5 µm long, typically spinose, often with spines arranged in longitudinal lines, evident in catenate conidia, sometimes finely spinulose or apparently smooth-walled.

Typification. Blaser (1975) designated CBS 123.53 as type of E. cristatum, but wrongly attributed the isolate to K.B. Raper, rather than H.J. Swart.

Distinguishing characteristics. The ascospores of E. cristatum have the general appearance of those of E. amstelodami, but are distinctive because crests are much prominent, and walls are distinctly spiked rather than rough. Some isolates produce almost smooth conidia, but the typical form is spinose, with the spines arranged in rows paralel to the conidium axis. Conidia of A. amstelodami are sometimes rather similar.

Taxonomy. When the variety A. chevalieri var. intermedius was originally introduced by Thom and Raper (1941), the most distinctive feature was stated to be the small, smooth walled conidia, and also ascospores intermediate between E. amstelodami, and E. chevalieri. Examination of the type isolate of this variety, raised by Blaser (1975) to species level, showed a proportion of conidia to be spinose, with a distinctive pattern of spines, and strikingly similar to those of E. spinulosum. The ascospores als proved to be similar, with the same very distinctive ornamentation. The presence of smooth-walled conidia in the type of E. intermedium is regarded here as an unstable variant, and all three species brought together. Eurotium cristatum has priority for the teleomorph. A. cristatus is unavailable: of the two names introduced by Blaser (1975), A. spiculosus has been chosen here, because it is more stable in culture and better represents the conidial type which I believe to be characteristic of this species.

EUROTIUM ECHINULATUM Delacr.- Bull. Soc. mycol. Fr. 9: 266, 1893

Anamorph: Aspergillus brunneus Delacr.- Bull. Soc. mycol. Fr. 9: 185, 1893.

= Aspergillus echinulatus (Delacr.) Thom & Church, Aspergilli: 107, 1926(nom.inval.,includes teleomorph)
= Eurotium appendiculatum Blaser, Sydowia 28: 38, 1975.
= Aspergillus appendiculatus Blaser, Sydowia 28: 38, 1975.

Colonies on CY20S after 7 days at 25 C variable, 20-45 mm diam, plane or sulcate, usually consisting of inconspicuous mycelium bearing densely packed dull to dark green conidial heads in patches or over the whole surface; reverse pale orange or brown; at 14 days, 60 mm or more diam, sometimes irregular in outline, deep; mycelium in a basal layer, Bright Yellow to Reddish Orange (4-7-A6-8), enveloping yellow cleistothecia, concealed generally or in patches beneath layers or radiate conidial heads, sometimes pale green, usually Dark Green to Olive (26E4-2F4), often with randomly distributed cleistothecia and sparse orange hyphae visible also; reverse pale to yellow or brown.

Cleistothecia yellow, spherical, 75-150 µm diam, maturing in 2.5 to 3 weeks; ascospores ellipsoidal, commonly 8-10 x 6-7 µm, smaller in occasional isolates, with a slight longitudinal furrow flanked by two crests, low, narrow and sinuous, sometimes discontinuous, with walls smooth or with minute spines near the crests.

Conidiophores 700-1000 µm long, with relatively thin, smooth walls, broadening towards the apex; vesicles only 1.5 to 2 times stipe width, 25-35 µm diam, bearing phialides only; phialides uncrowded, 12-15 µm long; conidia of variable shape and size, usually ellipsoidal to pyriform and 8-10 µm long, conspicuously spinose.

Typification. As neotype of *E. echinulatum*, Blaser (1975) designated CBS 524.65.

Distinguishing characteristics. Besides producing the largest ascospores in the genus, which have characteristic low, irregular crests, *E. echinulatum* grows relatively slowly on CY20S, with yellow to dull red mycelium, and produces conidial heads which turn dull yellow green to olive in age. The ascospores, which take a little longer to mature on CY20S than the common, small spored species, are the principal diagnostic feature.

Taxonomy. Delacroix (1893 a, b) validly described both the teleomorph and anamorph names for this species: *Eurotium echinulatum* and *Aspergillus brunneus*. Apart from its recent devalidation due to changes in Art. 50 of the International Botanical Code, the commonly used name *A. echinulatum* has always been a synonym of *A. brunneus*.

Blaser (1975) described *E. appendiculatum* as having large ascospores (5.5-6.9 µm long plus crests) with a shallow furrow, and hyaline, filiform, flexuous appendages attached irregularly to the spore body. He noted that culturally the species resembled *E. echinulatum*. Examination in this study showed that while a proportion of the ascospores produced very irregular spike-like ornamentations, a considerable proportion of them, ca 40%, produced the low irregular crests characteristic of *E. echinulatum*. My interpretation of these is that this particular isolate has a slightly defective system for adhesion of crests to the ascospores - a phenomenon occasionally observed with other Ascomycetes species. Colony appearance and colouring, and growth rate on CY20S are all indicative of *E. echinulatum*. Moreover the anamorph has the broadening stipes and small vesicles characteristic of *A. brunneus*. *E. appendiculatum* is placed in synonymy.

PROBLEM SPECIES

Lastly, I wish to discuss three species, all of which produce teleomorphs similar to, even indentical with, earlier teleomorph species, but produce distinctive anamorphs. Guidance on their nomenclature is sought.

Eurotium heterocaryoticum Christensen et al.- Mycologia 57: 535-542 (1965) is a curious fungus, because it produces brown conidia not found in any other *Eurotium* species. Moreover it also produces white conidia as sectors or as a stable character. Many of the conidia are smooth walled, or have very short spines, others are spinose.

The appearance of each kind of colony is very distinctive. However, both kinds produce ascospores characteristic of *E. amstelodami*, and colony growth rates and mycelial colouration are consistent with the species having a perfectly normal *E. amstelodami* teleomorph.

Two questions arise: first, should the teleomorph *E. heterocaryoticum* be placed in synonymy with *E. amstelodami*? And second, is the anamorph, which despite its very distinctive appearance probably differs in only a few genes from the anamorph of *E. amstelodami*, worthy of recognation as a species?

The first question hinges on interpretation of the meaning of the name *E. amstelodami*. If we refer only to the teleomorph, it seems logical that the teleomorph of *E. heterocaryoticum* is a synonym. However, if by the name *E. amstelodami* we are referring to the holomorph, then recognition of *E. heterocaryoticum* has some validity, as it clearly has some differences from the holomorph *E. amstelodami*. As to the second, the author feels that while taxonomy of these fungi remains based on visual appearance, the case for recognition of *A. heterocaryoticum* is adequate and it should be recognised.

Eurotium pseudoglaucum (Blochwitz) Malloch and Cain. Raper and Fennell (1965) and Blaser (1975) maintained *E. pseudoglaucum* as a separate species. However comparative cultural and microscopic studies indicate that the teleomorph, colony colours and colony growth rates make this species inseparable from *E. repens*. However, the anamorph, *Aspergillus pseudoglaucus* Blochwitz, has unusually small conidiophores, and is readily distinguishable from *A. repens*. There are two problems. First, it is justifiable to reduce the teleomorph to synonymy and regard the anamorph as distinct? Is this good, practical taxonomy? And second, *A. pseudoglaucus* was described inclusive of the teleomorph, and is an incorrect name. If this anamorph is maintained, it will have to be given a new name under the Sydney Code. Would, then, this quite rare morphological anamorph ever be recognised again?

Eurotium glabrum Blaser was distinguished from *E. repens* primarily on the basis of its smooth conidia (Blaser, 1975). Examination in this study confirmed his viewpoint: the teleomorph, colony colours and growth rates are typical of *E. repens*. Taxonomically, a similar problem occurs as with *E. pseudoglaucum*. Should *E. glabrum* be reduced to synonymy? If so, should *Aspergillus glaber* be regarded as a second anamorph of *E. repens*, along with *A. repens*? In this case, the taxonomic decision is simplified by the fact that the name *A. glaber* Blaser is valid.

REFERENCES

BLASER, P. 1975. Taxonomische und physiologische Untersuchungen uber die Gattung Eurotium Link ex Fries. Sydowia 28: 1-49.

DELACROIX, G. 1893 a. Espèces nouvelle observées au Laboratoire de Pathologie vegetale. Bull. Soc. mycol. Fr. 9: 184-188.

DELACROIX, G. 1893 b. Champignons parasites nouveaux. Bull. Soc. mycol. Fr. 9: 264-268.

DOMSCH, K.H., GAMS, W. and ANDERSON, T.H. 1980. "Compendium of Soil Fungi". London: Academic Press. 2 vol.

HAWKSWORTH, D.L. 1984. Recent changes in the international rules affecting the nomenclature of fungi. Microbiol. Sci. 1: 18-21.

HAWKSWORTH, D.L. and SUTTON, B.C. 1974. Article 59 and names of perfect state taxa in imperfect state genera. Taxon 23: 563-568.

KORF, R.P. 1982. Mycological and lichenological implications of changes in the code of nomenclature enacted in 1981. Mycotaxon 14: 476-490.

KORNERUP, A. and WANSCHER, J.H. 1978. "Methuen Handbook of Colour". 3rd edn. London: Eyre Methuen.

MALLOCH, D. and CAIN, R.F. 1972. The Trichocomataceae: Ascomyetes with Aspergillus, Paecilomyces, and Penicillium imperfect states. Can. J. Bot. 50: 2613-2628.

MANGIN, M.L. 1909. Qu'est-ce que l'Aspergillus glaucus? Etude critique et experimentale des formes groupees sous ce nom. Ann. Sci. nat. Bot., Ser. 9, 10: 303-71.

PITT, J.I. 1979. "The Genus Penicillium and its Teleomorphic States Eupenicillium and Talaromyces". London: Academic Press.

PITT, J.I. and HOCKING, A.D. 1985. "Fungi and Food Spoilage". Sydney: Academic Press (in press).

RAPER, K.B. and FENNELL, D.I. 1965. "The Genus Aspergillus". Baltimore: Williams and Wilkins.

SAMSON, R.A. 1979. A compilation of the Aspergilli described since 1965. Stud. Mycol., Baarn 18: 1-38.

STOLK, A.C. and SCOTT, D.B. 1967. Studies on the genus Eupenicillium Ludwig. 1. Taxonomy and nomenclature of the Penicillia in relation to their sclerotioid ascocarpic states. Persoonia 4: 391-405.

THOM, C. and CHURCH, M.B. 1926. "The Aspergilli". Baltimore: Williams and Wilkins.

THOM, C. and RAPER, K.B.. 1941. The Aspergillus glaucus group. U.S. Dep. Agric. Misc. Pub. 426: 1-46.

VOSS, E.G.et al.(eds.). 1983. "International Code of Botanical Nomenclature adopted by the Thirteenth International Botanical Congress, Sydney, August, 1981". Utrecht: Bohn, Scheltema and Holkema.

DIALOGUE FOLLOWING DR. PITT'S PAPER

PITT: My question to the group here is: If we have two Eurotium species with identical teleomorphs but different anamorphs, should we put the teleomorphic species together and keep the anamorphs separate, i.e. have one teleomorph species with two anamorphs, or should we keep the teleomorphs separate as well?

MALLOCH: If you have two similar teleomorphs with different anamorphs, you must keep them separate, because the holomorphs are different.

PITT: All right. If the holomorphs are different, the anamorphs logically must be kept separate.

SAMSON: If you have two different teleomorphs which appear to have the same anamorph, shouldn't we use the same anamorph name for both?

PITT: No, because the holomorphs are different.

GAMS: If they are really indistinguishable you are justified in lumping them if there is a need.

HAWKSWORTH: We have this problem, of course, in Trichoderma as well. M. Rifai came up with the solution of using "species aggregate" for some of these groups of anamorphs with different teleomorphs, which provided an interim solution to the problem pending further work.

PITT: It's funny how many taxonomists are more than happy to lump anamorphs together, but they never lump teleomorphs together.

GAMS: This is because teleomorph taxonomists are differentiating between different organisms.

PITT: Wait a moment. Strictly anamorphic species such as Penicillium crustosum are in no way an artificial organisms. They may not have teleomorphs but they are perfectly well developed organisms.

WICKLOW: Did you attempt to look at Aspergillus restrictus, searching for strains that might have lost their teleomorph stage?

PITT: No, I consider A. restrictus to be a long way from Eurotium. A. penicilloides appears to be more closely related to the Eurotia because it is a xerophile, but I don't think it had a teleomorph for a long time.

SAMSON: But sometimes, I have seen, that A. penicilloides strains produces sclerotia very similar to immature Eurotium cleistothecia, indicating its close relationship.

THE ASCOMYCETE GENUS PENICILLIOPSIS AND ITS ANAMORPHS

Robert A. SAMSON and Keith A. SEIFERT

Centraalbureau voor Schimmelcultures
POB 273, 3740 AG Baarn

SUMMARY

The ascomycete genus Penicilliopsis is revised, along with some associated anamorph genera. Sarophorum and Stilbodendron are accepted as anamorphs of Penicilliopsis. The anamorph genus Stilbothamnium, assumed to be a Penicilliopsis anamorph, is reduced to synonymy with Aspergillus and a new subgenus within Aspergillus is proposed to accommodate these species. The anamorph genus Pseudocordyceps, related to the above anamorphs but as yet unknown to have a teleomorph, is redescribed. New combinations proposed are Sarophorum palmicola (Henn.) Samson & Seifert, Stilbodendron cervinum (Cooke & Massee) Samson & Seifert, Aspergillus amazonensis (Henn.) Samson & Seifert, Aspergillus togoensis (Henn.) Samson & Seifert, and Aspergillus vitellinus (Ridley) Samson & Seifert. One new species, Penicilliopsis africana Samson & Seifert, is described. A key to accepted species is provided.

INTRODUCTION

The genus Penicilliopsis was described by Solms-Laubach (1887) on seeds of Diospyros macrophylla collected in the Botanical Garden of Bogor (formerly Buitenzorg) in Indonesia. The detailed description clearly depicts an ascomycete with a phialidic anamorph. Despite this, Penicilliopsis was for many years considered an anamorph genus, and most of the species included by subsequent authors were described without teleomorphs. In addition to the anamorphic "Penicilliopsis", here considered under the generic name Sarophorum Sydow, three other seminicolous anamorphic genera were monographed by Hauman (1936): Stilbodendron H. &. P. Sydow, Stilbothamnium Henn. and Pseudocordyceps Hauman. We have positively connected two of these genera, Sarophorum and Stilbodendron, with mature Penicilliopsis ascostromata, and have observed sterile, Penicilliopsis-like stromata on specimens of Stilbothamnium (here considered a synonym of Aspergillus).

Given the conspicuous synnemata produced by these species and their common occurrence in tropical rain forests, it is surprising that the group has not been critically reexamined in the past fifty years. Carmichael et al. (1980), in their compilation of Hyphomycete genera, did not illustrate any of these seminicolous fungi. It is possible that the yellow to yellow-green synnemata produced by these species are so large and conspicuous as to be considered mosses or other non-fungal cryptogams by most field mycologists. Indeed, for mycologists accustomed to studying the modest conidiomata produced by temperate Hyphomycetes, it may be difficult to believe that the gigantic and elaborate synnemata produced by Penicilliopsis species are actually anamorphic fungi.

It is tempting to speculate that the luxurious growth and large conidiomatal and conidiogenous structures of these species are similar to those of the ancestors of some of the mononematous Penicillia and Aspergilli. As the ancestors of these fungi moved from durable substrates and humid environments into more ephemeral substrates and unpredictable environments such as grains or soil, it presumably became less practical to produce such large, elaborate synnemata. The tendency of some Aspergilli inhabiting agriculturally important seeds to produce sterile stromata also suggests a relationship between these species and Penicilliopsis. These ideas are explored more fully in the section on Aspergillus subgenus Stilbothamnium below.

In addition to type specimens, we have studied many supplementary collections of tropical seminicolous fungi. This paper presents a revision of these spectacular taxa.

Key to species included in this paper

1. Mature ascomata produced..................Penicilliopsis 2

   Mature ascomata not produced.............................3

2. Ascospores 7-10 x 5.0-5.5 µm; ascogenous tissue yellow; anamorph Sarophorum ....P. clavariaeformis

   Ascospores 7.0-8.5 x 3.5-5.0 µm; ascogenous tissue brick-red; anamorph Stilbodendron..................P. africana

3. Conidiophores typically Aspergillus-like; conidiophore stipe sometimes covered with sheathing hyphae........... ....Aspergillus series Stilbothamnium..4

   Conidiophores lacking conspicuous Aspergillus vesicle.....8

4. Conidiophores mononematous..............A. erythrocephalus

   Conidiophores synnematous................................5

5. Aspergillus stipes granular, not covered by a sheath of hyphae...6

Aspergillus stipes smooth, covered by a sheath of hyphae when mature..................................A. dybowskii

6. Synnemata composed primarily of tightly packed Aspergillus conidiophore stipes which splay out to form a divergent head; metulae more than 40 µm long..........A. vitellinus

Synnemata racemosely branched, with a stipe composed of hyphae distinctly different from Aspergillus conidiophore stipes; metulae less than 30 µm long...................7

7. Conidia subglobose to ellipsoidal; metulae more than 15 µm long; distribution in South America........A. amazonensis

Conidia clavate to narrowly doliiform; metulae less than 15 µm long; distribution in Africa and Asia.. A. togoensis

8. Synnemata fertile along most of length in a hymenium-like layer..............................Sarophorum palmicola

Fertile area confined to apices of synnemata or their racemose branches..................................... 9

9. Synnemata racemosely branched; metulae not swollen...................................... Stilbodendron cervinum

Synnemata unbranched, or if branched not racemose; metulae swollen.......................Pseudocordyceps seminicola

PENICILLIOPSIS Solms-Laubach - Annls Jard. Bot. Buitenzorg 6: 53-72. 1887.

Holotype species: Penicilliopsis clavariaeformis Solms-Laubach.

Ascomata consisting of hard, stipitate stromata, single or clustered at the apex of the stipe, uni- or multiloculate, nonostiolate, yellow-green to brown, turning purple-red in KOH, yellow-green in lactic acid. Stromata with a wall of pigmented textura angularis and a cortex of hyaline textura intricata in which asci are embedded, no distinct ascocarp wall present. Asci ellipsoidal to globose, nonstipitate, evanescent, 8-spored. Ascospores ellipsoidal with hyaline longitudinal crests.

Anamorphs: Sarophorum, Stilbodendron, presumably also Aspergillus subgenus Stilbothamnium and Pseudocordyceps.

Penicilliopsis was classified in the Trichocomaceae (Eurotiales) by Malloch and Cain (1972), where it is closely related to such genera as Petromyces Malloch & Cain and Dichlaena Mont. & Durieu. All three genera produce ascospores in hard, slowly maturing ascostromata. However, the ascostromata of these two genera are nonstipitate, and hence are easily distinguished from those of *Penicilliopsis*.

Two species are accepted here in *Penicilliopsis*, although we anticipate that several more species will eventually be discovered. The teleomorphs are treated together with their anamorphs below. The majority of anamorphs included have not been connected to teleomorphs.

The presence of a pH-sensitive pigment in the stromatal walls of *Penicilliopsis* and the synnemata of its known anamorphs is worthy of note. Such KOH/lactic acid reactions are common in the Hypocreales, where they have been considered of considerable taxonomic importance (Rossman 1983), but they have not been noted to date in the Eurotiales. In *Penicilliopsis* anamorphs, these pigments occur in the hyphae of the stipe and as the synnemata age may accumulate in the metulae and phialides as well.

THE ANAMORPH GENERA

Although only two species of *Penicilliopsis* are known, the anamorphs of these species can be included in a group of similar anamorphs which are not yet known to produce teleomorphs. The anamorphs of known species of *Penicilliopsis* are referred to *Sarophorum* and *Stilbodendron*. Although we have not seen mature ascocarps associated with the anamorphs here treated in *Aspergillus* subgenus *Stilbothamnium*, immature *Penicilliopsis*-like stromata have been seen. Ascospores were observed, however, by Raciborski (1909) and by J. Nicot and R. Heim (Roquebert & Nicot, this publication). Of the anamorph genera included below, only *Pseudocordyceps* is not known to be associated with stromatic bodies.

Taken together, these anamorph genera share many similarities, and represent different combinations of several common characters (Fig. 1). The phialides of all these species are practically identical. They are light yellow, cylindrical or slightly clavate with an abruptly tapering apex, a cylindrical to slightly flaring collarette, and conspicuous periclinal thickening. Generally they are quite large, usually at least 15-20 µm long and 5 µm wide. The phialides are arranged on metulae which are cylindrical to slightly clavate in some of the genera, and remarkably swollen in others. The conidia are generally ellipsoidal, catenate, and with walls that may be minutely echinulate to conspicuously and irregularly roughened. The ornamentation of conidia is extremely variable. The most striking feature of most of these anamorphs is the formation of very long, often highly branched yellow-green synnemata. The hyphae of these synnemata are often 15-20 µm wide. Some species contain KOH-sensitive pigments, others do not. It is probable that these pigments change colour dramatically on drying. Colour photographs and paintings prepared from living material demonstrate that when freshly collected the synnemata are often bright yellow or yellow-green. Dried specimens lose most of the yellow coloration and are predominantly green or greenish brown.

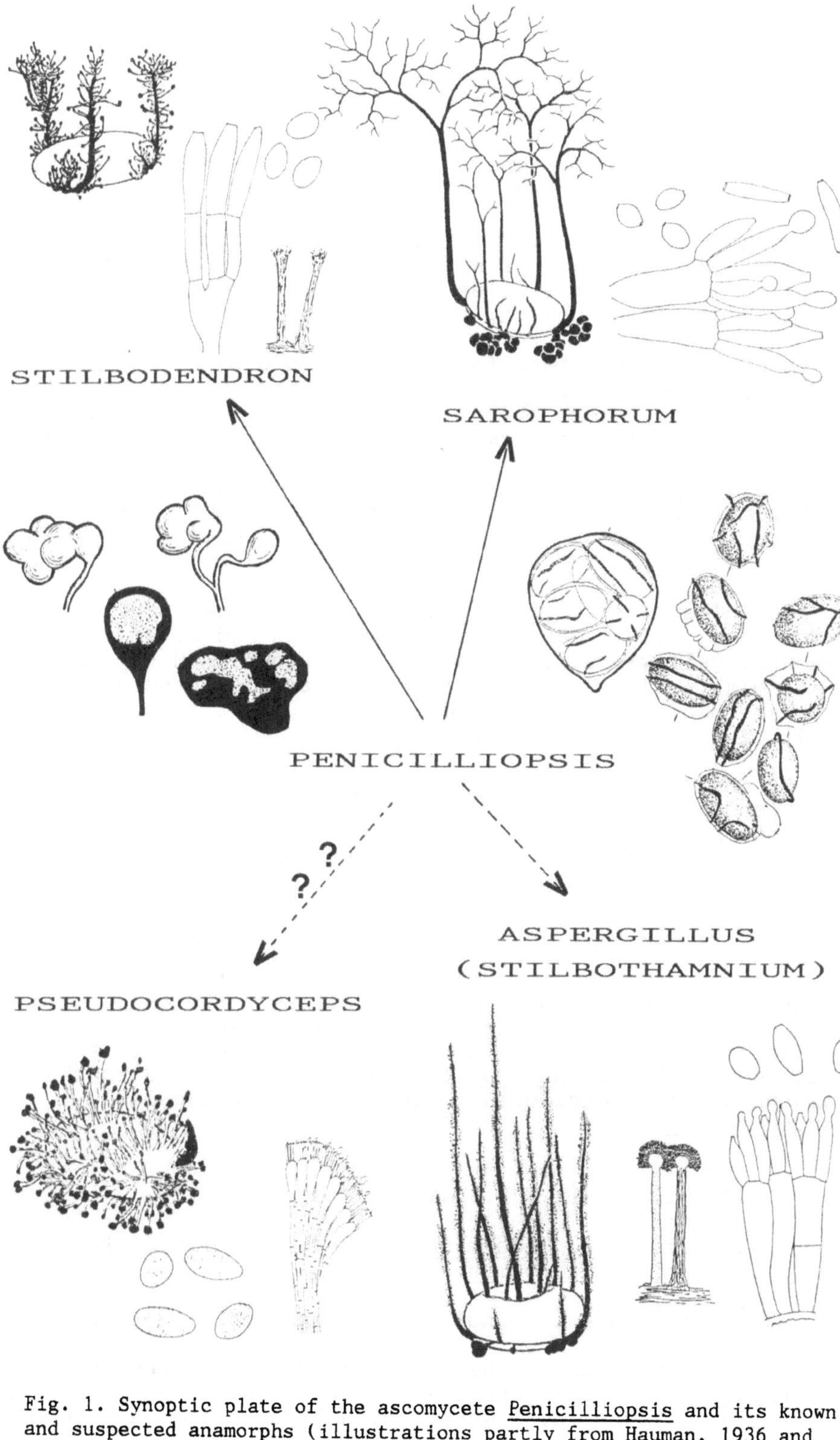

Fig. 1. Synoptic plate of the ascomycete Penicilliopsis and its known and suspected anamorphs (illustrations partly from Hauman, 1936 and Kobayasi and Indoh, 1943)

Although Sarophorum, Stilbodendron and Pseudocordyceps are not readily comparable with other anamorph genera of the Eurotiales, Stilbothamnium shows many similarities with Aspergillus (Roquebert & Abadie 1973, Samson 1979), and we have synonymized Stilbothamnium with Aspergillus below, retaining the species of Stilbothamnium in a separate subgenus.

The four anamorph-genera revised below can be recognized by the following combinations of characters:

Sarophorum - Synnemata indeterminate, unbranched, dichotomously branched, or racemosely branched, yellow-green to green, KOH+, phialides arranged on swollen metulae in a hymenium-like layer over much of the synnemata, conidia dimorphic.

Stilbodendron - Synnemata fertile only in capituli of branches, racemosely branched, with indeterminate central axis and determinate branches, greenish or reddish brown, KOH +, phialides on cylindrical metulae in terminal capituli, conidia monomorphic.

Pseudocordyceps - Synnemata determinate, unbranched or percurrently proliferating, white or light brown, KOH -, phialides arranged on swollen metulae in a terminal capitulum, conidia monomorphic.

Aspergillus subgenus Stilbothamnium - Synnemata unbranched and determinate, or with an indeterminate axis and determinate branches, yellow-green, green or light brown, KOH -, conidiophores with a wide stipe and terminal vesicle, generally with biseriate heads, conidia monomorphic. One mononematous species is also included in this series.

Of these genera, only Pseudocordyceps shows much resemblance to other known genera of synnematous Hyphomycetes. The genus Phialostele Deighton (Cejp & Deighton 1969) also produces determinate white to light brown KOH - synnemata on fruits or seeds, with phialides and dry masses of conidia in short chains. It is distinguished from Pseudocordyceps primarily by its conidiophore branching, which is more or less penicillate, without conspicuously swollen metulae. Our observations of type and authentic material of P. scytopetali Deighton (IMI) have shown that this species, the only one so far included in Phialostele, is sometimes associated with sterile stromata imbedded in the tissue of the seed. There is little doubt that P. scytopetali is closely related to the anamorphic fungi considered in this paper, particularly Pseudocordyceps.

SAROPHORUM Syd. -- Bot. Jb. 54: 260. 1916

Type species: S. ledermannii Sydow, anamorph of Penicilliopsis clavariaeformis.

=Bovetia Persiani & Onofri -- Mycotaxon 15: 256. 1982.

No type specimen of the type and only described species of Sarophorum, S. ledermannii Sydow, was located in B, L or S. The original diagnosis, however, clearly describes a racemosely branched seminicolous fungus and the illustration is compatible with the anamorph of Penicilliopsis clavariaeformis. The measurements given by Sydow for the conidia and phialides of S. ledermannii are smaller than those observed by us, but this may have been the result of a calibration error or the preservation of the specimen examined by Sydow in spirits.

The genus Bovetia, recently described by Persiani and Onofri (1982) from seeds of Diospyros and cultures originating from soil, is here considered a synonym of Sarophorum. Although these authors did not mention the dimorphic conidia of the species, cylindrical conidia are clearly illustrated along with ellipsoidal conidia (Persiani & Onofri, 1982, Fig. 1c, 2c). The occurrence of polyphialides and the clavate rather than swollen metulae reported by Persiani and Onofri are both features typical of cultures of P. clavariaeformis, although these features differ in herbarium specimens.

Penicilliopsis clavariaeformis Solms-Laubach -- Annls Jard. Bot. Buitenzorg 6: 53. 1887. Fig. 2

= Penicilliopsis brasiliensis Moller apud Schimper -- Bot. Mitt. Tropen 9: 296. 1901.

Anamorph: Sarophorum palmicola (Henn.) Seifert & Samson, comb. nov. Figs 3, 4.

= Penicilliopsis palmicola Henn. -- Hedwigia 43: 352. 1904 (basionym, described as anamorph).
= Penicilliopsis jurensis Henn. -- Hedwigia 44: 59. 1905 (described as anamorph).
= Sarophorum ledermannii Sydow -- Bot. Jb. 54: 260. 1916
  = Penicilliopsis ledermannii (Sydow) Hauman -- Bull. Soc. r. Bot. Belg. 69: 111. 1936.
= Penicilliopsis longissima Hauman -- Bull. Soc. r. Bot. Belg. 69: 110. 1936 (described as anamorph).
= Penicilliopsis microsequoia Hauman -- Bull. Soc. r. Bot. Belg. 69: 112. 1936 (described as anamorph).
= Penicilliopsis dichotomus Hauman -- Bull. Soc. r. Bot. Belg. 69: 113. 1936 (described as anamorph).
= Bovetia diospyri Persiani & Onofri -- Mycotaxon 15: 252. 1982.

Ascomata single to several on seed, accompanied by anamorph or not, stipitate with 1-3 stromata per stalk which usually meld together, hard, globose to ellipsoidal or slightly turbinate, dark green or reddish brown, dark red or purple in KOH, 1-4 mm diam, with a smooth, slightly warty, granulose or tomentose surface, the stipe 750-2000 µm long, 250-625 µm wide. Ascostromata uni- or multiloculate, at first hard and green but becoming yellowish brown and flaky as the asci mature; the wall 75-150 µm thick, composed of a textura angularis to textura epidermoidea of cells 6-16 µm wide with walls to 0.5 µm thick; the asci are embedded in a textura intricata of hyphae 3-6 µm wide. Asci globose to ellipsoidal, 8-spored, 11-17 x 9-11 µm. Ascospores ellipsoidal, 7-10 x 5-5.5 µm, with several hyaline longitudinal crests, slightly echinulate between crests.

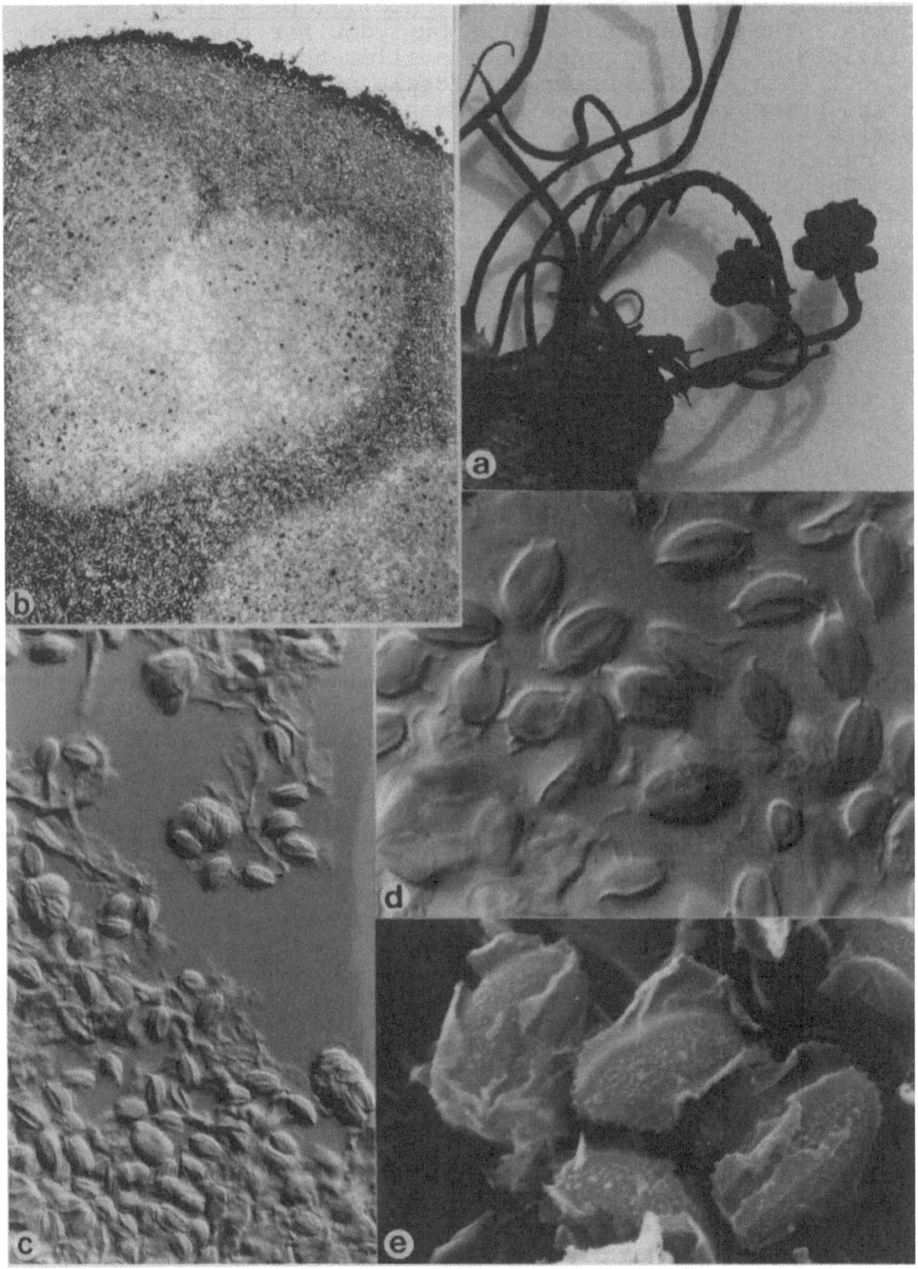

Fig. 2. Penicilliopsis clavariaeformis. a. Stipitate ascostroma emerging from seed of Diospyros macrophylla, van Overeem no. 11, herb. K (x 2.5). b. Section of ascostromata showing locules and imbedded asci, IMI 93061 (x 125). c-e. Ascospores from material collected in Botanical garden, Bogor (BO), c, d, differential phase contrast microscopy (x 650, x 1659 resp.), e, SEM (x 4200).

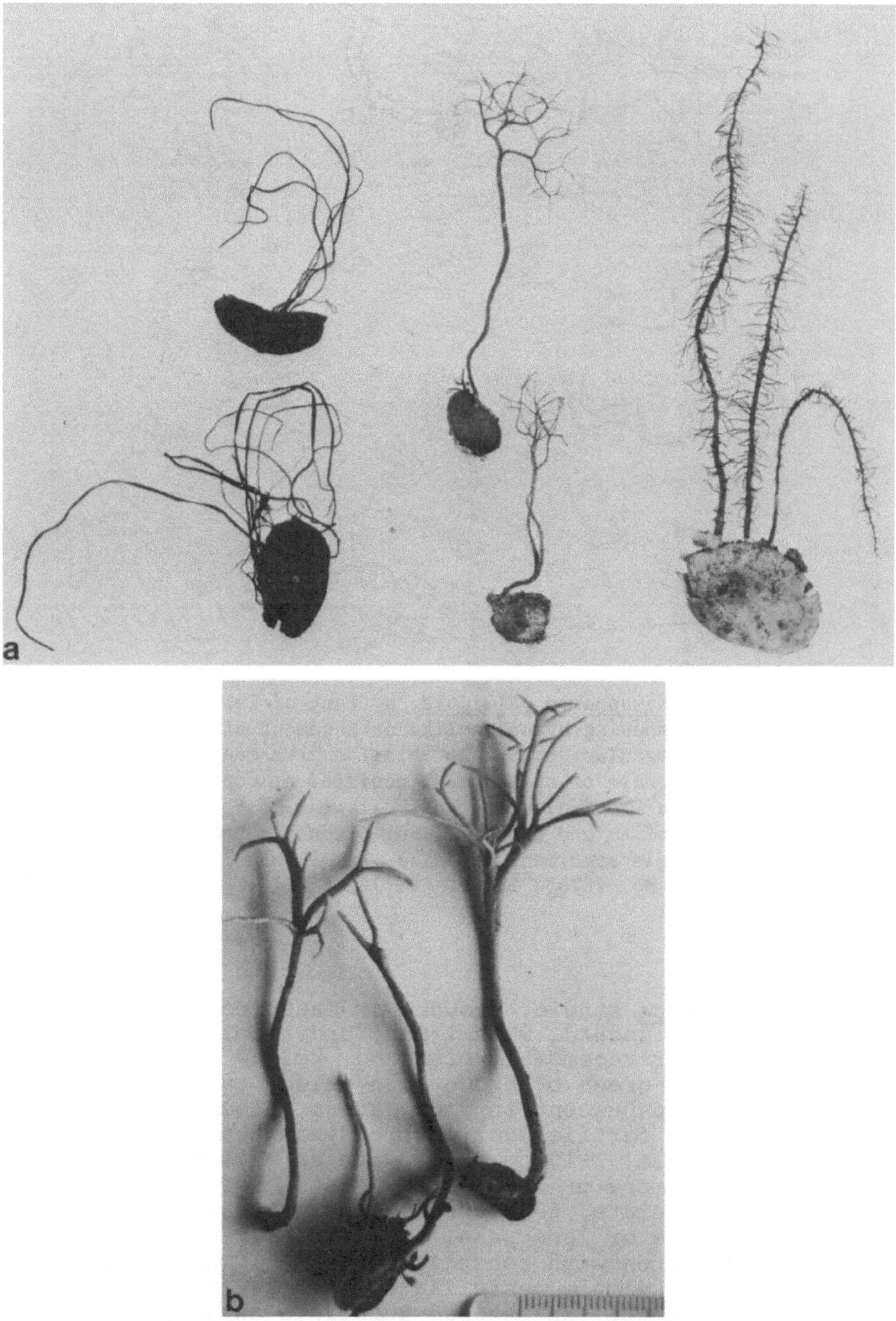

Fig. 3. Sarophorum palmicola. a. Dried herbarium specimens (x 0.6). Left, unbranched synnemata. Centre, dichotomously branching synnemata. Left, racemosely branching synnemata. b. Living specimen, photo. J. Rammeloo (x 0.9). (a. left, lectotype, P. longissimus, centre, holotype P. dichotomus BR, right, P. microsequoia, Hauman no. 581, BR. b. J. Rammeloo no. 6029, BR).

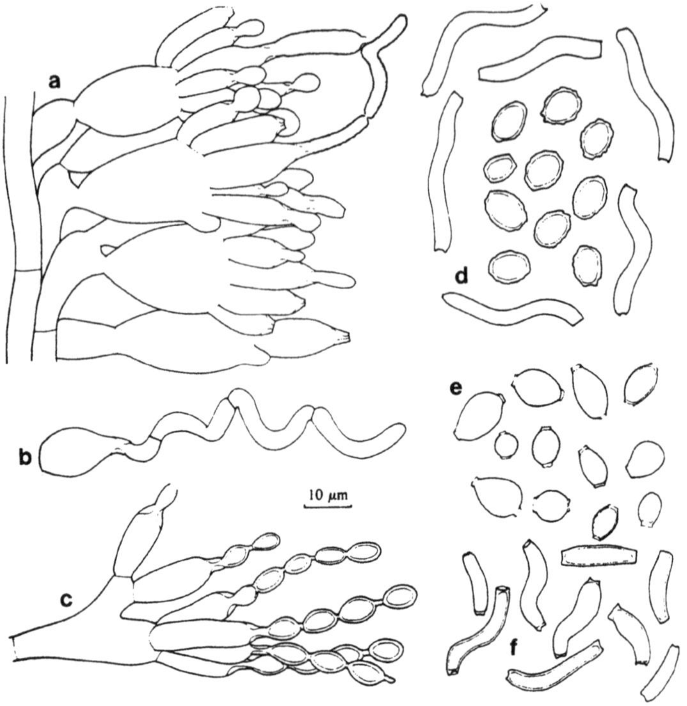

Fig. 4. Sarophorum palmicola. a. Part of lateral portion of synnema showing hymenium-like arrangement of conidiophores and swollen metulae. b. Single phialide from centre of whorl with helical chain of cylindrical conidia. c. Conidiophore from synnemata from culture on oatmeal agar. d-f. Conidia. d, mixture of A- and B-conidia. e, B-conidia and f, A-conidia from single specimen. (a, d, J. Rammeloo no. 6029, BR. b, e, J. Louis no. 16158, BR. c, CBS 761.68).

Synnemata single, crowded or caespitose, filiform, sinuous, unbranched, with 1-many orders of dichotomous branching, or racemosely branched, yellow to yellow-green when fresh, olive-green or brown when dried, turning red to purple-red in KOH, pubescent in young areas, floccose to lanose in older areas, fertile over entire synnema although basal 1 cm may be sterile, 2-10 cm tall, with an axis 650-1500 µm wide, tapering to more or less acute points; racemose branches, if present, subulate, 600-4000 µm long, 200-560 µm wide. Hyphae of stipe more or less parallel, 5-15 µm wide, with walls 1-2 µm thick, often with constricted septa. Conidiophore branching more or less simple penicillate, with 1-3 metulae terminal on a hypha of stipe, on natural substrate swollen, clavate, ellipsoidal or obovate, (8-)20-35 x (8-)10-20(-27) µm, in culture cylindrical or slightly clavate, 2-5 um wide. Phialides in terminal whorls of 5-13 on metulae, 14-27(-35) µm long, (3.5)5-8 µm wide, the central phialide usually wider, 6.5-11 um wide, with conidiogenous aperture 2-3.5 µm wide. Conidia of two types; A-conidia ellipsoidal, obovate, oblong ellipsoidal or subglobose, (5-)7.5-15(-17) x 3-8(-11) µm; B-conidia cylindrical, straight, curved, or sinuous, (8-)13-32.5 x 3-6 um, forming a helical chain arising from the central phialide, usually not formed in culture; both types of conidia initially hyaline, becoming light yellow, yellowish green in mass, base often truncate or with a hyaline frill 2-3 µm wide, to 1 µm deep, usually rounded at apex, at first smooth and thin-walled, becoming roughened and slightly thick-walled.

HABITAT: Seeds. Recorded from Diospyros spp., Strychnos melaecensis, Borassus flabellifer, Lecythis sp. Occasionally isolated from soil.

DISTRIBUTION: Pantropical, recorded from Africa (Belgian Congo, Burundi, Ivory Coast, Nigeria), SE Asia (Indonesia, Malaysia, Phillipines, Singapore), South Asia (India), West Indies (Trinidad), Pacific Archipeligo (New Britain Island), South America (Brazil).

SELECTED MATERIAL EXAMINED:

Herbarium specimens: Specimens marked * have teleomorph (immature or mature) present. Authentic material only is listed.

Penicilliopsis brasiliensis, isotype, E. Ule, Myc. Brasiliensis, on seed of Lecythis sp., leg. E. Ule, Amazonas Jurua, 1901 (L, BR)

Penicilliopsis clavariaeformis,*authentic material from Solms-Laubach collected in Botanical Garden at Bogor, ex herb. Hauman (BR).

Penicilliopsis dichotomus, lectotype here designated, on seeds, leg. P. Staner no. 711, Zaire, Bokuma-Bokatola, 1930 (BR).

Penicilliopsis juruensis, holotype, E. Ule, Herb. Brasiliense no. 2833, on seeds of ?Myrtaceae, Brazil, Estado de Amazonas, Jurua, June 1901 (HBG, BR).

Penicilliopsis longissimus, lectotype here designated, on seeds of Ouratea, leg. P. Staner no. 560, Zaire, Bamania near Eala, Sept. 1930 (BR). Syntype, on seed, leg. E. & M. Laurent no. 240, Zaire, Ubangi, 21 Dec. 1903 (BR)

Penicilliopsis microsequoia, holotype, on fallen seeds, leg. L. Hauman no. 581, Zaire, Bolafa, sur Lopari, Sept. 1932 (BR).

Penicilliopsis palmicola, isotype, E. Ule, Herb. Brasiliense no. 2834, on seed of palm, Jurua, Mory, Estado de Amazonas, July 1901 (L 910.219-145, HBG, BR).

Living cultures:

CBS 257.33, ex seed Diospyros sp., K.B. Boedijn, Bogor; CBS 392.50, ex seed Diopspyros celebica, K. B. Boedijn & J. Reitsma, Bogor.

CBS 323.83, Bovetia diospyri, ex-type culture isol. ex seed of Diospyros cf. ivorensis, Ivory Coast, Tai National Park.

and several others

Penicilliopsis clavariaeformis is recognized by the yellowish colour of the ascogenous tissues, its ascospores, which are slightly larger than those of P. africana, and its conspicuous Sarophorum anamorph. The species is a common parasite of seeds in the tropics, and may attack seeds while they are still on the tree. Descriptions of the species have also been presented by Hauman (1936) and by Kobayasi and Indoh (1943).

Sarophorum palmicola can be recognized by the hymenium-like arrangement of phialides on swollen metulae, and the helical chains of cylindrical conidia and straight chains of more or less ellipsoidal conidia emerging from each whorl of phialides. Hauman (1936) described a number of species based on the different patterns of synnema branching. The microscopic characters of these differently branched synnemata are identical. Synnemata formed in culture tend to be unbranched or sometimes dichotomously branched, irrespective of whether the original specimen is unbranched. dichotomously branched or racemosely branched. For example, IMI 62866 includes the original specimen, in which the synnema is racemosely branched, and a dried culture, in which synnemata are dichotomously branched. Lectotypes have been designated for two of Hauman's species above.

P. clavariaeformis is known by the vernacular name "Botawa-tora" in the Turumbi dialect of Zaire, according to annotations on one specimen collected by Louis (BR).

STILBODENDRON H. & P. Sydow - Annls mycol. 14: 260. 1916

Type species: Stilbodendron camerunense H. &. P. Sydow,
Anamorph of Penicilliopsis africana Samson & Seifert

Following the work of Morris (1963), Stilbodendron has been considered a synonym of Corallodendron Junghuhn. The type species of Corallodendron, C. leucocephalum Junghuhn, was described from submerged wood. The authentic material we have examined (L 910.222-3466, L 910.222-3464) consists of Anomoeoneis-like diatoms overgrowing coralloid structures of unknown composition. Corallodendron, therefore, is best considered a nomen dubium. Hasskarlindia O. Kuntze is a superfluous name for Corallodendron (Carmichael et al., 1980).

PENICILLIOPSIS AFRICANA Samson & Seifert, sp. nov. Fig. 5 d-f

Anamorph: Stilbodendron cervinum (Cooke & Massee) Samson & Seifert, comb. nov. Figs 5 a-c, 6.

= Corallodendron cervinum Cooke & Massee -- Grevillea 16: 71. 1888.
= Stilbodendron camerunense Sydow -- Annls mycol. 14: 260. 1916.
= Stilbodendron congoense Hauman -- Bull. Soc. r. Bot. Belg. 69: 115. 1936.

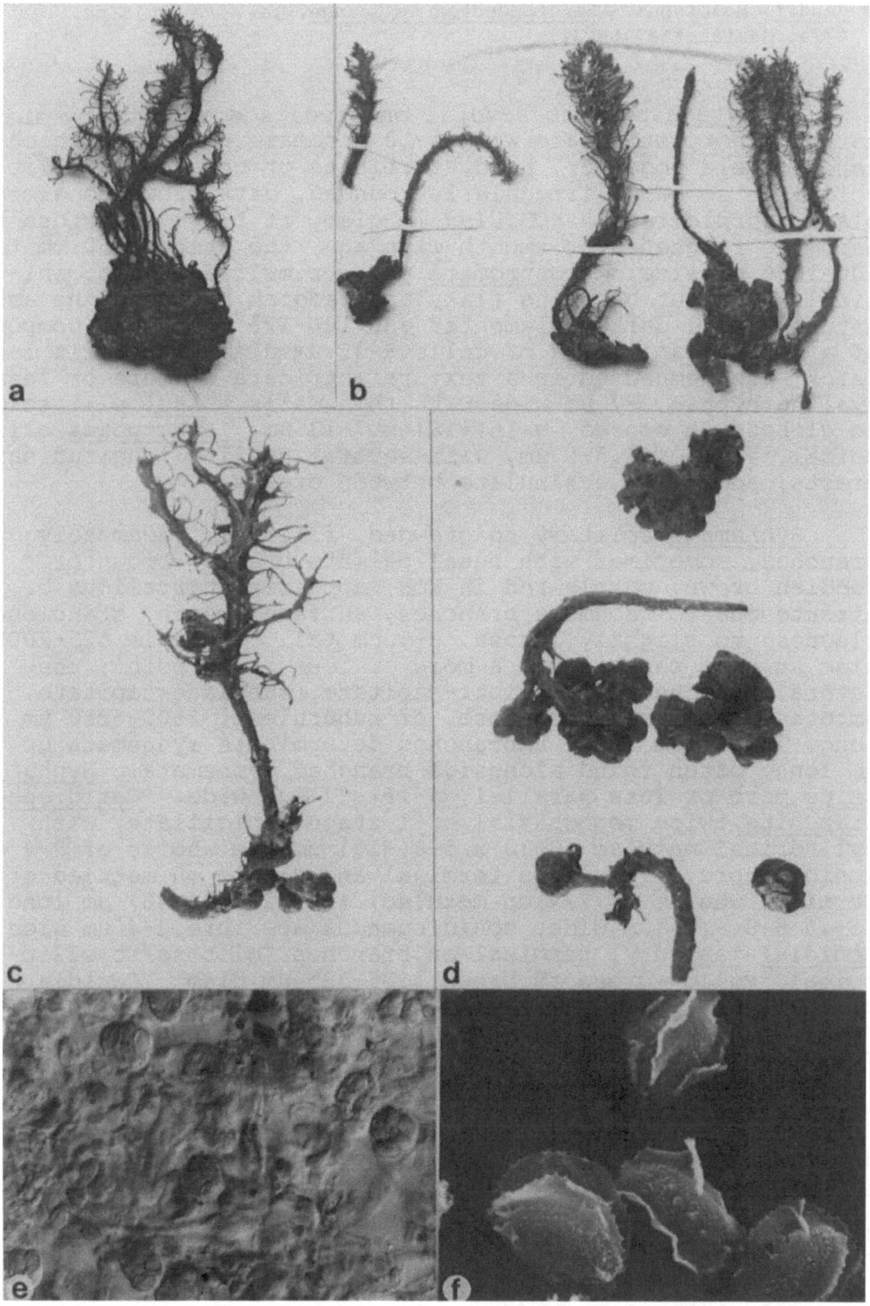

Fig. 5. Penicilliopsis africana and its anamorph, Stilbodendron cervinum. a-c. Habit of synnemata. a. BR no. 2436, B. herb. Hauman 10.1036, BR, c. no. A 7238 BR. d. Ascostromata of holotype P. africana (x 4.5). e-f. Asci and ascospores, e. differential phase contrast microscopy (x 500), f, SEM. (x 4200).

A *P. clavariaeformi* differt ascosporis 7.0-8.6 x 3.5-5.0 µm, contextu ascogeno lateritio et anamorphe typi *Stilbodendri*. Holotypus: "*Stilbodendron congoense*, leg. J. Louis no. 6275, Metiquette" (BR).

Ascomata single to several on seed, accompanied by anamorph or not, stipitate with 1-10 stromata per stipe which tend to meld together, hard, turbinate or compressed ellipsoidal or somewhat irregularly rounded, dark brown to almost black, purple-red in KOH, 1-3 mm diam, at first tomentose but becoming roughened or smooth with age, the stipe 2-10 mm tall, 500-1000 µm wide. Ascostromata uni- or multiloculate, initially hard but becoming flaky and reddish brown as the asci mature; wall darkly pigmented and 100-125 µm thick, composed of a textura angularis of cells 8-25 um diam with walls 1-2 µm thick; asci embedded in a textura intricata of more or less hyaline hyphae 3-7 µm wide with thin walls. Asci ellipsoidal to globose, 8-spored, 9-16(-18) x 7-13 µm. Ascospores ellipsoidal, 7-8.5 x 3.5-5 µm, with several hyaline longitudinal crests, sometimes echinulate between crests.

Synnemata solitary to crowded, filiform, racemosely branched, sometimes with basal palmate branch, brown or reddish brown, purple-red in KOH, appearing hirtellous or hirsute due to racemose branches, surface smooth, granulose, floccose to slightly lanose, 3-5 cm tall, the axis 625-2000 µm wide at base tapering to a more or less acute point, the lateral branches cylindrical-capitate, subulate-capitate, smooth, granulose, tomentose, or puberulent, 2500-6250 µm long, 75-175 µm wide; unbranched determinate synnemata up to 3 mm long, often found alongside branched synnemata. Hyphae of stipe more or less parallel, 5-15(-22) µm wide. Conidiophore branching twice monochasial or 1-stage penicillate, with cylindrical metulae 10-30 x 5-8(-10) µm, in whorls of 2-4 on conidiophore. Phialides terminal and single on metulae or in terminal whorls of 2-4 on metulae, (13-)20-30(-40) µm long, (5-)6.5-8(-10) µm wide, conidiogenous aperture 2-4 µm wide. Conidial mass dry, terminal on branches, globose to ellipsoidal, yellow-green to brown, 125-325 µm diam. Conidia ellipsoidal, oblong-ellipsoidal or obovate, often with polar scars or hyaline frill, base often truncate, apex usually rounded, olive brown, in chains with obvious connectives, (5.5-)8-12(-16) x 5.5-10(-11) µm, with granulose to verrucose walls which may be thin or slightly thickened.

HABITAT: Seeds, recorded from *Carpodinus* sp., *Landolfia lemdolphia*.

KNOWN DISTRIBUTION: Africa (Belgian Congo, Cameroon, Ivory Coast, Zaire).

SELECTED MATERIAL EXAMINED:

Herbarium specimens: Specimens marked * have mature or immature teleomorph present. Authentic specimens only are listed.

Corallodendron cervinum, holotype, on seeds of *Landolfia lemdolphia*, leg. E.M. Holmes, Africa (K).

Penicilliopsis africana, holotype, as Stilbodendron congoense, on seed, leg. Louis no. 6275, Metiquette (*BR).

Stilbodendron camerunense, holotype, on seed, leg. Ledermann no. 1156, Cameroon, Bajoki, Balbon, 20 Nov. 1908 (*B).

Stilbodendron congoense, lectotype here designated, on seeds, leg. Mme. Goosens-Fontana no. 379a, Zaire, Eala, Sept. 1925 (BR). Syntype, on seed, leg. L. Hauman no. 583, Zaire, forest between Mangana and Bolafa, Sept. 1932 (BR).

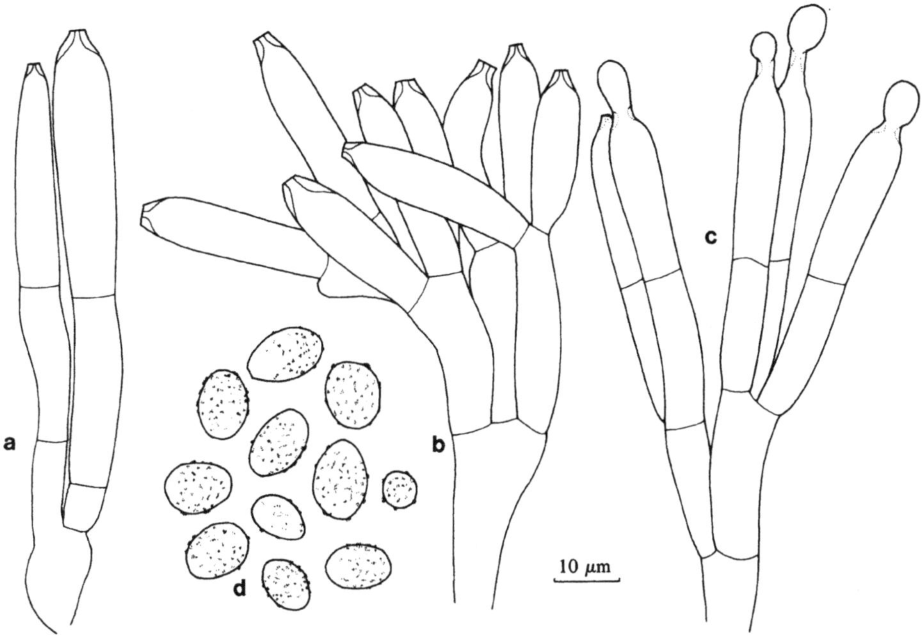

Fig. 6. Stilbodendron cervinum. a-c. Conidiophores from different specimens. d. Conidia. (a, Ghesquiere s.n., 10.1936, BR. b, "P. dybowskii", Vandereyst s.n., BR. c,d, Hauman 2936, BR.

Penicilliopsis africana can be differentiated from P. clavariaeformis most easily by its Stilbodendron anamorph, though the ascospores are smaller, and the colour of the mature ascogenous tissue is red rather than yellow.

A lectotype for Stilbodendron congoense Hauman has been designated above from available syntype material.

PSEUDOCORDYCEPS Hauman -- Bull. Soc. r. Bot. Belg. 69: 110. 1936.

Type and only species: Pseudocordyceps seminicola Hauman, teleomorph unknown.

Pseudocordyceps seminicola Hauman -- Bull. Soc. r. Bot. Belg. 69: 117. 1936. Fig. 7, 8.

Synnemata gregarious to caespitose, cylindrical-capitate, clavate, often flattened in one dimension, straight or sinuous, slender to robust, unbranched, sometimes sharing a common base, light brown, smooth, 1-10 mm tall, 150-875 µm wide. Hyphae of stipe parallel, 5-12 µm wide with granular cytoplasm. Conidiophore branching simple penicillate, with terminal clavate metulae 40-100 x 13-23 µm. Phialides in terminal whorls of 5-10 on metulae, light yellow, 40-65 x 6-10 µm, conidiogenous aperture 2.5-4 µm wide. Conidial mass terminal, orange, brown, orange-brown to dark grey, hemispherical to ellipsoidal, 250-500 µm diam. Conidia ellipsoidal, oblong ellipsoidal or ovate, sometimes with slightly truncate base, light yellow, 10-20 x 5-11 µm, walls roughened to inconspicuously horizontally striate. Stromata not observed.

HABITAT: Seeds of unknown identity.

KNOWN DISTRIBUTION: Africa (Zaire).

MATERIAL EXAMINED:

Pseudocordyceps seminicola, neotype here designated, on seeds, leg. Vrijdagh, Zaire, Bambesa, K2882 (BR). On seeds, leg. Louis no. 6519, Zaire, Yangambi, 6 Nov. 1937, K2894 (BR).

Pseudocordyceps seminicola is recognized by its white to light brown KOH- determinate synnemata, its occurrence on seeds, and the swollen clavate metulae with terminal whorls of phialides. It is most likely to be confused with Phialostele scytopetali, as has been discussed above. Hauman (1936) described percurrent proliferation of synnemata in P. seminicola, a character we have not observed on the specimens we examined. The possible occurrence of this feature should be noted by future researchers.

In the original description of Pseudocordyceps seminicola, Hauman (1936) mentioned only a single collection. We have been unable to locate this specimen among Hauman's collections in BR, PC, K and IMI and for this reason we have selected a neotype above, which apparently originates from Hauman's herbarium.

ASPERGILLUS Subgenus STILBOTHAMNIUM (Henn.) Samson & Seifert, comb. nov.

= Stilbothamnium Henn. -- Bot. Jb. 23: 542. 1897 (Basionym).

As previous authors have indicated (Roquebert and Abadi, 1973; Samson, 1979), the conidiophores of species described in

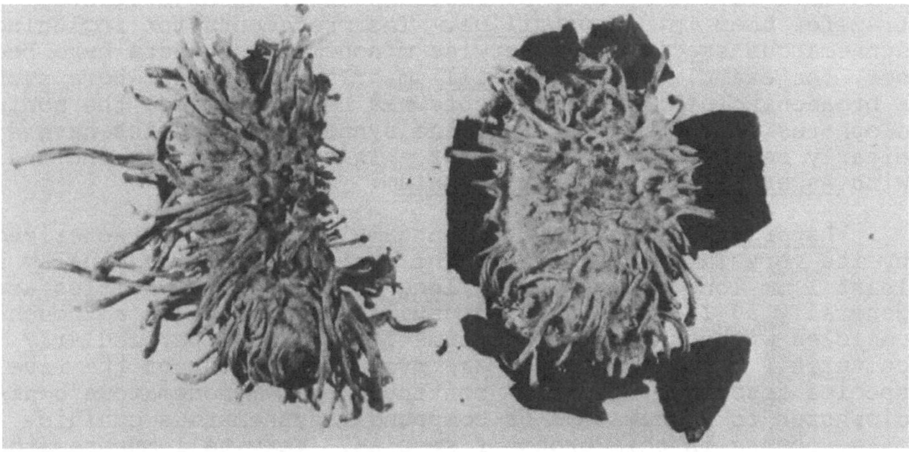

Fig. 7. Pseudocordyceps seminicola, habit photograph of neotype (x 4).

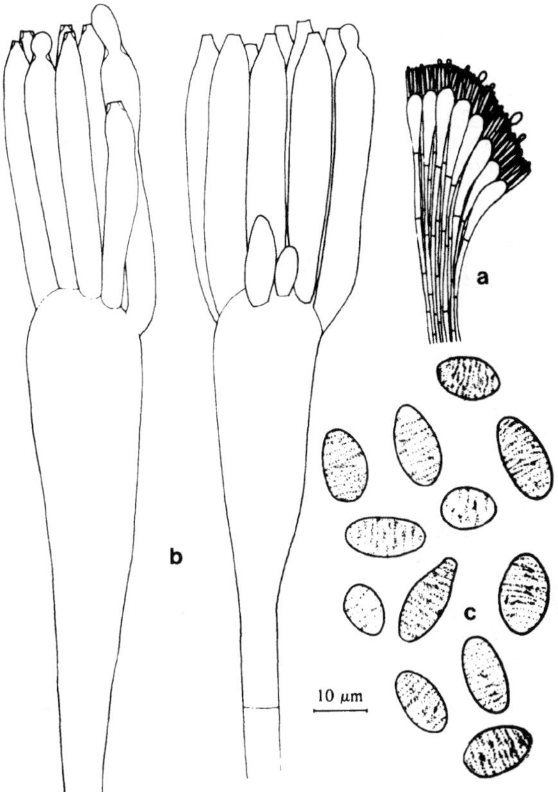

Fig. 8. Pseudocordyceps seminicola, neotype. a. Schematic section of apex of synnema showing arrangment of conidiophores. b. apex of conidiophores, showing swollen terminal metulae. c. conidia.

Stilbothamnium are Aspergillus-like and, despite the prominent synnemata formed by some of these species, we have elected to transfer them into Aspergillus. The precedents for including synnematous species in otherwise mononematous genera have been set, for example, with Penicillium. In most cases where such a broadening of generic concepts has been proposed, the conidiophores of the mononematous and synnematous species have clearly been shown to be similar. This is certainly the case with Aspergillus and Stilbothamnium.

Aspergillus subgenus Stilbothamnium can be characterized by its very large structures; the conidiophores are usually at least 1 mm long and may be as long as 2 cm. The vesicles are generally 50-100 µm diam, and usually have metulae, although vesicles without metulae are sometimes found, particularly in cultures. Large synnemata are produced by four of the five species described below. A continuum from mononematous conidiophores to synnematous or compoundly synnematous conidiomata, occurs in this group of species. Very tall mononematous conidiophores have been seen in some specimens.

Sterile Penicilliopsis-like KOH negative sclerotia have been found associated with two of the four synnematous anamorphs described below (see also Fig. 12), and sterile sessile Petromyces-like stromata with the mononematous species. We have not observed ascospores but Raciborski (1909) described ellipsoidal ascospores 11-13 x 6-7.5 µm, with equatorial ridges for Stilbothamnium penicilliopsis, here considered a nomen dubium. According to Roquebert and Nicot (this publication), French mycologists have also observed fertile ascomata in A. togoensis.

Aspergillus subgenus Stilbothamnium appears closely related to Aspergillus section Flavi and section Circumdati and may bear a great deal of similarity to the ancestors of these domesticated fungi. A. flavus often produces sclerotia on corn or peanuts, but no ascospores have so far been observed. The morphologically similar ascomata of Petromyces (anamorph: Aspergillus alliaceus) are usually sterile and the ascogenous apparatus probably matures late, as is also the case in ascostromata of Penicilliopsis. The occurrence of these temperate species on seeds and the production of stromatic bodies suggests that species such as A. flavus and A. ochraceus may represent domesticated relatives of series Stilbothamnium. In fact, we have noted a number of specimens of tropical seminicolous fungi which appear to represent A. ochraceus. These specimens have a strong tendency to produce fasciculate conidiophores, and the conidia are roughened, like those of many of the species described below.

Despite the large amount of material we have examined, the delimitation of species in subgenus Stilbothamnium remains problematic. Although A. erythrocephalus, A. dybowskii and A. vitellinus appear to be relatively well-defined species, the delimitation of A. amazonensis and A. togoensis is at present unsatisfactory. In addition, we have examined specimens which do not fit well into any of the species we describe below. The variability of these fungi presents a remarkable challenge to the taxonomist, and we feel we have been only partially successful in understanding this fascinating group of species.

We have been unable to locate type or authentic material of the type species of Stilbothamnium, S. togoense Henn. amongst Hennings' collections in B, S, BR or K. But from the diagnosis and illustration (see Roquebert & Nicot, this publication, Fig. 1) it is clearly identifiable as the common species with naked conidiophore stipes occurring in Africa. We have selected a specimen from herb. BR as the neotype and consequently the species is combined into Aspergillus below.

Aspergillus erythrocephalus Berk. & Curtis -- J. Linn. Soc. Bot., London 10: 362. 1869. Figs 9, 10 a-c.

Conidiophores gregarious, 1.5-4 mm tall, stipes hyaline, 50-70 um wide, with a granular wall 4-5 µm wide at base narrowing to 2-3 µm thick below the capitulum, vesicles globose, (100-)140-200 µm diam, capitulum orange-yellow to slightly reddish brown. Phialides terminal on intercalary cells, cylindrical with a tapered apex, 20-25 µm long, 6-8 um wide, with a conidiogenous aperture 2.5-3.5 µm wide, and a flaring collarette, intercalary cells cylindrical, 11-14 x 6-8 µm. Conidia subglobose to ovate, 8.5-13 x 6-8 µm, echinulate, slightly yellow. Stromata sterile, embedded in seed, crowded, more or less globose, sessile, almost black, light green inside, shiny, hard, 1-2 mm diam, with a smooth to slightly verrucose surface.

MATERIAL EXAMINED:

Aspergillus erythrocephalus, holotype, on seeds, Wright no. 764, Cuba, herb. Berk. (K).

Aspergillus erythrocephalus is a somewhat gigantic relative of A. ochraceus, and serves to connect the synnematous species of subgenus Stilbothamnium to the A. ochraceus and A. flavus series*. Apart from the absence of synnemata, A. erythrocephalus is a typical member of subgenus Stilbothamnium. The sterile stromata seen inside the seeds of the type specimen are sessile and Petromyces-like, offering further evidence of a link between Penicilliopsis and Petromyces.

Raper and Fennell (1965) suggested that A. erythrocephalus was a gigantic form of A. tamarii, and Subramanian (1971) subsequently replaced the name of the latter species by the former. Although there are similarities between the two species, the conidia of A. tamarii are globose and distinctly tuberculate, as opposed to the subglobose echinulate conidia of A. erythrocephalus. The ecologically specific niche of A. tamarii in koji makes it unlikely that it could be the same as A. erythrocephalus.

*Section Flavi and Circumdati, see page 60.

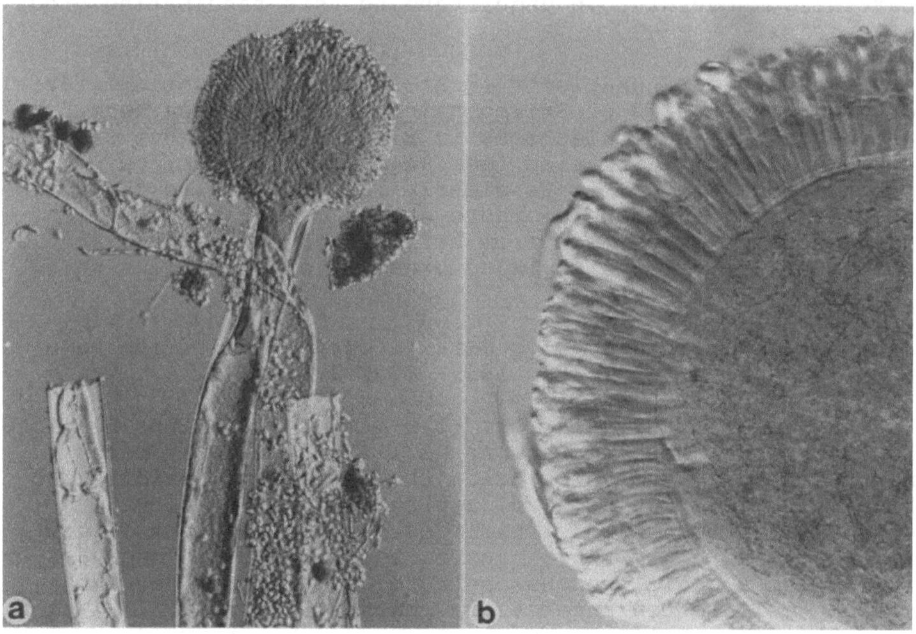

Fig. 9. Aspergillus erythrocephalus, holotype. a. Conidiophore (x 140). b. Detail of biseriate head (x 540).

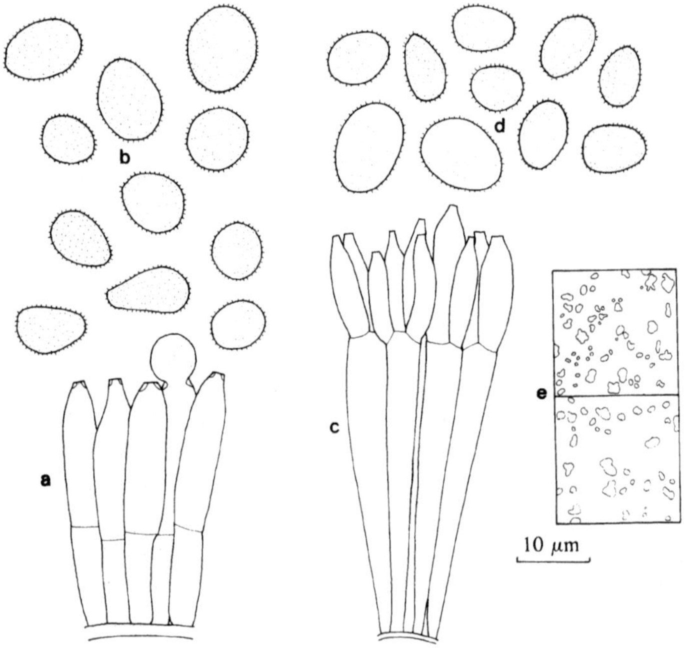

Fig. 10. a-b. Aspergillus erythrocephalus, holotype. a. Arrangement of phialides and metulae on vesicle. b. Conidia. c-e. Aspergillus vitellinus, c. Phialides and metulae on vesicle. d. Conidia. e. Roughening of conidiophore stipe from two different conidiophores on the same synnema.

Aspergillus vitellinus (Ridley) Samson & Seifert, comb. nov., Fig. 10 d-f, 11.

= Sterigmatocystis vitellina Ridley --J. Botany, Brit. & Fopr. 34: 152. 1896 (basionym).

Synnemata solitary or 2-3 on seed, with a stipe composed of tightly packed and anastomosing conidiophores which splay out to form a divergent fertile head, yellow brown, 1-3 cm tall, stipe 3-5 mm wide. Stipe of synnema composed of conidiophore stipes 35-70 um wide with granular walls 1-4 µm thick, bound together by a network of interweaving hyphae 5-16 um wide with smooth thin walls. Conidiophores 3-3.5 mm long, stipe 70-110 µm wide with a granulose wall, warts up to 2 um wide, often distinctly yellow, wall to 12 µm thick at the base, narrowing to about 2-5 µm thick below capitulum, vesicles globose 150-200 µm diam. Metulae clavate, 40-75 x 4-7 µm. Phialides in terminal whorls of 2-4 on metulae, cylindrical or slighty clavate with a tapered apex, 14-19 µm long, 3-6 µm wide, with a conidiogenous aperture 1.5-2 µm wide without a noticeable collarette. Conidia ellipsoidal, ovate, sometimes obpyriform or cylindrical, 6.5-13 x 5-10, echinulate. Stromata not observed.

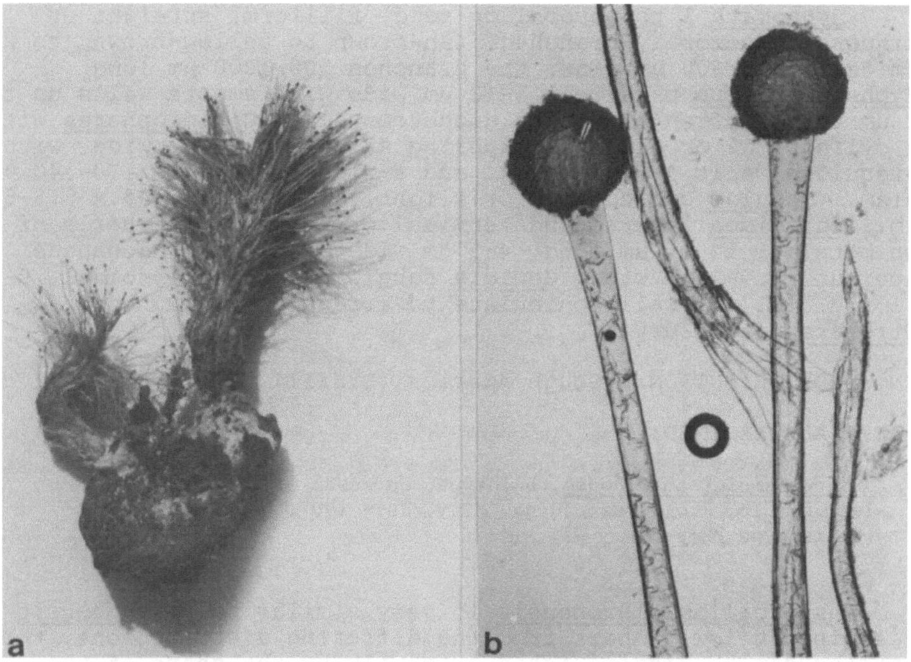

Fig. 11. Aspergillus vitellina, holotype. a. Synnemata emerging from seed (x 3). b. Conidiophores (x 40).

MATERIAL EXAMINED:

Sterigmatocystis vitellina, holotype, Flora of Singapore no. 2970, leg. H. Ridley, 1894 (K, isotype IMI).

Aspergillus vitellinus is distinguished from the similar A. amazonensis and A. togoensis primarily by the organization of its synnemata, in which the stipe is composed of conidiophore stipes woven together by a network of thinner hyphae. The synnemata tend to be shaped like a shaving brush, rather than racemosely branched as in the other two species. The very long metulae of A. vitellinus serve to further distinguish it from the other two species.

Petch (1944) suggested that A. vitellinus was a synonym of Sterigmatocystis monilifera (Berk. & Br.) Petch. The holotype of Periconia monilifera in K is similar to Aspergillus ochraceus but has roughened conidia. It differs from A. vitellinus in having smaller conidia and narrower conidiophore stipes. Although the conidiophores of P. monilifera are densely caespitose and occasionally fasciculate, they do not form stipitate synnema such as those produced by A. vitellinus.

Aspergillus amazonensis (Henn.) Samson & Seifert, comb. nov., Fig. 12.

= Stilbothamnium amazonense Henn. -- Hedwigia 43: 396. 1904 (Basionym).

Synnemata 1 to several on seed, filiform, straight or sinuous, racemosely branched, tan-brown to yellow-brown, to 2 cm tall, 300-600 µm wide, the branches 500-1000 µm long. Hyphae of stipe parallel, 5-13 µm wide with smooth walls up to 1 um thick, often branching dichotomously. Conidiophores with a cylindrical or slightly tapering stipe 60-100 um wide, with granulose walls 3-5 µm thick, and a globose vesicle 110-140 µm diam. Metulae cylindrical or slightly clavate, 15-20 x 3.5-5 µm. Phialides lateral and terminal or in terminal whorls of 3 on metulae, 14-15 µm long, 4-5 µm wide with a conidiogenous aperture 1.5-2 µm wide. Conidia subglobose to ellipsoidal, 8-11 x 7.5-11 µm, wall echinulate to irregularly roughened. Stromata not observed.

KNOWN DISTRIBUTION: South America (Brazil).

MATERIAL EXAMINED:

Stilbothamnium amazonense, holotype, on seeds of Lecythidea, leg. E. Ule, Rio Jurua, Jurua, Miry, July 1907. herb. Hennings (S, isotype BR).

Aspergillus amazonensis is very similar to A. togoensis described below. Apart from the differing distributions, the species can be distinguished primarily by the shape of the conidia, which in A. amazonensis are almost globose, and in A. togoensis tend to be oblong ellipsoidal to somewhat doliiform. Unfortunately, the one specimen of A. amazonensis we have seen was in poor condition, and it is probable that additional variation of macroscopic characters will be noted in future collections.

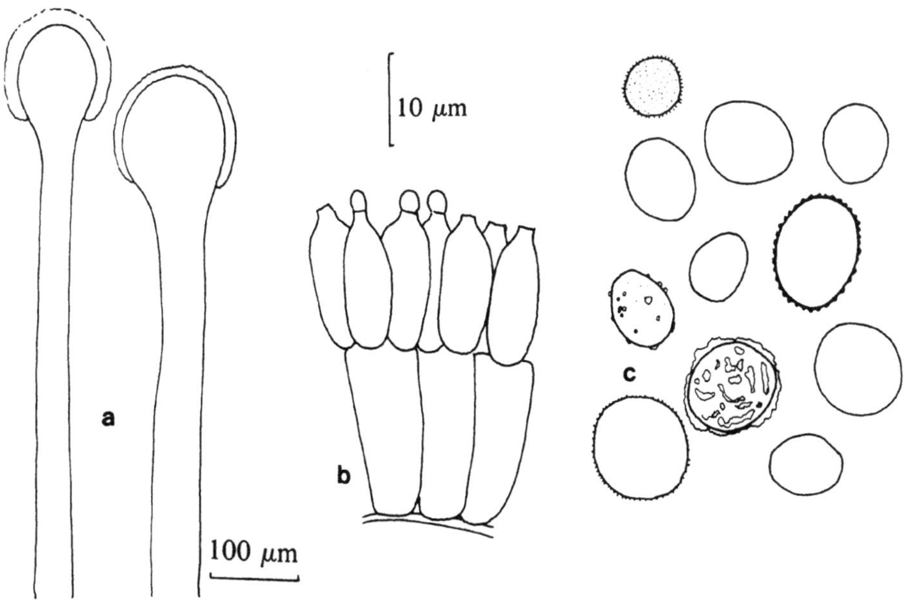

Fig. 12. Aspergillus amazonensis, holotype. a. Conidiophores. b. Arrangment of phialides and metulae on vesicle. c. Conidia.

Aspergillus togoensis (Henn.) Samson & Seifert, comb. nov.
Figs 13, 14.

=Stilbothamnium togoense Henn. -- Bot. Jb. 23: 542. 1897 (Basionym)
= Stilbothamnium novoguineense H. & P. Sydow -- Annls mycol. 14: 262. 1916.
= Stilbothamnium nudipes Hauman -- Bull. Soc. r. Bot. Belg. 69: 123. 1936.
= Aspergillus coremiiformis Bartoli & Maggi -- Trans. Br. mycol. Soc. 71: 386. 1978.

Synnemata solitary or crowded, filiform, straight or sinuous, racemosely branched, appearing hirsute because of branching, surface smooth, floccose or granulose, yellow - brown, tan-brown or greenish brown, often dark green at base, 2-8 cm tall, 625-3000 μm wide at base, 300-900 μm wide in fertile region. Hyphae of stipe more or less parallel, 10-21 μm wide, with walls 0.5-2 μm thick. Conidiophores concentrated on apical part of stipe, emerging at nearly right angles or slightly antrorse, cylindrical-capitate or subulate-capitate, the stipe light yellow-brown, 1000-2000 μm long, 20-60 μm wide, with granulose walls 1-5 μm thick near base, 0.5-2 μm thick below vesicle, vesicle globose, 35-100 μm diam, with or without metulae, or with 'septate phialides'. Metulae cylindrical to slightly capitate, 10-15 x 3-5 μm. Phialides arising directly from vesicles, single and terminal on inter-

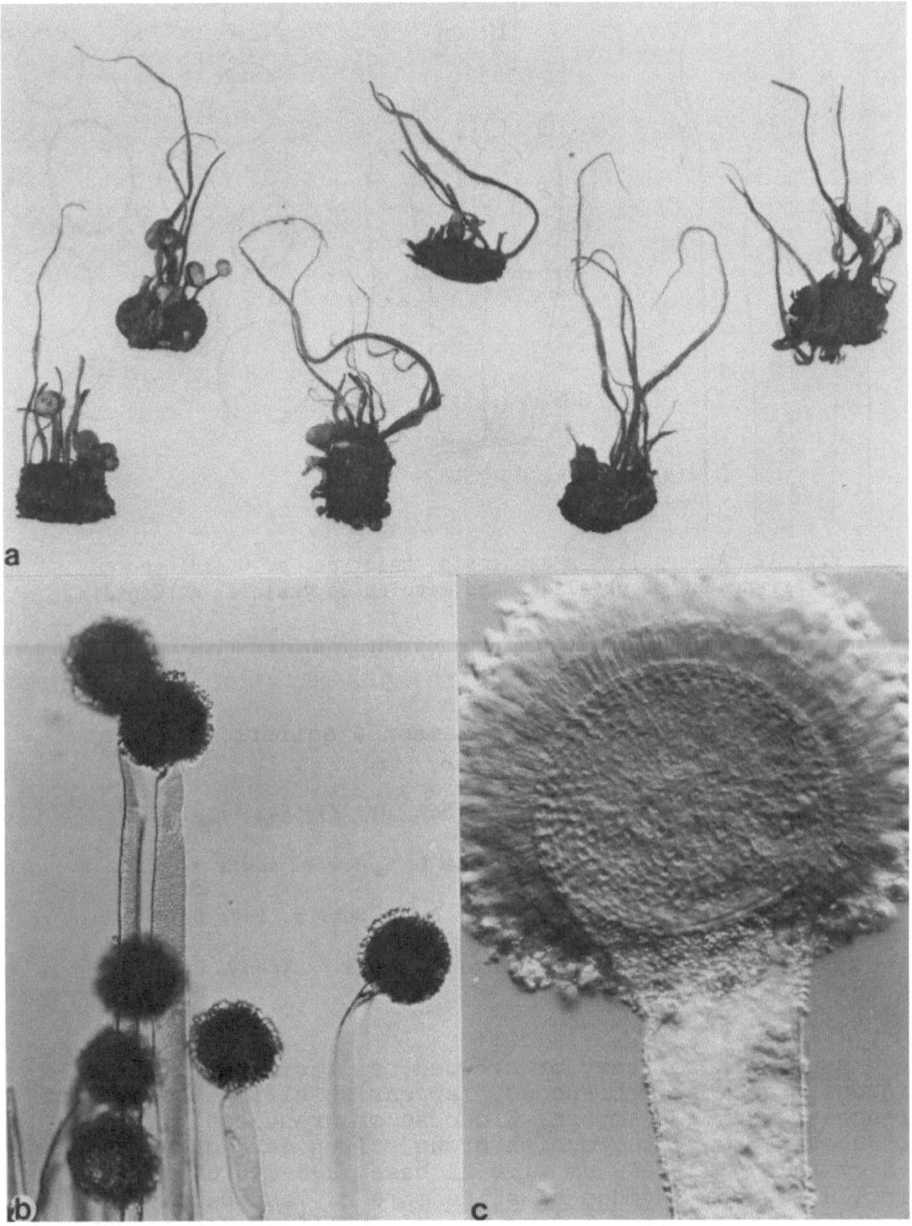

Fig. 13. Aspergillus togoensis, neotype. Synnemata on seeds. (x 0.6). b. conidiophores showing roughening of stipe (x 80), c. Conidiophore head showing metulae and phialides (x 1250)

calary cells, or in terminal pairs or whorls of 3 on metulae, 11-20 x 6-7 µm, the conidiogenous aperture 1.5-3 µm wide.

Conidia oblong-ellipsoidal, ellipsoidal or ovate, often appearing somewhat doliiform, yellow, yellow-brown to orange-brown in mass, (5-)6-11 x (2.5-)4-8 um, with smooth, echinulate or granulose walls which are thin to slightly thickened. Stromata sometimes present, sterile, hard, smooth, ovate, ellipsoidal, grey-green to brown, 2.5-3.3 mm diam, dark green inside.

HABITAT: Seeds of Landolphia, Strychnos.

KNOWN DISTRIBUTION: Southeast Asia (Borneo, New Guinea, Thailand, Singapore), Africa (Sierra Leone, West Africa, Uganda).

MATERIAL EXAMINED:

Herbarium specimens: Specimens marked * have sterile stromata present. Authentic specimens only are listed.

Stilbothamnium novoguineense, *holotype, on seed, leg. C. Ledermann no. 8065, New Guinea, Urwald, 27 July 1912 (B).

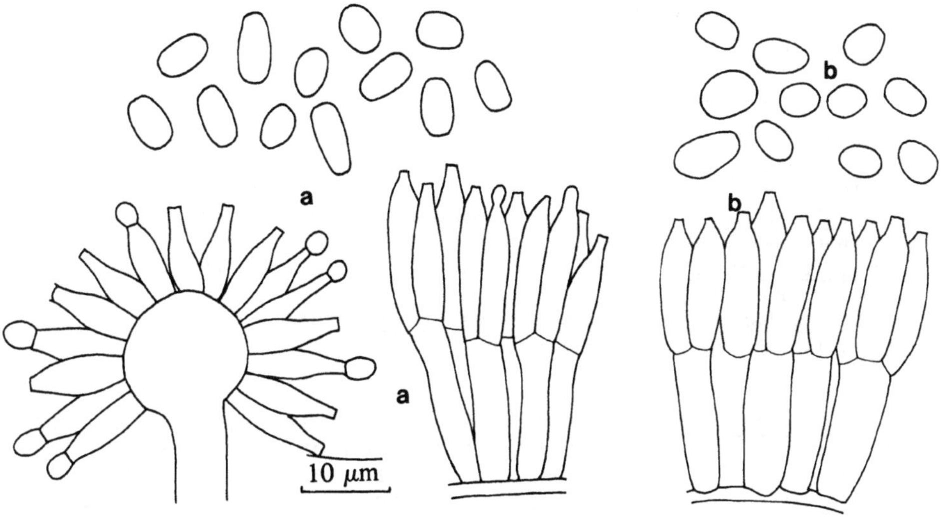

Fig. 13 Aspergillus togoensis. Camera lucida drawings of conidial heads and conidia. a. Specimen "P. clavariaeformis" no. 41, Singapore, K., b. S. novoguineense, holotype.

Stilbothamnium nudipes, lectotype here designated, on seed, leg. L. Hauman no. 586, Zaire, Forest between Bolafa and Mongana, Sept. 1932 (BR). Syntype, on seed, leg. L. Hauman no. 587, same time and place as no. 586 (BR).

Stilbothamnium togoense, neotype here designated, as Stilbothamnium nudipes, leg. J. Louis no. 6196, Zaire, s. dat., B 1009 (BR).

Living cultures:

CBS 205.75, received as Stilbothamnium nudipes, isol. ex seed of Landolphia sp., M.F. Roquebert.

CBS 553.77, ex-type culture of Aspergillus coremiiformis, isol. ex soil, Ivory Coast, A. Rambelli.

The characters which distinguish Aspergillus togoensis from the similar A. amazonensis are discussed above under the description of the latter species. A. togoensis is notable for the extreme variability of the arrangement of phialides. On any one synnema, capitula may be found with or without metulae, or with 'septate phialides'. This latter type has been examined in some detail by Roquebert and Nicot (this publication).

Samson (1979) suggested that Aspergillus coremiiformis was actually an isolate of Stilbothamnium. Our comparison of the ex-type culture of A. coremiiformis with a culture originating from a specimen of A. togoensis (CBS 205.75, received as S. nudipes from M.-F. Roquebert) confirms that cultures of these two species are identical, and A. coremiiformis is hence regarded a synonym of A. togoensis here.

Aspergillus dybowskii (Pat.) Samson & Seifert, comb. nov., Figs 15, 16.

= Penicilliopsis dybowski Pat. -- Bull. Soc. mycol. Fr. 7: 54. 1892 (basionym, anamorph only described) = Stilbothamnium dybowskii (Pat.) Henn. -- Monsunia 1: 7. 1899.
= Stilbothamnium usneoides Sydow -- Annls mycol. 14: 262. 1916.
= Stilbothamnium dybowskii var. macrospora Beeli -- Bull. Soc. r. Bot. Belg. 59: 16. 1927.
= Stilbothamnium dybowskii var. laevispora Hauman -- Bull. Soc. r. Bot. Belg. 69: 120. 1936.
= Stilbothamnium usneoides var. verrucosum Hauman -- Bull. Soc. r. Bot. Belg. 69: 122. 1936.

Synnemata solitary, crowded or caespitose, 1-12 synnemata per seed, filiform, sinuous, unbranched or dichotomously branched at base, then racemosely branched, appearing hirsute because of racemose branching, surface smooth but sometimes lanose at base, greenish brown to dark brown, 2-13 cm tall, 0.5-1(-4) mm wide at base, 300-625 µm wide throughout most of length. Hyphae of stipe more or less parallel, light brown, (2-)4.5-23 um wide, with walls thin to 1 µm thick. Conidio-

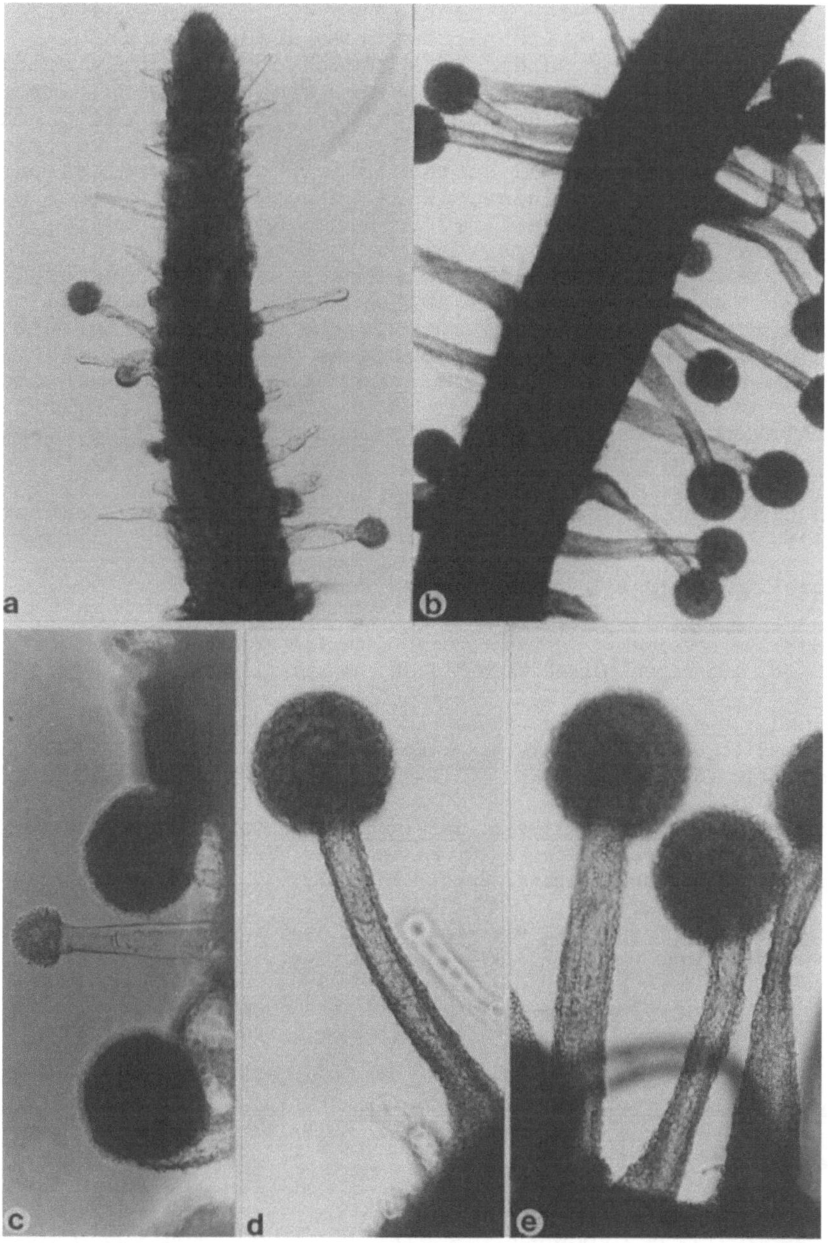

Fig. 15. Aspergillus dybowskii, development of sheathing hyphae on a single synnema. a. Apex of synnema with conidiophore stipes mostly naked (x 50). b. Lower down on synnema with conidiophore stipes covered by sheathing hyphae (x 50). c. Detail of naked conidiophore stipe near synnema apex. d. Detail of conidiophore stipe covered in the lower part by sheathing hyhae, and becoming covered in upper part. e. Detail of conidiophore stipes completely covered by sheathing hyphae. (c-e, all x 140)

phores scattered over lower part of stipe, concentrated in upper part, emerging at more or less right angles to the stipe, cylindrical-capitate or subulate-capitate, 375-2000 um long, 25-75 um wide with a central core 15-50 um wide with smooth walls 3-4 um thick narrowing to about 1 um thick beneath vesicle, surrounded by a sheath of frequently septate interweaving hyphae 2-6(-10) um wide, sometimes with naked stipes near apex of synnemata, vesicle globose to ellipsoidal, 50-100 um diam, metulae 7-21(-35), normally unbranched, but often developing internal septa. Phialides in terminal pairs or whorls of 3-5 on metulae, 12-22 um long, 3-7 um wide, with a conidiogenous aperture 1.5-3 um wide. Conidia obovate, ellipsoidal, oblong-ellipsoidal, or cylindrical, often with a truncate base with a visible detachment scar, olivaceous, (3-) 5-10 (-12) x (2.5-) 3.5-7 um, with smooth to verruculose walls. Stromata sterile, stipitate, hard, globose to turbinate, brown, 2-3.5 mm diam, the inside grey or green, the peridium 10-25 um thick, stipe 1-2 mm tall, 500-600 um wide.

HABITAT: Seeds of Anondium, Carpodinus lingustifolius, Landolphia dubreucquinana, L. florida.

KNOWN DISTRIBUTION: Africa (Belgian Congo, Burundi, Cameroon, Zaire).

MATERIAL EXAMINED:

Herbarium specimens: Only authentic material examined is listed. Specimen marked with * contains sterile stromata.

Penicilliopsis dybowskii, authentic material, on seed, leg. Dybowski, Congo, Jan. 1894, ex herb. Patouillard in herb. Bresadola (S).

Stilbothamnium dybowskii var. laevispora, *lectotype here designated, on seed, leg. L. Hauman no. 585, Zaire, forest between Bolafa and Mangana, Sept. 1932 (BR).

Stilbothamnium dybowskii var. macrospora, holotype, leg. Mme Goosens-Fontana no. 379B, Zaire, Lisala, Sept. 1925, A 599, BR).

Stilbothamnium usneoides, holotype, on seed of Anonidium sp., leg. J. Mildbraea no. 4560, Cameroon, between Lokoma, Bumba and Bange, Feb. 1911 (S).

Stilbothamnium usneoides var. verrucosum, holotype, on seeds, leg. J. Lebrun no. 4183, Zaire, forest between Irumi and Mambasa, Oct. 1931 (BR).

Aspergillus dybowskii is distinguished from the similar A. amazonensis and A. togoensis by the sheathing hyphae which cover the conidiophore and the smooth rather than roughened Aspergillus-stipes. These sheathing hyphae are unusual in Aspergillus, although less extensively developed sheathing hyphae are sometimes seen in A. togoensis.

Hauman (1936) considered S. dybowskii var. macrospora a nomen dubium because the type specimen was a mixture of S. dybowskii and Stilbodendron cervinum. Our observations confirm

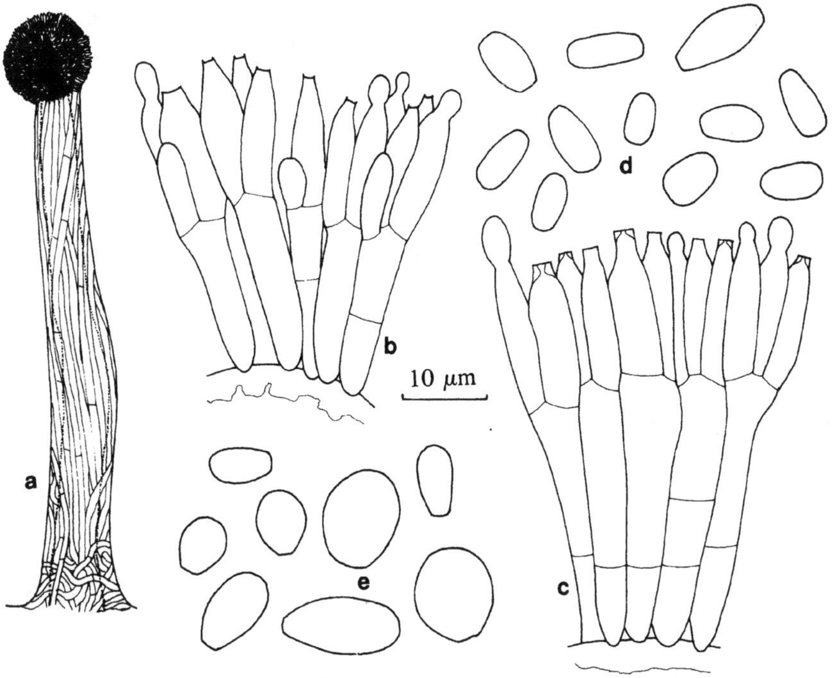

Fig. 16. Aspergillus dybowskii. a. Conidiophore covered with sheathing hyphae. b, c. Arrangement of metulae and phialides on vesicle. d, Conidia. (a, b, Rammeloo 6677, BR. c, d, authentic material, S).e. Conidia (Goosens, no. 379, BR).

this, but we include this variety in the synonymy above because the protologue clearly describes the Stilbothamnium to the exclusion of the Stilbodendron.

EXCLUDED SPECIES

Penicilliopsis bambusae Nag Raj & Govindu -- Sydowia 23: 112. 1969. The type specimen, IMI 89197, is not referable to any of the genera treated here. Its affinities are uncertain. No teleomorph was mentioned in the diagnosis.

Penicilliopsis togoense Henn. -- Bot. Jb. 30: 40. 1901. Type not in B, S. The diagnosis suggests this may be Stilbodendron cervinum or Sarophorum palmicola, but the habit on leaves rather than seeds makes the identity uncertain.

Stilbothamnium javanicum Henn. & E. Nym -- Monsunia 1: 37. 1899. Type not in B, BO, S. A specimen collected by Bakhuizen v. d. Brink labelled with this name deposited as BO 2306 contains no synnemata. Although the diagnosis appears to describe an Aspergillus, Stilbodendron cannot be ruled out. Nomen dubium.

Stilbothamnium penicillopsis Henn. & E. Nym. -- Monsunia 1: 37. 1899 = Aspergillus penicillopsis (Henn. & E. Nym.) Raciborski -- Bull. Acad. Sci. Cracovie 1909: 372. Type not in B, BO, S. Two specimens with this name collected by Van

Overeem after the publication of the name and deposited as BO 655 and BO 656 with this name are *Sarophorum palmicola*. This does not fit the diagnosis of *S. penicillopsis*. Nomen dubium.

Acknowledgments

We are particularly grateful to Dr. Jan Rammeloo (BR) for his efforts in locating specimens from Hauman's herbarium and for the loan of many additional collections of seminicolous fungi. We are also indebted to the curators of the other herbaria mentioned in the text for the loan of type and suplimentary specimens. The junior author thanks the Natural Sciences and Engineering Research Council of Canada for the award of a postgraduate scholarship.

REFERENCES

CARMICHAEL, J. W., KENDRICK, W. B., CONNORS, I. L. & SIGLER, L. 1980. "Genera of Hyphomycetes". Edmonton: University of Alberta Press.

CEJP, K. & DEIGHTON, F. 1969. Microfungi III. Mycol. Pap. 117: 1-31.

HAUMAN, L. 1936. Les champignons seminicole des forêts tropicales. Bull. Soc. r. Bot. Belg. 69: 96-129.

KOBAYASI, Y. & INDOH, H. 1943. On the genus *Penicilliopsis*. Bull. Centr. Natl. Mus. Machuokuo 5: 15-34 (in Japanese).

MALLOCH, D. & CAIN, R. F. 1972. The Trichocomataceae: Ascomycetes with *Aspergillus*, *Paecilomyces*, and *Penicillium* imperfect states. Can. J. Bot. 50: 2613-2628.

MORRIS, E. F. 1963. The synnematous genera of fungi imperfecti. Western Illinois Univ., Ser. Biol. Sci. 3: 1-143.

PERSIANI, A. M. & ONOFRI, S. 1982. A new genus of synnematous hyphomycetes from tropical rain forest. Mycotaxon 15: 254-260.

PETCH, T. 1944. Ceylon fungi, new and old. Trans. Br. mycol. Soc. 27: 137-147.

RACIBORSKI, M. 1909. Parazitische und epiphytische Pilzen Java's. Bull. Acad. Sci. Cracovie, 1p. 346-394.

RAPER, K. B. & FENNELL, D. I. 1965. "The genus *Aspergillus*". Baltimore: Williams & Wilkins.

ROQUEBERT, M-F. & ABADI, M. 1973. Etude ultrastructurale de la sporogenese chez un micromycete: *Stilbothamnium nudipes* Hauman. Comp. R. Acad. sci. 276: 281 - 285.

ROSSMAN, A. Y. 1983. The phragmosporous species of *Nectria* and related genera. Mycol. Pap. 150: 1-164.

SAMSON, R. A. 1979. A compilation of the Aspergilli described since 1965. Stud. mycol, Baarn 18: 1-40.

SOLMS-LAUBACH, H. G. 1887. *Penicilliopsis claviaeformis*, ein neuer javanischer Ascomycet. Ann. Jard. Bot. Buitenzorg 6: 53-72.

SUBRAMANIAN, C. V. 1971. "Hyphomycetes: an account of Indian species except Cercosporae". New Delhi: Indian Council of Agricultural Research.

DIALOGUE FOLLOWING DR. SAMSON'S PRESENTATION

MALLOCH: I don't think *Petromyces* should ever be considered a synonym of *Penicilliopsis* because *Petromyces* has a distinct cleistothecial wall, whereas *Penicilliopsis*, so far as I know, has simply stromatic tissue and no cleistothecial wall surrounding the locules. I do agree that they are closely related.

SAMSON: Most of the *Penicilliopsis* ascocarps we saw were either unripe or overmature. It was very difficult to decide whether or not there was a distinct layer of cells around the locules separating them from the stroma.

MALLOCH: Have you germinated any of those ascospores? Were they bivalved?

SAMSON: No. We tried, but these are all old herbarium specimens. I think the ascospores are somewhat bivalved.

PITT: I question the wisdom of putting *Stilbothamnium* into *Aspergillus*. The forms producing synnemata which you have shown us may well represent species ancestral to *Aspergillus*, but they are not the same thing. *Aspergillus* is a modern, well defined genus. These genera are certainly related, but they are not the same.

SAMSON: What would you do with *Aspergillus erythrocephalus*?

PITT: I don't mind what you do with it. Put it where it fits best.

SAMSON: Bartoli & Maggi isolated *Aspergillus coremiiformis* from soil not knowing it was *Stilbothamnium*, and in my opinion, placed it in the correct genus. Probably within a few years all of these fungi will turn up in culture.

WICKLOW: We wouldn't want to take the synnematous Penicillia, such as *P. claviforme*, out of *Penicillium*.

SAMSON: *Penicillium isariiforme* may be a similar situation to these *Stilbothamnium* species. It was isolated from soil in a tropical rain forest. Is the synnematous habit your only objection to combining *Stilbothamnium* with *Aspergillus*?

PITT: I think the synnematous habit comes from a whole different world and a whole different era in fungal evolution.

MALLOCH: There's a whole gradation between mononematous and synnematous forms.

SEIFERT: I've looked at a great deal of material of *Stilbothamnium* over the past two years and I am convinced that it. cannot be separated from *Aspergillus*. Sometimes you even find mononematous conidiophores, synnematous conidiophores and everything in between on a single specimen.

PITT: All right. I'm convinced.

WICKLOW: I think this is a beautiful evolutionary story. You have this wet tropical forest environment which supports that sort of structure, and then you have all these tropical Aspergilli which are found in arid tropical environments, where time does not permit the production of these elaborate structures because it is only wet for brief periods. These species tend to have sclerotia which germinate by directly producing mononematous conidiophores. These may have been the forms which eventually come to colonize foods.

CONTRIBUTION TO OUR KNOWLEDGE OF ASPERGILLUS TELEOMORPHS: SOME TAXONOMIC PROBLEMS

Shun-ichi UDAGAWA and M. TAKADA

National Institute of Hygienic Sciences
Tokyo, Japan and Toyo Jozo Co., Ltd., Ohito
Shizuoka, Japan

SUMMARY

Based on recent knowledge of Aspergillus teleomorphs, some species belonging to the genera Eurotium, Neosartorya and Emericella are treated taxonomically. The second heterothallic species of Neosartorya, N. spathulata, is described. Results on interspecific pairings between another heterothallic species, N. fennelliae, and A. fumigatus are also noted.

INTRODUCTION

The numerous species of Aspergillus are distributed widely as environmental contaminants of air, foods, dust, etc. Although only a relatively small proportion of the whole Aspergillus genome is responsible for ascomatal formation in their life cycle, those which do form asci, e.g. the species belonging to the genera Eurotium, Neosartorya and Emericella, are as common and as cosmopolitan as the Aspergillus anamorphs. The taxonomy of Aspergillus species has been elaborated for years; however, only recently has the significance of teleomorph-anamorph connections for Trichocomaceae systematics as well as the other genera with a Penicillium or Paecilomyces anamorph been of concern (Malloch and Cain, 1972; Subramanian, 1972; Malloch, 1979; Subramanian, 1979). In their monograph Raper and Fennell (1965) listed 39 species and 8 varieties of Aspergillus as having teleomorphs. Information on the Aspergillus teleomorphs has considerably expanded within the last two decades. Attention is focussed on some species of Eurotium, Neosartorya, and Emericella in this study.

EUROTIUM

The revision by Blaser (1976) includes 19 species in the genus Eurotium. Besides Blaser's taxa, one additional species, E. xerophilum Samson et Mouchacca (1975) has been isolated from soil samples collected in the Western Desert of Egypt.

Samson (1979) has disagreed with Blaser at the following points: E. repens and E. acutum are considered as synonyms of E. herbariorum; and E. heterocaryoticum is conspecific with E. amstelodami. Concurrently, he stated that the genera Edyuilla and Gymnoeurotium, both based on Aspergillus athecius in the A. glaucus group, should be regarded as synonymous with Eurotium.

A variety of Eurotium echinulatum Delacr. (Fig. 1)

During a survey of the mycota in the domestic environment, an atypical strain of E. echinulatum was frequently encountered on the dilution plates of carpet dust prepared with DG 18 agar (Hocking and Pitt, 1980). This fungus could be identified as similar to E. echinulatum, but some of its morphological characters are distinguished from those of the type strain as described below.

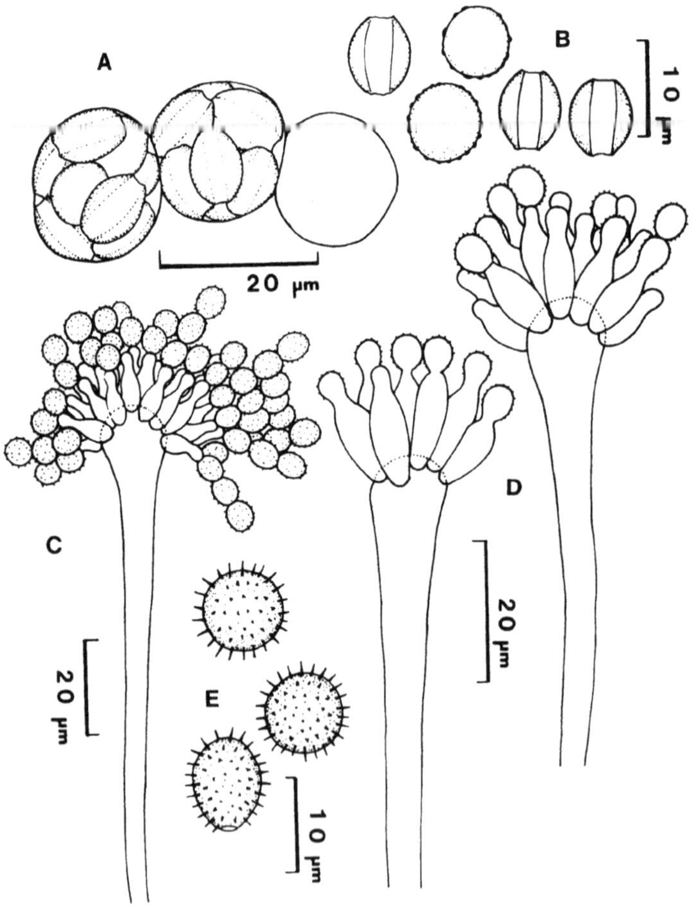

Fig.1. Eurotium sp. (NHL 2938). A. asci, B. ascospores, C-D. conidiophores. E. conidia.

Colonies on M40Y agar spreading, greyish green; reverse uncolored to olive brown. Cleistothecia pale yellow, 115-160 µm diam. Ascospores broadly lenticular, 8-9 x 6-7 µm, with convex surfaces smooth or very finely roughened, with two equatorial crests. Conidial heads radiate. Conidiophores mostly 40-200 x 5-6 µm, sometimes up to 450 µm long. Vesicles clavate to dome-shaped, 10-20(-30) µm diam. Phialides cylindrical to flask-shaped, (10-)12-16 x 5-6 µm. Conidia subglobose to ovoid or ellipsoidal to pyriform, (7-)8-14(-16) x (5.5-)7-12(-14)) µm, distinctly echinulate.

Material examined: NHL 2938, isolated from house dust of carpet, Bunkyo-ku, Tokyo, Japan, 16. 3, 1984, coll. S. Udagawa.

The variety strain differs from the type species of E. echinulatum in producing greyish green colonies, short conidiophores and larger conidia. It is our impression that the fungus is assignable to a variety of E. echinulatum, though more exact placement must await further comparative study.

NEOSARTORYA

Within the genus five species and two varieties have been recognized by Raper and Fennell (1965) and are widely distributed in tropical to temperate soils. Since 1965, only one species N. fennelliae Kwon-Chung et Kim (1974) has been described as a new species. It was isolated from eyeballs of a rabbit in the U.S.A.

N. spathulata Takada et Udagawa, a new heterothallic Neosartorya*

During an investigation of the mycota of Formosan soils, a heterothallic fungus with an Aspergillus anamorph (assignable to the A. fumigatus group) was isolated and appeared to be sufficiently different from N. fennelliae to warrant its proposal as a second heterothallic new species of Neosartorya. The brief description of this fungus is as follows:

Colonies on malt extract agar spreading, floccose, producing abundant dull green to greyish green conidial heads, with pale yellow cleistothecia developed only after appropriate pairings; reverse uncolored to somewhat greyish brown. Colonies on oatmeal agar growing moderately, floccose, greyish green; conidial heads and cleistothecia as described on malt; reverse uncolored. Colonies on both malt extract and oatmeal agar at 37°C growing more rapidly than at 23°C.

Heterothallic. Cleistothecia pale yellow to light yellow, globose to subglobose, 100-260 µm diam, covered by hyphal strands; peridium membranaceous, composed of several layers of flattened cells; outer layer of "textura epidermoidea" and inner of "textura angularis". Asci borne in short chains, 8-spored, globose to subglobose, 12-15.5 x 10-12 µm, evanescent.

---

* Formal description published elsewhere (Udagawa and Takada, Mycotaxon, in press).

Ascospores hyaline, lenticular, 5-7 x 3-4.5 µm including crests, with smooth convex surfaces, and bearing two widely spaced equatorial crests. Conidial heads uniseriate, loosely columnar, 120-300 x 40-80 µm. Conidiophores usually arising from the basal felt, sometimes from aerial hyphae, of two kinds: (a) large type, more or less scattered, 500-1500 x 11-18(-25) µm, projecting above the layer of smaller conidiophores, verrucose, thick-walled, broadening to dome- or flask-shaped vesicles measuring (22-)25-52 µm diam; (b) small type, 60-250 x 4-10 µm, smooth, thick-walled, hyaline, broadening to vesicles measuring 8-15(-24) µm diam. Phialides borne on the upper 1/2 to 3/4 surface of the vesicle, flask-shaped, 6-10 x 2-4 um. Conidia faintly yellow-green, mostly ellipsoidal, 3-5.5 x 2-4.5 µm, smooth.

Material examined: NHL 2947, dried culture of NHL 2948 (A) x 2949 (a), both strains isolated from cultivated soil of *Alocasia macrorrhiza* fields in Tai-Ping, Ilan, Formosa, 13.10.1982, coll. M. Takada.

The yellow cleistothecia, catenate asci, and ascospores with nearly smooth convex surfaces seen in *N. spathulata* are in contrast to the white cleistothecia, single asci, and ascospores with reticulate convex surfaces of *N. fennelliae*. The two species can also be separated on the basis of the differences in their anamorphs. *N. fennelliae* always forms one type of conidial structures and small globose conidia, while in *N. spathulata* the conidial structures are of two kinds and the conidia are large and ellipsoidal.

All the mating experiments were made on malt extract and oatmeal agars at 25°C, 14 days or at 37°C, 8-14 days. Two kinds of mating types, "A" and "a", were recognized in the isolates of this fungus from cultivated soil in Formosa. When monoconidial strains from the above isolates were mated on agar plates two bands of fertile cleistothecia near the junction of the two colonies appeared between, but only with compatible strains. None of the crosses between this fungus and each strain (NRRL 5534, 5535; NHL 2951, 2952) of *N. fennelliae*, even between "A" and "a", was fertile.

Neosartorya fennelliae Kwon-Chung et Kim (Fig. 2)

St. anam.: *Aspergillus fennelliae* Kwon-Chung & Kim

Colonies on malt extract agar spreading, floccose, producing white cleistothecia after appropriate pairings; reverse reddish yellow to greyish yellow. Heterothallic. Cleistothecia globose, 90-240 µm diam, peridium thin, membranaceous. Asci 8-spored, globose to ovoid, 12-14(-16) x 10-13 µm. Ascospores hyaline, lenticular, 5-7 x 4-4.5 µm including crests, with two equatorial crests and reticulate surfaces. Conidial heads greyish green, loosely columnar, uniseriate. Conidiophores 95-450 x 3.5-7 µm, smooth-walled.

vesicles flask-shaped, (6-)8-12 µm diam. Phialides 6-7 x 2-3 µm. Conidia mostly globose to subglobose, 2-3.5 µm diam, smooth.

Material examined: NHL 2950, dried culture of NHL 2951 (A) x 2952 (a), isolated from marine sludge, Fuji City, Shizuoka, Japan, 29.10.1981, coll. M. Takada

A fungus which appeared to be a heterothallic _Neosartorya_ from marine sludge in Japan was crossed with monoconidial isolates of a pair of tester strains of _N. fennelliae_, NRRL 5534 "A" and NRRL 5535 "a", on oatmeal agar plates. Among the pairings, our isolate NHL 2951 "A" developed two bands of fertile cleistothecia near the junction of the two colonies when mated with the strain NRRL 5535 "a" and the strain NHL 2952 "a" developed similar fertile cleistothecia when mated with the strain NRRL 5534 "A". The results indicate that the Japanese heterothallic _Neosartorya_ is identical with _N. fennelliae_, not only morphologically but in matings.

This isolation, to our knowledge, represents the second record of _N. fennelliae_. With only minor differences such as sizes of cleistothecia and conidiophores, the microscopical features of the Japanese isolate are quite identical with those of the type strains from NRRL.

Interspecific pairings between NHL strains of Neosartorya fennelliae and Aspergillus fumigatus

Kwon-Chung and Kim (1974) reported that interspecific pairings of _Neosartorya fennelliae_ and _Aspergillus fumigatus_ not only failed to cross but usually produced clear zones between the strains. However, as an exception, they found that strains of _N. fennelliae_ stimulated hyphal growth of _N. fischeri_ var. _glabra_ along the line of contact.

Their findings led us to examine the reaction of the Japanese strains of _N. fennelliae_ in pairings with _A. fumigatus_. A total of 43 isolates of _A. fumigatus_ were tested. In pairings between the "A" mating type tester strain (NHL 2951) of _N. fennelliae_ and almost all isolates of _A. fumigatus_, many abortive cleistothecia were produced in a band on the side of the tester strain. Stimulation of abortive cleistothecial production of _N. fennelliae_ occurred at the zone of contact between cultures of NHL 2951 and each reactive isolate of _A. fumigatus_ but never occurred on the "a" mating type strain (NHL 2952) when paired with all tested _A. fumigatus_ isolates. These results mean that the tested isolates of _A. fumigatus_ potentially behaved only as "a" mating type. If unequal distribution of this sex function is not restricted within certain narrow geographic areas (e.g. Japan), it seems to be the phylogenic nature of sexual characteristics in _A. fumigatus_. The possibility cannot be excluded that an ancestoral "_N. fennelliae_" lost the "A" characters and that only the "a" type persists to the present time as _A. fumigatus_.

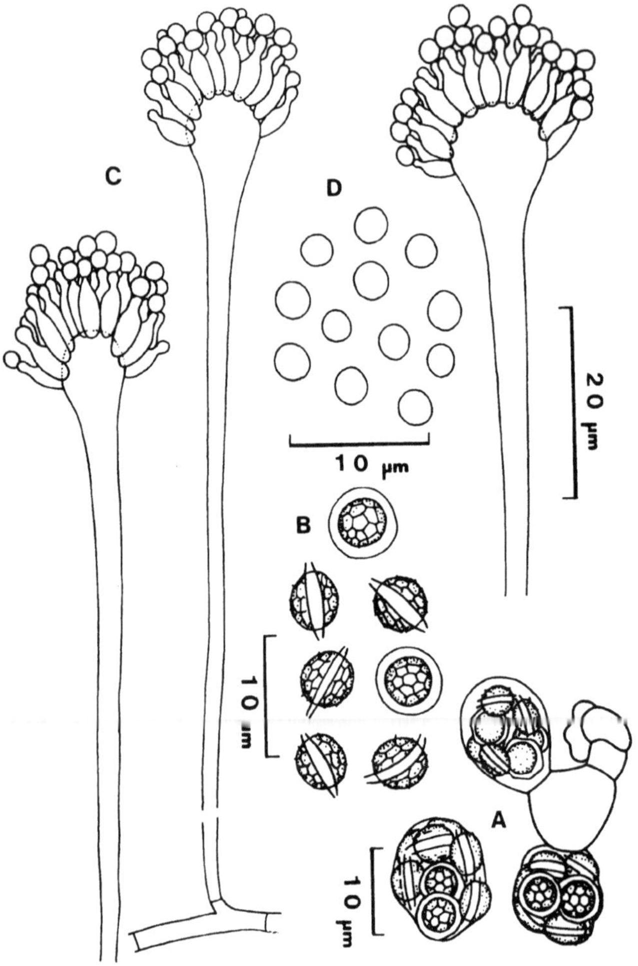

Fig. 2. Neosartorya fennelliae (NHL 2950). A. asci. B. ascospores. C. conidiophores, D. conidia.

EMERICELLA

The genus Emericella has been reviewed recently by Christensen and States (1982), who listed 29 species and 5 varieties in their synoptic key, considerably expanding our knowledge of the genus. One additional species has been described by Horie (1979). Horie (1980) made SEM observations on ascospore ornamentation of 21 species and 5 varieties. Emericella species are mainly distributed in tropical and subtropical zones, as shown by frequent isolations of E. striata from India and Nepal (Raper and Fennell, 1965; Udagawa and Sugiyama, 1981, 1982).

REFERENCES

BLASER, P. 1976. Taxonomische und physiologische Untersuchungen über die Gattung Eurotium Link ex Fr. Sydowia 28: 1-49.

CHRISTENSEN, M. and STATES, J.S. 1982. Aspergillus nidulans

group: *Aspergillus navahoensis*, and a revised synoptic key. Mycologia 74: 226-235.
HOCKING, A.D. and PITT, J. I. 1980. Dichloran-glycerol medium for enumeration of xerophilic fungi from low-moisture foods. Appl. Environ. Microbiol. 39: 488-492.
HORIE, Y. 1979. New or interesting *Emericella* from herbal drugs. Trans. mycol. Soc. Japan 20: 481-491.
HORIE, Y. 1980. Ascospore ornamentation and its application to the taxonomic re-evaluation in *Emericella*. Trans. mycol. Soc. Japan 21: 483-493.
KWON-CHUNG, K.J. and KIM, S.J. 1974. A second heterothallic *Aspergillus*. Mycologia 66: 628-638.
MALLOCH, D. 1979. Plectomycetes and their anamorphs. In "The Whole Fungus, Vol. 1" B. Kendrick, ed. Ottawa: National Museums of Canada. pp. 153-165.
MALLOCH, D. and CAIN, R.F. 1972. The Trichomataceae: Ascomycetes with *Aspergillus*, *Paecilomyces*, and *Penicillium* imperfect states. Can. J. Bot. 50: 2613-2628.
RAPER, K.B. and FENNELL, D.I. 1965. The genus *Aspergillus*. Baltimore: Williams and Wilkins, 686 pp.
SAMSON, R.A. 1979. A compilation of the Aspergilli described since 1965. Stud. Mycol., Baarn 18: 1-38.
SAMSON, R.A. and MOUCHACCA, J. 1975. Additional notes on species of *Aspergillus*, *Eurotium* and *Emericella* from Egyptian desert soil. Antonie van Leeuwenhoek 41: 343-351.
SUBRAMANIAN, C.V. 1972. The perfect states of *Aspergillus*. Curr. Sci.41: 755-761.
SUBRAMANIAN, C.V. 1979. Phialidic Hyphomycetes and their teleomorphs -- an analysis. In "The Whole Fungus", Vol. 1" B. Kendrick, ed. Ottawa: National Museums of Canada. pp. 125-151.
UDAGAWA, S. and SUGIYAMA, Y. 1981. Additions to the interesting species of Ascomycetes from imported spices. Trans. mycol. Soc. Japan 22: 197-212.
UDAGAWA, S. and SUGIYAMA, Y. 1982. New records and new species of ascomycetous microfungi from Nepal, a preliminary report on the expedition of 1980. In "Reports on the Cryptogamic Study in Nepal" Y. Otani, ed. Tokyo: National Science Museum. pp. 11-46.

DIALOGUE FOLLOWING DR. UDAGAWA'S PAPER

WICKLOW: In some of the crosses Dr. Kurtzman did with yeasts, he found that the ascospores produced in apparantly good matings would not germinate. Did you do germination studies of the ascospores from some of your crosses?

UDAGAWA: No, we did not test this.

HAWKSWORTH: How do you prepare your specimens for scanning electron microscopy? Your photographs are very beautiful.

UDAGAWA: We use the normal proceedure with osmium tetroxide using a suspension of ascospores. Sometimes we use a sonicator to settle the ascospores into an even layer.

SECONDARY METABOLITES AS AN AID TO EMERICELLA CLASSIFICATION

Jens C. FRISVAD

Food Technology Laboratory
Technical University of Denmark
2800 Lyngby, Denmark

SUMMARY

The genus Emericella contains thirty species and five varieties, including eight related and strictly anamorphic Aspergillus species. Chemotaxonomic results support the established taxonomy in the genus. Some species appeared to be only morphological variants of the type species E. nidulans, E. nidulans and its varieties (lata, echinulata, dentata, and acristata), E. quadrilineata, E. corrugata, E. foveolata, E. rugulosa, E. cleistominuta and maybe E. violacea. E. navahoensis and A. recurvatus had very similar mycotoxin profiles to E. fruticulosa. The species mentioned above, except E. violacea and A. recurvatus, produced the carcinogenic mycotoxin sterigmatocystin. Other producers of sterigmatocystin were E. spectabilis, E. bicolor, E. heterothallica and A. multicolor. E. nidulans (including the varieties listed above), E. striata, E. variecolor var. variecolor and E. variecolor var. astellata all produced asperthecin, a compound related to teleomorph formation. Other species in Emericella and related Aspergilli produced specific profiles of unknown secondary metabolites.

INTRODUCTION

Members of the Aspergillus nidulans group and Emericella are found in soils from subtropical and tropical regions, and sometimes in foods, especially E. nidulans and E. variecolor. They are also important pathogens (Raper and Fennell, 1965). Together with species in the A. versicolor and A. ustus groups, these species produce a number of secondary metabolites : pentostatin, emerin, asperlin, quadrilineatin, asperugin, siderin, emericellins, asteltoxin, sterigmatocystins, nornoidulins, austins, astellolid, siderin, andibenin, echinocandin, asperthecin, nidulotoxin and aspertetronin. All of these species of Aspergillus produce green conidia, and often Hülle cells and brown biseriate conidiophores. Data on the production of secondary metabolites by these fungi (Turner and

Aldridge, 1983) indicate that species like E. nidulans and E. rugulosa are closely related as both produce specific groups of biosynthetically different secondary metabolites: pentostatin, asperthecin, emericellin, sterigmatocystin and echinocandin. Other species, such as E. variecolor var. variecolor, produce a great number of other secondary metabolites.

Recent genetic studies on members of Emericella support the chemotaxonomic indications mentioned above. Earl et al. (1981) showed, that the mitochondrial genomes of E. nidulans var. dentata and E. nidulans var. nidulans are nearly identical, except six small inserted base sequences in the former variety. The insertions did not appear to influence the phenotype. Kevei and Peberdy (1984) showed that after protoplast fusion allodiploids could be recovered from crosses of E. nidulans var. nidulans with E. rugulosa, E. quadrilineata, E. nidulans var. echinulata and E. violacea. These results show that the latter species are very alike. E. violacea seemed to have more pronounced boundaries than the other species however.

Apart from ascospore morphology (Horie, 1980), morphological differences between the varieties of E. nidulans, E. rugulosa and E. quadrilineata are rather small (Raper and Fennell, 1965; Christensen and States, 1982). Even though ascospore morphology is considered important by taxonomists (Horie, 1980; Stolk and Samson, 1983) fungal taxonomy should include many other phenotypic expressions of the genome (Pitt, 1984). The aim of this study was to find out if chemotaxonomy, based on simple thin layer chromatography methods, could contribute to the taxonomy of this well known telemorphic genus.

MATERIAL AND METHODS

The isolates investigated are listed in Table I. Each taxon include the available type cultures. These isolates were grown on Czapek yeast extract agar (CYA), malt extract agar (MEA), yeast extract sucrose agar (NO2) (Frisvad, 1981; Pitt, 1979) and oat meal agar (Gams et al., 1980). The fungi were incubated at 25°, 30° and 37°C and observed after 7, 14, 21 and 28 days of incubation. Colony diameters were measured after 7 days of growth at 25°C. The profiles of secondary metabolites were determined by the simple thin-layer chromatographic methods of Filtenborg and Frisvad (1980) and Filtenborg et al. (1983). Extracellular metabolites were detected on YES agar (Filtenborg and Frisvad, 1980), and intracellular metabolites were detected on CYA, YES and oat meal agar after 14 to 28 days of growth using chloroform-methanol (2:1, vol/vol) as the solvent. Where possible agar plugs were taken both from purely conidial areas in the colonies and purely cleistothecial areas and analyzed separately. The developing systems used were toluene-ethyl acetate-90% formic acid (5:4:1, vol/vol/vol) and Chloroform-acetone-isopropanol (85:15:20, vol/vol/vol). Griseofulvin, sterigmatocystin and asperthecin were used as external and internal standards. All TLC plates were examined under visible and UV light (254 and 366 nm) before and after treatment with spray reagents. Anisaldehyde spray was used for extracellular metabolites (Filtenborg et al., 1983) and $AlCl_3$ spray and heating for 5 min at

120°C was used for intracellular metabolites (Filtenborg et al., 1983).

RESULTS AND DISCUSSION

Some of the secondary metabolites produced by species of Emericella and the Aspergillus nidulans group are listed in Table I. E. nidulans and its varieties, E. quadrilineata, E. rugulosa, E. corrugata and E. foveolata produced both sterigmatocystin and asperthecin except for E.nidulans var. acristata IMI 90326 and E. nidulans IMI 85473. The latter isolate was the first known producer of nidulin and nornidulin (Dean et al., 1954) and it was first identified as A. ustus. The isolate received from IMI was closer to E. unguis. Subsequent isolation of nornidulin and similar metabolites from E. unguis (Turner and Aldridge, 1983) suggest that IMI 85473 is indeed an isolate of the latter species. Sterigmatocystin was also detected in E. spectabilis, E. bicolor, E. fruticulosa, E. navahoensis, E. heterothallica, A. multicolor and E. cleistominuta. Asperthecin was also detected in E. violacea, E. striata, E. variecolor var. variecolor and E. variecolor var. astellata. Sterigmatocystin was detected in both anamorph and teleomorph structures, while asperthecin was detected in teleomorph structures only.

The following species in Emericella and the A. nidulans group produced distinct profiles of known and unknown secondary metabolites: E. variecolor var. variecolor, E. unguis, E. aurantiobrunnea, E. heterothallica, E. desertorum, E. foeniculicola, E. purpurea, A. egyptiacus, and A. multicolor. These distinct profiles provide evidence for the validity of these species. Two other groups of species had many common secondary metabolites. E. fruticulosa, A. recurvatus and E. navahoensis produced a very distinct TLC pattern in common. The most conspicuous of these metabolites were sterigmatocystin and three bluish turquoise fluorescing metabolites. Morphological characteristics also overlap in these three species (Christensen and States, 1982), and their habitats are all in desert soil. Perhaps the three species are synonymous.

The most important group of species is E. nidulans and its varieties, E. quadrilineata, E. rugulosa, E. cleistominuta, E. corrugata and E. quadrilineata, E. foveolata. These taxa produced the same profile of secondary metabolites and they were alike macromorphologically. Some isolates in this group grew restrictedly on all media: E. nidulans var. echinulata IMI 193022, E. rugulosa IMI 136775 and IMI 153724 and E. cleistominuta IMI 131554. This may be due to production of self inhibiting secondary metabolites in high amounts. All members of the group grew well at 37°C in contrast to E. aurantiobrunnea, E. bicolor, E. spectabilis, E. foeniculicola, E. heterothallica, and E. purpurea (Horie, 1980). Isolates in the subgroup mentioned first in this paragraph are not geographically isolated and they are widespread on similar substrates, such as soil, hay, plants, basidiomyceteous fungi and mycotic lesions. Isolates of E. nidulans, E. nidulans var. lata and E. quadrilineata were even found in the same sample of gum arabic by E. Yuill (CMI culture collection catalogue, 1982). Morpho-

Table 1. Species of *Emericella* and the *Aspergillus nidulans* group and their mycotoxin production.

Species	No. of isolates	Sterigma-tocystin	Asper thecin	Specific profile[a]
E. nidulans var. *nidulans*	11	+	+	A
E. nidulans var. *lata*	2	+	+	A
E. nidulans var. *acristata*	2	+/-	+	A
E. nidulans var. *dentata*	2	+	+	A
E. nidulans var. *echinulata*	2	+	+	A
E. quadrilineata	3	+	+	A
E. rugulosa	2	+	+	A
E. cleistominuta	1	+	-	(A)
E. corrugata	1	+	+	A
E. foveolata	1	+	+	A
E. violacea	2	-	+	
E. striata	1	-	+	
E. variecolor var. *variecolor*	2	-	+	B/(B)
E. variecolor var. *astellata*	1	-	+	
E. spectabilis	1	+	-	
E. bicolor	1	+	-	
E. fruticulosa	1	+	-	C
E. navahoensis	1	+	-	C
A. recurvatus	1	-	-	C
E. unguis	2	-	-	D
E. nidulans var. *nidulans*	1	-	-	D?
E. aurantiobrunnea	1	-	-	E
E. parvathecia	1	-	-	(F)
E. heterothallica	2	+	-	G
E. desertorum	1	-	-	H
E. foeniculicola	1	-	-	I
E. purpurea	1	-	-	J
A. egyptiacus	2	-	-	K
A. aureolatus	1	-	-	
A. crustosus	1	-	-	
A. aeneus	2	-	-	
A. kassunensis	1	-	-	(L)
A. subsessilis	2	-	-	-/(L)
A. multicolor	1	(+)	-	M
A. eburneocremeus		Not tested		

[a]The profiles of secondary metabolites listed as B, C, D, F, G, H, I, J, K, and M were distinctive. The A profile was most distinctive in *E. nidulans* and varieties, *E. quadrilineata*, *E. rugulosa*, *E. cleistominuta*, *E. corrugata* and *E. foveolata*. *E. spectabilis* and *E. bicolor* produced two distinctive metabolites in common.

logical differences, apart from ascospore ornamentation, are small or overlapping in these taxa (see the synoptic key in Christensen and States, 1982). Thus, most characters indicate that the taxa in the subgroup mentioned above are conspecific. This conclusion is supported by recent genetic experiments. Earl et al. (1981) found that mitochondrial DNA of *E. nidulans* and *E. nidulans* var. *echinulata* were homologous, except from six small inserted nucleotid fragments in the latter variety.

The additional DNA fragments in the variety did not seem to be expressed phenotypically. Kevei and Peberdy (1984) showed that after somatic hybridisation of protoplasts, E. nidulans formed allodiploids with E. quadrilineata, E. rugulosa, E. nidulans var. echinulata and (with much lower fusion frequency) E. violacea. Crosses between E. nidulans and E. unguis, E. variecolor or E. heterothallica were unsuccesfull.

The suggested conspecificity of E. nidulans with its varieties, E. quadrilineata, E. rugulosa, E. cleistominuta, E. foveolata and E. corrugata raises the question as to the validity of spore ornamentation as an exclusive taxonomic character and the species concept in Aspergillus and its teleomorph. In most cases spore ornamentation is an excellent and diagnostic character, but only if it is correlated with other positive characters of different types. The rough-walled and smooth walled conidia of P. purpurugenum and P. rubrum respectively were used to separate the two species by Raper and Thom (1949), for example, but Pitt (1979) synonymized these species, because they has many other characters in common. This synonymy is further supported by the profile of secondary metabolites: both species produce rubratoxins and phoenicin (Turner and Aldridge, 1983). Pitt (179) also synonymized P. corymbiferum (smooth conidia) and P. hordei (slightly rough-walled conidia). Both species produce terrestric acid and roquefortine C, but they also differ in a number of secondary metabolites. This difference between the taxa is supported by the difference in habitats: P. corymbiferum is a pathogen of onions and bulbs in soil, whereas P. hordei is found on barley (or in the air surrounding barley kernels). The latter two species probably has a common ancestor (they are rather closely related), but the difference in conidium morphology is correlated to other characters and the taxa should be separated, among other for practical reasons. E. violacea is also closely related to E. nidulans, but correlated differences dictate a separation of the species, maybe on the variety level.

E. spectabilis and E. bicolor both produced sterigmatocystin and two extracellular metabolites. These taxa may also prove to be conspecific. They are also rather similar in morphological characters. As mentioned by Christensen and Raper (1978), two strains of E. violacea differed in having echinulate and smooth ascospore walls, and differences in E. bicolor and E. spectabilis may also prove to be too slender to separate them. More detailed studies of the morphology, physiology and chemotaxonomy of such related strains may reveal their correct status.

Rather weak profiles of secondary metabolites indicated that A. kassunensis and A. subsessilis are conspecific, as suggested by Samson (1979).

E. sublata Horie was not investigated, but the description (Horie, 1979) indicates, that it, too, is conspecific with E. nidulans.

Thus the original 35 taxa in Emericella and the A. nidulans group seem to represent only 21 taxa based on all kinds of characters, with emphazis on morphology and profiles of secondary metabolites. Further studies, for example using gel electrophoresis and other chemotaxonomical polyphenic methods, may show if the reduction in the number of species in Emericella and the A. nidulans group suggested here is reasonable. Such studies should preferably also include the closely related species in A. versicolor group and the A. ustus group of Raper and Fennell (1965).

Further studies on the connection between different morphological structures, e.g. conidiophore stipes and conidia, Hulle cells, cleistothecia, ascospores and encrusted hyphae, and profiles of secondary metabolites may also be very rewarding. The colours of these structures are traditionally very important characters (Raper and Fennell, 1965); Christensen and States, 1982) and the recovery of trace amounts of secondary metabolites related to teleomorph formation in the hyphae may help in the search for teleomorphic phases in anamorph species. In this study asperthecin was found in the cleistothecia of several species. The recovery of asperthecin in a sclerotial strain in the A. nidulans group may indicate that the fungus has a teleomorphic state in Emericella. Wicklow and Cole (1982) found that aflatrem and aflavinine were found in sclerotia but not in the conidia of Aspergillus flavus. The recovery of these metabolites in teleomorphs of Aspergillus would be very interesting.

REFERENCES

CHRISTENSEN M. & RAPER, K.B. 1978. Synoptic key to Aspergillus nidulans group species and related Emericella species. Trans. Br. mycol. Soc. 71: 177-191.

CHRISTENSEN, M. and STATES, J.S. 1982. Aspergillus nidulans group: Aspergillus navahoensis, and a revised synoptic key. Mycologia 74: 226-235.

DEAN, F.M., ROBERTS, J.C. and ROBERTSON, A. 1954. The Chemistry of fungi. XXII. Nidulin and nornidulin ("ustin"): chlorine-containing compounds of Aspergillus nidulans. J. Chem. Soc. 1954: 1432-1439.

EARL, A.J., TURNER, G., CROFT, J.H., DALES, B.G., LAZARUS, C.M., LUNSDORF, H. and KUNTZEL, H. 1981. High frequency transfer of species specific mitochondrial DNA sequences between members of the Aspergillaceae. Curr. Genet. 3: 221-228.

FILTENBORG, O. and FRISVAD, J.C. 1980. A simple screening method for toxigenic moulds in pure cultures. Lebensm. Wiss. Technol. 13: 128-130.

FILTENBORG, O., FRISVAD, J.C. and SVENDSEN, J.A. 1983. Simple screening method for molds producing intracellular mycotoxins in pure cultures. Appl. environ. Microbiol. 45: 581-585.

FRISVAD, J.C. 1981. Physiological criteria and mycotoxin production as aids in identification of common asymmetric Penicillia. Appl. environ. Microbiol. 41: 568-579.

GAMS, W., AA, H.A. van der, PLAATS-NITERINK, A.J. van der, SAMSON, R.A. and STALPERS, J.A. 1980. "CBS Course of Mycology" Baarn: Centraalbureau voor Schimmelcultures.

HORIE, Y. 1979. New or interesting Emericella from herbal drugs. Trans. mycol. Soc. Japan. 20: 481-491.

HORIE, Y. 1980. Ascospore ornamentation and its application to the taxonomic re-evaluation in Emericella. Trans. mycol. Soc. Japan. 21: 483-493.

KEVEI, F. and PEBERDY, J.F. 1984. Further studies on protoplast fusion and interspecific hybridization within the Aspergillus nidulans group. J. gen. Microbiol. 130: 2229-2236.

PITT, J.I. 1979. "The Genus Penicillium and its Teleomorphic States Eupenicillium and Talaromyces". London: Academic Press.

PITT, J.I. 1984. The value of physiological characters in the taxonomy of Penicillium. In "Toxigenic fungi - their toxins and health hazard, H. Kurata, and Y. Ueno, eds. Amsterdam: Elsevier. pp. 107-118.

RAPER, K.B. and FENNELL, D.I. 1965. "The Genus Aspergillus" Baltimore: Williams and Wilkins.

RAPER, K.B. and THOM, C. 1949. "A Manual of the Penicillia". Baltimore: Williams and Wilkins.

SAMSON, R.A. 1979. A compilation of the Aspergilli described since 1965. Stud. Mycol., Baarn 18: 1-38.

STOLK, A.C. and SAMSON, R.A. 1983 The Ascomycete genus Eupenicillium and related Penicillium anamorphs. Stud. Mycol., Baarn 23: 1-149.

TURNER, W.B. and ALDRIDGE, D.C. 1983. "Fungal metabolites II". London: Academic Press.

WICKLOW, D.T. and COLE, R.J. 1982. Tremorgenic indole metabolites and aflaxotoxins in sclerotia of Aspergillus flavus: an evolutionary perspective. Can. J. Bot. 60: 525-528.

DIALOGUE FOLLOWING DR. FRISVAD'S PAPER

MOUCHACCA: I disagree with lumping all the Emericella species together. In fact, I think that the morphological differences separating some of the varieties of E. nidulans are quite stable, particularly characters of the ascospores, and that these varieties should be raised to species rank.

FRISVAD: I can understand your reservations.

PITT: I wouldn't expect that Emericella species would have the wide variety of metabolites you find in Penicillium. If there is no biochemical differentiation within the genus, then we have to trust the morphological differentiation.

PATERSON: But these morphological differences are only differences in ascospore ornamentation, aren't they? But there are three metabolites and they're all the same. Who's to say that one should be weighted higher than the other?

CHRISTENSEN: The ascospore ornamentation is the character that allows us to separate these species. How much DNA is involved in determining the morphology of the ascospore? Is there a different rate of evolutionary change between morphological characters and metabolic characters? We don't know this. I wonder if these metabolites you are detecting are an indicator of their similar ecological habitats.

FRISVAD: This is possible. I can accept that these are different species, but they are certainly closely related.

MALLOCH: The different ascospore morphologies may represent different dispersal mechanisms. The species may not be physiologically isolated, but they must be isolated in some way, or you wouldn't have constant differences in ascospores all over the world. Each shape must have its adaptational advantage for the organism.

FRISVAD: If you consider E. violacea, some isolates have ascospores with many spines, but sometimes you find isolates from the same ecological niche which lack this ornamentation.

MALLOCH: But other species are very constant in morphology, such as E. rugulosa.

ARE THE SPECIES OF PENICILLIUM AS IMPERFECT AS WE THINK?

Carlos RAMÍREZ

Laboratory of General and Applied Mycology
Instituto "Jaime Ferrán" de Microbiología
2806-Madrid, Spain

INTRODUCTION

Brefeld (1874) was the first to report the development of a Penicillium teleomorph, giving detailed descriptions of the formation of sclerotioid ascomata and their maturation from the centre outwards. Later studies on Penicillium teleomorphs were carried out by Emmons (1935), Benjamin (1955), Stolk and Scott (1967) and Scott (1968). Malloch and Cain (1972) proposed the family Trichocomaceae Fischer as the appropiate one for species of cleistothecial Ascomycetes with Aspergillus, Paecilomyces, and Penicillium anamorphs. Udagawa and Horie (1973) suggested the division of the genus Eupenicillium on the basis of ascospore ornamentation, as viewed with the scanning electron microscope, without including the anamorphs in their considerations.

Pitt (1974) pointed out some similarities between Eupenicillium species and sclerotial Penicillia. Later Pitt (1979) linked together for the first time a nonsclerotial Penicillium anamorph with an Eupenicillium teleomorph when identifying P. phoeniceum van Beyma as the anamorph of E. cinnamopurpureum Stolk & Scott.

But in my opinion, a great advance in the clarification of this matter was recently made by Stolk and Samson (1983) in connecting many Penicillium anamorphs with Eupenicillium (or Talaromyces) teleomorphs, and adopting a broader species concept in Penicillium than those of previous authors.

In Eupenicillium ascomata are enclosed within pseudoparenchymatous or sclerotioid cleistothecia. Cleistothecia originate from dendritically ramified hyphal networks and asci are irregularly disposed, ripening from the centre outwards, as described by Emmons (1935). However, a spirally coiled archicarp initiates ascomata development in at least the following genera of the Family Trichocomaceae: Byssochlamys Westling, Dactylomyces Videns, Dichlaena Montagne & Durieu, Eurotium Link, Gymnoeurotium Malloch & Cain, Hamigera Stolk & Samson,

Harpezomyces Malloch & Cain, and *Neosartorya* Malloch & Cain.

Sclerotial species of *Penicillium* have been excluded for a long time from *Eupenicillium* by their failure to produce asci, The same criterium can be applied to the non-sclerotial species of *Penicillium*.

In some species the structures which appear to function as an ascogonium are merely slightly differentiated portions of the vegetative mycelium and are probably structures simplified through reduction. In other species the ascogonium is highly differentiated. In still others, a simple substitute type of ascogonium reappears, but is borne upon a specialized system of branched hyphae.

Species of *Eupenicillium* and *Talaromyces* are always homothallic, as proved by many single spore isolations. Presumably, therefore, the lack of a teleomorph is not caused by segregation of two complementary strains. Perhaps, the great majority of species of *Penicillium* are found in their anamorphic state not because they are haplonts, but because some variation or mutation has given rise to a loss of fertility. Moreover, the fact that some species do not develop their entire life cycle in some media but do in others could be a proof of a very delicate balance of nutritive conditions which might well dominate conidium production as well as ascomata formation. Without attempting any explanation of the mechanism of this loss, we are justified in assuming it occurs, for such loss of fertility can be observed in variants arising in the laboratory. It is then reasonable to assume that the same forces may operate in nature to produce strictly conidial species of *Penicillium*.

The development of ascomata in genera in the family Trichocomaceae may be divided into the following stages:

a) Differentiation of some vegetative hyphae to form various types of ascomatal initials.

b) Development of these initials into various structures, which, in the case of *Eupenicillium*, are stromata, firm and pseudoparenchymatous, or very hard and sclerotioid.

c) Development of the ascogenous hyphae from the centre of the structure. These ascogenous hyphae develop directly and consecutively into asci.

Certain features in the development of the ascocarp are of considerable interest because of the information as to relationships which they may give.

A review of the morphological studies by Emmons (1935) of the reproductive structures involved in the teleomorphic state of several species of *Talaromyces* will show a surprising diversity in ascocarpic initials. In *T. trachyspermum* and *T. wortmanii*, initials are barely differentiated segments of the vegetative hyphae; in *T. bacillisporus* and *T. luteus* they are paired coiled organs; in *T. stipitatus* and *T. flavus* they are true ascogonia and antheridia. In *Eupenicillium crustaceum* (and other species) initials appear as upright branched hyphal networks from the centre of which ascocarps arise directly.

In my own experience, while preparing a monograph on Penicillia (Ramírez, 1982), many Penicillium anamorphs have been observed showing what appeared to be ascocarpic initials of the coiled type encountered in the family Trichocomaceae. However, no further development into sclerotia or ascomata was observed in these isolates.

MATERIAL AND METHODS

For fast growing isolates, microscopical examination was performed on three day old, undisturbed, living slide cultures, grown on MEA 25°C (Ramírez, 1982). Slower growing isolates were incubated for up to six days. Before observation, the slide culture was dipped in 95% ethanol for a few seconds in order to expel all air bubbles, wetting efficiently the specimen before staining. The simplest method of staining was to mix the stain with the mounting fluid. BDH methyl blue (0.05% in lactophenol) was used throughout this work with excellent results. With a staining time of 3-5 minutes, young structures stain faily deeply. After staining, slides were quickly rinsed with distilled water and then into 95% ethanol. The culture was gently covered with a coverglass, using 95% ethanol as mounting fluid for immediate observation and subsequent photogrpahy. A Reichert "NEOPAN" microscope (Austria), equipped with a Reichert "KAM ES" electronic microphotographic equipment, using a SPL 40/=,65 objective and a PK 8 x eyepiece was utilized for all observations.

RESULTS

Observations of coiled hyphae were made in the following isolates: P. cinereoatrum Chalabuda (= P. citreonigrum Dierckx) fide Stolk and Samson (1983); Ramírez, 1982, Fig. 41); P. tarraconense Ramírez et al. (Ramírez, 1982, Fig. 17); P. albocinerascens Chalabuda(= P. citreonigrum Dierckx fide Stolk and Samson 1983; Ramírez, 1982, Fig 63). P. moldavicum Milko & Beljakova (=P. verruculosum Peyronel fide Stolk and Samson, 1983; Ramirez, 1982, Fig. 123); P. giganteum Roi & Singh (Ramirez, 1982, Fig. 190); P. rubicundum Miller et al. (Ramirez, 1982, Fig. 202); P. variabile Sopp (Ramírez, 1982, Fig. 209) and P. novae-zeelandiae van Beyma (non-sclerotial strain) (Ramirez, 1982, Fig. 184).

More recently we have observed these coiled structures in about 30 isolates. Some are shown in figures 1, 2, and 3.

All these ascomata initials are coiled, directly produced on vegetative hyphae more or less profusely, depending on the strain.

I consider that all these isolates are potentially capable to complete their entire life cycle by starting to produce the first step of it: the ascomata initials. However, some unknown deficiencies (genetic blocks, environmental factors, etc.), stop the process at its early stage.

Perhaps some day we may find that all species of Penicillium are actually "perfect".

REFERENCES

BENJAMIN, C.R. 1955. Ascocarps of Aspergillus and Penicillium. Mycologia 47: 669-687.

BREFELD, O. 1874. Botanische Untersuchungen über Schimmelpilze. heft 2:"Die Entwicklungsgeschichte von Penicillium". Leipzig

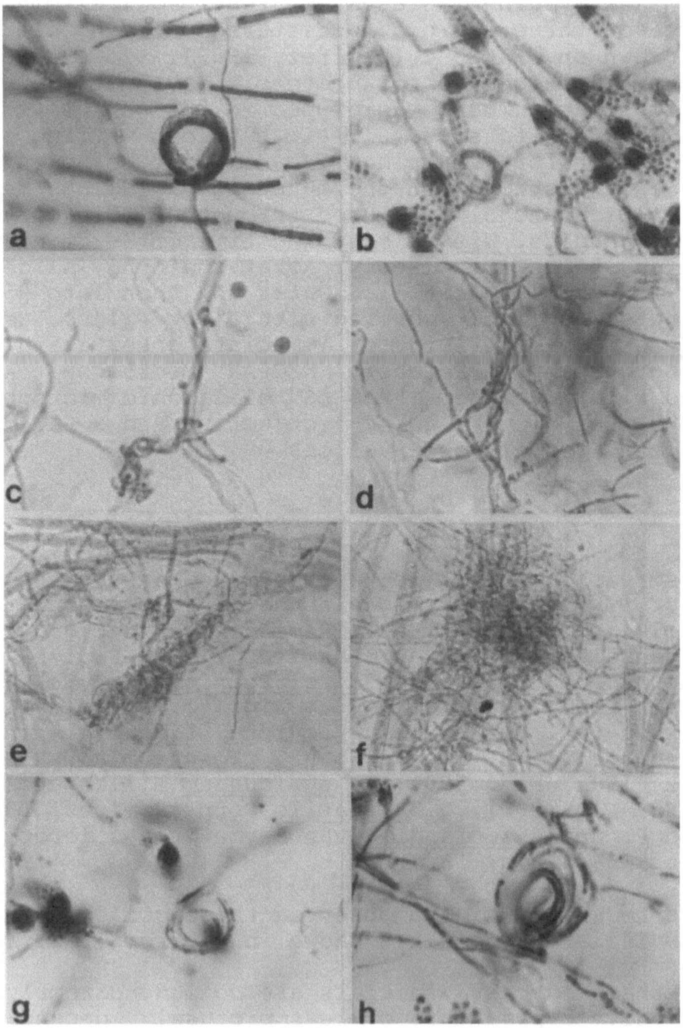

Fig. 1. Coiled ascoma initials observed in different isolates of Penicillium. a. P. valentinum CBS 172.81 (sclerotial); b. P. purpurescens IJFM 1661 (conidial); c, d. P. thomii IJFM 3705 (sclerotial); e, f. P. thomii IJFM 3705, knot of irregularly wound hyphae and loosely ramified tuft-like hyphal system preceding the development of sclerotia; g. P. lapatayae IJFM 19.012 (sclerotial); h. P. mclennaniae IMI 253.561 (conidial). x 250.

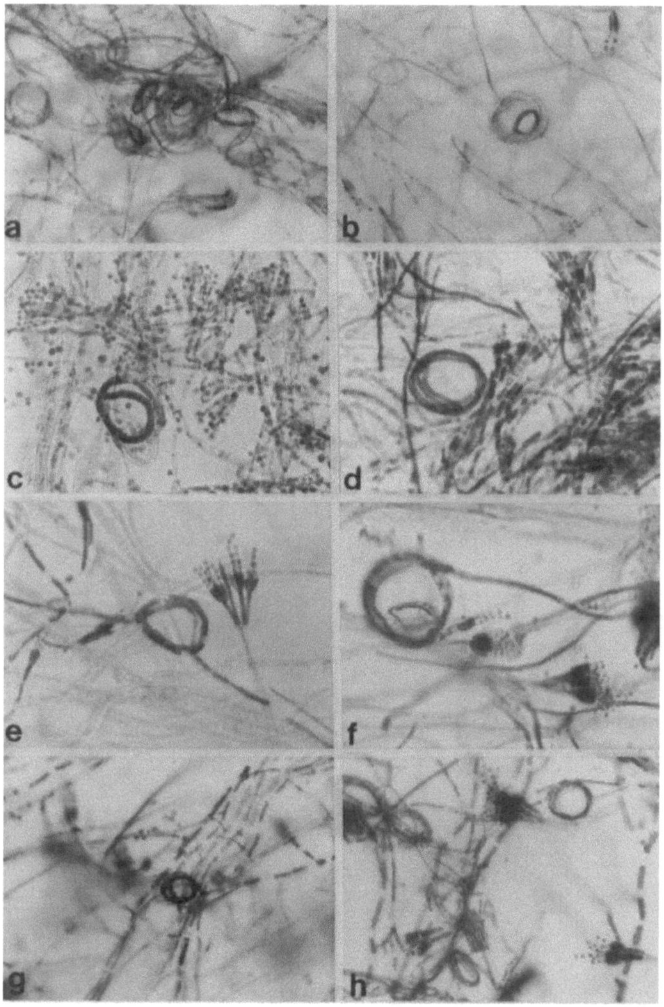

Fig. 2. Coiled ascoma initials observed in different isolates of Penicillium. a. P. novaecaledoniae IJFM 140 441 (conidial); b. P. novaecaledoniae IJFM 7673 (sclerotial); c. P. chrysogenum IJFM 1586 (conidial); d. P. chrysogenum IJFM 7116 (conidial); e. P. raciborskii CBS 689.77 (conidial); f. P. raciborskii IJFM 19.115 (conidial); g. P. sajarovii IJFM 7674 (conidial); h. P. smithii IJFM 7675 (conidial). x 250.

EMMONS, C.W. 1935. The ascocarps in species of Penicillium Mycologia 27: 128-150.

MALLOCH, D. & CAIN, R.F. 1972. The Trichocomataceae: Ascomycetes with Aspergillus, Paecilomyces and Penicillium imperfect states. Can. J. Bot. 50: 2613-2628.

PITT, J. 1974. A synoptic key to the genus Eupenicillium and to sclerotigenic Penicillium species. Can. J. Bot. 52: 2231-2236.

PITT, J. 1979. "The genus Penicillium and its telemorphic states Eupenicillium and Talaromyces". London: Academic Press.

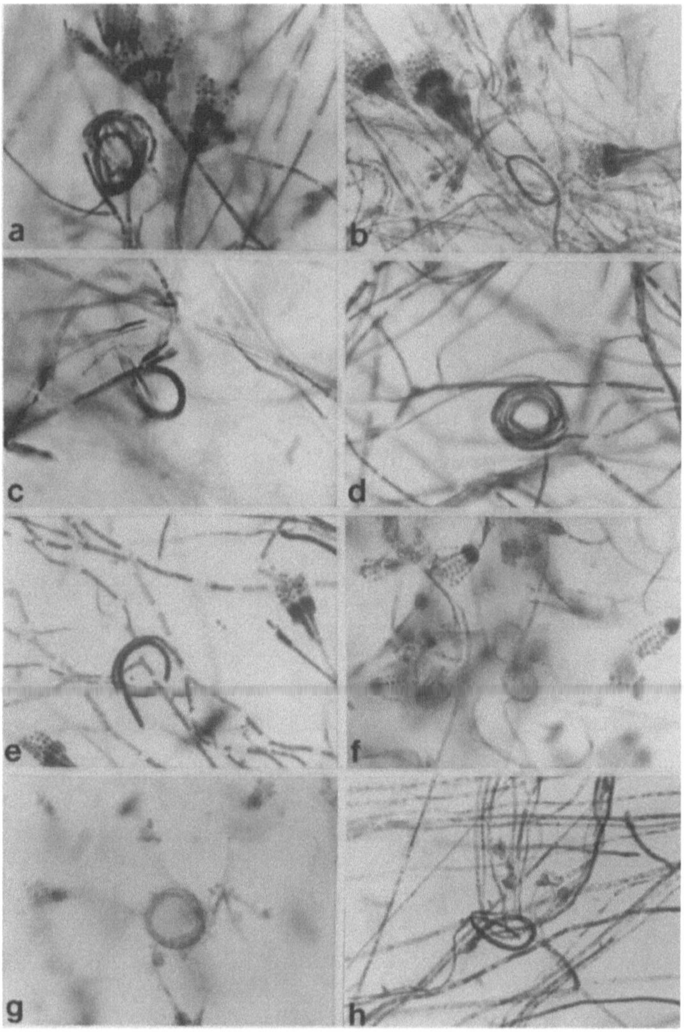

Fig. 3. Coiled ascoma initials observed in different strains of *Penicillium* anamorphs. a. *P. rolfsii* IJFM 19.052 (sclerotial); b. *P. soppii* IJFM 19.106 (conidial); c. *P. soppii* IJFM 19.107 (sclerotial); d. *P. albidum* IJFM 19.062 (conidial); e. *P. walsksmanii* IJFM 7830 (conidial); f. *P. murcianum* CBS 161.81 (conidial); g. *P. camemberti* IJFM 19.116 (conidial); h. *P. purpurogenum* IJFM 19.112 (conidial). x 250.

RAMIREZ, C. 1982. "Manual and Atlas of the Penicillia". Amsterdam: Elsevier Biomedical.

SCOTT, D.B. 1968. "The Genus *Eupenicillium* Ludwig". Res. Rep. Counc. scient. ind. Res. Pretoria. 272: 1-150.

STOLK, A.C. and SAMSON, R.A. 1983. The ascomycete genus *Eupenicillium* and related *Penicillium* anamorphs. Stud. Mycol., Baarn 23: 1-149.

STOLK, A.C. and SCOTT, D.B. 1967. Studies on the genus *Eupenicillium* Ludwig. I. Taxonomy and nomenclature of Penicillia in relation to their sclerotial ascocarpic states. Persoonia 4: 391-405.

UDAGAWA, S. & Horie, Y. 1973. Surface ornamentation of ascospores in _Eupenicillium_ species. Antonie van Leeuwenhoek 39: 313-319.

DIALOGUE FOLLOWING DR. RAMÍREZ'S PAPER

SAMSON: In one of your pictures you showed a penicillus arising from a coil. Doesn't this indicate that it cannot be an ascoma initial?

RAMÍREZ: No, the coil is coming from the bottom of the stipe. I'm just guessing that these are ascoma initials.

SAMSON: If these are really ascoma initials, they cannot be those of _Talaromyces_ or _Eupenicillium_ which are both quite different. They would have to belong to another genus.

RAMÍREZ: Yes. They might be _Eurotium_.

MALLOCH: I don't recall ever seeing a coil like that initiating an _Eupenicillium_ ascostroma. If there are coils in _Eupenicillium_ they must be inside the stroma; I've never seen any.

SAMSON: I have never encountered such an abundance of these coils in _Penicillium_ cultures.

NIRENBERG: I see these coils in old and deteriorated _Fusarium_ strains, so I don't think they have anything to do with sexual reproduction.

RAMÍREZ: On the contrary, I often see these coils in fresh, healthy cultures and I can find them within 3 days on a slide culture.

GAMS: I have seen such structures in many soil fungi, but they are independent from ascoma initials in the same cultures. In _Eurotium_, ascoma initials are also coils, but they are of quite a different pattern than the coils you have shown us here. I have sometimes seen conidiophores of several Hyphomycete genera formed from the type of coils you have shown us.

MALLOCH: These coils are also present in cultures of Basidiomycetes. Isn't it possible that this is some kind of tropism, perhaps a coiling around a concentration of nutrients?

Chapter 7:

RECOMMENDATIONS

GENERAL RECOMMENDATIONS

The following recommendations were agreed to by the participants at the workshop and represent a consensus on various aspects of the systematics of these genera. In cases where no complete agreement was reached these points are cited as an appendix at the end of this chapter.

COOPERATION

Recommendation 1:

That this workshop recognizes the value of international and interdisciplinary approaches to systematics and recommends that such cooperation be encouraged.

NOMENCLATURE

Recommendation 2:

(a) That old and unfamiliar names for which no holotype or material suitable for lectotypification exists are not re-introduced by neotypification other than in exceptional circumstances.

(b) That the International Commission on the Taxonomy of Fungi (ICTF) be encouraged to prepare a "Guide to good taxonomic practice in the fungi".

Recommendation 3:

(a) That dried cultures be preserved as nomenclatural types in accordance with the International Code of Botanical Nomenclature accepted in Sydney in 1981 (Art. 9.1). Holotypes should be preserved from 7 day old cultures on CYA in the case of anamorphs; additional cultures incubated for 10 and 14 days also should be preserved. Where teleomorphs are formed, cultures including mature cleistothecia should be dried also.

(b) That, in addition, living cultures derived from the material dried to provide the nomenclatural type of a name should be preserved in at least two major service culture collections using a long-term storage technique (e.g. lyophilization, liquid nitrogen).

(c) That living cultures so preserved are to be referred to as "ex-type" or "derived from the type", i.e. originating from the material that is the nomenclatural type of the name. "Ex-holotype", "ex-neotype" and "ex-lectotype" etc. are also acceptable where appropriate.

(d) That names where only a living type was designated in the original diagnosis of a species and where no dried material was preserved, should not be treated as invalidly published on that basis alone under the Sydney Code. A dried culture should be prepared as the nomenclatural type and referred to as "lectotype" for the name in accordance with the procedures adopted by Pitt (1979).

Recommendation 4:

(a) That where well-established names of *Aspergillus* and *Penicillium* species of economic importance are threatened by earlier names which should be adopted under the Code, the established species names should first be proposed for conservation in *Taxon* and considered by the Special Committee for Fungi as provided for in Article 14.1.

(b) That where well-established names of *Aspergillus* and *Penicillium* species are found to have been applied in a sense other than their types, such names be proposed for rejection under Art. 69.1 of the Code, published in *Taxon* and considered by the special Committee for Fungi before they are taken up in a changed sense.

Recommendation 5:

(a) That nomenclaturally correct teleomorph and anamorph names should be introduced wherever such states are both known, in accordance with Art. 59.1-.6 as adopted in the Sydney Code.

(b) That while some participants of the Workshop regret the consequences of the changes made in Art. 59 in 1981, and that procedures which could have resulted in the saving of many well-known names in *Aspergillus* and *Penicillium* (Hawksworth and Sutton-- *Taxon* 23: 563-568, 1974) were not adopted, this rule should not now be remodified in view of the further confusion that could then result.

Recommendation 6:

That where it is felt appropriate to recognize infrageneric groupings within *Aspergillus* and *Penicillium*, names in formal taxonomic ranks (e.g. subgenus, section, and series as appropriate), should be adopted in accordance with the Code.

LOCALITY DATA

Recommendation 7:

That in referring to publications dealing with descriptions of new species of *Aspergillus* and *Penicillium*, full details should be provided as to their origin including precise locality, habitat, substratum, soil profile, depth, date and method of isolation, etc., as appropriate.

TAXONOMY

Recommendation 8:

(a) That as wide a range of characters as possible be considered in the formulation of species concepts in *Aspergillus* and *Penicillium*, including morphology, physiology, biochemistry, metabolites, growth conditions, ultrastructure, serology, DNA and RNA hybridization, etc.

(b) That in keys and identification aids, characters which are reproducible and suitable for use by non-specialists should be adopted wherever possible.

PRESERVATION

Recommendation 9:

(a) That major culture collections are urged to obtain fresh isolates from as wide a range of geographical and ecological situations as possible, with the object of maintaining the maximum genetic diversity as this may assume major importance in the future.

(b) That workers obtaining isolates showing unusual features are urged to deposit them in a major culture collection wherever possible.

(c) That isolates that have been characterized biochemically or otherwise documented should also be so preserved.

MEDIA AND CULTIVATION

Recommendation 10:

That Czapek agar (Cz) and/or Czapek Yeast Extract Agar (CYA), and Malt Extract Agar (MEA) be used as the standard media for descriptions of *Penicillium* and *Aspergillus* species. The standard formulations for these media are as follows:

Czapek agar (Cz)

$NaNO_3$	3.0 g
K_2HPO_4	1.0 g
KCl	0.5 g
$MgSO_4 \cdot 7H_2O$	0.5 g
$FeSO_4 \cdot 7H_2O$	0.01 g
Sucrose	30 g
Agar	15 g
Distilled water to	1 litre

Autoclave at 121°C for 10 min. Final pH 6.0-6.5.

Czapek Yeast Extract Agar (CYA)

$NaNO_3$	3.0 g
K_2HPO_4	1.0 g
KCl	0.5 g
$MgSO_4 7H_2O$	0.5 g
$FeSO_4 7H_2O$	0.01 g
Yeast extract	5.0 g
Sucrose	30 g
Agar	15 g
Distilled water to	1 litre

Autoclave at 121°C for 10 min. Final pH 6.0-6.5

Malt Extract Agar (MEA)

Malt extract, powdered	20 g
Peptone, bacteriological	1 g
Glucose, A.R.	20 g
Agar	15 g
Distilled water to	1 litre

Autoclave at 121°C for 10 min. Final pH 5.0-5.5

Recommendation 11:

That for teleomorphs, one or more of the following media be used which have provided useful results: for *Eupenicillium* and *Talaromyces*, Cornmeal agar, Oatmeal agar and CBS malt agar (MA); for *Eurotium*, Czapek Agar with 20% sucrose (C20S); Czapek Yeast Extract Agar with 20% sucrose (CY20S); and Malt Yeast 40% Sucrose Agar (M40Y).

Formulations are as follows:

Corn meal agar (CMA)

Add 60 g freshly ground cornmeal to 1 l water, heat to boiling and simmer gently for 1 h. Strain through cloth and sterilize for 1 h at 1 atm overpressure. Fill up to 1 l.

Oat meal Agar (OA)

Heat 30 g oat flakes in 1 l water to hoiling and simmer gently for 2 h. Filter through cloth and sterilize 1 h at 1 atm. Fill up to 1. When using powdered oatmeal, less powder has to be used and filtering is superfluous.

CBS Malt Agar (2%) (MA)

Add water to malt extract from the brewery until it contains 10% sugar (measurement with areometer). Mix 200 ml of this solution with 15 g agar and 800 ml water. Malt agar may also conveniently be prepared with malt syrup (10-40 g/l) or malt powder (10-20 g/l).

Czapek Agar with 20% Sucrose (C20S)

$NaNO_3$	3.0 g
K_2HPO_4	1.0 g
KCl	0.5 g
$MgSo_4 7H_2O$	0.5 g
$FeSo_4 7H_2O$	0.01 g
Sucrose	200 g
Agar	15 g
Distilled water to	1 litre

Autoclave at 121°C for 10 min. Final pH 6.0-6.5

Czapek Yeast Extract Agar with 20% Sucrose (CY20S) is formulated as C20S, with the addition of 5 g yeast extract per litre.

Malt Yeast 40% Sucrose Agar (M40Y)

Malt extract, powdered	20 g
Yeast extract	5 g
Sucrose	400 g
Agar	15 g
Distilled water to	1 litre

Autoclave at 121 C for 10 min.

Chemicals for media

Inorganic chemicals should be analytical grade. Sucrose may be analytical grade or refined table grade, provided it does not contain sulphur dioxide. Glucose may be analytical grade or food grade glucose monohydrate; in the latter case it should be used at 1.1 x the stated concentration to allow for the water of crystallisation.

Yeast extract, malt extract and agar should be obtained from reliable media manufactures. However, light or pale malt extract sold for home brewing is also suitable and much less expensive. Hydrogen ion concentration can influence results with any of these media, but there is no published information defining the extent of this influence.

INCUBATION

Recommendation 12:

a) That the standard recommended incubation temperature be 25°± 1°C, where such a precise temperature control is impractical, the acceptable limits are 22-27°C.

b) That the standard incubation time for anamorph species be 7 days; for species with teleomorphs longer periods will usually be required for teleomorph maturation. In describing new species, attention should be given to reporting colony growth under conditions other than 25°C.

CHECK LIST OF COMMON PENICILLIUM SPECIES

R. A. Samson & J. I. Pitt

The following is a check list of *Penicillium* species where we are in agreement both in the correct epithet and on the circumscription. It is emphasized that this list was jointly agreed only by the two of us (RAS and JIP) and does not imply general agreement by all those present at the workshop. Note that varieties of *P. verrucosum* established by Samson et al. (1976) have been raised to species level by Stolk and Samson (this publication).

Genus EUPENICILLIUM

E. cinnamopurpureum Scott & Stolk
E. ochrosalmoneum Scott & Stolk
E. pinetorum Stolk
E. javanicum (van Beyma) Stolk & Scott (Samson much broader that Pitt)
E. lapidosum Scott & Stolk
E. crustaceum Ludwig
E. shearii Stolk & Scott

Genus PENICILLIUM

subgenus (section) *Aspergilloides*

P. glabrum (Wehmer) Westling
P. spinulosum Thom
P. lividum Westling
P. thomii Maire
P. sclerotiorum van Beyma
P. citreonigrum Dierckx
P. decumbens Thom

Section Divaricata

P. janthinellum Biourge = *P. simplicissimum* sensu Stolk & Samson
P. janczewskii Zaleski
P. canescens Sopp

Section Furcatum

P. oxalicum Currie & Thom
P. raciborskii G. Smith = P. raistrickii sensu Pitt
P. citrinum Thom
P. corylophilum Dierckx
P. simplicissimum (Oudem.) Thom

Section Penicillium

P. expansum Link
P. chrysogenum Thom
P. viridicatum Westling) P. verrucosum sensu Samson et al.
P. verrucosum Dierckx)
P. crustosum Thom
P. hirsutum Dierckx
P. hordei Stolk
P. echinulatum Raper & Thom ex Fassatiová
P. aurantiogriseum Dierckx
P. puberulum Bain. = P. aurantiogriseum sensu Stolk and Samson
P. camemberti Thom
P. roqueforti Thom
P. griseofulvum Dierckx
P. olivicolor Pitt
P. brevicompactum Dierckx
P. granulatum Bain. (= P. glandicola see Seifert & Samson, this publication, p. 147).
P. italicum Wehmer
P. concentricum Samson & al. (= P. coprophilum, see Seifert & Samson, this publication, p. 145).
P. digitatum (Pers. :Fr.) Sacc.
P. olsonii Bain. & Sartory
P. arenicola Chalabuda

Subgenus Biverticillium

P. duclauxii Delacr.
P. claviforme Bain. (= P. vulpinum, see Seifert & Samson, this publication, p. 144).
P. minioluteum Dierckx
P. pinophilum Hedgcock
P. funiculosum Thom
P. purpurogenum Stoll
P. verruculosum Peyronel
P. islandicum Sopp
P. variabile Sopp
P. rugulosum Thom
P. piceum Raper & Fennell

Genus TALAROMYCES

T. flavus (Kloecker) Stolk & Samson
T. stipitatus C.R. Benjamin
T. luteus (Sacc.) Stolk & Samson
T. wortmanii C.R. Benjamin
T. trachyspermus (Shear) Stolk & Samson
T. thermophilus Stolk
T. striatus (Raper & Fennell) C. R. Benjamin (*Hamigera striata* sensu Stolk & Samson)

GLOSSARY OF TERMS FOR THE DESCRIPTIONS OF ASPERGILLI AND PENICILLIA

(compiled by R. A. Samson & J. I. Pitt)

ACEROSE: cylindrical, then tapering rapidly to a point, like leaves of *Pinus* spp.

AMPULLIFORM: swelling from a narrow base, then tapering to a relatively broad apex, like a Greek ampulla or a ten pin bowling pin.

ANAMORPHIC: refers to the state of a fungus which bears asexual spores (Hennebert and Weresub, 1977)

APICULATE: pointed at both ends; lemon-shaped.

ASCOCARP, ASCOMA: the general term for the sexual fruiting structure produced by Ascomycetes; for *Talaromyces* Pitt (1979) uses gymnothecium.

ASCOGENOUS HYPHAE: the hyphae on which asci are borne within an ascocarp.

ASCOSPORE: the sexual spore characteristic of Ascomycetes; it results from meiosis and is borne in an ascus.

ASCUS: produced within an ascocarp, the ascus is the sac within which ascospores are formed by a characteristic process known as "free cell formation".

BISERIATE: in *Aspergillus*, a head with phialides borne on metulae; also "metulate".

BIVERTICILLATE: refers to a penicillus with two branch points between stipe and conidium, i.e. one bearing metulae and phialides; compare with Samson's one-stage branched.

BRANCH: an element in the penicillus between metula and stipe; see also ramus and ramulus

CLAVATE: club-shaped, enlarging towards the apex.

CLEISTOTHECIUM: a nonostiolate ascocarp with a well defined peridium; see also Malloch's paper on Trichocomaceae (p. 378).

COLLULUM: the neck of a phialide (Zaleski, 1927)

CONIDIAL HEAD: the terminal conidium-bearing structure in *Aspergillus*.

CONCURRENT: borne from the same point.

CONIDIOGENOUS CELL: a cell, from which conidia are borne; in the present context a phialide.

CONIDIOPHORE: the complete fruiting structure which bears conidia.

CONIDIUM: a specialised, nonmotile spore not developing by cytoplasmic cleavage or free cell formation (Kendrick, 1971).

EXUDATE: droplets of liquid exuded by colonies on the obverse.

FASCICLE: a closely packed bundle of conidiophores, often with a common origin.

FERTILE HYPHA: a hypha bearing one or more conidiophores.
FIMBRIATE: fringed with hyphae of irregular length and density.
FLANGED: bearing flanges or crests as on a pulley wheel.
FLOCCOSE: of loose texture, similar to that of cotton wool.
FUNICLE: an aerial rope of fertile hyphae from which conidiophores are borne at irregular locations.
FUNICULOSE: composed of, or bearing, funicles.
FUSIFORM: pointed at both ends, spindle-shaped.
HULLE CELLS: terminal and intercalary very thick-walled cells occurring in _Aspergillus_ species.
INTERCALARY: positioned between base and apex.
LECTOTYPE: a type selected from the original specimens on which a taxon was based, when no holotype exists, to serve as the nomenclatural type for the taxon; in the present context, also a specimen derived from a culture ex type at a time later than the original publication.
LENTICULAR: lens-shaped; biconvex.
METULA: a specialised cell giving rise to a verticil of phialides.
MONOVERTICILLATE: refers to a penicillus with one branch point between stipe and conidium, i.e. composed of phialides only; Samson et al. (1976) use the term "simple".
NUTRITIONALLY DEFICIENT: unable to grow on a minimal medium of mineral salts and a simple carbon source; in the present context, unable to grow on Czapek agar.
PENICILLUS: the structure which bears conidia in _Penicillium_ and similar genera. It is borne on a well defined stipe and consists of phialides or other conidiogenous cells alone or in combination with metulae and/or other supportive elements.
PERIDIUM: the wall of an ascocarp or other fruiting body.
PHIALIDE: a conidiogenous cell which produce, from a fixed conidiogenous locus, a basipetal successsion of enteroblastic conidia whose walls arise _de novo_ (Kendrick, 1971; Cole and Samson, 1979).
PYRIFORM: pear-shaped.
QUATERVERTICILLATE: refers to a penicillus with four branch points between stipe and conidium; i.e. bearing rami, ramuli, metulae and phialides; Samson et al. (1976) use "three stage branched".
RAMULUS: an element similar to a ramus, interposed between ramus and metulae in a quaterverticillate penicillus.
RAMUS: a specilised cell giving rise to a verticil of metulae and phialides.
RETICULATE: ornamented with a network.
REVERSE: the backside of a colony viewed through the medium.
RUGOSE: with surface wrinkled or roughened.
RUGULOSE: with surface finely wrinkled or roughened.
SCLEROTIGENIC: bearing sclerotia.
SCLEROTIOID: very hard, stone-like.
SCLEROTIUM: in _Penicillium_ and _Aspergillus_, a body indistinguishable from an immature cleistothecium, but which is not known to produce asci and ascospores.
SOLUBLE PIGMENT: pigment diffusing into the medium from the reverse of a colony.
SPATHULATE: spatula- or spoon-shaped; of a stipe, apically enlarged, then tapering towards the base.
SPINOSE: spiny.

SPINULOSE: with minute spines.

STIPE: a stalk, in the present context a specialised hypha arising from a fertile hypha and supporting a penicillus or conidial apparatus of Aspergillus

STRIATE: marked with ridges.

SULCATE: furrowed or grooved.

SYNNEMA: a cluster of appressed or fused fertile hyphae bearing conidiophores along length.

TELEOMORPH: refers to the state of a fungus which bears sexual spores such as ascospores or basidiospores (Hennebert and Weresub, 1977).

TERVERTICILLATE: refers to a penicillus with three branch points between stipe and conidium, i.e. bearing rami, metulae and phialides; Samson et al. (1976) use "two stage branched".

UNISERIATE: in Aspergillus, a head with phialides borne directly on the vesicle.

VELUTINOUS: refers to a colony in which most or all conidiophores are borne as a subsurface or surface layer, so that the surface texture is like velvet.

VERRUCOSE: covered with small wart-like outgrows; in the present context, sometimes indicating a distinctly roughened surface.

VERTICIL: a cluster of more than two metulae or phialides with a common origin.

VESICLE: a swelling; in the present context, limited to the apical swelling of a stipe where the diameter of the swelling is at least twice that of the stipe (Pitt, 1973).

VESICULATE: terminating in a vesicle.

APPENDIX

On colour photographs:

In discussion, the suggestion was made by Dr. Ramírez that colour photographs of type cultures accompany type specimens, and that copies of negatives be distributed with the culture when it is sent to the major culture collections. It was felt that colour reproduction is not yet sufficiently accurate and stable to allow this procedure to be useful.

On living types:

Prof. Hawksworth asked the participants if they were of the opinion that living types should be accepted or whether the Code as amended in 1981 met their requirements for *Aspergillus* and *Penicillium* species. The general feeling was that dried types were necessary and that the rules were now satisfactory, but that carefully preserved living cultures are also highly desirable and should be preserved wherever possible (see also Recommendation 3b).

On single spore cultures:

Some participants would recommend that authors of new species consider use of single spore cultures in view of the heterokaryotic nature of mycelia in cultures of *Aspergillus* and *Penicillium*. However, a number of participants were unconvinced of the necessity of single spore techniques for the taxonomy of these genera.

On the naming of chemical races

Dr. Pitt questioned the desirability of naming chemically distinguished races formally. This met with general approbation and recommendations are embodied in the contribution in chapter 2 by Pitt and Hawksworth, which was prepared during the workshop to address the particular problem.

LIST OF THE WORKSHOP PARTICIPANTS

Dr. J. ANNÉ
Rega instituut, Microbiologische afdeling.
Minderbroederstraat 10,
B-2000 Leuven, Belgie

Dr. N. ARAN
Dept. Food Science, TUBITAK,
Marmara Scientific and Industrial Research Institute,
Gebze/Istanbul, Turkey.

Dr. L. R. BEUCHAT
Dept. of Food Science, University of Georgia,
College of Agriculture, Experiment Station,
Georgia 30212, USA

Dr. P. D. BRIDGE
Commonwealth Mycological Institute,
Ferry Lane, KEW, Surrey TW9 3AF, UK

Prof. Dr. M. CHRISTENSEN
Dept. of Botany, Aven Nelson Building.
University of Wyoming, LARAMIE, Wyoming 82071, U.S.A.

Mll. Joelle DUPONT
Laboratoire de Cryptogamie,
Rue de Buffon 5, 75005 Paris, France

Dr. O. FASSATIOVÁ
Dept. of Cryptogamic Botany,
Benatska 2, 128 01 Praha 2, Czechoslovakia

Dr. J. C. FRISVAD
The Technical University of Denmark,
Food Technologu Lab., Building 221,
DK-2800 Lyngby, Denmark

Prof. Dr. K. W. GAMS
Centraalbureau voor Schimmelcultures,
P. O. Box 273, 3740 AG BAARN, the Netherlands

Dr. S. GOCHENAUR
Dept. of Biology, Adelphi University,
Garden City, Long Island, NY 11530, USA

Prof. Dr. D. L HAWKSWORTH
Commonwealth Mycological Institute,
Ferry Lane, KEW, Surrey TW9 3AF, UK

Prof. Dr. G. L. HENNEBERT
Laboratoire de Mycologie systématic et appliquée
Université Catholique de Louvain,Place de Croix du Sud 3
B-1348 Louvain-la-Neuve, Belgique

Dr. R. A. HILL
Ruakura Soil and Plant Research Station
Ministry of Agriculture and Fischeries, Private Bag,
Hamilton, New Zealand

Dr. Maren KLICH
USDA Southern Regional Research Center,
P. O. Box 19687, New Orleans, LA 70179, USA

Dr. Z. KOZAKIEWICZ
Commonwealth Mycological Institute,
Ferry Lane, KEW, Surrey TW9 3AF, UK

Dr. D. MALLOCH
Department of Botany, University of Toronto,
TORONTO, Ont. M5S 1A1, Canada

Dr. J. MOUCHACCA
Laboratoire de Cryptogamie,
Rue de Buffon 5, 75005 Paris, France

Dr. Helgard I. NIRENBERG
Institut fur Mikrobiologie,
Biologische Bundesanstalt fur Land- und Forstwirtschaft,
Konigin Luise Str. 19, D-1000 Berlin 33, FRG.

Dr. A. H. S. ONIONS
Commonwealth Mycological Institute,
Ferry Lane, KEW, Surrey TW9 3AF, UK

Dr. R. R. PATTERSON
Commonwealth Mycological Institute,
Ferry Lane, KEW, Surrey TW9 3AF, UK

Dott. L. POLONELLI
Istituto di Microbiologia,
Università Cattolica del Sacre Cuore,
Via delle Pineta Sacchetti, 00168 ROMA, Italy

Dr. J. I. PITT
C.S.I.R.O., Division of Food Research,
P.O. Box 52, North RYDE, N.S.W. 2113, Australia.

Prof. Dr. C. RAMÍREZ
Instituto "Jaime Ferrán" de Microbiología,
Joaquín Costa 32, MADRID-10, Spain.

Dr. M-F. ROQUEBERT
Laboratoire de Cryptogamie,
Rue de Buffon 5, 75005 Paris, France

Dr. R. A. SAMSON
Centraalbureau voor Schimmelcultures,
P. O. Box 273, 3740 AG BAARN, the Netherlands

Dr. K. A. SEIFERT
Centraalbureau voor Schimmelcultures,
P. O. Box 273, 3740 AG BAARN, the Netherlands

Drs. A. C. STOLK
Wittelaan 19, BAARN, the Netherlands

Drs. E. S. VAN REENEN- HOEKSTRA
Centraalbureau voor Schimmelcultures,
P. O. Box 273, 3740 AG BAARN, the Netherlands

Dr. Shun-ichi UDAGAWA
National Institute of Hygienic Sciences,
Laboratory of Mycology, 1-18-1, Kamiyoga,
Setagaya, Tokyo, Japan.

Dr. D. T. WICKLOW
United States Department of Agriculture, NRRL
1815 North University Street,
PEORIA, Illinois 61604, U.S.A.

Mr. A. P. WILLIAMS
Leatherhead Food Research Association.
Randalls Road, Leatherhead, Surrey KT 22 7RY, UK

INDEX

MIX
Papier aus verantwortungsvollen Quellen
Paper from responsible sources
FSC® C105338

If you have any concerns about our products,
you can contact us on
ProductSafety@springernature.com

In case Publisher is established outside the EU,
the EU authorized representative is:
Springer Nature Customer Service Center GmbH
Europaplatz 3, 69115 Heidelberg, Germany

Printed by Libri Plureos GmbH
in Hamburg, Germany